Monitoring and Modelling of Geological Disasters Based on InSAR Observations

Monitoring and Modelling of Geological Disasters Based on InSAR Observations

Editors

Chisheng Wang
Daqing Ge
Guohong Zhang
Wu Zhu
Siting Xiong

MDPI • Basel • Beijing • Wuhan • Barcelona • Belgrade • Manchester • Tokyo • Cluj • Tianjin

Editors
Chisheng Wang
Shenzhen University
China

Daqing Ge
China Geophysical Surveying and Remote Sensing Center for Natural Resources (AGRS)
China

Guohong Zhang
China Earthquake Administration
China

Wu Zhu
Chang'an University
China

Siting Xiong
Guangdong Laboratory of Artificial Intelligence and Digital Economy (SZ)
China

Editorial Office
MDPI
St. Alban-Anlage 66
4052 Basel, Switzerland

This is a reprint of articles from the Special Issue published online in the open access journal *Remote Sensing* (ISSN 2072-4292) (available at: https://www.mdpi.com/journal/remotesensing/special_issues/InSAR_Geological_Disasters).

For citation purposes, cite each article independently as indicated on the article page online and as indicated below:

LastName, A.A.; LastName, B.B.; LastName, C.C. Article Title. *Journal Name* **Year**, *Volume Number*, Page Range.

ISBN 978-3-0365-6380-0 (Hbk)
ISBN 978-3-0365-6381-7 (PDF)

Cover image courtesy of Chisheng Wang

Contents

About the Editors

Chisheng Wang

Chisheng Wang is currently an associate professor with the School of Architecture and Urban Planning, Shenzhen University, Shenzhen, Guangdong, China. He received a B.S. degree from Beijing Normal University, Beijing, China, in 2007; an M.S. degree from the Institute of Applied Remote Sensing, Chinese Academy of Science, Beijing, in 2010; and a Ph.D. degree from the Department of Land Surveying and Geo-Informatics, Hong Kong Polytechnic University, Kowloon, Hong Kong. His research interests include disaster and infrastructure monitoring, InSAR, point cloud processing, and photogrammetry.

Daqing Ge

Daqing Ge is currently a Senior Engineering (Professorship) with the China Aero Geophysical Survey and Remote Sensing Center for Land and Resources, Beijing, China. He was born in Shaanxi, China. He received a B.S. degree in engineering of surveying and mapping from the China University of Mining and Technology (CUMT), Xuzhou, China, in 2002; an M.S. degree in photogrammetry and remote sensing from the China University of Mining and Technology (Beijing) (CUMTB), Beijing, China, in 2005; and a Ph.D. degree from the China University of Geosciences (Beijing) (CUGB), Beijing, China, in 2013. He has served as the Principal Investigator of several Chinese national projects. His research interests mainly include radar interferometry technology, geohazards monitoring applications, and the development of SAR satellite application systems.

Guohong Zhang

Guohong Zhang is a research professor in the Institute of Geology, China Earthquake Administration. He received his PhD degree in Geophysics from Institute of Geology, China Earthquake Administration. His current research interests include the history of earthquake ruptures using joint inversion of InSAR and seismic data, co- and post-seismic crustal deformation using InSAR technology and its modeling, the active deformation of the continents, and the rheology of the lithosphere. He has more than 80 journal and conference publications.

Wu Zhu

Wu Zhu is currently a professor with School of Geology Engineering and Geomatics, Chang'an University, Xi'an, China. He received an M.Sc. degree in geodesy and geodynamics from Chang'an University, Xi'an, China, in 2010 and a Ph.D. degree from the Department of Land Surveying and Geo-Informatics (LSGI), Hong Kong Polytechnic University, Kowloon, Hong Kong, in 2015. His research interests include the estimation and correction of ionospheric effects on interferometric synthetic aperture radar (InSAR) and analyses of InSAR-measured surface deformation associated with geophysical hazards.

Siting Xiong

Siting Xiong is currently an associate researcher with the Guangdong Laboratory of Artificial Intelligence and Digital Economy (SZ), Shenzhen, China. She received a B.Sc. degree from the School of Geodesy and Geomatics, Wuhan University, Wuhan, China, in 2011; an M.Sc. degree from the Department of Earth and Space Sciences, Peking University, Beijing, China, in 2014; and a Ph.D. degree from the Department of Space and Climate Physics, University College London, London, U.K., in 2019. She was a post-doctoral researcher with the College of Civil and Transportation Engineering, Shenzhen University, Shenzhen, China, from 2019 to 2022. Her research interests include image processing, radar subsurface mapping, interferometric SAR, and planetary radar data processing. She has more than 30 journal and conference publications.

remote sensing

MDPI

Article

Rupture Models of the 2016 Central Italy Earthquake Sequence from Joint Inversion of Strong-Motion and InSAR Datasets: Implications for Fault Behavior

Chuanhua Zhu [1,2,3], Chisheng Wang [1,2,*], Xinjian Shan [3], Guohong Zhang [3], Qingquan Li [1,2], Jiasong Zhu [1,4], Bochen Zhang [1,4] and Peng Liu [5]

1 Ministry of Natural Resources (MNR) Key Laboratory for Geo-Environmental Monitoring of Great Bay Area, Guangdong Key Laboratory of Urban Informatics, Shenzhen Key Laboratory of Spatial Smart Sensing and Services, Shenzhen University, Shenzhen 518060, China; czhu@szu.edu.cn (C.Z.); liqq@szu.edu.cn (Q.L.); zjsong@szu.edu.cn (J.Z.); zhangbc@szu.edu.cn (B.Z.)
2 School of Architecture and Urban Planning, Shenzhen University, Shenzhen 518060, China
3 State Key Laboratory of Earthquake Dynamics, Institute of Geology, China Earthquake Administration, Beijing 100029, China; xjshan@ies.ac.cn (X.S.); zhanggh@ies.ac.cn (G.Z.)
4 College of Civil and Transportation Engineering, Shenzhen University, Shenzhen 518060, China
5 Department of Earth and Space Sciences, Southern University of Science and Technology, Shenzhen 518055, China; liup3@sustech.edu.cn
* Correspondence: wangchisheng@szu.edu.cn

Abstract: We derived the joint slip models of the three major events in the 2016 Central Italy earthquake sequence by inverting strong-motion and InSAR datasets. *b*-values and the historic earthquake scarp offset were also investigated after processing the earthquake catalog and near-field digital elevation model data. The three major events gradually released seismic moments of 1.6×10^{18} Nm (M_w 6.1), 1.5×10^{18} Nm (M_w 6.1), and 1.1×10^{19} Nm (M_w 6.7), respectively. All the ruptures exhibit both updip and along-strike directivity, but differ in the along-strike propagation direction. The high *b*-value found beneath three mainshock hypocenters suggests possible fluid intrusions, explaining the cascading earthquake behavior. The cumulative surface scarp from past earthquakes shows rupturing features that are consistent with the 2016 earthquake sequence, suggesting a characteristic fault behavior. Under the assumption of the Gutenberg–Richter law, the slip budget closure test gives a maximum magnitude of M_w 6.7 and implies the seismic hazard from the largest event has been released in this sequence.

Keywords: source inversion; rupture model; strong motion; InSAR; seismic hazard

Citation: Zhu, C.; Wang, C.; Shan, X.; Zhang, G.; Li, Q.; Zhu, J.; Zhang, B.; Liu, P. Rupture Models of the 2016 Central Italy Earthquake Sequence from Joint Inversion of Strong-Motion and InSAR Datasets: Implications for Fault Behavior. *Remote Sens.* **2022**, *14*, 1819. https://doi.org/10.3390/rs14081819

Academic Editor: Salvatore Stramondo

Received: 20 March 2022
Accepted: 8 April 2022
Published: 10 April 2022

1. Introduction

Earthquakes have sustainedly caused a large number of casualties and damages worldwide. Seismological research relies heavily on the inversion for the earthquake source, including the source location, fault geometry, slip distribution, rupture directions, etc. Using that, we can explore fault behaviors and seismic hazards [1–4]. As such, inversion for earthquake sources has long been one of the most popular topics after a destructive earthquake. Benefitting from the algorithm progressive and data coverage, joint inversion of seismic and geodetic data have been successfully applied for detailed source rupture processes of earthquakes worldwide, e.g., the 1999 Mw 7.1 Duzce, the 2011 Mw 9.0 Tohoku, and the 2012 Mw 7.6 Nicoya, Costa Rica, earthquake, etc. Additionally, it has been proved to be able to provide a stabler and higher resolution source model than a single-data inversion [5–8].

On 24 August 2016 (UTC 01:36, local time 03:36), a destructive earthquake (M_w 6.2) occurred in central Italy (the Amatrice earthquake). The USGS reported that the earthquake originated at a depth of 4.4 km, with a normal faulting mechanism. The epicenter was located at 42.70°N, 13.23°E, between the towns of Norcia and Amatrice. According to the official figures of the Protezione Civile, this event caused the death of 297 people, with 234 of the casualties occurring in the town of Amatrice. Two months later, two major earthquakes were triggered northwest of the Amatrice earthquake on 26 October (19:18 UTC) and 30 October (19:18 UTC), with individual moment magnitudes of 6.1 (Visso earthquake) and 6.6 (Norcia earthquake). There were no reports of serious injuries in the October earthquakes.

The ground motion associated with the Amatrice earthquake was recorded by many geodetic and seismic instruments, including space-borne synthetic aperture radar (SAR) sensors and ground seismometers. These datasets can be used to constrain a fault slip model, which can help us to understand the earthquake mechanics and seismic hazard. To date, source models have been inferred using one or several datasets [9–16]. For example, Tinti et al. (2016) presented the first kinematic model of this event by inverting the waveforms from 26 3-component strong-motion accelerometers [14]. Lavecchia et al. (2016) derived a static slip model constrained by differential interferometric SAR (DInSAR) measurements from several SAR satellites [12]. However, the features of these models differ from each other because different datasets were used in these studies. As InSAR datasets and strong-motion datasets have complementary strengths in earthquake source inversion [17], it is possible to invert both datasets simultaneously for a more comprehensive fault slip model.

The future hazard of a fault after an earthquake is of great importance for human life and property safety. The 2016 central Italy earthquake sequence is located in a segmented fault system mixed with modern and ancient structures. Conventional seismic hazard models assume earthquake ruptures are controlled by fault segmentation, where the rupture is believed to be unlikely propagate from one segment to another. In the 2016 sequence, the heterogeneous crust has broken the lateral continuity of the seismogenic fault, prohibiting it to grow to a large devastating earthquake. However, increasing observations, e.g., the 2010 EI Mayor-Cucapah earthquake [18] and the 2016 Kaikoura earthquake [19], show that multi fault segments are possible to simultaneously rupture in a single earthquake. In addition, the segmented faults may become more continuous and mature after many earthquakes repeatedly ruptured, and consequently, inclined to host a larger earthquake. Therefore, the future seismic hazards in this region need to be re-examined based on more observations. Besides the comprehensive rupture model, we obtained as mentioned above, the complete Italian earthquake catalogue and available near-field high-resolution topography enable us to retrieve more fault information including *b*-value, historical events trace, and seismic activity. A detailed interpretation of the causative fault by combining the information can therefore benefit the earthquake hazard evaluation.

In this paper, we first present the rupture processes of the three major earthquakes in the central Italy earthquake sequence by inverting joint datasets, including both InSAR and strong-motion data. We then explore the *b*-values and historic earthquake scarp offsets through processing the earthquake catalog and near-field digital elevation model (DEM) data. Based on these results, we interpret the fault behavior of the earthquake sequence and discuss the future seismic hazard in this area.

2. Tectonics

The 2016 central Italy earthquake sequence took place in the northern and central Apennines of Italy, which were formed due to the collision between the Adria microplate and the Eurasian plate. The collision induced compression in the front arc and extension in the back arc. The back arc of the Apennines chain is thus characterized by a set of NW–SE-striking normal fault systems with a current overall east–west extension rate of 1.5–3 mm/yr [20,21]. The earthquake sequence occurred across several fault systems (Figure 1), including the Mt. Vettore–Mt. Bove Fault system (VBFS) and the Mt. Gorzano Fault system (GFS).

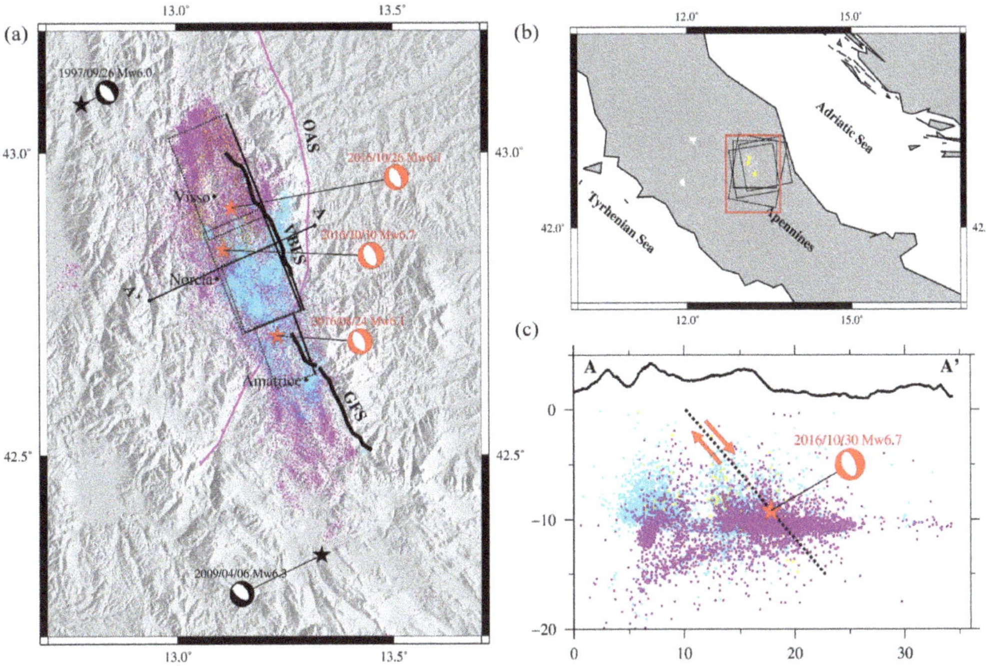

Figure 1. Overview of seismicity, faults and SAR data coverage for the 2016 central Italy earthquake sequence. (**a**) Seismotectonic setting. The three major earthquakes in the sequence are denoted by the red stars and beach ball symbols, and the two recent large historical events are colored in black. The dashed black rectangles represent the surface projection of the fault planes adopted in this study. The bold black lines are the two seismogenic normal fault systems, namely, the Mt. Vettore–Mt. Bove Fault (VBFS) system and the Mt. Gorzano Fault (GFS) system. The pink line shows the simplified trace of the preexisting compressional front named the Olevano–Antrodoco–Sibillini (OAS) thrust. Aftershocks are marked by the small dots with different colors. The blue, purple, and yellow points represent the events taking place between 24 August 2016–26 October 2016, 26 October 2016–30 October 2016, and 30 October 2016–8 October 2017, respectively. (**b**) Map view of the SAR data coverage. The black rectangles represent the coverage of the Sentinel-1A and ALOS 2 SAR data for generating the interferograms. The red rectangle denotes the area shown in (**a**). (**c**) Cross section through A-A′ shown in Figure 1a. The dashed black line represents the fault plane of the 30 October 2016 M_W 6.7 Norcia earthquake.

The VBFS is a SW-dipping fault system, composed of different splays and segments. The fault system scarps are exposed along the SW foothills of Mt. Vettore, Mt. Porche, and Mt. Bove, with a length of ~27 km. The GFS is a SW-dipping fault, with a ~26 km long fault scarp developing along the foothills of Mt. Gorzano. These two faults are segmented by existing tectonic structures inherited from pre-Quaternary compressional tectonics [22]. A ~3 km thick layer in which small events and some large extensional aftershocks occur is found below the seismogenic fault, limiting the seismicity to the first 8 km of the crust [10].

A series of moderate earthquakes have repeatedly struck this area in the last 400 years. The largest one occurred near Norcia town in 1703, with a magnitude of 6.8. The closest large historical event was the 1639 M_W 6.2 earthquake that took place near Amatrice town. According to instrumental records, two M_W > 6 earthquakes have also struck the area in modern times. One was the 1997 M_W 6.0 Colfiorito earthquake that occurred to the northwest, and the other was the 2009 M_W 6.3 L'Aquila earthquake to the south. The 2016 earthquake sequence occurred in a seismic gap which is located between the areas hit by the 1997 Colfiorito earthquake and the 2009 L'Aquila earthquake.

3. Rupture Model

3.1. Data

We used four SAR image pairs to measure the ground displacement relative to the 24 August M_W 6.1 Amatrice earthquake, the 26 October M_W 5.9 Visso earthquake, and the 30 October M_W 6.5 Norcia earthquake (Figure A1 and Table A1). To better serve the rupture model inversion, each selected SAR pair only covered one event mentioned above. Among the image pairs, three were from the Sentinel-1 satellite and the other was from the Advanced Land Observing Satellite (ALOS) satellite. Many institutions and researchers have created interferograms for the Amatrice and Visso events using Sentinel-1 images, and their results are quite similar. In this study, we directly used the InSAR interferograms from the European Space Agency's InSARap program for the analysis (provided free online at http://insarap.org/, last accessed 20 March 2022). For the 24 August event, the deformation pattern is characterized by two NNW–SSE striking main distinctive lobes with different shapes, but having almost the same maximum negative line-of-sight (LOS) deformation of around 20 cm. Regarding the 26 October event, the interferogram reveals an ear-shaped deformation pattern, striking NNW–SSE, similar to the 24 August event.

The 30 October M_W 6.5 Norcia earthquake generated much larger ground displacement than the previous two events, making the InSAR measurements more challenging. Therefore, we adopted the L-band ALOS 2 SAR image pair, which has better resistance to phase incoherence, to retrieve the coseismic deformation. A two-pass technique was used to process the data with Gamma software. The TanDEM-X DEM with a 12-m resolution was used to remove the topographic components in the interferogram. A baseline refinement step was carried out to remove the ionospheric disturbance. Finally, we obtained the displacement map after unwrapping the interferogram by the use of the minimum cost flow (MCF) algorithm. The InSAR interferogram reveals a similar ear-shaped deformation pattern to the 26 October event, with around ~90 cm maximum negative LOS displacement. To reduce the quantity of InSAR data, we first applied uniform downsampling to reduce the displacement map to a size of ~500 × 500, and we then used the equation-based quadtree downsampling algorithm to select ~1000 LOS measurements from each interferogram [23].

In addition to the InSAR datasets, we also collected three-component strong-motion datasets of the three earthquakes for joint inversion [24]. These data were obtained from the Italian National Accelerometric Network operated by the Istituto Nazionale di Geofisica e Vulcanologia (INGV) and the Rete Accelerometrica Nazionale (RAN). We selected strong-motion stations with good azimuthal coverage and epicentral distances of less than 50 km. Under these criteria, we separately adopted strong-motion data from 30, 24, and 19 stations for the 24 August, 26 October, and 30 October earthquakes (Figure A2). To better represent the longer period features of strong motion, the velocity waveform data were used for the inversion. We first removed the mean offset and instrument response from the original accelerogram, then filtered the local site effects, and finally, integrated the accelerogram in time. The frequency of each time-series velocity waveform was resampled to 2 Hz to reduce the computational burden during the joint inversion. More details on the inversion strategy and setting are provided in the Supporting Information Texts S1 and S2 [9,25–31].

3.2. Slip Models

We tested three different inversion scenarios: InSAR data inverted alone, strong-motion data inverted alone, and a joint inversion with both geodetic and seismic data (Figures A2 and A3). In the three events, the InSAR-only model differs from the strong-motion data-only model in slip distribution. However, the joint inverted slip distributions seem to make a compromise, absorbing the characteristics from both models. Overall, the joint slip model is closer to the InSAR-only results, which is consistent with the fact that the near-field InSAR data have a better ability to constrain the fault slip pattern. The InSAR and strong-motion prediction from the joint model fits quite well with the observations (Figures A2 and A7, Figures A8 and A9).

The joint model of the 24 August earthquake shows two separate major slip concentrations with a maximum slip of 0.76 m and 0.72 m, respectively, locating at depths between 5 km and 3 km (Figures 2 and A4). The characteristics of the rupture model are in general accordance with previous models [13,14], exhibiting a normal faulting mechanism, bilateral rupture directivity, and a relatively fast rupture velocity. A more detailed comparison with previous models is provided in supporting information Text S3 [9–14,16,32]. Assuming a shear modulus of 30 GPa, the overall seismic moment of the two fault segments is 1.6×10^{18} Nm, equivalent to a moment magnitude of M_w 6.1. The slip pattern is mostly constrained by the near-field InSAR data, but the relative far-field strong motion still fits quite well. During the first 6 s, the majority of the seismic moment was released. The moment rate rapidly increased in the initial stage and reached 3.9×10^{17} Nm/s at 2.8 s, and then decreased rapidly with time. The rupture process took place in a relative manner, propagating gradually from the epicenter to distant patches, and no delayed slip patches were observed.

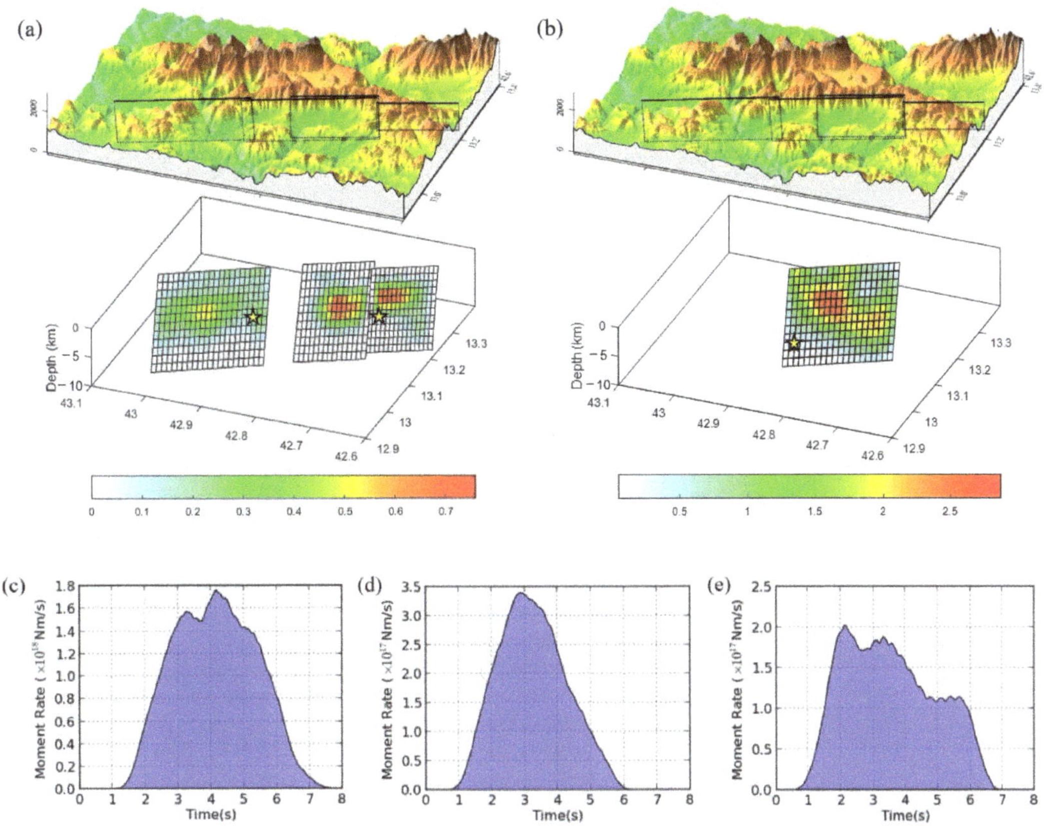

Figure 2. The inverted joint slip models for the 2016 central Italy earthquake sequence. (**a**) Distributions of slip at depth for the Visso (left) and Amatrice (right) earthquakes. The yellow stars denote the start point for the ruptures. (**b**) Distributions of slip at depth for the Norcia earthquake. (**c**–**e**) The moment rate functions of the Visso, Amatrice, and Norcia earthquakes, respectively.

For the 26 October earthquake, our joint model suggests a single elongated normal faulting slip concentration in the northern segment of the VBFS (Figures 2 and A5). The rupture started from the southeast part of the fault plane and propagated mostly unilaterally toward the northwest. The moment rate rapidly increased at the initial stage, reaching 2×10^{17} Nm/s at 2.1 s, but the relatively high moment rate ($>1 \times 10^{17}$ Nm/s) lasted for about 4 s, and then decreased rapidly. This event released a seismic moment of 1.5×10^{18} Nm, corresponding to a M_W 6.1 event.

The 30 October earthquake was the largest event in the earthquake sequence. The inverted rupture model suggests a normal faulting mechanism, consistent with the moment tensor solution, as well as the long-term behavior of the VBFS system. Two slip patches were found, the larger one located in the north with a maximum slip of 2.8 m at a depth between 4 km and 6 km, and the other one peaking at 2.3 m at a depth between 5 km and 7 km (Figures 2 and A6). It can be noted that the latter slip patch is almost below the northwest slip concentration of the 26 August event. The slip history shows that the largest moment rate reached 1.8×10^{18} Nm/s at 4.2 s, and almost all of the coseismic moment was released in the first 7 s. The overall seismic moment was 1.1×10^{19} Nm, equivalent to a M_W 6.7 event.

4. Mapping of the *b*-Values and Scarp Offsets

4.1. *b*-Values

The Gutenberg–Richter (G–R) law is the commonly used statistical model when describing the size distribution of earthquakes. In G–R law, the number (N) of earthquakes having a magnitude $\geq$M follows a logarithmic relationship with magnitude M, expressed as $\log_{10} N = a - bM$, where a and b are constants. Parameter b describes the occurrence ratio of small to large earthquakes. The variance of the *b*-value is thought to be related to local conditions, e.g., stress applied to the material, the strength heterogeneity of the material, the crack density, and the thermal gradient [33]. Among these potential factors, applied stress is usually cited. It is believed to have a negative linear relationship with the *b*-value, which has been proved by a number of laboratory experiments and earthquake observations. In this section, we investigate the *b*-value variation of the 2016 central Italy earthquake sequence area, attempting to reveal the potential stress heterogeneity of the seismogenic fault. ZMAP software [34] was used to calculate the *b*-values. An event catalog covering the 12 recent years of 3 April 2005–8 October 2017 with a depth below 30 km was downloaded from the INGV (http://cnt.rm.ingv.it/en, last accessed 20 March 2022) and adopted as the data source (Figure 3a).

The results give a cut-off magnitude (MC) for this region of 1.3. The average *b*-value for this sequence is estimated to be equal to 1.03 ± 0.02, with a 90% goodness-of-fit level. The average *b*-value is slightly larger than the global mean value of 1.0, which is consistent with previous findings of normal fault-related earthquakes having relatively high *b*-values compared with strike-slip and reverse-slip faulting mechanisms [35]. We further mapped the spatial distribution of the *b*-value in a $0.05° \times 0.05°$ grid (Figure 3a). Each grid selected 300 neighboring events and required at least 50 events above the local value of MC. The estimated *b*-values vary from 0.55 to 2.05. The three events are found to be located in the low *b*-value region, suggesting high-stress regimes. High *b*-values are also observed to the southwest of the 24 August earthquake hypocenter. This event happened in the connecting area between the Gorzano fault and Vettore fault, where the Olevano–Antrodoco–Sibillini (OAS) thrusting structure is thought to intersect. The complex structure may generate a highly fractured rock mass, which is thought to correspond to large *b*-values.

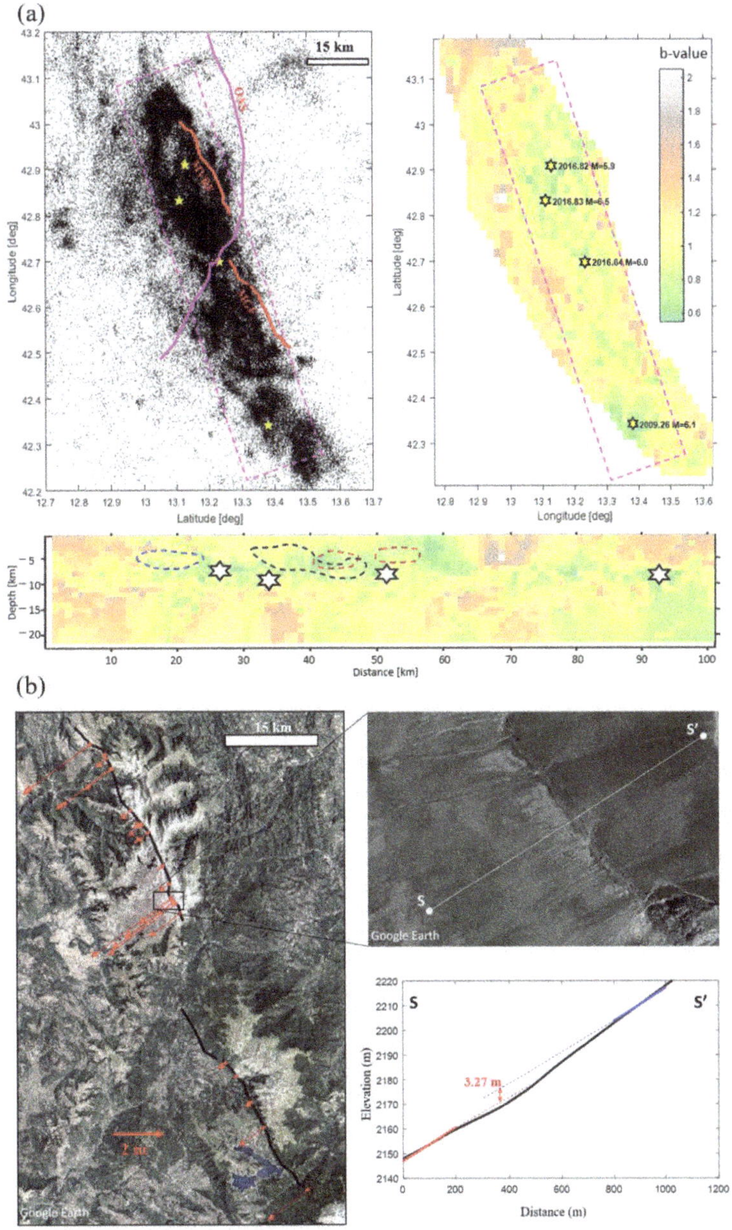

Figure 3. *b*-value and cumulative vertical displacements for the 2016 central Italy earthquake sequence. (**a**) Seismic activity, *b*-value distribution map, and cross section using the INGV earthquake catalog covering the period from 3 April 2005 to 8 October 2017. The pink dashed rectangles indicate the location of the *b*-value swath profile. The red, blue, and black dashed lines indicate the 60% maximum slip area of the Amatrice, the Visso, and the Norcia earthquakes. (**b**) Cumulative vertical displacements (red arrows) along the VBFS/GFS fault systems, and an example of scarp offset measurement. Red arrows represent vertical offsets of ruptures, and they are orthogonal to fault strikes.

We also mapped the depth variation of the *b*-values across the rupture zone (Figure 3a), which was performed in a 1 km × 1 km grid on the swath profile, with 300 neighboring events and a minimum of 50 events above MC. The cross section clearly reveals a low *b*-value layer above 10 km, suggesting that high differential stress exists in the first 10 km of the upper crust. The lower part has low stress possibly due to the presence of fluids in the rock matrix, and the energy is released by a series of small events, as observed. However, these features are not held to the southeastern end of the GFS, where the number of fault branches increases followed by rotating counterclockwise from NNW–SSE to NWW–SEE trending [36], indicating that the magnitude and direction of tectonic stress changed considerably.

4.2. Scarp Offsets

Scarps are universal along the GFS and VBFS systems. Measuring the scarp offset through topography analysis is a common approach in paleoseismology for slip rate estimation. The diversity of the offset values along the fault trace can, therefore, represent the rupture history on the fault. We adopted the 12 m resolution TanDEM-X DEM obtained before the earthquake sequence to extract the scarp offsets along the seismogenic fault trace. Guided by the active fault map reported in Falcucci et al. (2016) [37], we identified the fault trace through visual analysis of high-resolution Google imagery. We then exacted the elevation profile from the TanDEM-X DEM across the fault trace. Two parallel lines orthogonal to the fault strike were generated by least-squares fitting on each side of the fault. As the height profile across the fault is complex, we designed an interactive approach where the user can manually select a profile section with a relatively constant slope for linear fitting. The distance between the lines is considered as the vertical offset of the scarp. As the long-term interseismic slip may not generate a near-fault scarp, we assume that the measured topographic offsets were contributed by large historic earthquakes.

We located 28 sites in the Google imagery and TanDEM-X DEM where obvious scarp features can be observed (Figure 3b and Table A2). The scarp features distribute along the surface trace of the Norcia and Visso earthquakes and to the south of the Amatrice earthquake, while no obvious scarp is found on the south fault segment of the Amatrice earthquake. This agrees with the results reported in Falcucci et al. (2016) [37], i.e., there is no evidence at the surface of late Quaternary fault activity in this area. The scarp features are consistent with the mechanisms of the 2016 central Italy earthquake sequence. Vertical offsets in the height profile are obvious at these sites, while no horizontal displacement is visible in the high-resolution satellite imagery. The average of the measured offsets is about 1.4 m, with the largest value exceeding 3.8 m, which is approximately 1.5 times larger than the maximum slip found for the 30 October Norcia earthquake.

5. Discussion

5.1. Fault Behavior

Historical earthquakes in the central Apennines of Italy show an obvious space-time clustering behavior, where one main shock triggers a series of subsequent events in a relatively short period. Two recent earthquakes nearby this sequence share the same cascading character. To the north, the 1997 M_W 6.0 Colfiorito earthquake triggered six $M > 5$ events in 20 days. To the south, the 2009 M_W 6.3 L'Aquila earthquake started a strong sequence of aftershocks. The 2016 sequence is a typical cascading event in the central Apennines, which offers us another opportunity to investigate the fault behavior in this region.

One notable feature in our joint slip models is the complex multi-fault segments and heterogeneous coseismic slip distribution. Several slip concentrations with different maximum slips are located in different fault patches. Such heterogeneity is also evidenced by spatial *b*-value mapping, where the estimated *b*-value varies largely in different locations (Figure 3a). The slip distributions of three events connect quite well, with few overlaps or gaps, implying an almost complete release of the accumulated interseismic energy. An exception is the conjunction location between the two fault segments of the 26 August event. Such discontinuity is thought to be related to the inherited compressional thrust fault. The interaction between the inherited thrust fault and the active normal fault controls the seismicity very obviously, by which the aftershocks are divided into two clusters, separately distributing on each side of the inherited thrust fault. Fault segmentation is a common features found in earthquakes worldwide, where the fault is composed of discrete segments divided by geometrical discontinuities [38]. It normally acts as a structural control on earthquake magnitude and rupture progress. The central Apennines is dominated by a mix of modern and ancient structures, resulting in heterogeneous crust and segmented faults. Comparing some mature faults which have up to 10 million years of history (e.g., most faults in the San Andreas fault system), the modern extending faults in the central Apennines starting from about a half million years ago are still very young. In an immature fault system, it is mechanically difficult for rupture propagate from one segment to another. The moderate magnitude and slip heterogeneity as observed in our joint slip model is thus a result of the segmentation and immaturity of the seismogenic fault system. Research on the 2016 Kaikoura earthquake [19] warns that multi-fault segments are likely to rupture simultaneously to generate a large earthquake, but its seismogenic faults within the southern island of New Zealand are relatively mature. The immature fragmented fault systems in the central Apennines may still favor moderate magnitude earthquakes at the current stage.

Normally, it is believed that the heterogeneous fault slip distribution is mainly controlled by the fault strength variation [39]. As reported by the joint slip model, several asperities were found in this fault system, with the middle one corresponding to a much larger fault strength. From the observation of the *b*-value mapping, we can also find that low *b*-values are associated with these strong patches. The *b*-value represents the relative occurrence of large and small events, where a low value indicates a larger proportion of large earthquakes [35]. The *b*-value is normally believed to be related to the stress condition, as large earthquakes are thought to be caused by highly loaded stress. Beneath the three mainshock hypocenters, we can observe a high *b*-value region associated with low strain energy (Figure 3a). It is probably due to the presence of fluids in the rock matrix, which is also proposed to exist in the nearby 2009 L'Aquila earthquake [40]. The hypothesis of deep fluid intrusion may also give a good explanation for the space-time clustering behavior of earthquakes in the central Apennines [41]. Pore fluid pressure diffusion after an earthquake can dramatically reduce the shear strength of adjacent fault segments, and consequently induce a cascade of multiple events on them even they are not fully loaded.

Another notable feature of this earthquake sequence is the rupture directivity of the slip models. The retrieved rupture progress shows that the nucleation of three major events all started at the bottom of the fault and then propagated to the upper patches. The along-strike unilateral propagation is also very obvious. The Amatrice and Visso events propagated mostly toward the NNW, and the Norcia event was toward the SSE. The preference for unilateral propagation is observed in many earthquakes, and a potential explanation is the fault segmentation [42], as the earthquake prefers to propagate unilaterally along strike until reach discontinuities. The up-dip rupture direction might result from the material property contrast along with the depth. As denoted above, high *b*-value is observed at the deeper depth from the *b*-value cross section, suggesting the existence of fractured rock mass saturated with fluids over there. The rupture may thus prefer to initiate at the more compliant down-dip part of the rupture zone.

Laboratory experiments have shown that non-uniform normal stress characterized by a high amplitude single-stress asperity favors the occurrence of strong characteristic microquakes, which share similar locations, magnitudes, and return periods [43]. Both rupture model and *b*-value mapping suggest heterogeneous stress in terms of several asperities in this fault system, and the middle one associated with the M_W 6.7 Norcia earthquake has the largest amplitude. We can expect that the M_W 6.7 Norcia earthquake was a characteristic earthquake in this fault system, and that it may rupture again within a certain period. An effective way to test the long-term characteristic behavior of an active fault is through a comparison between historical fault scarps and recent earthquakes [44]. We can refer to such independent observations to verify this characteristic earthquake assumption. In this case, we obtained the cumulative scarp features of historic earthquakes through measuring the near-fault topographic offset. The scarp characteristics agree quite well with the joint slip model in this fault system, where the vertical displacement is obvious in the seismogenic fault of the Norcia event and is several times larger than the coseismic slip. This suggests that long-term constant strong patches locate in this area, and this fault may have displayed characteristic slip behavior.

5.2. *Slip Budget Closure Test*

If we assume a characteristic fault behavior in this fault system for this earthquake sequence, an important issue related to the seismic hazard assessment is whether the Norcia earthquake is the largest earthquake in this fault and whether there is any unruptured fault segment that has the potential to generate a larger earthquake. For a characteristic earthquake, it is possible to derive information about the maximum-magnitude earthquake and its return period from the earthquake catalog by assuming that the frequency magnitude follows G–R law. The relationship between the maximum magnitude (M_w) and its corresponding frequency (N_{max}) can be expressed as [45]:

$$\log N_{max} = -\frac{3}{2}M_w - 9 + logM_0 + log\left[\alpha\left(1 - \frac{2b}{3}\right)\right] \quad (1)$$

where, α is the fraction of transient slip that is seismic, b is the b-value in G–R law, and M_0 is the seismic moment deficit.

The slip budget closure test can therefore be run with the maximum earthquake frequency equation and G–R law. This test has been found to be quite successful in several faulting systems, such as the longitudinal valley fault in Taiwan and the Sumatra Megathrust fault [45]. It is thus interesting to examine the fault slip budget in this complex normal faulting system under frequency–magnitude law.

We tested the seismicity in the area between the 1997 Colfiorito and 2009 L'Aquila earthquakes. We assumed a 75 km length (L) and 16 km width (W) fault plane and half-area ($S_L = 0.5 \times L \times W$) is locked according to the observed distribution of the earthquakes. The long-term slip rate (V) was set to be 2 mm/yr, referring to previous studies [20,21]. Using a shear modulus (G) of 30 GPa, we obtained a rough seismic moment deficit of 7.06×10^{16} Nm/yr, according to the simplified equation $M_0 = GS_LW$.

We combined three catalogs for the frequency–magnitude linear fitting. The INGV catalog includes a wide range of earthquakes with different magnitudes, but the span period is only from 2005 to 2022. The USGS catalog has a longer cover period from 1950 to 2022, but is incomplete for small events. The DBMI catalog contains historical events between 1005 to 2014, but also lacks small and recent earthquakes (http://emidius.mi.ingv.it/DBMI11/, last accessed 20 March 2022). As earthquakes in the central Italy often occurred in terms of cascading earthquake storm, the earthquake frequency in a short-period earthquake catalog is significantly biased. In order to obtain a more comprehensive catalog, we formed a combined DBMI-USGS catalog, where earthquakes before 2014 are from DBMI and events between 2014 and 2017 are from the USGS. The three catalogs give a similar b-value, while the a-values differ a lot. This is because the seismic rates are overestimated in the previous two catalogs. Therefore, we prefer to use the combined DBMI-USGS catalog. If we assume that the seismic moments are released partly seismically with $\alpha = 0.8$, estimated according to the area proportion of asperities on fault planes, the return period line given by Equation (1) will intersect with the frequency–magnitude line at point ($M_w = 6.7$, Lg N = –3.02), suggesting a maximum magnitude of 6.7 and a predicted return period of 1064 years (Figures 4 and A10).

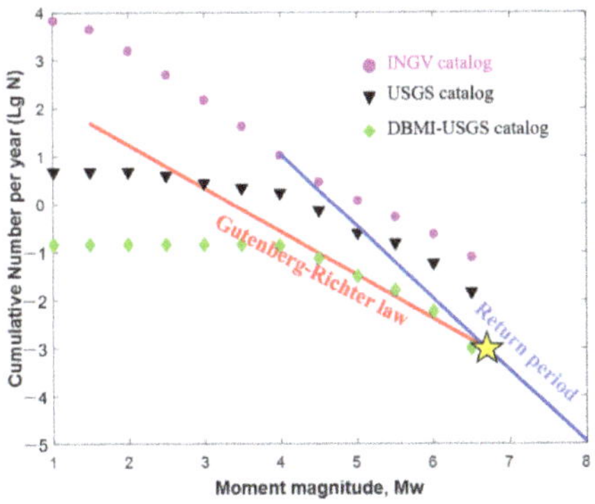

Figure 4. Seismic slip budget closure test on the 2016 central Italy earthquake sequence. The red line represents the G–R law fitted by the DBMI-USGS catalog. The blue line shows the return period of the maximum magnitude event given in Equation (1). The yellow star depicts the intersection of the blue line and red line, giving a maximum magnitude of 6.7 and a return period of 1064 years.

We can also directly estimate the earthquake recurrence interval T by T = S/V, as suggested by Shen et al. (2009) [46], where S is the mean coseismic slip on the fault segment and V is the secular slip rate. Our joint slip model suggests 1.8 m average slip on the slip concentration for the Norcia event (~0.16-m and ~0.24-m average slips for the Amatrice and Visso events). Together with the predetermined ~2 mm/yr slip rate [20,21], we can obtain a recurrence interval of around 900 years for the Norcia fault segment (~80- and ~120-year recurrence intervals for the Amatrice and Visso fault segments). This is in general agreement with the slip budget closure test, implying a low seismic hazard for this area in the future. However, we should also note that both results are rough estimations because they rely on many assumptions, such as the characteristic earthquake, the frequency–magnitude law, the secular fault slip rate, and the catalog completeness. Meanwhile, although the return period for a maximum-magnitude earthquake is long (~1000 yr), such normal faulting areas usually have relatively high *b*-values, which means that the return period of an earthquake drops sharply with the decrease of the magnitude, according to the magnitude–frequency law. It should also be noted that the risk of smaller earthquakes (M_W 5–6) still exists, because they can rupture again with a much shorter return period.

6. Conclusions

This paper has provided a complete rupture history of the three main shocks in the 2016 central Italy earthquake sequence. The 24 August Amatrice earthquake occurred on two fault segments divided by an inherited compressional thrust fault, with a total seismic moment of 1.6×10^{18} Nm (M_W 6.1). The 26 October and 30 October events both ruptured in the VBFS system, releasing seismic moments of 1.5×10^{18} Nm (M_W 6.1) and 1.1×10^{19} Nm (M_W 6.7), respectively. The *b*-value mapping reveals a complex and non-uniform stress condition in this area. The complex segmented faults and heterogeneous coseismic slip distribution suggest an immature fault behavior over there. Possible fluid intrusion implied by high *b*-value at deep depth may be the cause of the rupture of multi-fault segments in a short period. The cumulative surface scarp displacement from past earthquakes shows faulting that is consistent with the 2016 earthquake sequence, suggesting a characteristic fault behavior. Under the slip budget closure test for characteristic earthquakes, we obtained a maximum magnitude of 6.7 in this area. It is likely that the maximum earthquake has already been triggered in this sequence, and the seismic hazard from large earthquakes (M 6+) in this region will be at a low level for a long period (~1000 years).

Supplementary Materials: The following supporting information can be downloaded at: https://www.mdpi.com/article/10.3390/rs14081819/s1, Text S1: inversion strategy; Text S2: inversion setting; Text S3: slip model comparison.

Author Contributions: Conceptualization, C.W.; methodology, C.Z., C.W. and G.Z.; software, C.Z. and C.W.; validation, X.S., Q.L. and J.Z.; formal analysis, B.Z. and P.L.; writing—original draft preparation, C.Z. and C.W.; writing—review and editing, C.Z., C.W., B.Z. and P.L.; supervision, Q.L. and J.Z.; funding acquisition, C.Z., C.W. and Q.L. All authors have read and agreed to the published version of the manuscript.

Funding: This research was funded by the National Key Research and Development Program of China (Grant No. 2019YFC1509203), the National Natural Science Foundation of China (Grant No. 41974006), the Shenzhen Science and Technology Program (Grant No. RCYX20210706092140076 and 20200812164904001), the state Key Laboratory of Earthquake Dynamics (Grant No. LED2021B05), and the Lhasa National Geophysical Observation and Research Station (Grant No. NORSLS20-08).

Institutional Review Board Statement: Not applicable.

Informed Consent Statement: Not applicable.

Data Availability Statement: The S1A InSAR interferograms were obtained from the European Space Agency's InSARap program (http://insarap.org/, last accessed 20 March 2022). ALOS-2 data are copyright of JAXA (2016), and they have been collected through the RA6 project (ID: 3078). The strong-motion data were obtained from the INGV and the RAN (available from the Engineering Strong-Motion database website: http://esm.mi.ingv.it/, last accessed 20 March 2022).

Acknowledgments: The authors would like to thank the European Space Agency, JAXA (2016), the INGV and the RAN for providing SAR images and strong-motion data. The Multi-Data Source Modeling and Inversion Toolkit was used for the strong-motion data processing and mapping (https://github.com/dmelgarm/MudPy, last accessed 20 March 2022). Some of the figures were generated by Generic Mapping Tools (GMT 4.5 from http://gmt.soest.hawaii.edu/, last accessed 20 March 2022). We are grateful to the three anonymous reviewers for constructive comments, that helped us to improve this manuscript.

Conflicts of Interest: The authors declare no conflict of interest. The funders had no role in the design of the study; in the collection, analyses, or interpretation of data; in the writing of the manuscript, or in the decision to publish the results.

Appendix A

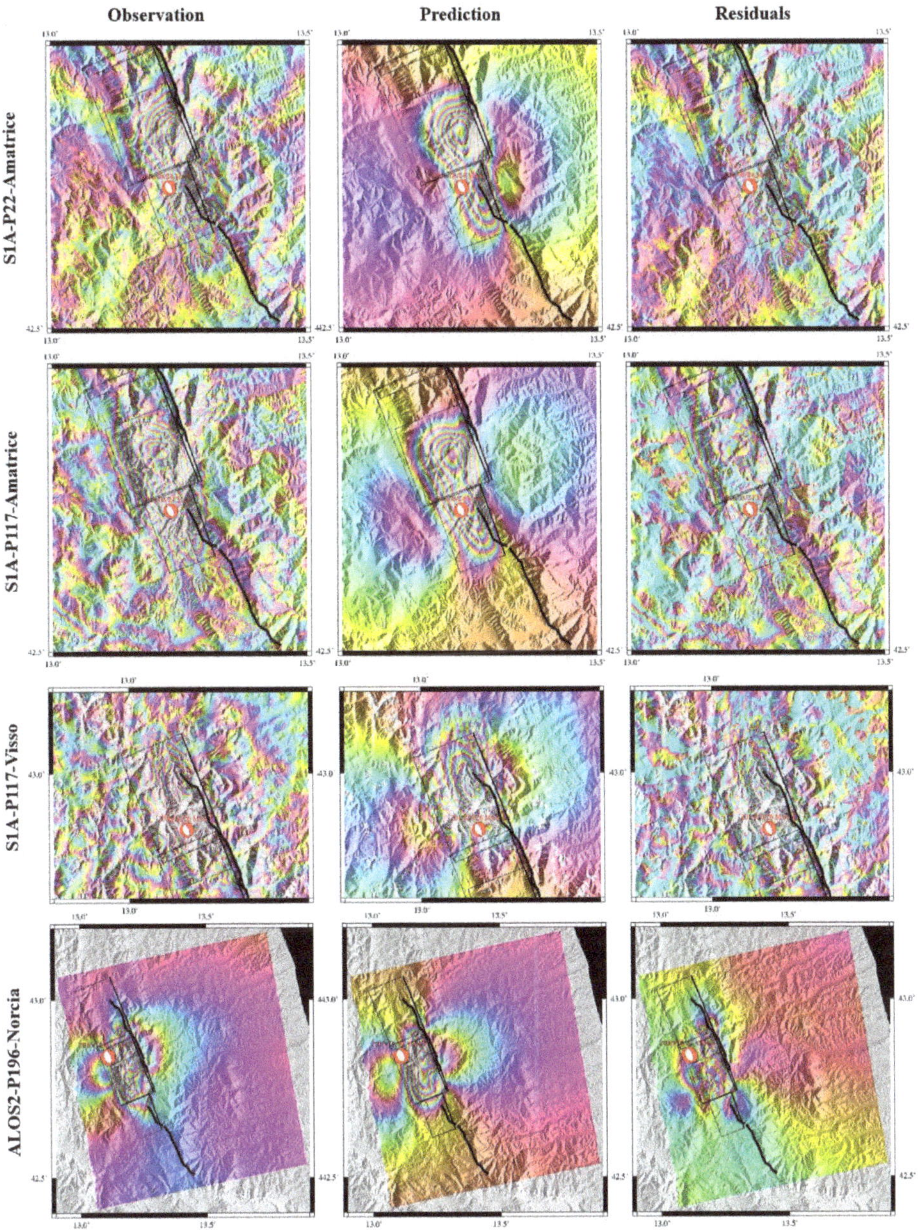

Figure A1. SAR Interferograms, predictions, and residuals based on the joint slip models derived in this study for the Central Italy earthquake sequence. The S1A interferograms are wrapped to 2.8 cm, and the ALOS interferogram is wrapped to 11.45 cm. The red beach ball in each interferogram represents the location and focal mechanism of the corresponding event.

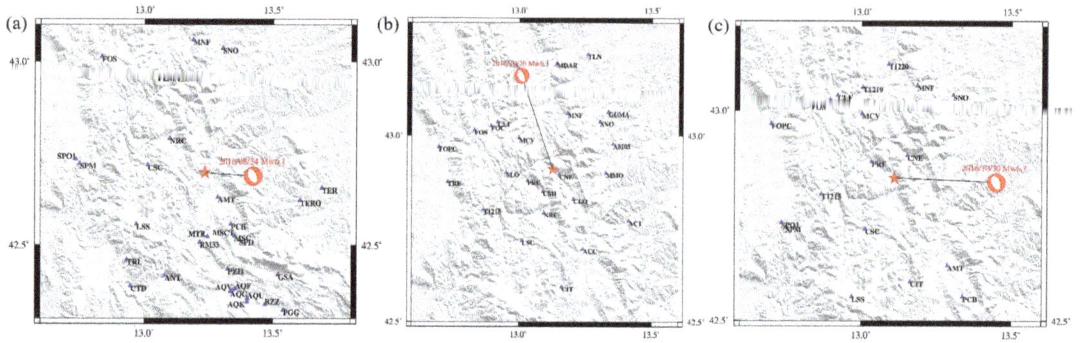

Figure A2. Distribution of the strong-motion stations used to retrieve the slip models of the Amatrice (**a**), Visso (**b**), and Norcia (**c**) earthquakes.

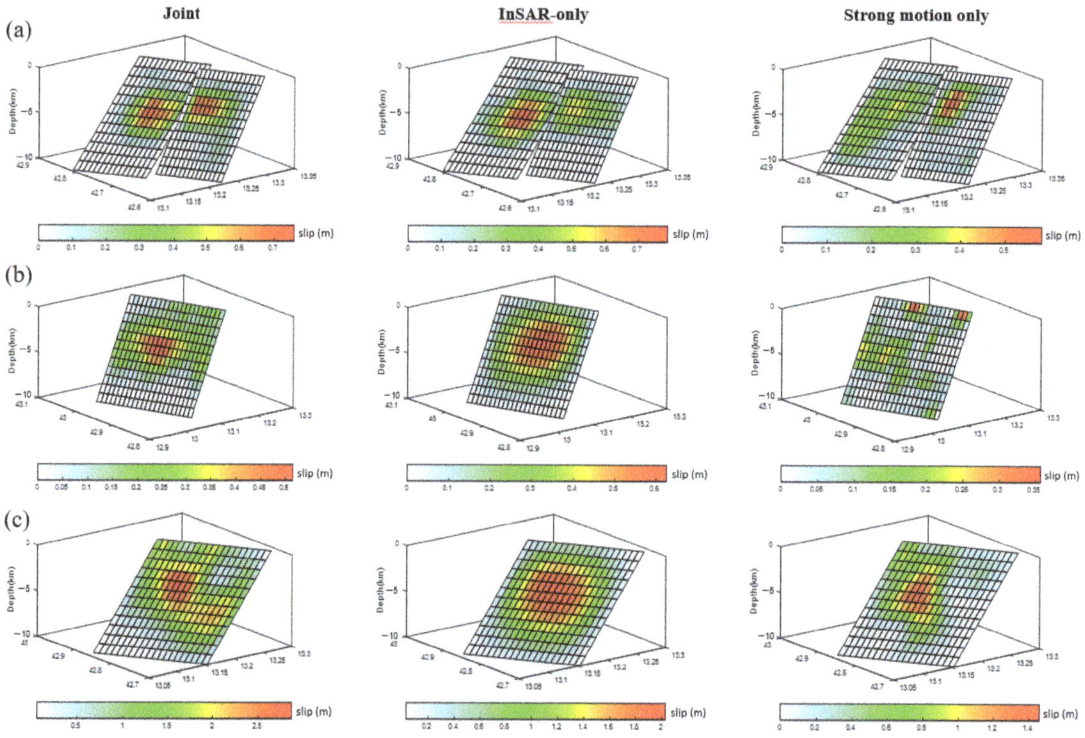

Figure A3. Comparison of slip models using different data constraints. (**a**) The Amatrice earthquake. (**b**) The Visso earthquake. (**c**) The Norcia earthquake.

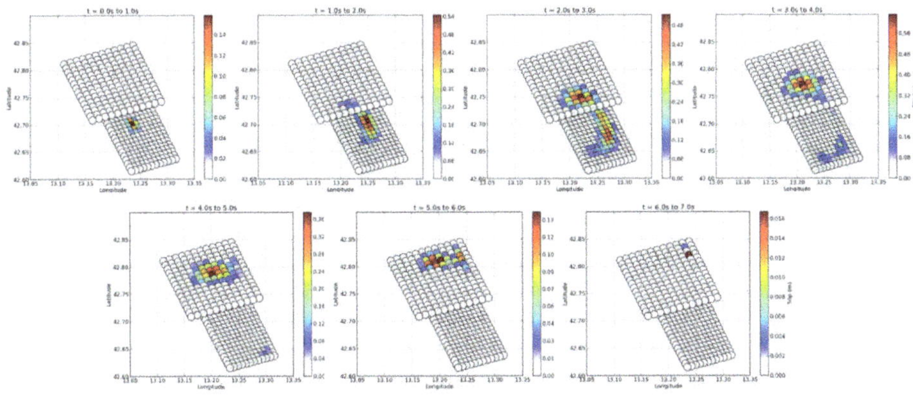

Figure A4. Rupture snapshots in a time interval of 1 s for the Amatrice earthquake.

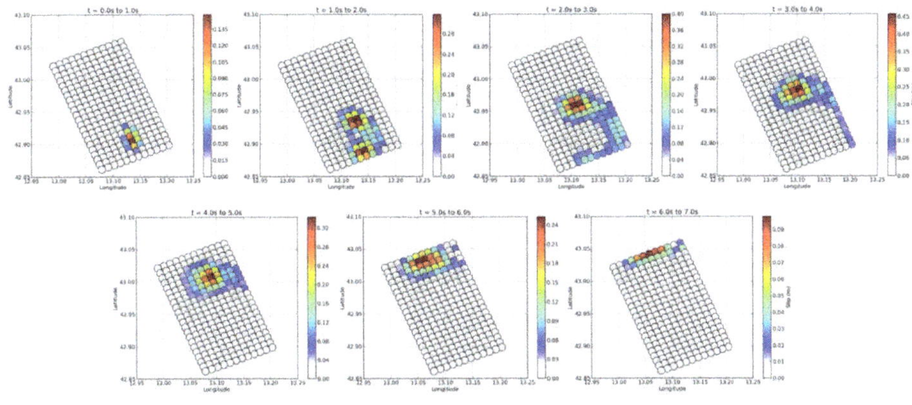

Figure A5. Rupture snapshots in a time interval of 1 s for the Visso earthquake.

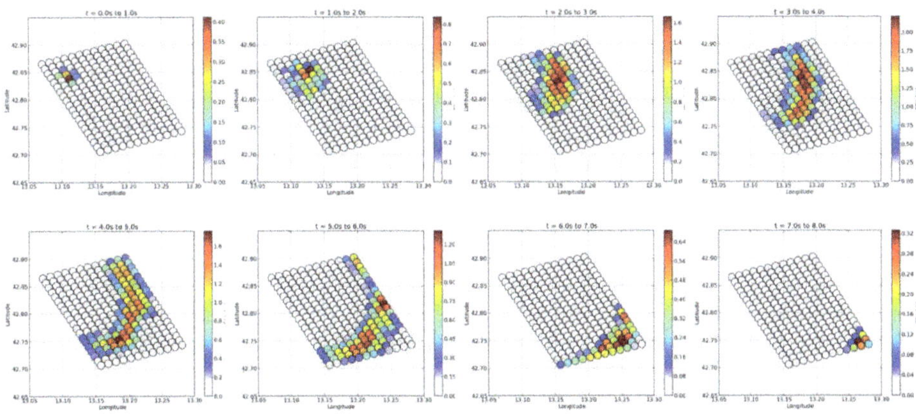

Figure A6. Rupture snapshots in a time interval of 1 s for the Norcia earthquake.

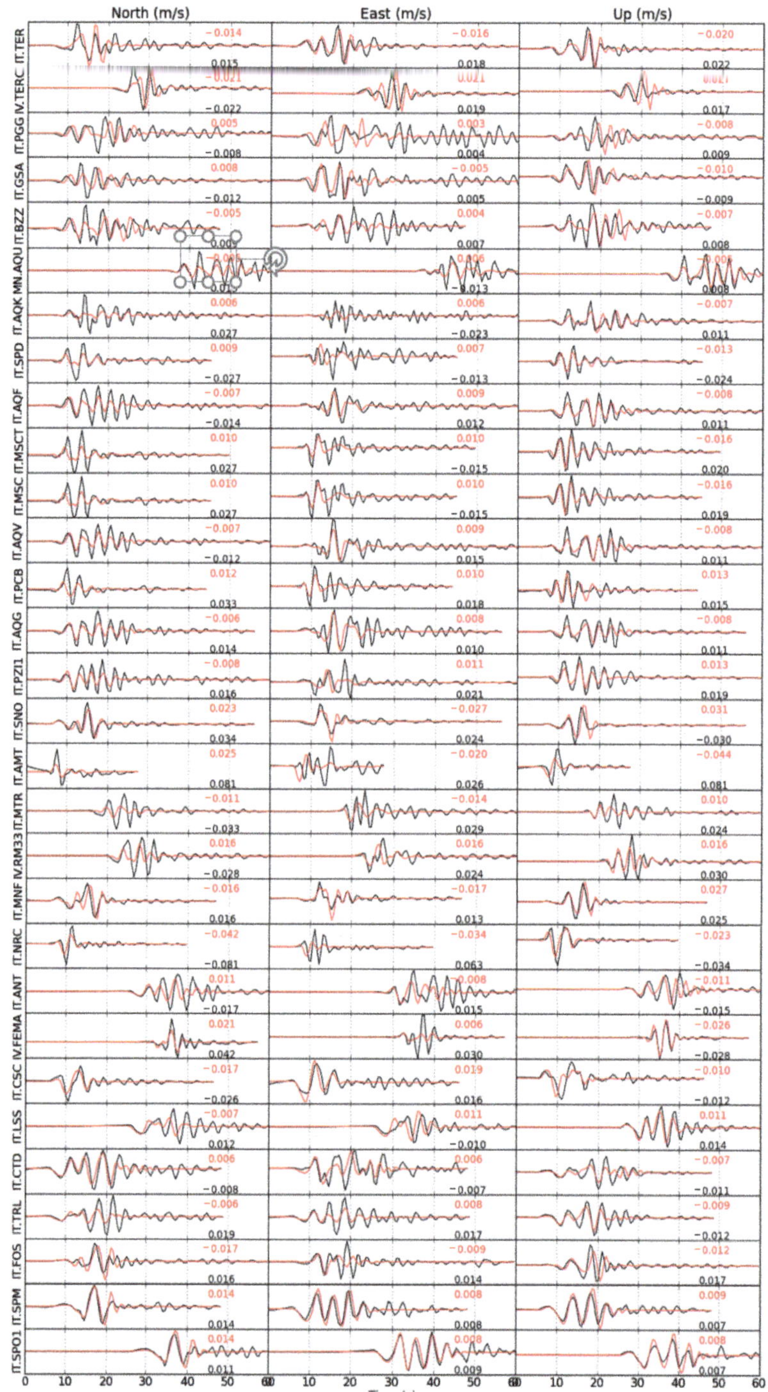

Figure A7. Comparison of strong-motion velocity records (black) and the synthetic seismograms (red) predicted by the joint slip model of the Amatrice earthquake.

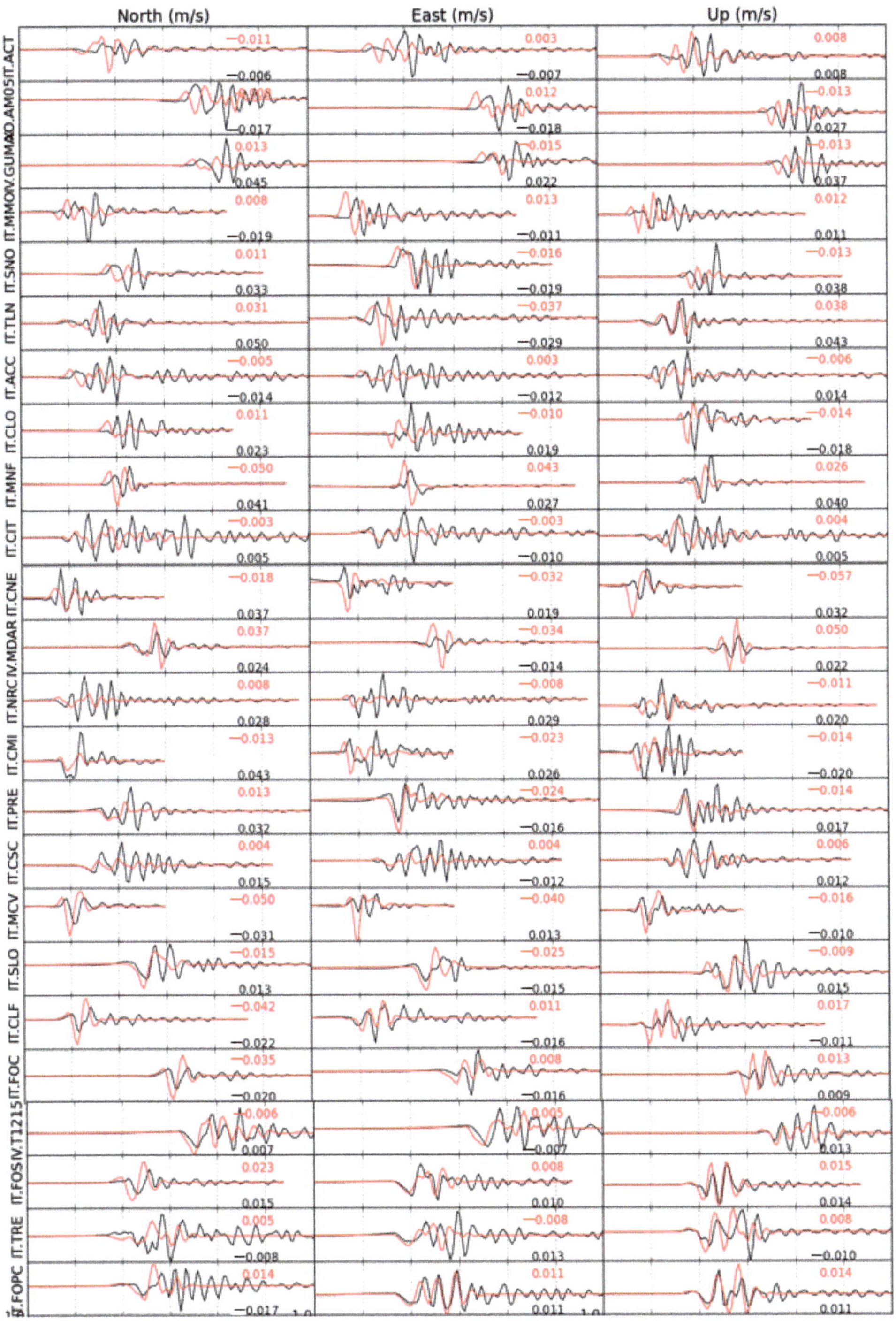

Figure A8. Comparison of strong-motion velocity records (black) and the synthetic seismograms (red) predicted by the joint slip model of the Visso earthquake.

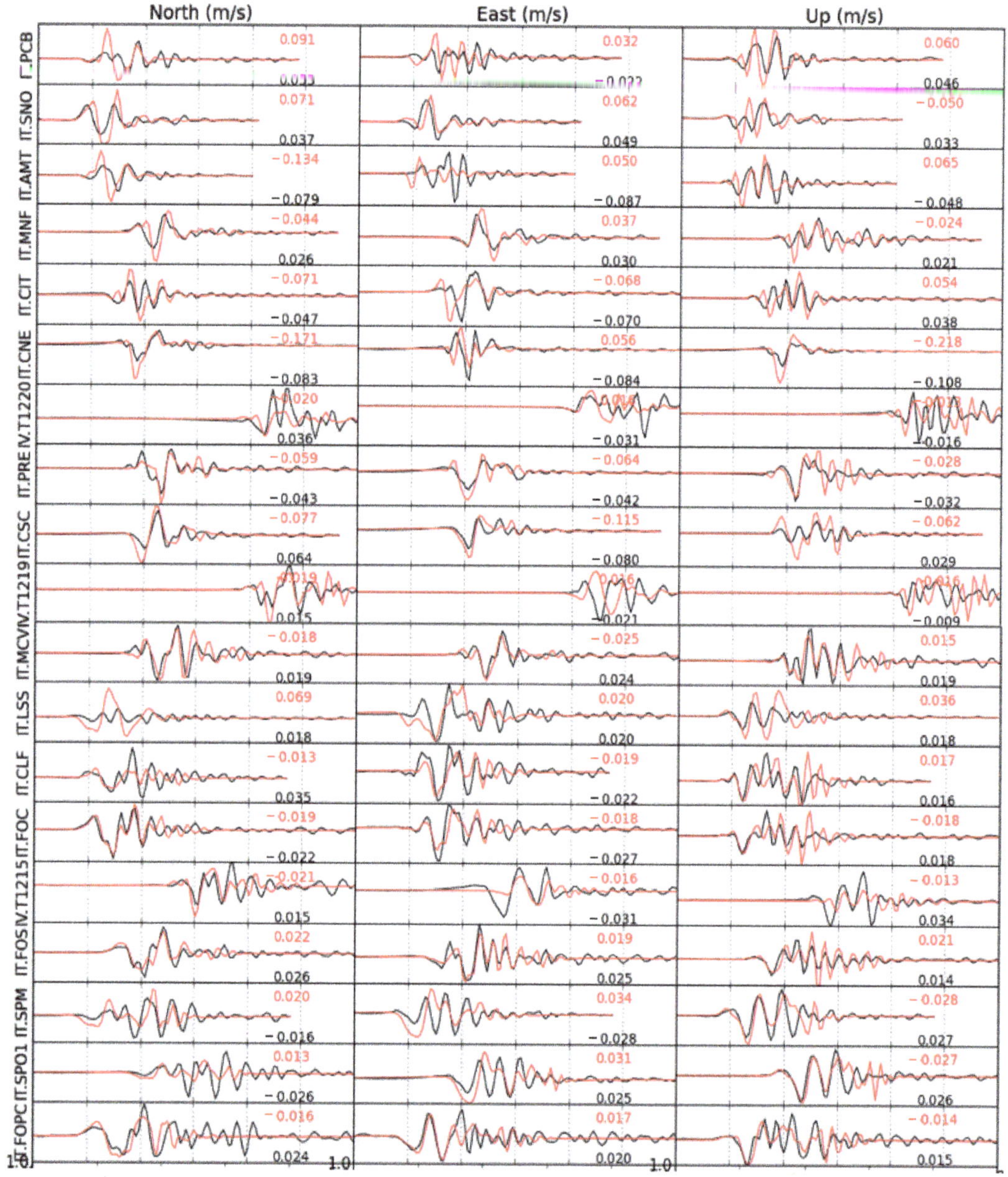

Figure A9. Comparison of strong-motion velocity records (black) and the synthetic seismograms (red) predicted by the joint slip model of the Norcia earthquake.

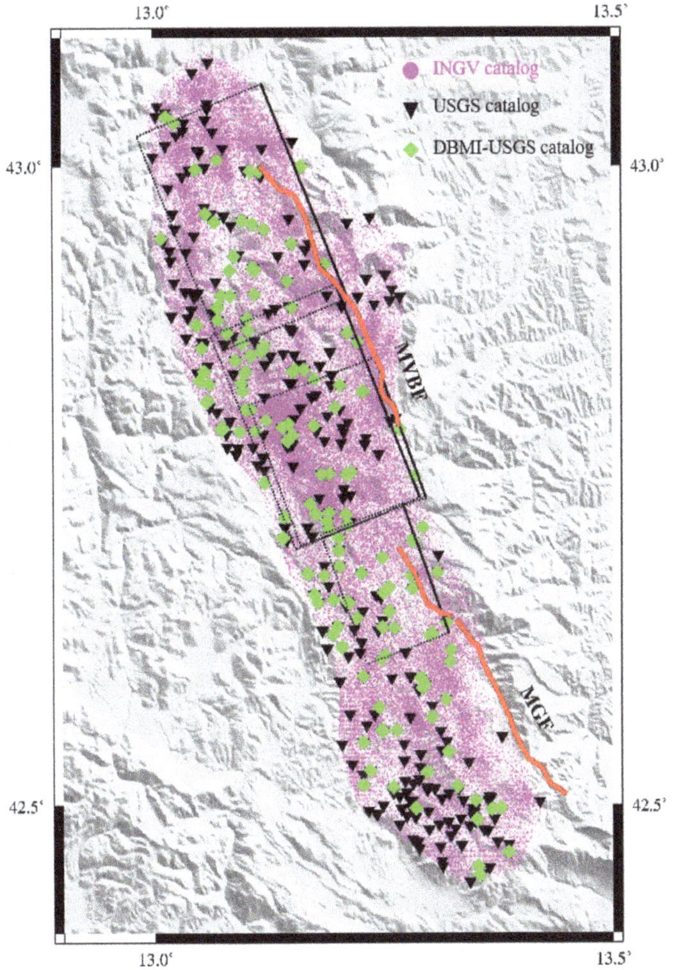

Figure A10. Distribution of seismic activity along the VBFS/GFS systems from the INGV, USGS and DBMI-USGS catalogs. The dashed black rectangles represent the surface projection of the fault planes adopted in this study. The bold red lines are the two seismogenic normal fault systems, namely, the Mt. Vet-tore–Mt. Bove Fault (VBFS) system and the Mt. Gorzano Fault (GFS) system.

Table A1. SAR scenes used for generating the coseismic interferograms.

No	Start Date	End Date	Satellite	Path	Heading	Event
1	0821	0827	S1A	22	D	24 August
2	0815	0827	S1A	117	A	24 August
3	1015	1027	S1A	117	A	26 October
4	1028	1111	ALOS2	196	A	30 October

Table A2. Scarp offset measurements through topography analysis.

No	Latitude	Longitude	Offset (m)	No	Latitude	Longitude	Offset (m)
1	42.983	13.137	3.0562	14	13.249	42.835	0.56817
2	42.983	13.137	3.1736	15	13.251	42.819	1.5543
3	42.972	13.158	0.19934	16	13.251	42.821	3.7760
4	42.968	13.159	0.53237	17	13.255	42.817	0.85712
5	42.957	13.167	2.3998	18	13.255	42.817	2.3591
6	42.957	13.167	2.6133	19	13.259	42.814	3.2770
7	42.956	13.168	1.3242	20	13.256	42.816	2.9960
8	42.913	13.194	0.53158	21	13.264	42.804	1.5217
9	42.897	13.210	0.69758	22	13.329	42.646	0.39271
10	42.899	13.207	0.32701	23	13.344	42.632	0.17142
11	42.890	13.219	0.69320	24	13.364	42.604	0.30947
12	42.855	13.239	0.22003	25	13.382	42.577	1.1996
13	42.854	13.239	0.16111	26	13.445	42.512	1.9561

References

1. Khattri, K. Great earthquakes, seismicity gaps and potential for earthquake disaster along the Himalaya plate boundary. *Tectonophysics* **1987**, *138*, 79–92. [CrossRef]
2. Li, Y.; Shan, X.; Zhu, C.; Qiao, X.; Zhao, L.; Qu, C. Geodetic Model of the 2018 M w 7.2 Pinotepa, Mexico, Earthquake Inferred from InSAR and GPS Data. *Bull. Seismol. Soc. Am.* **2020**, *110*, 1115–1124. [CrossRef]
3. Sykes, L.R.; Menke, W. Repeat times of large earthquakes: Implications for earthquake mechanics and long-term prediction. *Bull. Seismol. Soc. Am.* **2006**, *96*, 1569–1596. [CrossRef]
4. Zhu, C.; Wang, C.; Zhang, B.; Qin, X.; Shan, X. Differential Interferometric Synthetic Aperture Radar data for more accurate earthquake catalogs. *Remote Sens. Environ.* **2021**, *266*, 112690. [CrossRef]
5. Ji, C.; Wald, D.J.; Helmberger, D.V. Source description of the 1999 Hector Mine, California, earthquake, part I: Wavelet domain inversion theory and resolution analysis. *Bull. Seismol. Soc. Am.* **2002**, *92*, 1192–1207. [CrossRef]
6. Konca, A.O.; Leprince, S.; Avouac, J.-P.; Helmberger, D.V. Rupture process of the 1999 M W 7.1 Duzce earthquake from joint analysis of SPOT, GPS, InSAR, strong-motion, and teleseismic data: A supershear rupture with variable rupture velocity. *Bull. Seismol. Soc. Am.* **2010**, *100*, 267–288. [CrossRef]
7. Yokota, Y.; Koketsu, K.; Fujii, Y.; Satake, K.; Sakai, S.I.; Shinohara, M.; Kanazawa, T. Joint inversion of strong motion, teleseismic, geodetic, and tsunami datasets for the rupture process of the 2011 Tohoku earthquake. *Geophys. Res. Lett.* **2011**, *38*, L00G21. [CrossRef]
8. Yue, H.; Lay, T.; Schwartz, S.Y.; Rivera, L.; Protti, M.; Dixon, T.H.; Owen, S.; Newman, A.V. The 5 September 2012 Nicoya, Costa Rica Mw 7.6 earthquake rupture process from joint inversion of high-rate GPS, strong-motion, and teleseismic P wave data and its relationship to adjacent plate boundary interface properties. *J. Geophys. Res.-Earth* **2013**, *118*, 5453–5466. [CrossRef]
9. Cheloni, D.; De Novellis, V.; Albano, M.; Antonioli, A.; Anzidei, M.; Atzori, S.; Avallone, A.; Bignami, C.; Bonano, M.; Calcaterra, S. Geodetic model of the 2016 Central Italy earthquake sequence inferred from InSAR and GPS data. *Geophys. Res. Lett.* **2017**, *44*, 6778–6787. [CrossRef]
10. Chiaraluce, L.; Di Stefano, R.; Tinti, E.; Scognamiglio, L.; Michele, M.; Casarotti, E.; Cattaneo, M.; De Gori, P.; Chiarabba, C.; Monachesi, G. The 2016 central Italy seismic sequence: A first look at the mainshocks, aftershocks, and source models. *Seismol. Res. Lett.* **2017**, *88*, 757–771. [CrossRef]
11. Huang, M.H.; Fielding, E.J.; Liang, C.; Milillo, P.; Bekaert, D.; Dreger, D.; Salzer, J. Coseismic deformation and triggered landslides of the 2016 Mw 6.2 Amatrice earthquake in Italy. *Geophys. Res. Lett.* **2017**, *44*, 1266–1274. [CrossRef]
12. Lavecchia, G.; Castaldo, R.; De Nardis, R.; De Novellis, V.; Ferrarini, F.; Pepe, S.; Brozzetti, F.; Solaro, G.; Cirillo, D.; Bonano, M. Ground deformation and source geometry of the 24 August 2016 Amatrice earthquake (Central Italy) investigated through analytical and numerical modeling of DInSAR measurements and structural-geological data. *Geophys. Res. Lett.* **2016**, *43*, 12389–12398. [CrossRef]
13. Liu, C.; Zheng, Y.; Xie, Z.; Xiong, X. Rupture features of the 2016 Mw 6.2 Norcia earthquake and its possible relationship with strong seismic hazards. *Geophys. Res. Lett.* **2017**, *44*, 1320–1328. [CrossRef]
14. Tinti, E.; Scognamiglio, L.; Michelini, A.; Cocco, M. Slip heterogeneity and directivity of the ML 6.0, 2016, Amatrice earthquake estimated with rapid finite-fault inversion. *Geophys. Res. Lett.* **2016**, *43*, 10745–10752. [CrossRef]
15. Walters, R.J.; Gregory, L.C.; Wedmore, L.N.; Craig, T.J.; McCaffrey, K.; Wilkinson, M.; Chen, J.; Li, Z.; Elliott, J.R.; Goodall, H. Dual control of fault intersections on stop-start rupture in the 2016 Central Italy seismic sequence. *Earth Planet. Sci. Lett.* **2018**, *500*, 1–14. [CrossRef]
16. Xu, G.; Xu, C.; Wen, Y.; Jiang, G. Source parameters of the 2016–2017 Central Italy earthquake sequence from the Sentinel-1, ALOS-2 and GPS data. *Remote Sens.* **2017**, *9*, 1182. [CrossRef]

17. Delouis, B.; Giardini, D.; Lundgren, P.; Salichon, J. Joint inversion of InSAR, GPS, teleseismic, and strong-motion data for the spatial and temporal distribution of earthquake slip: Application to the 1999 Izmit mainshock. *Bull. Seismol. Soc. Am.* **2002**, *92*, 278–299. [CrossRef]
18. Fletcher, J.M.; Oskin, M.E.; Teran, O.J. The role of a keystone fault in triggering the complex El Mayor–Cucapah earthquake rupture. *Nat. Geosci.* **2016**, *9*, 303–307. [CrossRef]
19. Hamling, I.J.; Hreinsdóttir, S.; Clark, K.; Elliott, J.; Liang, C.; Fielding, E.; Litchfield, N.; Villamor, P.; Wallace, L.; Wright, T.J. Complex multifault rupture during the 2016 M w 7.8 Kaikōura earthquake, New Zealand. *Science* **2017**, *356*, eaam7194. [CrossRef]
20. D'Agostino, N.; Mantenuto, S.; D'Anastasio, E.; Avallone, A.; Barchi, M.; Collettini, C.; Radicioni, F.; Stoppini, A.; Fastellini, G. Contemporary crustal extension in the Umbria–Marche Apennines from regional CGPS networks and comparison between geodetic and seismic deformation. *Tectonophysics* **2009**, *476*, 3–12. [CrossRef]
21. Serpelloni, E.; Anzidei, M.; Baldi, P.; Casula, G.; Galvani, A. Crustal velocity and strain-rate fields in Italy and surrounding regions: New results from the analysis of permanent and non-permanent GPS networks. *Geophys. J. Int.* **2005**, *161*, 861–880. [CrossRef]
22. Pizzi, A.; Galadini, F. Pre-existing cross-structures and active fault segmentation in the northern-central Apennines (Italy). *Tectonophysics* **2009**, *476*, 304–319. [CrossRef]
23. Wang, C.; Ding, X.; Li, Q.; Jiang, M. Equation-based InSAR data quadtree downsampling for earthquake slip distribution inversion. *IEEE Geosci. Remote Sens. Lett.* **2014**, *11*, 2060–2064. [CrossRef]
24. Luzi, L.; Puglia, R.; Russo, E.; ORFEUS, W. *Engineering Strong Motion Database, version 1.0*; Istituto Nazionale di Geofisica e Vulcanologia: Rome, Italy, 2016.
25. Fukahata, Y.; Wright, T.J. A non-linear geodetic data inversion using ABIC for slip distribution on a fault with an unknown dip angle. *Geophys. J. Int.* **2008**, *173*, 353–364. [CrossRef]
26. Herrmann, R.B.; Malagnini, L.; Munafò, I. Regional moment tensors of the 2009 L'Aquila earthquake sequence. *Bull. Seismol. Soc. Am.* **2011**, *101*, 975–993. [CrossRef]
27. Kikuchi, M.; Kanamori, H. Inversion of complex body waves. *Bull. Seismol. Soc. Am.* **1982**, *72*, 491–506. [CrossRef]
28. Funning, G.J.; Fukahata, Y.; Yagi, Y.; Parsons, B. A method for the joint inversion of geodetic and seismic waveform data using ABIC: application to the 1997 Manyi, Tibet, earthquake. *Geophys. J. Int.* **2014**, *196*, 1564–1579. [CrossRef]
29. Okada, Y. Internal deformation due to shear and tensile faults in a half-space. *Bull. Seismol. Soc. Am.* **1992**, *82*, 1018–1040. [CrossRef]
30. Olson, A.H.; Apsel, R.J. Finite faults and inverse theory with applications to the 1979 Imperial Valley earthquake. *Bull. Seismol. Soc. Am.* **1982**, *72*, 1969–2001. [CrossRef]
31. Zhu, L.; Rivera, L.A. A note on the dynamic and static displacements from a point source in multilayered media. *Geophys. J. Int.* **2002**, *148*, 619–627. [CrossRef]
32. Scognamiglio, L.; Tinti, E.; Casarotti, E.; Pucci, S.; Villani, F.; Cocco, M.; Magnoni, F.; Michelini, A.; Dreger, D. Complex fault geometry and rupture dynamics of the MW 6.5, 30 October 2016, Central Italy earthquake. *J. Geophys. Res.-Earth* **2018**, *123*, 2943–2964. [CrossRef]
33. Scholz, C. The frequency-magnitude relation of microfracturing in rock and its relation to earthquakes. *Bull. Seismol. Soc. Am.* **1968**, *58*, 399–415. [CrossRef]
34. Wiemer, S. A software package to analyze seismicity: ZMAP. *Seism. Res. Lett.* **2001**, *72*, 373–382. [CrossRef]
35. Schorlemmer, D.; Wiemer, S.; Wyss, M. Variations in earthquake-size distribution across different stress regimes. *Nature* **2005**, *437*, 539–542. [CrossRef] [PubMed]
36. Benedetti, L.; Manighetti, I.; Gaudemer, Y.; Finkel, R.; Malavieille, J.; Pou, K.; Arnold, M.; Aumaitre, G.; Bourles, D.; Keddadouche, K. Earthquake synchrony and clustering on Fucino faults (Central Italy) as revealed from in situ 36Cl exposure dating. *J. Geophys. Res.-Earth* **2013**, *118*, 4948–4974. [CrossRef]
37. Falcucci, E.; Gori, S.; Galadini, F.; Fubelli, G.; Moro, M.; Saroli, M. Active faults in the epi-central and mesoseismal Ml 6.0 24, 2016 Amatrice earthquake region, central Italy. Methodological and seismotectonic issues. *Ann. Geophys.* **2016**, *59*, Fast Track 5. [CrossRef]
38. Manighetti, I.; Caulet, C.; De Barros, L.; Perrin, C.; Cappa, F.; Gaudemer, Y. Generic along-strike segmentation of A far normal faults, E ast A frica: Implications on fault growth and stress heterogeneity on seismogenic fault planes. *Geochem. Geophys. Geosyst.* **2015**, *16*, 443–467. [CrossRef]
39. Wesnousky, S.G. Displacement and geometrical characteristics of earthquake surface ruptures: Issues and implications for seismic-hazard analysis and the process of earthquake rupture. *Bull. Seismol. Soc. Am.* **2008**, *98*, 1609–1632. [CrossRef]
40. Malagnini, L.; Lucente, F.P.; De Gori, P.; Akinci, A.; Munafo, I. Control of pore fluid pressure diffusion on fault failure mode: Insights from the 2009 L'Aquila seismic sequence. *J. Geophys. Res.-Earth* **2012**, *117*, B05302. [CrossRef]
41. Chiarabba, C.; De Gori, P.; Amato, A. Do earthquake storms repeat in the Apennines of Italy? *Terra Nova* **2011**, *23*, 300–306. [CrossRef]
42. McGuire, J.J.; Zhao, L.; Jordan, T.H. Predominance of unilateral rupture for a global catalog of large earthquakes. *Bull. Seismol. Soc. Am.* **2002**, *92*, 3309–3317. [CrossRef]

43. Caniven, Y.; Dominguez, S.; Soliva, R.; Peyret, M.; Cattin, R.; Maerten, F. Relationships between along-fault heterogeneous normal stress and fault slip patterns during the seismic cycle: Insights from a strike-slip fault laboratory model. *Earth Planet. Sci. Lett.* **2017**, *480*, 147–157. [CrossRef]
44. Zhou, Y.; Elliott, J.R.; Parsons, B.; Walker, R.T. The 2013 Balochistan earthquake: An extraordinary or completely ordinary event? *Geophys. Res. Lett.* **2015**, *42*, 6236–6243. [CrossRef]
45. Avouac, J.-P. From geodetic imaging of seismic and aseismic fault slip to dynamic modeling of the seismic cycle. *Annu. Rev. Earth Planet. Sci.* **2015**, *43*, 233–271. [CrossRef]
46. Shen, Z.-K.; Sun, J.; Zhang, P.; Wan, Y.; Wang, M.; Bürgmann, R.; Zeng, Y.; Gan, W.; Liao, H.; Wang, Q. Slip maxima at fault junctions and rupturing of barriers during the 2008 Wenchuan earthquake. *Nat. Geosci.* **2009**, *2*, 718–724. [CrossRef]

remote sensing

MDPI

Article

InSAR Atmospheric Delay Correction Model Integrated from Multi-Source Data Based on VCE

Xiaobo Li [1,2,3], Xiaoya Wang [1,2,4,*] and Yanling Chen [1,2,4]

1 Shanghai Astronomical Observatory, Chinese Academy of Sciences, Shanghai 200030, China
2 School of Astronomy and Space Science, University of Chinese Academy of Sciences, Beijing 100049, China
3 Institute of Earthquake Forecasting, CEA, Beijing 100036, China
4 Shanghai Key Laboratory of Space Navigation and Positioning Techniques, Shanghai Astronomical Observatory, Chinese Academy of Sciences, Shanghai 200030, China
* Correspondence: wxy@shao.ac.cn; Tel.: +86-021-3477-5234

Abstract: With the rapid development of interferometric synthetic aperture radar (InSAR) measurement technology, its measurement accuracy requirements are increasing. Atmospheric delay errors must be corrected, especially in the case of crustal deformation monitoring, the 20% variation of tropospheric water vapor among InSAR pairs generally produces range from 10 cm to 14 cm deformation errors. Such errors can be of the same magnitude as the annual changes in crustal deformation, or even greater, masking crustal deformation information and seriously affecting the results of crustal deformation monitoring. Therefore, in order to obtain a more accurate InSAR atmospheric delay correction model, this paper calculated and integrated atmospheric delays that were estimated by different sources, including the 37 pressure levels of the fifth generation of the European Centre for Medium-Range Weather Forecasts (ECMWF)) numerical weather prediction model, ECMWF Reanalysis v5 (ERA5), and Global Navigation Satellite System (GNSS) measurement data from the crustal movement observation network of China, based on the variance component estimation (VCE) weighting method. The results showed that the integrated model, based on the VCE method, is better than the generic atmospheric correction online service (GACOS) model for InSAR measuring of crustal deformation. The precision in monitoring crustal deformations was improved by approximately 5 mm, the correlation coefficient of atmospheric delay errors and crustal deformations improved from 0.287 to 0.347, and accuracy improved by approximately 25%. However, the improvement in accuracy was limited because of system error decoherence that was induced by atmospheric noise caused by abundant vegetation or snow cover. Therefore, in order to achieve more accurate results, we recommend the adoption of the multi-source integrated atmospheric delay correction model, based on the VCE method, for InSAR high-precision measuring of crustal deformation and seismic activities.

Keywords: InSAR; ERA5; the crustal movement observation network of China; variance component estimation (VCE) weighting; accuracy analysis

Citation: Li, X.; Wang, X.; Chen, Y. InSAR Atmospheric Delay Correction Model Integrated from Multi-Source Data Based on VCE. *Remote Sens.* **2022**, *14*, 4329. https://doi.org/10.3390/rs14174329

Academic Editors: Guohong Zhang, Chisheng Wang, Wu Zhu, Daqing Ge and Siting Xiong

Received: 4 July 2022
Accepted: 26 August 2022
Published: 1 September 2022

1. Introduction

Interferometric synthetic aperture radar (InSAR) is an earth observation (EO) technique that combines synthetic aperture radar (SAR) technology and interferometry. After periodically obtaining and aligning images of the same area, interferometric images are generated based on the imaging geometry relationship between the SAR satellite and the ground targets; then, the monitoring of crustal deformation and seismic activities can be carried out [1]. InSAR technology has attracted much attention, due to its advantages of large coverage and high spatial resolution; in addition, it can provide all-weather, all-day, high-precision deformation monitoring, especially in some geographical areas that are not covered by the traditional measurement methods. Therefore, the data processing accuracy of InSAR technology for atmospheric delays is improving and the technology is receiving increased attention (see the Supplementary Materials). In 1997, Zebker et al. [2] indicated that the deformation errors caused by atmospheric delays attributed to water vapor alone can reach 10 cm to 14 cm, a degree of error that is difficult to ignore in monitoring millimeter-scale crustal deformation. Therefore, atmospheric delay correction must be carried out for InSAR data processing (see the Supplementary Materials).

We investigated the developmental history of atmospheric delay correction techniques via InSAR measurement, from the use of InSAR's own image data to the use of external data to correct the atmospheric delay effect on deformation monitoring. We also examined single-technology and multi-technology fusion correction methods. In each step, we witnessed the persistence of researchers with respect to the atmospheric delay correction of InSAR measurements and the gradual improvement of accuracy.

In this paper, we built a new atmospheric delay correction model by fusing the newer numerical meteorological forecasting model and the GNSS stations of the crustal movement observation network of China, based on a more reasonable weighting approach, and verified the accuracy of the new model by comparing it with the generally used generic atmospheric correction online service (GACOS) model for InSAR (http://www.gacos.net/, accessed on 11 May 2021), under the same experimental conditions and performing the deformation validation on GNSS stations.

In 1994, Massonnet [3] studied the Landers earthquake and identified the presence of atmospheric signals in the interferogram of repeated tracks. After that, there were more and more studies on the correction of atmospheric delay errors in InSAR measurements. The methods that were used can be divided into two main categories: one category is the correction of errors by on the basis of a method's own image data without the involvement of meteorological parameters; the other category is the use of external data in making corrections. The first category includes the phase accumulation method (stacking) [4], the permanent scatterer (PS) technique [5], and the small baseline subset (SBAS) technique [6]. Among these three methods, the phase accumulation method may reduce the temporal resolution of InSAR measurements; the PS and SBAS techniques may waste a large amount of SAR image data, and the processor's subjective judgment is arbitrary. The second category of methods utilizes external data from the ground-based global navigation satellite system (GNSS), satellite-based medium resolution imaging spectrometer (MERIS) and moderate-resolution imaging spectrometer (MIRIS) data, moderate-resolution imaging spectroradiometer (MODIS) observations, numerical weather prediction models, ground-based meteorological station observations [7], and wireless sounder observations [8]. Among these methods, the use of GPS data is the primary source of information.

In 1997, Zebker et al. [2] pointed out that when spatial and temporal variations in atmosphere relative to humidity reaches 20%, deformation measurement errors of 10 cm to 14 cm and elevation measurement errors of 80 m to 290 m can result. In 1998, Williams et al. [9] used GPS water vapor data obtained from the GPS monitoring network in southern California to reduce the influence of the atmosphere on InSAR interferogram maps. Their results confirmed that GNSS-derived water vapor data could provide a more thorough and rigorous modeling of the effects of water vapor on InSAR interferometric data after a certain spatial interpolation, and that the GPS zenith delay was correlated with elevation, which was consistent with the conclusion of Zebker [2] and the statistical model of Treuhaft-Lanyi [10,11]. Li et al. [12,13] and Luo et al. [14] proposed different interpolation models and investigated the relationship between atmospheric delay and topography. They found that the use of the topography-dependent turbulence model (GTTM) [15,16] could result in the interpolation correction map being more accurate.

In 2011, Jolivet et al. [17] conducted InSAR atmospheric delay correction in the Kunlun Mountains region, using reanalysis information from the European Centre for Medium-Range Weather Forecasts (ECMWF). They calculated the ratio of phase/elevation before and after atmospheric delay correction within a certain displacement window range. They found the monitoring accuracy was improved by 2.0 rad/km. However, their study also found that the atmospheric delay values obtained from meteorological models alone were not necessarily reliable [18].

In 2014, Wang et al. [19] used the ECMWF meteorological forecast model to correct the effects of INSAR atmospheric delay errors and found that the ECMWF model could not accurately reflect the details of local atmospheric changes, especially in regions with large changes in relative humidity. The ECMWF model was more suitable for regions with relatively small changes in large-scale meteorological conditions [20]. Regardless of whether GNSS data or other external data, such as data derived from numerical weather prediction models, are used, they need to match the temporal and spatial resolution of InSAR-obtained image data. However, there are no data that can achieve complete agreement with InSAR-obtained image data, in terms of spatial and temporal resolution. Therefore, it is necessary to combine multiple data to develop a better InSAR atmospheric delay correction model.

In 2007, Li et al. [21] analyzed the Gaussian morphology, anisotropy, and energy spectrum characteristics of atmospheric noise by four interferograms in the Shanghai area, and theoretically confirmed that an external data spatial resolution of at least 0.3 km is required in order to correct 90% of atmospheric delay errors in InSAR measurements. However, GNSS stations are discrete and unevenly distributed, requiring the use of other data to achieve higher spatial resolution. Thereafter, many scholars [22–28] studied the integration of GPS and MODIS/MERIS water vapor data for better atmospheric delay correction. However, both MODIS and MERIS are near-infrared water vapor measurements, which can only describe the amounts of column water vapor on clouds in clear daytime land areas, as well as in cloud-free land and ocean areas. Currently, MERIS water vapor data are not freely available.

Therefore, we built an atmospheric delay correction model for InSAR measurements by integrating newer numerical weather prediction model reanalysis data and ground-based GNSS observation data, based on a variance component estimation (VCE) weighting method. The model was validated and compared with the current widely used generic atmospheric correction online service (GACOS) model for InSAR measurements [29], and then applied to deformation monitoring of the 2019 Changning M6.0 earthquake that occurred in Sichuan, China, to evaluate its monitoring accuracy.

We emphasize that atmospheric delay errors exist in a variety of measurement techniques; such errors are among the important sources of error that affect the accuracy of crustal deformation monitoring. Our atmospheric delay correction model can be extended and used in the data processing of various types of crustal deformation monitoring. Our model is not limited seismic deformation applications. Although the magnitude of the seismic deformations is generally larger than that of atmospheric deformations, atmospheric delay is significant in monitoring crustal deformation via InSAR measurements, especially in regions with a large tropospheric influence, such as those regions near the equator. Atmospheric delay is also significant in monitoring urban surface subsidence.

2. Study Area and InSAR Data

Changning is a region of complex terrain in the Sichuan province of China. It is located on a secondary fault near the Changning-Shuanghe back-slope tectonics on the edge of the Sichuan Basin, with a northwest–southeast trending stress field that is prone to triggering landslide hazards [30]. On 17 June 2019, there was an M6.0 earthquake with an epicenter located at 28.34°N (north latitude) and 104.90°E (east longitude). It had a 16 km source depth (http://www.ceic.ac.cn/history, accessed on 29 January 2021). The study area of interest is near the epicenter. Figure 1 shows the range and the topography of the study area. We monitored co-seismic deformation by InSAR measurements from the Sentinel-1A satellite and calculated the atmospheric delay correction model for InSAR measurements in the study area in order to improve the accuracy of crustal deformation monitoring.

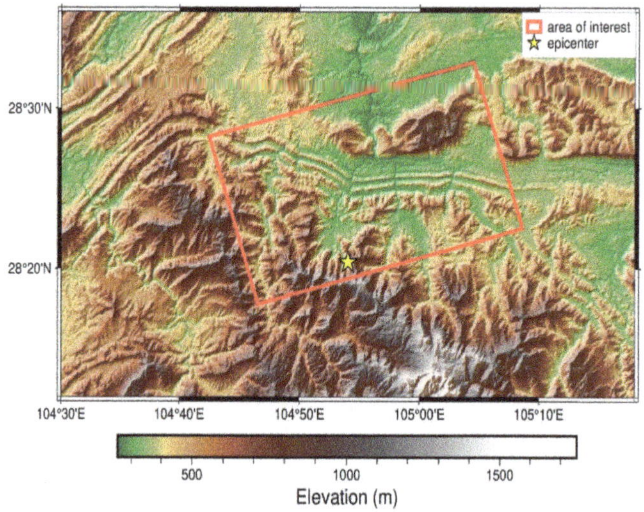

Figure 1. The range of the area of interest and the topography from SRTM3 v4.1 DEM [31]. The red rectangular represents the area of interest, the yellow pentagram represents the earthquake location as identified by the China Earthquake Networks Center. The color-coded strip represents the elevation of the topography.

The Sentinel satellite is another epoch-making C-band SAR satellite, after ERS and ENVISAT satellites. It can provide all-weather, all-day, high spatial resolution remote sensing data with good spectral quality. Due to its short wavelength, the C-band can be ignored in considering the effect of ionospheric delay in atmospheric delay, and only the tropospheric delay in the signal propagation process needs to be considered. The Sentinel-1 series satellite opens a new era of free data accessibility [32]. The measurement data of the Sentinel-1 satellite can be downloaded freely from European Space Agency (ESA) data (https://scihub.copernicus.eu/dhus/#/home, accessed on 24 September 2019) by user registration and login; the orbit file, with measurement data, can be downloaded freely in the other website of the ESA (https://scihub.copernicus.eu/gnss/#/home, accessed on 27 October 2019) via different user information. The Sentinel-1 satellite has two-track data, three-track data, and four-track data, the two-track data is generally used in different InSAR data, i.e., D-InSAR (differential interferometric synthetic aperture radar).

Due to the short wavelength and high frequency of the C-band, the effect of ionospheric delay is very small and the effect of atmospheric delay is mainly that of the tropospheric delay in the signal propagation process.

Usually, InSAR data processing is used to extract DEM with high accuracy and D-InSAR data processing is used to obtain the surface deformation displacement caused by an earthquake, a volcano, or city surface subsidence, and to carry out interference stake time series analyses, such as persistent scatters (PS) and Small baseline subsets (SBAS). For simplicity, in this paper, we refer to D-InSAR as InSAR.

In processing the InSAR measurements for crustal deformation displacement, co-polarization is usually used, because its penetrability is better than that of cross-polarization. Sngle look complex (SLC) radar image data, before and after the interested event, is taken as input data (e.g., Sentinel-1A VV polarization image data). One image is used as the master image and another is used as the slave image. The quality of the interference image pairs is evaluated by a baseline estimation and multi-look processing by azimuth resolution and range resolution to suppress the speckle noise of the SAR images. Then, an interferogram could be obtained, based on STRM DEM [31], in order to reduce the influence of terrain error. This paper adopted a DEM of 90 m $\times$ 90 m resolution. The phase unwrapping is carried out by the Goldstein filtering method, using the Delaunay minimum cost flow theory to solve the problem of phase ambiguity. The corresponding precision orbit is used to adjust the orbit and re-flatten it by selected ground control points (GCP). Finally, the line-of-sight (LOS) co-seismic deformation displacement in the WGS84 coordinate system is obtained via geocoding, and the initial or raw displacement field before atmospheric delay correction is obtained.

We chose the repeat track data of the Sentinel-1A satellite for a 12-day interval period before and after the Changning M6.0 earthquake in the interest area, i.e., SLC image data for 9 June 2019 (day of year, DOY 160) and 21 June 2019 (DOY 172) were downloaded, respectively; we obtained the co-seismic displacement field of the area of interest by D-InSAR measurements.

3. Methodology

3.1. Numerical Meteorological Forecasting Atmospheric Delay Calculation Model

In 2016, the European Centre for Medium-Range Weather Forecasts (ECMWF) [33] (http://www.ecmwf.int, accessed on 8 May 2021) released a revised version of the fifth generation of the global atmospheric numerical forecast reanalysis (ECMWF Reanalysis v5, ERA5); the data has been updated since 2019 [34]. This information provides 137 model levels of information on atmospheric, terrestrial, and oceanic climate changes within 80 km of height and 37 pressure levels of the relevant atmospheric data, stratified by hourly pressure values from 1 hPa to 1000 hPa on an 0.25° × 0.25° grid. The latter information is freely available. Therefore, this paper applied the pressure, temperature, specific humidity, potential volume, and relative humidity information provided from the 37 pressure levels.

There are two steps in ERA5 atmospheric zenith total delay (ZTD) calculations (subsequently referred to as ERA5 ZTDs). First, the ERA5 ZTDs are calculated for each grid corresponding to the 37 pressure levels' atmospheric data, by integral methods, as follows [33]:

$$ZTD_{\text{int\_}sfc} = 10^{-6} \cdot \int_{h_{sfc}}^{h_{top}} N \cdot \mathrm{d}h \tag{1}$$

In addition,

$$N = k_1 \cdot \frac{(P-e)}{T} + k_2 \cdot \frac{e}{T} + k_3 \cdot \frac{e}{T^2}$$

$$e = sh \times P/0.622$$

$$k_1 = 77.604\mathrm{K/Pa};\ k_2 = 64.79\mathrm{K/Pa};\ k_3 = 377600.0\mathrm{K/Pa}$$

where $ZTD_{\text{int\_}sfc}$ denotes the atmospheric delay from the bottom to the top of the 37 pressure levels' meteorological values; h_{sfc} and h_{top} denote the bottom and top heights of the ERA5 meteorological data, respectively; N is the total refractive index; P denotes the atmospheric pressure in hPa; e denotes the water vapor pressure in hPa; sh denotes the specific humidity; T denotes the temperature of the nth level at the corresponding grid point; the unit is K; $\mathrm{d}h = h_{n-1} - h_n$; and h_n, the potential height, indicates the height of the nth level on the grid point, and is obtained by the potential divided by the local acceleration of gravity named, the unit is km; n decreases with the increase in height.

After deriving the ZTD value by the integration method from the 37 pressure levels' meteorological data, at approximately 47 km in height, the zenith delay above the top level must be added in order to make it comparable with the ZTD of GNSS stations, based on GNSS's high-precision data processing solution [33]. However, there is no meteorological data above the top level of approximately 47 km in height. Fortunately, the effect of the wet delay is almost negligible, due to the reduced water vapor content above the top level [35].

The ZTD value above the top level could be calculated by the Saastamoinen delay model, as in the following equation [36]:

$$ZTD_{top} = 0.002277 \times \frac{\left[P_{top} + \left(0.05 + \frac{1255}{T_{top}+273.15}\right) \cdot e_{top}\right]}{f(\varphi, H)} \quad (2)$$

In addition,

$$e_{top} = rh \times 6.11 \times 10^{\frac{7.5 \cdot T_{top}}{T_{top}+273.15}}$$

$$f(\varphi, H) = 1 - 0.00266 \cdot \cos(2\varphi) - 0.00028 \cdot H$$

where P_{top} denotes the integrated top pressure level in hPa, T_{top} denotes the integrated top level Celsius temperature in °C, e_{top} denotes the integrated top level water vapor pressure; rh denotes the integrated top level relative humidity (0~1); φ is the geocentric geodetic latitude of the station or grid point in rad; and H is the grid height in km above the sea level.

By adding the integrated atmospheric delay and the corresponding grid atmospheric delay obtained via the Saastamoinen model, the atmospheric zenith delay derived by the numerical weather prediction model can be expressed as follows [33]:

$$ZTD = ZTD_{\text{int\_}sfc} + ZTD_{top} \quad (3)$$

This ZTD value is the atmospheric delay at the grid position with the pressure of 1000 hPa. In fact, due to the existence of terrain undulations, the obtained atmospheric delay is not necessarily the surface elevation observed by InSAR measurements. Therefore, it is necessary to convert the ZTD value to the surface elevation-by-elevation correction, and the conversion formula used is the following [37,38]:

$$P = P_0 \cdot \left[1 + \left(\frac{8.419 \times 10^{-5} \cdot (H_0 - H)}{P_0{}^{0.190284}}\right)\right]^{5.255303} \quad (4)$$

$$T = T_0 - 6.5 \cdot (H - H_0) \quad (5)$$

where P is the surface air pressure and the unit is hPa; P_0, T_0, H_0 are the air pressure, temperature, and elevation at the known points of the corresponding grid in hPa, K, and km, respectively; H is the elevation of the surface; T is the surface temperature; and the unit is K.

3.2. GNSS Data Estimated Atmospheric Delay Model

Using GAMIT/GLOBK software, GNSS ZTDs with high accuracy could be obtained [39]. This paper applied the continuously observed GNSS stations of the crustal movement observation network of China (i.e., the land state network, CMONC) and the IGS stations around the study area to obtain absolute ZTD values, for GNSS network stations with a maximum baseline of more than 500 km. All models, including receiver antenna files, antenna phase center models, earth orientation parameters, moon ephemeris tables, sun ephemeris tables, and ocean tide models, were updated. ITRF2014 provided the prior coordinates, with constraints of 0.1 m for CMONC stations and 0.01 m for IGS stations. The GMF mapping function was used for the tropospheric delay model and the zenith wet delays, as the unknown parameters, were solved once per hour, together with other parameters, such as station coordinates and satellite orbits.

However, ZTDs determined by the GNSS network or by ERA5 reanalysis data are often different from ZTDs that are required at the time of InSAR satellite transit in the spatio-temporal distribution. They must be interpolated and matched. This paper applied Lagrange's interpolation to obtain the required ZTDs of the InSAR observation area at the time of SAR satellite transit.

3.3. GACOS Model

In 2018, CHEN et al. [29] proposed taking advantage of the 0.125° × 0.125° horizontal resolution global coverage and the 137 levels of vertical resolution data information provided by the 6-h ECMWF numerical meteorological model, as well as the high temporal resolution (5 min) and high accuracy measurements of discrete fixed GPS stations, to form an atmospheric delay correction model for InSAR measurements. The optimum weight between ECMWF and GPS is the lowest cross-validation RMS value of the GPS network stations within 150 km of the decorrelation distance. The test range of the ECMEF/ GPS relative weight is from 0.0 to 10 by steps of 0.1. In [29], the atmospheric delay was separated into the elevation-related components and the turbulent components by an iterative approach. Moreover, the atmospheric delay correction model for InSAR measurements was built by interpolation and exponential coefficients for the considered pixel. This GACOS model is generally applicable in both flat and upland areas. Users can download the atmospheric delay correction model for a certain region at a certain time on the GACOS website (http://www.gacos.net/, accessed on 11 May 2021) and load it into their own study area. The GACOS model is a commonly used method to correct the atmospheric delay for InSAR measurements. In this paper, it is used as the comparative atmospheric delay model.

We calculated the GNSS ZTDs by using the more than 60 GNSS stations on the Qinghai-Tibet Plateau of the Crustal Movement Observation Network of China (CMONC) and eight surrounding IGS stations. Figure 2 shows the distribution of the selected GNSS stations in the CMONC, and the location of the epicenter.

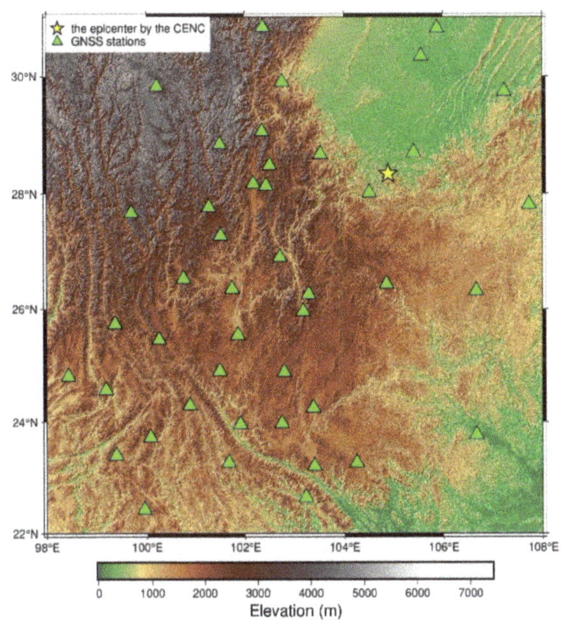

Figure 2. Distribution of selected GNSS stations in the crustal movement observation network of China near the area of interest. The green triangle represents the selected GNSS stations of the Crustal Movement Observation Network of China (CMONC). The yellow pentagram represents the location of the epicenter by the China earthquake networks center (CENC).

3.4. Multi-Source Data Integrated Atmospheric Delay Modeling Based on VCE

With the weather forecasting model update and the increase in the number of GNSS stations, and most importantly with the need for a more reasonable weighting method, this paper established a new multi-source data integration atmospheric delay correction model for InSAR measurements, based on the variance component estimation (VCE) weighting method [40–44], by GNSS high-precision real ZTD observations and high spatial and temporal resolution ERA5 ZTDs.

In order to integrate the same volume for the different sources, we converted the surface elevation of the ERA5 ZTDs to the corresponding elevation position of the GNSS station by the elevation-correction-of-the-neighborhood-grid-points method and compared them with the ZTDs at the GNSS station. The ERA5 ZTDs and the GNSS ZTDs were discussed in Sections 3.1 and 3.2.

The integrated ZTD is a sum of the elevation-related stratified component and the turbulence component, in that the stratified component can be expressed as the exponential function. The formula is as follows [29].

$$ZTD = S + T + \varepsilon \tag{6}$$

In addition,

$$S = L \cdot e^{-\beta h} \tag{7}$$

where ZTD is the integrated atmospheric delay value, T is the turbulence component delay, S is the stratified component delay value, ε is the remaining residual, L is the stratified delay at sea level, β is the exponential coefficient, and h is the height.

Because the accuracy and spatio-temporal resolution of the ZTD obtained by different techniques are different, the integration from different sources needs to consider the weight assignment of each source to obtain the high-precision measurement result. In data processing, the variance component estimation (VCE) is a widely used weighting method [40–44] that can set an arbitrary initial weight value to provide a pre-adjustment and calculate the variance of observation; then, the new variance estimation weight is calculated to improve the previous weight value. These steps were performed repeatedly until the weight factor was convergent. It was most important to establish the error equations for the VCE method. Suppose there are two technologies; then, the VCE method is as set out below.

The functional model is as follows.

$$l_1 = B_1 \cdot x + \Delta_1 \tag{8}$$

$$l_2 = B_2 \cdot x + \Delta_2 \tag{9}$$

Then,

$$D(l_1) = D(\Delta_1) = {\sigma_{01}}^2 \cdot {P_1}^{-1} \tag{10}$$

$$D(l_2) = D(\Delta_2) = {\sigma_{02}}^2 \cdot {P_2}^{-1} \tag{11}$$

where l_1 and l_2 are the observations of two technologies, respectively; Δ_1 and Δ_2 are observation errors of two technologies, respectively; x is the unknown matrix; B_1 and B_2 are the coefficient matrices of the two technologies, respectively; $D(\Delta_2)$ and $D(\Delta_1)$ are the pretest variances of the two technologies, respectively; ${\sigma_{01}}^2$ and ${\sigma_{02}}^2$ are the variances in the unit weight of the two technologies, respectively; and P_1 and P_2 are the observation weights.

The established error equations are as follows.

$$v_1 = B_1 \cdot \hat{x} - l_1 \tag{12}$$

$$v_2 = B_2 \cdot \hat{x} - l_2 \tag{13}$$

Generally, the initial weight is not appropriate, while the variances in unit weight are identical for the different technologies, as follows.

$${\sigma_{01}}^2 \approx {\sigma_{02}}^2 \tag{14}$$

This means that we need to establish the relationship between the residual error and the variance in unit weight. The specific formula can be found in references [40–45].

In this study, we supposed, first, that the turbulence components were zero and then we established the error equation by derivation of the exponential function. The error equation was as follows.

$$v = \ln L - \beta h - \ln S \tag{15}$$

We calculated the coefficient β and L of the stratified component by the VCE method. The iteration stopped until the β and L were convergent., During the process, the stratified and turbulence components were updated constantly by new β and L and the new stratified component was again weighted by the variance component estimation. These steps were repeated. This iteration stopped until the weight factor was convergent [45,46] and each component was also convergent.

The threshold setting of the convergence condition affects the iteration number and the accuracy of the results. the lower the stop threshold, the greater the number of iterations and the longer the procedure operation time; however, higher accuracy is not necessarily obtained. We chose different thresholds of weight factor change for our experiments to facilitate building the more accurate and appropriate atmospheric delay correction model based on the VCE. We found that the correction result was not more highly accurate when the threshold was lower, but the result was consistent with the accuracy obtained by setting the threshold at 0.001. Moreover, the coefficient convergence condition of 0.001 was sufficient to obtain the accuracy of the atmospheric delay correction values at sub-millimeter levels.

The threshold referred to was the convergence condition weight factor. In order to separate the stratification associated with the elevation and turbulent components, each component was required to converge simultaneously with the iterative process in calculating the exponential function coefficients.

We used the atmospheric delays in the study area on DOY 160 and DOY 172 in 2019 as the basis for calculating the daily atmospheric delays based on the VCE method (Figure 3c,d), and compared those results with the daily results produced by the GACOS model in the same areas at the same time (Figure 3a,b). In Figure 3, the vertical axis and horizontal axis represent the latitudinal and longitudinal ranges of the study areas, respectively, and the color-coded strips represent the calculated ZTD values. It can be seen that the two atmospheric delay models were almost the same at the same DOY times.

We found that the ERA5 weight value relative to the GNSS weight value of each considered pixel did not change significantly and was stable on DOY 160 prior to the earthquake; however, there was a significant change on DOY 172 after the earthquake, at which time the weight values varied abruptly from 0.049 to 0.053 between adjacent pixel points, The sizes of the weights on these two days were significantly different. The weight ratio between ERA5 and GNSS on DOY 172 was larger than the weight ratio before the earthquake, but it was far less than 1.0. This might have been because the ERA5 meteorological model had a certain forecast error, so that its accuracy was lower than the accuracy of the GNSS ZTD. However, the increase in the weight ratio of ERA5 relative to GNSS, as well as the weight ratio between some adjacent pixels, was unstable after the earthquake, which might be related to the co-seismic deformation of GNSS stations that led to an increase in the atmospheric delay error and a decrease in their weights.

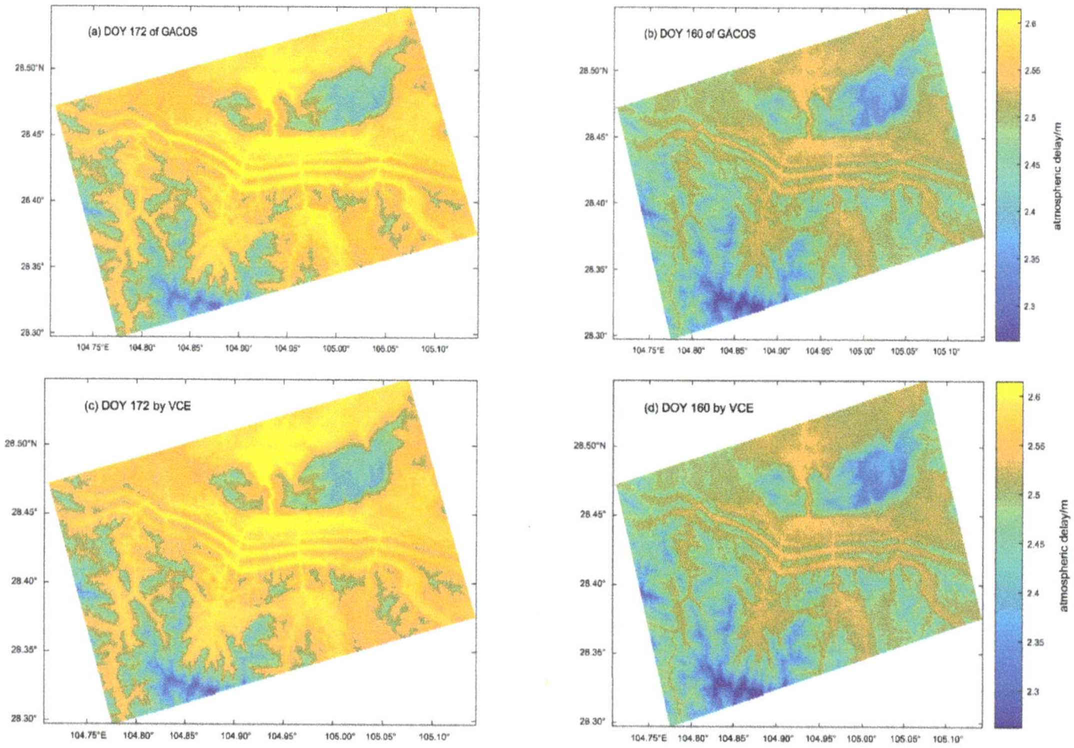

Figure 3. Daily atmospheric delay obtained from different models: (**a**) the atmospheric delay of the InSAR signals on DOY 172 according to the GACOS model; (**b**) the atmospheric delay of the InSAR signals on DOY 160 according to the GACOS model; (**c**) the atmospheric delay of the InSAR signals on DOY 172 according to the variance component weighting model; and (**d**) the atmospheric delay of the InSAR signals on DOY 160 according to the variance component weighting model. The color−coded strips represent the calculated ZTD; the vertical axis and horizontal axis represent the north latitudinal range and the east longitudinal range of the study area, respectively.

The difference between the atmospheric delay correction model on DOY 172 and on DOY 160, obtained by multi-source data integration based on the estimated variance components and the corresponding correction by the GACOS model, is provided in Figure 4 (top). The corresponding pixel point ZTD corrected difference between the two models is provided in Figure 4 (bottom). It can be seen that the maximum difference in value between the two models was about 15 mm, which occurred in the northeast part of the study area, which had large changes in topography elevation and was far from the epicenter. The big differences were perhaps because the GNSS stations were not sufficiently dense in places with large terrain changes, leading to ZTD grid calculation errors. The other error caused by the large variations in topography may be due to the fact that the ERA5 meteorological data used in this paper included a $0.25° \times 0.25°$ grid, while the ECMWF meteorological data used in the GACOS model included a $0.125° \times 0.125°$ grid.

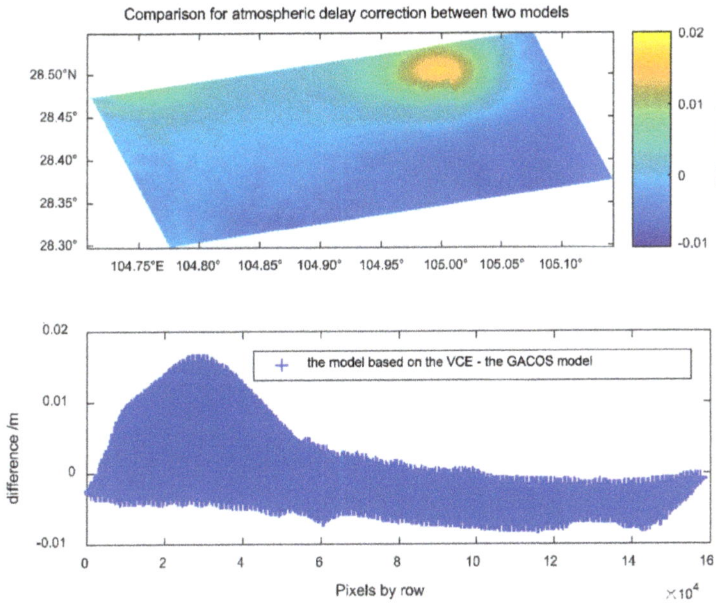

Figure 4. Comparison for atmospheric delay correction between the variance component weighting model and the GACOS model: the top shows the difference between the two models in the area of interest; the color−coded strips represent the difference between the atmospheric delay correction model, based on the VCE method, and the GACOS model; the vertical axis and the horizontal axis represent the northern latitudinal range and the eastern longitudinal range of the study area, respectively; at the bottom is the difference between the two atmospheric delay correction models in the corresponding pixels: the horizontal axis represents the pixels extracted by row, and the vertical axis represents the difference values; the unit is meter.

4. Results of the Atmospheric Delay Correction Model Applied to Crustal Deformation Monitoring

In order to verify that the precision of the deformation can be improved by the atmospheric delay model, based on the VCE method, the Sentinel-1A satellite data were used to obtain the raw crustal deformation information for the area of interest. The raw co-seismic deformation displacement is shown in Figure 5, which depicts an obvious regional variability; the deformation near the epicenter was drastic. It is shown as striped.

The InSAR measurement provides the line-of-sight (LOS) direction displacement of the satellite in the study area, so that the atmospheric delay correction model needs to be mapped to the LOS of the satellite by a trigonometric function on the mean elevation angle at the moment of the satellite's transmitting signal. Then, the corrected displacement of the raw displacement field minus the multi-source data integrated atmospheric delay correction model, based on the VCE method, and the GACOS model are provided in Figures 6 and 7, respectively. Generally, it is believed that the crustal deformation in the area far from the epicenter should be small or even nil. Therefore, we emphasized the eastern part of the image, which is far from the epicenter, and compared it with the GACOS model.

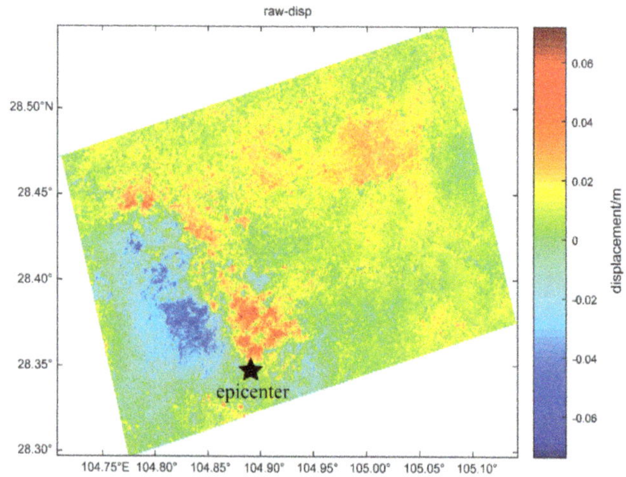

Figure 5. InSAR co−seismic deformation for the area of interest, corresponding to Figure 1. The red rectangle represents the area of interest; the black pentagram represents the earthquake location according to the China Earthquake Networks Center; the color−coded strip represents the displacement of co-seismic deformation for InSAR.

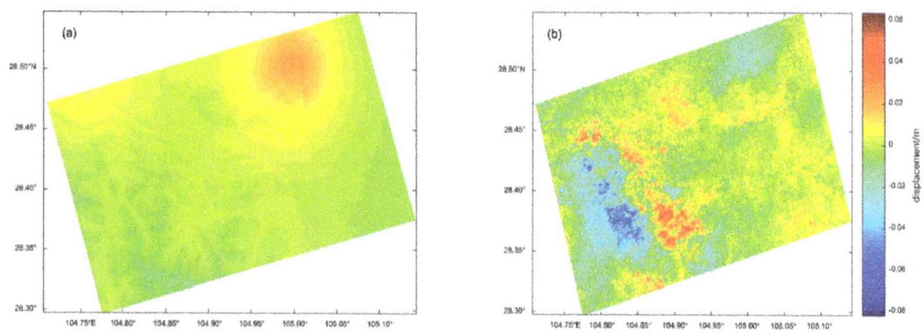

Figure 6. The corrected deformation displacement according to the atmospheric delay model based on the VCE method: (**a**) the multi−source data integrated atmospheric delay correction model based on the variance component estimation (VCE) method; (**b**) the corrected co−seismic deformation displacement according to the model based on the VCE. The color−coded strip represents the atmospheric delay correction model and the displacement in (**a**,**b**), respectively.

As shown in Figures 6 and 7, the results were consistent in most regions; however, there were some differences in certain regions. On the one hand, this might be because of the uneven distribution of the GNSS stations that were used in the GACOS model, which were mainly distributed in Europe. Few GNSS stations in China could affect the accuracy of the GACOS atmospheric delay model. On the other hand, GACOS model weighting by a test step size of 0.1 could affect its accuracy in integrating the meteorological forecasting model and the GNSS data. Although this weighting method is much improved, compared with the previous equal-weight integration method, the choice of its step size was not a good method for optimal weights [47].

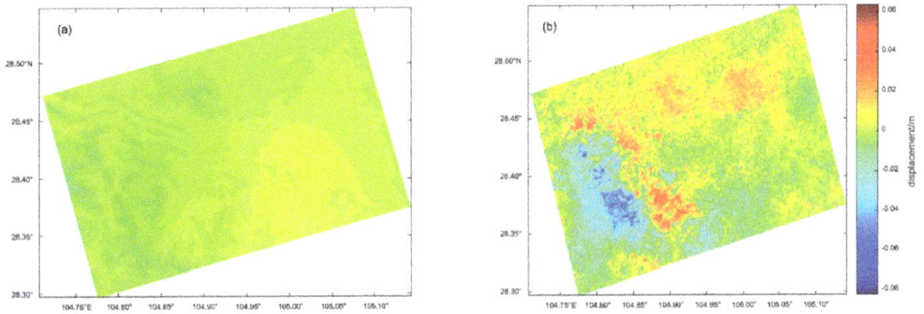

Figure 7. The corrected deformation displacement according to the GACOS model: (**a**) the multi-source data integrated atmospheric delay correction based on the GACOS model; (**b**) the corrected co−seismic deformation displacement according to the GACOS model. The color−coded strip represents the atmospheric delay correction model and the displacement in (**a**,**b**), respectively.

5. Discussion

In order to determine the main factor affecting the precision of the atmospheric delay correction model for the uneven GNSS stations and the weighting method, we built a new atmospheric delay correction model by using the same sources of the GNSS stations' ZTD and the ERA5 ZTD as those of the model based on the VCE method, and the same weighting method as that of the GACOS model. This was the test model. The results were as follows.

As shown in Figure 8, the atmospheric delay correction model was no better than the GACOS model, especially in areas where the elevation change in the northeast corner was dramatic and far from the epicenter. There was reason to believe that the weights method accounted for the more dominant factor influencing the accuracy of the atmospheric delay correction model.

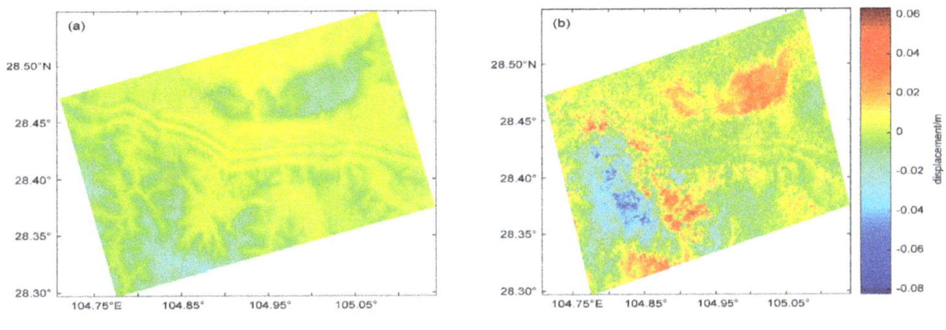

Figure 8. The corrected deformation displacement by the test model: (**a**) the atmospheric delay correction according to the test model; (**b**) the corrected co−seismic deformation displacement according to the test model. The color−coded strip represents the atmospheric delay correction model and the displacement in (**a**,**b**), respectively.

The RMS could be used as an indicator to evaluate the accuracy of the atmospheric delay correction models. Therefore, the RMS values of the overall displacement fields in the study area were calculated on the raw displacement and the corrected displacement by different atmospheric delay correction models that were based on the VCE method and the GACOS model, respectively. The results showed that the RMS value was 19.8 mm on the raw displacement field, without atmospheric delay correction, in the study area. The RMS value was 22.7 mm on the corrected displacement field according to the GACOS model, with deterioration of accuracy. The RMS value was 21.4 mm on that corrected displacement according to the test model, which was almost identical to that of the GACOS model. The RMS value of the corrected displacement according to the multi-source data integrated atmospheric delay model based on the VCE method was 17.4 mm, which was a 5.3 mm improvement on that of the corrected displacement by the GACOS model, with accuracy improvement of approximately 25%, and a 2.4 mm improvement on that of the raw displacement, with accuracy improvement of approximately 25%. These were important improvements for highly accurate deformation monitoring.

Generally, the crustal deformation was small in most the regions, except in circumstances of sudden change. For example, near the earthquake center there was large displacement. In order to reduce the effect of the larger deformation by the earthquake, we tested the accuracy of the regions far from the epicenter. The results showed that the RMS was 21.4 mm, 24.4 mm, and 15.1 mm for the raw displacement, the displacement that was corrected via the GACOS model, and the displacement according to the new model, respectively. The results showed that the accuracy of the model based on the VCE method was better by providing 6 mm improvement compared with that of the raw displacement and provided 9 mm improvement in comparison with the displacement corrected by the GACOS model.

Because of the rich vegetation and complex topographic features of the area of interest, the raw phase measurement has low coherence, and other noise masked some of the accuracy of the atmospheric delay correction model. Accordingly, while the precision of the corrected result according to the integrated atmospheric delay correction model based on the VCE method was improved, the improvement was limited.

Another factor to verify the quality of the atmospheric delay correction model is the correlation between the raw displacement and the atmospheric delay correction [29]. The InSAR phase measurement is highly correlated with the atmospheric delay correction. This means that the atmospheric delay model is good when the atmospheric effect can be most reduced from the raw displacement. Therefore, the higher the correlation coefficient between the raw displacement and the atmospheric delay correction model, the better the accuracy of the atmospheric delay correction model. Generally, the calculation of the correlation coefficient is based on the degree of linear correlation between the atmospheric delay X and the displacement Y, which can be expressed as

$$Y = a \cdot X + b \tag{16}$$

where a is the slope and b is the intercept.

The correlation coefficient between the atmospheric delay and the raw displacement can be expressed as follows:

$$\rho = \frac{\mathrm{cov}(X,Y)}{\sqrt{\mathrm{var}(X)\mathrm{var}(Y)}} = \frac{\sum[(x_i - \overline{x}) \cdot (y_i - \overline{y})]}{\sqrt{\sum(x_i - \overline{x})^2 \cdot \sum(y_i - \overline{y})^2}} \tag{17}$$

where ρ is the correlation coefficient between X and Y, $\mathrm{cov}(X,Y)$ is the covariance of X and Y, and $\mathrm{var}(X)$ and $\mathrm{var}(Y)$ represent the variance of X and Y, respectively. In this study, X and Y represent the atmospheric delay correction model and the raw displacement, respectively. x_i is the atmospheric delay correction of each pixel in the study area, $\overline{x}$ is the average of the atmospheric delay correction for all pixels, y_i is the raw displacement of each pixel, and $\overline{y}$ is the average of the raw displacement for all pixels.

The correlation coefficients between the raw displacement and the atmospheric delay correction model based on VCE, as well as the coefficient between the raw displacement and the GACOS model, respectively, are shown in Table 1. It can be seen that the correlation coefficient between the displacement and the model based on the VCE method was higher than that between the displacement and the GACOS model. The former improved the correlation coefficient by approximately 25%, compared with the latter.

Table 1. Performance indicators for the GACOS model and the atmospheric delay correction model based on the VCE correlation coefficients.

Model	Linear Fit between Raw Displacement and Atmospheric Delay	Correlation Coefficient
GACOS model	Y = 1.6692X − 12.8042	0.2870
multi-source data integrated atmospheric delay correction model based on VCE	Y = 0.9495X − 6.8884	0.3465

In order to intuitively express the correction effect, the displacement of the test line was analyzed for raw displacement and corrected displacement by different atmospheric delay correction models. The results are shown in Figure 9.

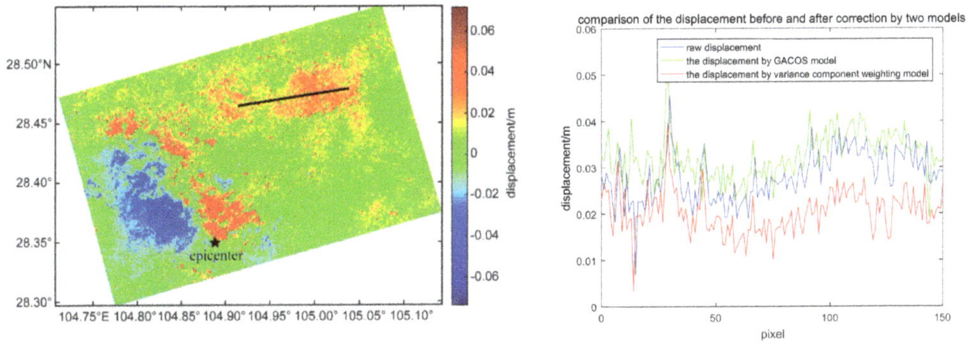

(**a**) (**b**)

Figure 9. A comparison of the test line displacement for the before and after correction with different atmospheric delay models. Part (**a**) shows a raw interference displacement of the study area. We can see the position of the test line; the black pentagram is the epicenter according to the China Earthquake Networks Center; the color−coded strip represents the displacement. Part (**b**) shows the displacement of the black test line located in (**a**); the bule curve is the raw displacement, the green curve is the corrected displacement according to the GAOCS model, and the red curve is the corrected displacement according to the model based on the VCE, respectively.

Figure 9a shows the raw deformation that was not corrected by the atmospheric delay correction model. Based on the test line in Figure 9a, we extracted the raw displacement and corrected it in different models, as expressed in Figure 9b. Figure 9b showed the displacement of the test line in Figure 9a. In Figure 9b, the blue area represents raw displacement, the green area represents the corrected displacement by the GACOS model, and the red area represents the corrected displacement according to the model based on the VCE method. Based on the principle that the smaller the displacement in a region far from the epicenter, the better the correction effect, we concluded that the red curve, representing the correction model, was the best, and that its curve contour shape but not its value was almost the same as the blue one, i.e., the curve trend of the corrected test line displacement based on the VCE method was similar to the raw displacement, and the smaller displacement than the raw one indicated that there was a systematic constant difference between the displacement corrected by the atmospheric delay correction model based on the VCE method and the raw displacement. Moreover, the corrected displacement according to the model based on the VCE method was lower. The existence of systematic error between the corrected displacement of the VCE model and the raw displacement could be the noise factors, such as rich vegetation in the region.

Given the high precision deformation monitoring capability of the GNSS stations, we considered using a GNSS station to verify the external accuracy of the results achieved.

Because the GNSS station was sparse in the area of interest, we chose and performed the deformation of the SCJU station that was close to the study area by the different models; then, we compared it with the solution of the GAMIT/GLOBK. The derived deformations against the SCJU station from the GAMIT/GLOBK solution, the correction model base on the VCE, and the GACOS model, were 0.016, 0.0377, and 0.0695, respectively. This indicated that the result of the model based on the VCE method was better than that of the GACOS model, due to the smaller deformation difference to the GAMIT/GLOBK. As there was still a distance of about 10 km between the considered pixel and the SCJU station, this external accuracy was used as the reference.

6. Conclusions

This paper's proposed atmospheric delay correction model is better than the GACOS model in separating the elevation-related stratified components from the water vapor components, based on the VCE method. It can be applied to various geographic situations for crustal deformation monitoring. For example, the selected Changning region has large terrain fluctuations and large changes in temperature and relative humidity. Therefore, its atmospheric delays, and especially its water vapor variations, are very complex. The experimental results showed that the model based on the VCE method is better than the GACOS model and that it is also suitable for flat terrain conditions.

In this paper, the denser GNSS data for the study area and the newer meteorological weather forecasting model, ERA5, were adopted. Even if the degree of granularity of the ERA5 37-layer pressure level is poorer than that of the GACOS model's ECMWF 137-layer spatial resolution, the accuracy of the new model, based on the VCE method, is much higher than that of the GACOS model. the model's corrected seismic deformation displacement RMS, based on the VCE method, decreased by 5 mm and 2.4 mm, compared with the GACOS model and the raw displacement. This is important for high-precision crustal deformation monitoring. Given the large deformations due to earthquakes, we tested the corresponding RMS in an area far from the epicenter, the precision improvement of the corrected deformation by the new model was 9.3 mm and 6.3 mm, compared with that of the GACOS model and the raw displacement.

The results indicate that with variance component estimation weighting it is feasible to build the multi-source data integrated atmospheric delay model, and the correction effect of the model based on the VCE method is better than that of the GACOS model. Given the more refined meteorological numerical information that can be obtaine, with an auxiliary to denser GNSS stations and other source data, it is possible to obtain a more accurate atmospheric delay correction model for InSAR by the variance component estimation (VCE) method.

In addition, the vegetation cover in the Changning area was very rich in the study period. For the C-band Sentinel satellite, the penetration of vegetation was not as good as the longer wavelength, such as L-band; thus, there was noise in the interferogram. However, the resulting analysis showed that the multi-source data integration atmospheric delay correction model based on the VCE method had higher precision than the GACOS model. The fault was consistent with the results of the analysis by Yu et al. [48] of the Changning M6.0 earthquake. The strike of the fault was northwest to southeast and the regional characteristics of the two sides of the fault were obviously different. The uplift of the southeastern side was up to 6 cm and the subsidence of the northwestern side was up to 8 cm. Compared with the GACOS model, the major improvement of the multi-source data integrated atmospheric delay correction model based on VCE was in the northeastern area of the study area, which had the largest elevation change. This indicates that accurate atmospheric delay correction is good for the correction effect of the large topographic elevation change.

We recommend the integration of the different multi-sources by the VCE method to obtain a more accurate atmospheric delay correction model to monitor crustal deformation. With the establishment of the China Seismic Science Experimental Site, the GNSS network in Sichuan and Yunnan will become more dense. This paper provides ideas and methods for the future integration of multi-source data for InSAR atmospheric delay correction. Such ideas and methods could be used as a reference for conducting atmospheric delay correction scientific research in different regions and different terrain conditions.

7. Summary

The atmospheric delay model for InSAR was established based on the VCE method; by integrating GNSS high-precision observation data and the high spatial—temporal resolution meteorological numerical forecasting model. ERA5. The model's reliability and advantages were evaluated by comparing the correction results with those of the GACOS model. We found that the VCE method is an effective way to improve the accuracy of crustal deformation monitoring in the multi-source data integrated atmospheric delay correction model.

Supplementary Materials: The Sentinel-1A SAR data is available at https://scihub.copernicus.eu/dhus/#/home, accessed on 24 September 2019; The GACOS model can be found in http://www.gacos.net/, accessed on 11 May 2021; The orbit file can be downloaded in https://scihub.copernicus.eu/gnss/#/home, accessed on 27 October 2019.

Author Contributions: X.W. proposed the conceptualization for the study, supervised the progress of the study, provided advice on issues that arose, and writing—review and editing; X.W. was also responsible for project administration and funding acquisition; X.L. established the study's methods, analyzed the results, and wrote the manuscript; X.L. was also responsible for validation, formal analysis, investigation, writing—original draft preparation; Y.C. provided suggestions for the model, writing—review and editing. All authors have read and agreed to the published version of the manuscript.

Funding: This research was funded by the National Natural Science Foundation of China, grant number 11973073; the National Key Research and Development Program of China, grant number 2016YFB0501405; the Basic project of Ministry of Science and Technology of China, grant number 2015FY310200; and the Shanghai Key Laboratory of Space Navigation and Position Techniques, grant number 06DZ22101.

Data Availability Statement: Data used to support the findings of this study are included within the article.

Acknowledgments: We want to thank the European Space Agency (ESA) for providing free Sentinel-1A SAR data, referred to in the Supplementary Materials.

Conflicts of Interest: The authors declare no conflict of interest.

References

1. Wang, Z.B. Research and Application of InSAR Atmospheric Delay Correction Based on Spatial Interpolation Algorithm. Ph.D. Thesis, Anhui University of Science and Technology, Anhui, China, 2018.
2. Zebker, H.A.; Rosen, P.A.; Hensley, S. Atmospheric effects in interferometric synthetic aperture radar surface deformation and topographic maps. *J. Geophys. Res.* **1997**, *102*, 7547–7563. [CrossRef]
3. Massonnet, D.; Feigl, K.L.; Rossi, M.; Adragna, F. Radar interferometry mapping of deformation in the year after the Landers earthquake. *Nature* **1994**, *369*, 227–230. [CrossRef]
4. Lanari, R.; Lundgren, P.; Manzo, M.; Casu, F. Satellite Radar Interferometry Time Series Analysis of Surface Deformation for Los Angeles, California. *Geophys. Res. Lett.* **2004**, *31*, 613–620. [CrossRef]
5. Ferretti, A.; Prati, C.; Rocca, F. Permanent Scatters in SAR Interferometry. *IEEE Trans. Geosci. Remote Sens.* **2001**, *39*, 8–20. [CrossRef]
6. Beradino, P.; Fornaro, G.; Lanari, R.; Sansosti, E. A new algorithm for surface deformation monitoring based on small baseline differential SAR interferograms. *IEEE Trans. Geosci. Remote Sens.* **2002**, *40*, 2375–2383. [CrossRef]
7. Li, Z.W.; Ding, X.L.; Liu, G.X. Modeling atmospheric effects on InSAR with meteorological and continuous GPS observations, algorithms and some test results. *J. Atmos. Sol.-Terr. Phys.* **2004**, *66*, 907–917. [CrossRef]
8. Song, X.G.; Li, D.R.; Shan, X.J.; Liao, M.S.; Cheng, L. Correction of atmospheric effect in repeat-pass InSAR measurements based on GPS and atmospheric transport model. *Chin. J. Geophys.* **2009**, *52*, 1156–1164.
9. Williams, S.; Bock, Y.; Fang, P. Integrated satellite interferometry: Tropospheric noise, GPS estimates and implications for interferometric synthetic aperture radar products. *J. Geophys. Res.* **1998**, *103*, 27051–27067. [CrossRef]
10. Treuhaft, R.N.; Lanyi, G.E. The effect of the dynamic wet troposphere on radio interferometric measurements. *Radio Sci.* **1987**, *22*, 251–265. [CrossRef]

11. Li, Z.W.; Ding, X.L.; Huang, C.; Zheng, D.W. Modeling of atmospheric effects on InSAR measurements by incorporating terrain elevation information. *J. Atmos. Sol.-Terr. Phys.* **2006**, *68*, 1189–1194. [CrossRef]
12. Li, Z.H.; Fielding, E.J.; Cross, P.; Muller, J.P. Interferometric synthetic aperture radar atmospheric correction: GPS topography-dependent turbulence model. *J. Geophys. Res.* **2006**, *111*, B02404. [CrossRef]
13. Luo, H.B.; He, X.F. A spatial interpolation for GPS-derived atmospheric delay corrections to InSAR. *J. Geod. Geodyn.* **2007**, *27*, 35–38.
14. Jarlemark, P.O.J.; Emardson, T.R. Strategies for spatial and temporal extrapolation and interpolation of wet delay. *J. Geod.* **1998**, *72*, 350–355. [CrossRef]
15. Jolivet, R.; Grandin, R.; Lasserre, C.; Doin, M.P.; Peltzer, G. Systematic InSAR tropospheric phase delay corrections from global meteorological reanalysis data. *Geophys. Res. Lett.* **2011**, *38*, L17311. [CrossRef]
16. Chen, Y.; Penna, N.T.; Li, Z.H. Generation of real-time mode high-resolution water vapor filed from GPS observations. *J. Geophys. Res. Atmos.* **2017**, *122*, 2008–2025.
17. Garthwaite, M.C.; Wang, H.; Wright, T.J. Broadscale interseismic deformation and fault slip rates in the central Tibetan Plateau observed using InSAR. *J. Geophys. Res.* **2013**, *118*, 5071–5083. [CrossRef]
18. Wang, H.; Peng, J.H. Test on InSAR atmospheric delay correction using ECMWF model. *J. Guangdong Univ. Technol.* **2014**, *31*, 74–77.
19. Li, Z.W.; Ding, X.L.; Huang, C.; Zou, Z.R. Atmospheric effects on repeat-pass InSAR measurements over Shanghai region. *J. Atmos. Sol.-Terr. Phys.* **2007**, *69*, 1344–1356. [CrossRef]
20. Li, Z.H.; Muller, J.P.; Cross, P.; Fielding, E.J. Interferometric synthetic aperture radar(lnSAR) atmospheric correction: GPS, Moderate Resolution Imaging Spectroradiometer (MODIS), and InSAR integration. *J. Geophys. Res. Solid Earth* **2005**, *110*, B30410.
21. Chen, M.; Chen, J.P.; Hu, C.W. Numerical Weather Models Applied in GNSS Precise Point Positioning. In Proceedings of the China Satellite Navigation Academic Conference, Changsha, China, 18–20 May 2016.
22. Li, Z.H. Correction of Atmospheric Water Vapour Effects on Repeat-Pass SAR Interferometry Using GPS, MODIS and MERIS Data. Ph.D. Thesis, University College London, London, UK, 2006.
23. Li, Z.H.; Fielding, E.J.; Cross, P.; Muller, J.P. Interferometric synthetic aperture radar atmospheric correction: Medium Resolution Imaging Spectrometer and Advanced Synthetic Aperture Radar integration. *Geophys. Res. Lett.* **2006**, *33*, L06816. [CrossRef]
24. Xu, W.B.; Li, Z.W.; Ding, X.L.; Feng, G.C. Correcting atmospheric effects in ASAR interferogram with MERIS integrated water vapor data. *Chin. J. Geophys.* **2010**, *53*, 1073–1084.
25. Li, X.F.; Li, Y.; Zeng, Q.M.; Zhao, Y.H. Correction of atmospheric effects on repeat-pass interferometric SAR using MERIS and ASAR synchronous data. *Acta Sci. Nat. Univ. Pekin.* **2009**, *45*, 1012–1018.
26. Hu, J.; Li, Z.W.; Zhu, J.J.; Ding, X.L.; Qian, S. Measuring three-dimensional surface displacements from combined InSAR and GPS data based on BFGS method. *Chin. J. Geophys.* **2013**, *56*, 117–126.
27. Zhang, X.Q.; Zhou, Y.H. Research on D-InSAR Atmospheric Delay Correction with CORS and MODIS. *Bull. Surv. Mapp.* **2015**, *8*, 13–18.
28. Zhou, W.B.; Xu, W.B.; Li, Z.W.; Wang, C.C. Elevation-dependent MERIS water vapor interpolation and its Application to Atmospheric Correction on ASAR Interferogram. *Geomat. Inf. Sci. Wuhan Univ.* **2012**, *37*, 963–977.
29. Chen, Y.; Li, Z.H.; Penna, N.T.; Paola, C. Generic atmospheric correction model for interferometric synthetic aperture radar observation. *JGR Solid Earth* **2018**, *123*, 8295–9285.
30. Qin, Z.P.; Liu, S.G.; Deng, B.; Li, Z.W.; Sun, W. Multiphase structural features and evolution of southeast Sichuan tectonic belt in China. *J. Chengdu Univ. Technol. Sci. Technol. Ed.* **2013**, *40*, 703–711.
31. Farr, T.G.; Rosen, P.A.; Caro, E.; Crippen, R.; Duren, R.; Hensley, S.; Alsdorf, D. The Shuttle Radar Topography Mission. *Rev. Geophys.* **2007**, *45*, 2. [CrossRef]
32. Rocca, F. Perspectives of Sentinel-1 for InSAR applications. In Proceedings of the 2012 IEEE International Geoscience and Remote Sensing Symposium (IGARSS 2012), Munich, Germany, 22–27 July 2012.
33. Chen, Q.M.; Song, S.L.; Heise, S.; Liou, Y. Assessment of ZTD derived from ECMWF/NCEP data with GPS ZTD over China. *GPS Solut.* **2011**, *15*, 415–425. [CrossRef]
34. Ma, Z.Q.; Chen, Q.M.; Gao, D.Z. Study on accuracy of ZTD and ZWD calculated from ERA-Interim data over China. *J. Geod. Geodyn.* **2012**, *32*, 100–104.
35. Meng, X.G.; Guo, J.J.; Han, Y.Q. Preliminarily assessment of ERA5 reanalysis data. *J. Mar. Meteorol.* **2018**, *38*, 91–99.
36. Saastamoinen, J. Atmospheric Correction for the Troposphere and Stratosphere in Radio Ranging Satellites. *Use Artif. Satell. Geodesy.* **1972**, *15*, 247–251.
37. Sheng, P.X.; Mao, J.T.; Li, J.G.; Ge, Z.M.; Zhang, A.C.; Sang, J.G.; Pan, N.X.; Zhang, H.S. *Atmospheric Physics*; Beijing University Press: Beijing, China, 2003.
38. Zhao, J.Y.; Song, S.L.; Zhu, Y.W. Accuracy Assessment of Applying ERA-Interim Reanalysis Data to Calculate Ground-based GPS/PWV over China. *Geomat. Inf. Sci. Wuhan Univ.* **2014**, *39*, 935–939, 1008.
39. Jia, W.; Pan, Y.J.; He, M.L.; Qin, Y.H. Calculating Analysis of Troposphere Delay Quantity Based on GAMIT. *Geospat. Inf.* **2013**, *11*, 46–48.
40. Cui, X.Z.; Yu, Z.C.; Tao, B.Z. *Generalized Surveying Adjustment*; Wuhan University Press: Wuhan, China, 2009.
41. Nie, Y.; Shen, Y.; Pail, R.; Chen, Q. Efficient variance component estimation for large-scale least-squares problems in satellite geodesy. *J. Geod.* **2022**, *96*, 1–15. [CrossRef]

42. Xu, P.; Shen, Y.; Fukuda, Y.; Liu, Y. Variance component estimation in linear inverse ill-posed models. *J. Geod.* **2006**, *80*, 69–81. [CrossRef]
43. Xu, P.; Liu, Y.; Shen, Y.; Fukuda, Y. Estimability analysis of variance and covariance components. *J. Geod.* **2007**, *81*, 593–602. [CrossRef]
44. Zhao, J.; Guo, J.F. A universal Formula of Variance Component Estimation. *Geomat. Inf. Sci. Wuhan Univ.* **2013**, *38*, 580–583, 588.
45. Shao, F. Research on high precision SLR and ITS astronomical and Geodetic Applications. Ph.D. Thesis, Shanghai Astronomical Observatory, Chinese Academy of Sciences, Shanghai, China, 2019; pp. 75–79.
46. Wu, X.Q. The Method of Weight Factor for Variance Component Estimation. *Geomat. Inf. Sci. Wuhan Univ.* **1989**, *14*, 8–15.
47. Löfgren, J. Tropospheric correction for InSAR using interpolated ECMWF data and GPS Zenith Total Delay from the Southern California Integrated GPS Network. In Proceedings of the 2010 IEEE International Geoscience and Remote Sensing Symposium (IGARSS 2010), Honolulu, HI, USA, 25–30 July 2010.
48. Yu, X.W.; Xue, D.J.; Wang, H.F. Deformation field extraction and simulation of Changning earthquake based on D-InSAR. *Bull. Surv. Mapp.* **2020**, *5*, 59–63.

remote sensing

MDPI

Article

Bayesian Estimation of Land Deformation Combining Persistent and Distributed Scatterers

Gen Li [1], Zegang Ding [1,2,3,*], Mofan Li [1], Zihan Hu [1], Xiaotian Jia [1], Han Li [1] and Tao Zeng [1,2,3]

1 School of Information and Electronics, Beijing Institute of Technology, Beijing 100081, China; 3120195416@bit.edu.cn (G.L.); 3120200781@bit.edu.cn (M.L.); 3120220696@bit.edu.cn (Z.H.); 3220200560@bit.edu.cn (X.J.); 7520210109@bit.edu.cn (H.L.); zengtao@bit.edu.cn (T.Z.)

2 Beijing Key Laboratory of Embedded Real-Time Information Processing Technology, Beijing Institute of Technology, Beijing 100081, China

3 Beijing Institute of Technology Chongqing Innovation Center, Chongqing 401147, China

* Correspondence: z.ding@bit.edu.cn

Abstract: Persistent Scatterer Interferometry (PSI) has been widely used for monitoring land deformation in urban areas with millimeter accuracy. In natural terrain, combining persistent scatterers (PSs) and distributed scatterers (DSs) to jointly estimate deformation, such as SqueeSAR, can enhance PSI results for denser and better coverage. However, the phase quality of a large number of DSs is far inferior to that of PSs, which deteriorates the deformation measurement accuracy. To solve the contradiction between measurement accuracy and coverage, a Bayesian estimation method of land deformation combining PSs and DSs is proposed in this paper. First, a two-level network is introduced into the traditional PSI to deal with PSs and DSs. In the first-level network, the Maximum Likelihood Estimation (MLE) of deformation parameters at PSs and high-quality DSs is obtained accurately. In the secondary-level network, the remaining DSs are connected to the nearest PSs or high-quality DSs, and the deformation parameters are estimated by Maximum A Posteriori (MAP) based on Bayesian theory. Due to the poor phase quality of the remaining DSs, MAP can achieve better estimation results than the MLE based on the spatial correlation of the deformation field. Simulation and Sentinel-1A satellite data results verified the feasibility and reliability of the proposed method. Regularized by the spatial deformation field derived from the high-quality PSs and DSs, the proposed method is expected to achieve robust results even in low-coherence areas, such as rural areas, vegetation coverage areas, or deserts.

Keywords: Persistent Scatter Interferometry (PSI); persistent scatterers (PSs); distributed scatterers (DSs); Bayesian theory; land deformation

Citation: Li, G.; Ding, Z.; Li, M.; Hu, Z.; Jia, X.; Li, H.; Zeng, T. Bayesian Estimation of Land Deformation Combining Persistent and Distributed Scatterers. *Remote Sens.* **2022**, *14*, 3471. https://doi.org/10.3390/rs14143471

Academic Editors: Chisheng Wang, Daqing Ge, Guohong Zhang, Wu Zhu and Siting Xiong

Received: 6 July 2022
Accepted: 16 July 2022
Published: 19 July 2022

1. Introduction

Differential Interferometry Synthetic Aperture Radar (DInSAR) is a powerful technique to precisely monitor land deformation, such as seismological activities, volcanism, landslides, and ground subsidence [1,2]. However, it has limitations due to atmospheric artifacts and temporal and geometric decoherence. Under unfavorable conditions, the deformation signal is obscured by atmospheric interference [3], temporal decorrelation noise (introduced by the changes in the scattering characteristics of the target and incoherent movement of individual scattering elements) [4], and geometric decorrelation noise (introduced by changes in radar viewing angles) [5,6].

Persistent Scatterer Interferometry (PSI) technology is an extension of DInSAR techniques that overcomes the limitations of decoherence and atmospheric problems through time-series methods [7–9]. PSI technology relies on persistent scatterers (PSs), whose amplitude and phase values are stable over time and imaging geometry [10]. Therefore, such technology has been widely used for monitoring land deformation with millimeter accuracy in urban areas where man-made structures behave like strongly reflecting corner reflectors [11].

However, PSI technology failed to perform satisfactorily in the absence of PSs in non-urban areas [9]. Especially in natural terrain or desert areas, the scattering mechanisms are dominated by distributed scatterers (DSs), whose back-scattered energy is less strong, but shares similar reflectivity values within a wide area. Contrary to PSs, DSs are seriously affected by temporal and spatial decorrelation. Therefore, the observed phases of DSs are much noisier than those of PSs [12]. This makes PSI an opportunistic approach that functionally limits the density of the useful scatterers [13].

Recently, several advanced techniques have been proposed to extract information from DSs and increase the spatial density of measured points. Among them, the most representative techniques are the Small Baseline Subset Algorithm (SBAS) [14] and SqueeSAR [15]. Unlike PSI, the SBAS method relies on multilooked and unwrapped differential interferograms with short spatial and temporal baselines to minimize the effects of decorrelation [16–18]. However, the error-prone phase-unwrapping and multi-looking operations in SBAS often lead to poor accuracy and low resolution of the estimated deformation [12]. Moreover, the rectangular multi-looking window in SBAS results in a superposition of different objects on the ground and then a loss of the deformation information contained in the isolated pixels [19]. In addition, the phase unwrapping often results in discontinuous fringes, especially in natural terrains.

In contrast, to reduce the decorrelation noise of DSs, the SqueeSAR method obtains the best possible estimates of the phase history over spatially statistically homogeneous pixels (SHPs) using all possible wrapped interferograms using a phase triangulation algorithm (PTA) [20,21]. The PTA can reduce the stochastic noise in DSs. Then, the DSs and PSs are jointly processed indiscriminately under the traditional PSI processing framework [15,22]. The optimum phase of DSs is estimated from the coherence matrix [23]. Recently, a number of works have recognized that coherence errors due to limited ensembles in data statistics affect the accuracy of InSAR measurements for DSs, especially over low coherence scenarios [24–28]. Moreover, although preprocessing of DSs is often simplistically depicted as transforming DSs into PSs, preprocessed DSs are statistically not equivalent to PSs [29]. In other words, the phase quality of a large number of DSs is far inferior to that of PSs, which deteriorates the deformation measurement accuracy [30–32]. In addition, the quality of the estimated phase of DSs can be indicated by the temporal coherence γ_{PTA}, which can be used effectively for the final selection of DSs with a reliable phase estimation [21,33]. However, a high value of temporal coherence indicates a large agreement between the observations and estimated parameters but fewer DS candidates. A low value of temporal coherence can involve more DS candidates but a large disagreement between the observations and estimated parameters. To sum up, the contradiction between measurement accuracy and coverage has not been fundamentally solved in both SBAS and SqueeSAR.

To solve the contradiction between measurement accuracy and coverage, a Bayesian estimation method of land deformation combining PSs and DSs is proposed in this paper. First, a two-level network is introduced into the traditional PSI to restrain the error propagation of low-quality DSs in the adjustment network. In the first-level network, the deformation parameters at PSs and high-quality DSs are obtained accurately based on the Maximum Likelihood Estimation (MLE). Then, the probability density function (PDF) of the deformation parameters at the remaining DSs can be obtained according to the spatial correlation of the deformation field. Finally, in the secondary-level network, the remaining DSs are connected to the nearest PSs or high-quality DSs and the deformation is estimated using Maximum A Posteriori (MAP) based on Bayesian theory.

This study aimed to achieve robust estimation results of land deformation even in low coherence areas, such as rural areas, vegetation coverage areas, or deserts. A total of 31 Sentinel-1A SAR data acquired between 2017 and 2019 were exploited to detect the land deformation at Remah in the United Arab Emirates (UAE) caused by the overexploitation of the aquifers. The proposed method not only greatly increased the density of the measuring points, but also ensured measurement accuracy and contained more detailed information, which reflected the depression cone correlating with the groundwater level. Even in low coherence areas, the deformation rate estimated using the proposed method is in good agreement with that of the groundwater level. Finally, the strengths, limitations, and application prospects of the proposed method are discussed.

2. Methodology

2.1. Maximum Likelihood Estimation of Land Deformation Combining PSs and DSs

2.1.1. MLE of Land Deformation in PSI

Assuming that the deterministic signal of PS y is corrupted by additive independent and identically complex circular Gaussian (CCG) noise, the likelihood function of the unknown geophysical parameters x can be expressed as [25,34]

$$\mathrm{P}(y|x) = \frac{1}{(2\pi\sigma^2)^N}\exp\left(-\frac{1}{2\sigma^2} \parallel y - \overline{y}(x) \parallel_2^2\right), \tag{1}$$

where N is the number of interferograms, σ denotes the standard deviation of the CCG noise, and $\overline{y}(\mathbf{x})$ is the modeled PS signal. Then, the MLE of the parameters in x can be written as

$$\underset{x}{\mathrm{argmax}}\mathrm{P}(y|x) = \underset{\Delta h,\ \Delta v}{\mathrm{argmax}}\left|\frac{1}{N}\sum_{n=1}^{N}\exp(-j\varphi_n(\Delta h, \Delta v))\right|, \tag{2}$$

and

$$\varphi_n(\Delta h, \Delta v) = \Phi_n - \frac{4\pi B_\perp^n}{\lambda R\sin\theta}\Delta h - \frac{4\pi}{\lambda}T_n\Delta v \tag{3}$$

is the modeled phase of the nth interferogram; Φ_n, Δh, and Δv are the differential phase difference, residual elevation difference, and deformation rate difference in the arcs between two adjacent PSs; $B_\perp^n$ is the perpendicular baseline; λ is the radar wavelength; R is the point-to-satellite distance; θ is the incidence angle; and T_n is the temporal baseline length. Under the MLE, the well-known maximum temporal coherence (MTC) can be rewritten as [8],

$$\gamma_{MTC} = \left|\frac{1}{N}\sum_{n=1}^{N}\exp\left(-j\varphi_n\left(\Delta\hat{h}_{MLE}, \Delta\hat{v}_{MLE}\right)\right)\right|. \tag{4}$$

Due to the common phase errors, such as atmospheric phase error, orbital phase error, nonlinear deformation phase error, and so on, which can be eliminated in the PS pair between two adjacent PSs, the modeled phase φ_n represents the noise level for the PS pair. Moreover, since PSs are hardly affected by temporal and spatial decorrelation, the noise level for the PS pair is usually very small

$$|\varphi_n| \ll \pi. \tag{5}$$

The MTC can be used as an indicator of noise level and estimation quality, and the higher the value of the MTC, the lower the noise level and the higher the estimation accuracy. Finally, the deformation parameters in the PSs can be solved using the weighted least-square adjustment method [35].

Ferretti has shown that the MLE of the deformation parameters can achieve an accuracy of some mm/year in urban areas [7–9]. To extract information from DSs and increase the spatial density of the measured points, the DSs and PSs can be jointly processed indiscriminately under the traditional PSI processing framework. Although preprocessing of DSs is often simplistically depicted as transforming DSs into PSs, preprocessed DSs are statistically not equivalent to PSs [29]. In the next subsection, the quality and CRB of the preprocessed phase of DSs are investigated in detail.

2.1.2. Cramér–Rao Bound (CRB) of the Optimum Phase of DSs

The optimum phase of DSs is estimated from the coherence matrix using PTA [20,21]. The CRB evaluates the highest achievable precision for the optimum phase. Under the assumption of scattering with a multivariate circular Gaussian distribution with absolute coherence matrix Y and atmospheric noise α as a zero-mean normal stochastic process, the variance matrix of the unbiased estimation of the interferometric phase $\hat{\phi}$ is bounded by

$$D\{\hat{\phi}\} = E\left\{(\phi - \hat{\phi})(\phi - \hat{\phi})^T\right\} \geq \left(\Theta^T\left(X^{-1} + Q_\alpha\right)^{-1}\Theta\right)^{-1}, \quad (6)$$

where Q_α is the $N \times N$ covariance matrix of the atmospheric signal, $\Theta = [0\ I_{N-1}]^T$ is the $N \times N - 1$ Jacobian matrix of the first-order partial derivatives of the interferometric phases with respect to the unknown parameters, and X is the Fisher Information Matrix (FIM) defined as

$$X = 2L\left(\mathrm{Y} \circ \mathrm{Y}^{-1} - I_N\right), \quad (7)$$

where $\circ$ indicates the Hadamard product, L is the number of looks, and I_N is an $N \times N$ identify matrix. Assuming the atmospheric effect is deterministic and the atmosphere covariance matrix is zero, the CRB of the estimated phase can be rewritten as

$$Q_{\hat{\phi}} \geq \left(\Theta^T X \Theta\right)^{-1}. \quad (8)$$

The coherence matrix can be simulated according to the exponential decorrelation model

$$\mathrm{Y}_{i,k} = (\gamma_0 - \gamma_\infty)exp\left(-\frac{t_{i,k}}{\tau}\right) + \gamma_\infty, \quad (9)$$

where γ_0 and γ_∞ are the short-term and long-term coherences, respectively, $t_{i,k}$ is the temporal baseline between the ith and kth acquisitions, and $1/\tau$ denotes the decorrelation rate. The parameters γ_0, γ_∞, and τ are related to types of land use.

As shown in Figure 1a, the temporal decorrelation behavior depends on land use [36,37]. The urban area tends to have higher γ_0 and γ_∞ but a middle τ, which indicates that the urban area keeps adequate coherence semi-permanently. The pasture (rural) area has a very low γ_∞ but a middle γ_0 and a higher τ, which indicates that the pasture (rural) area becomes almost completely decorrelated sooner or later. In the water area, we never have a coherent signal. Therefore, the γ_0, γ_∞, and τ are always small. According to Equation (5), assuming a stack of $N = 25$ SLC images containing an ensemble of $L = 30$ homogeneous pixels and the repeat cycle is 12 days in C-band, the CRB standard deviations of the preprocessed phase of DSs are simulated as shown in Figure 1b. It can be seen that the processed DS phase in the urban area can achieve a much better quality than that of PSs. However, in the pasture (rural) area, the processed DS phase quality does not reach the PS standard deviation, indicating that the information content of DSs is not sufficient for precise estimation of phase time series. In the water area, the processing of DSs is no longer valid because there is almost no coherent information in the stacks.

It should be noted that the CRB is achieved under the assumption that the coherence matrix is unbiased. The typical estimation methods, such as maximum likelihood (ML), eigenvalue decomposition (EVD), and integer least squares (ILS), usually cannot achieve such high precision in practice. The traditional preprocessing of DSs is often simplistically depicted as transforming DSs into PSs, and the DSs and PSs can be jointly processed indiscriminately under the traditional PSI processing framework discussed in Section 2.1. However, the phase quality of DSs, even after optimization, is far inferior to that of PSs, especially in low coherence areas (pasture or rural), which will lead to poor estimation accuracy of ($\Delta\hat{h}_{MLE}$, $\Delta\hat{v}_{MLE}$) in arcs connected to DSs, and errors will further propagate in the subsequent adjustment network. Moreover, by constraining γ_{MTC} and γ_{PTA}, some DSs with poor estimation can be removed, but this is bound to reduce the number of estimated points. Therefore, the contradiction between measurement accuracy and coverage has not been fundamentally solved.

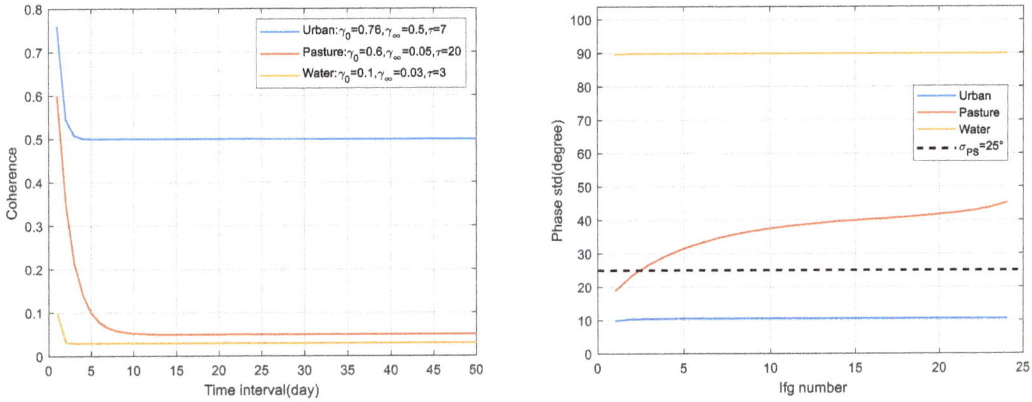

(a) (b)

Figure 1. Temporal coherence and CRB of the preprocessed phase of DSs. (**a**) Temporal coherence model in urban, pasture, and water areas; (**b**) The CRB standard deviations of the preprocessed phase of DSs and the standard deviation of a typical PS (~25°) are also plotted.

2.2. Bayesian Estimation of Land Deformation Combining Persistent and Distributed Scatterers

2.2.1. MAP Estimation of Deformation Parameters Based on Bayesian Theory

The main problem with the MLE is that the accuracy of the parameter estimation is greatly affected by the size and quality of the observation data, which means that the result of the parameter estimation is sensitive to random variations in the observation data. To deal with this, prior probabilities $\mathrm{P}(\boldsymbol{x})$ are introduced to improve the accuracy of the MLE. According to Bayesian theory, the posterior can be written as a product of likelihood and prior

$$\mathrm{P}(\boldsymbol{x}|\boldsymbol{y}) = \frac{\mathrm{P}(\boldsymbol{y}|\boldsymbol{x})\mathrm{P}(\boldsymbol{x})}{\mathrm{P}(\boldsymbol{y})}, \tag{10}$$

where $\mathrm{P}(\boldsymbol{y})$, called the probability of the observed data, is a normalizing constant ensuring that the posterior distribution has an area of 1. Then, we replace the likelihood with the posterior in the MLE in Equation (2), and we obtain the MAP estimation as in [38] by

$$\underset{x}{\mathrm{argmax}}\mathrm{P}(\boldsymbol{y}|\boldsymbol{x})\mathrm{P}(\boldsymbol{x}) = \underset{\Delta h,\,\Delta v}{\mathrm{argmax}} \left| \frac{1}{N} \sum_{n=1}^{N} \exp(-j\varphi_n(\Delta h, \Delta v)) \right| \mathrm{P}(\Delta h,\, \Delta v), \tag{11}$$

where $\mathrm{P}(\Delta h,\, \Delta v)$ is the prior joint probabilities of Δh and Δv.

The MAP estimation provides a framework for regularizing the tendency to match exactly with the observed data thus leading to a less noisy estimation. Moreover, MAP pulls the estimation towards the prior to an extent, which depends on the strength and bias of the prior relative to the likelihood.

As shown in Figure 2a, the MLE can achieve an unbiased estimation of Δh and Δv at (0, 0) without noise. However, as the noise increases, the temporal coherence decreases gradually and achieves the maximum of 0.6 at (2.8, −1.4) as shown in Figure 2b, and the MLE of Δh and Δv becomes worse. To improve the estimation accuracy of the MLE, the prior probabilities of Δh and Δv are given to achieve the MAP estimation. As shown in Figure 2c, the temporal coherence has greater kurtosis, and the MAP estimation is much better than the MLE in the case of unbiased Gaussian distribution for the priori probabilities. When the priori probabilities are biased slightly, the estimation accuracy of MAP will decline, but it is still better than the MLE as shown in Figure 2d. Thus, the priori probabilities of Δh and Δv determine the accuracy of the MAP estimation, and the PDF estimation method is given in the next subsection.

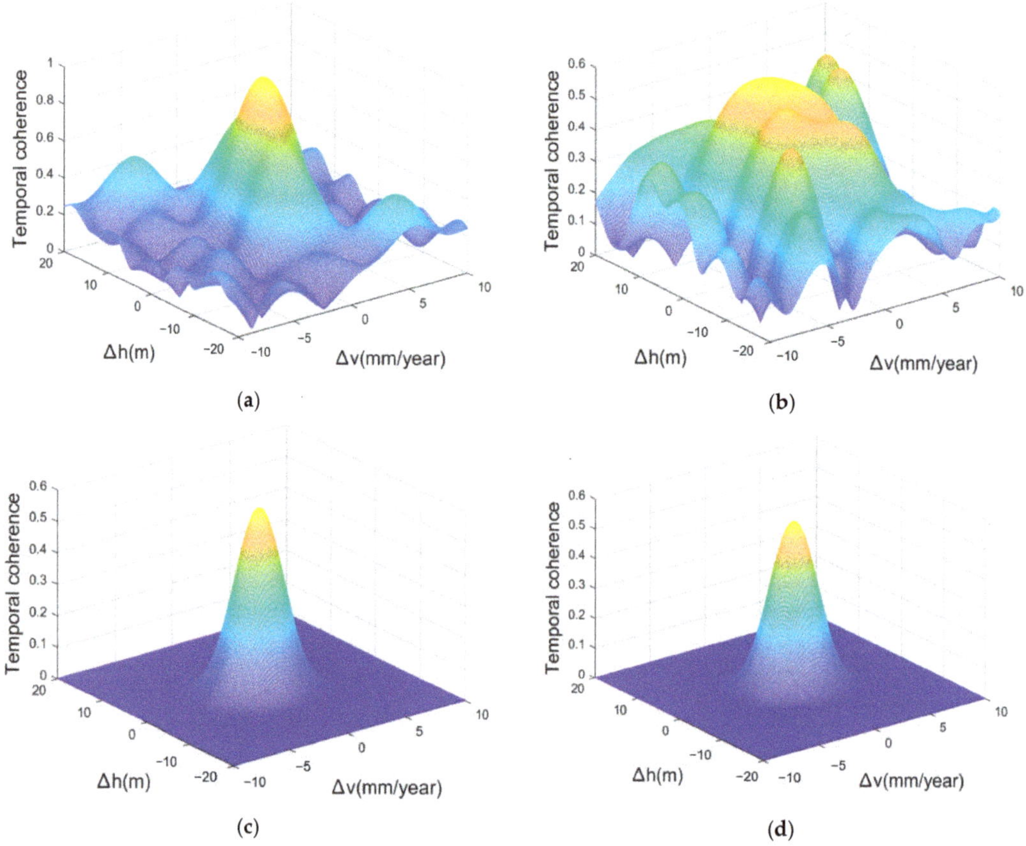

Figure 2. Estimation of Δh and Δv: (**a**) MLE without noise: $\gamma_{MTC} = 1$, $\Delta h = 0$, $\Delta v = 0$; (**b**) MLE with noise: $\gamma_{MTC} = 0.6$, $\Delta \hat{h} = 2.8$, $\Delta \hat{v} = -1.4$; (**c**) MAP estimation with noise: $\Delta \hat{h} = 0.2$, $\Delta \hat{v} = 0$ and unbiased prior: Δh~N(0,9), Δv~N(0,4); (**d**) MAP estimation with noise: $\Delta \hat{h} = 0.8$, $\Delta \hat{v} = 0.6$ and biased prior: Δh~N(1,9), Δv~N(1,4).

2.2.2. Two-Level Network for Deformation Parameters Estimation

- The first-level network for reliable scatterers

A two-level network is introduced into the traditional PSI to deal with PSs and DSs. In the first-level network, the deformation parameters at PSs and high-quality DSs are obtained accurately based on the Maximum Likelihood Estimation (MLE).

Given a stack of $N + 1$ coregistered SAR images, PSs are first selected using the amplitude dispersion method [7–9]. Meanwhile, DSs are selected to increase the density of the measuring points. The DSs, affected by temporal and spatial decorrelation, are much noisier than the PSs, so the phase history of DSs is optimized using the PTA [15].

Unlike the traditional PS–DS algorithm, the DSs are divided into high-quality DSs and low-quality DSs according to the posterior coherence factors ρ_{PTA_H} and ρ_{PTA_L}, respectively. The high-quality DSs and PSs are jointly processed in the Delaney triangle network. Due to the noise of these points being very small, the MLE can obtain reliable estimation results in the arcs. Then, the linear deformation rate can be solved using the weighted least-square adjustment method [32]

$$X = \left(B^T PB\right)^{-1} B^T PL, \tag{12}$$

where X is the vector of the deformation rates for the M reliable points, $L = \left[\Delta\hat{v}_1 \ \Delta\hat{v}_2 \cdots \ \Delta\hat{v}_Q\right]$ is the vector of the deformation rate difference of the Q arcs, and B is the coefficient matrix composed of 1 and -1 and has the form

$$B = \begin{bmatrix} 10{-1} \ 0 \ \ 0 \\ 00\ldots \ 1 \ -1 \\ \vdots\vdots \ \vdots \ \ \vdots \ \ \vdots \\ 01 \ 0 \ -1 \ 0 \end{bmatrix}_{Q\times M}, \tag{13}$$

where 1 represents the starting point and -1 corresponds to the ending point along each arc and P is the weight matrix composed of the MTC

$$P = \begin{bmatrix} \gamma^2_{MTC\_1} & 0 & 0 & 0 \\ 0 & \gamma^2_{MTC\_2}\cdots & & 0 \\ \vdots & \vdots & \vdots & \vdots \\ 0 & 0 & 0 & \gamma^2_{MTC\_Q} \end{bmatrix}_{Q\times Q}, \tag{14}$$

The residual elevation for reliable points can also be solved using the same process as above and $L = \left[\Delta\hat{h}_1 \ \Delta\hat{h}_2 \cdots \ \Delta\hat{h}_Q\right]$ is the vector of the residual elevation difference of the Q arcs.

- PDF estimation based on the Kriging model

Based on the results estimated from the first-level network and the spatial correlation of deformation field and elevation, the PDF of the deformation and elevation parameters at the low-quality DSs can be obtained using the Kriging model.

Kriging method, namely the Gaussian process regression method, is a geostatistical interpolation technology, which can give the best linear unbiased prediction at the unsampled locations according to the empirical observations [39]. In addition to a predicted value, it also provides the attached prediction error variance. The variance increases for predictions away from the data and as the spatial correlation weakens. Assuming that the output predicted values are h_{prior} and v_{prior} and the variances are σ_h and σ_v, respectively, the prior joint Gaussian distribution for h and v can be expressed as

$$P(h,\ v) = \frac{1}{2\pi\sigma_h\sigma_v} exp\left[-\frac{1}{2}\left(\frac{\left(h - h_{prior}\right)^2}{\sigma_h^2} + \frac{\left(v - v_{prior}\right)^2}{\sigma_v^2}\right)\right]. \tag{15}$$

- The second-level network for the remaining DSs

In the second-level network, to restrain the error propagation of low-quality DSs in the adjustment network, the low-quality DSs are connected to the nearest PSs or high-quality DSs in the second-level network. Additionally, to improve the estimation accuracy, the deformation parameters of the low-quality DSs are estimated using MAP based on Bayesian theory.

Assuming that the deformation parameters of the nearest reliable scatterers are h_r and v_r according to Equations (11) and (15), the MAP estimation can be written as

$$\Delta h_{MAP}, \Delta v_{MAP} = \underset{\Delta h,\ \Delta v}{\operatorname{argmax}} \left\{ \left| \frac{1}{N} \sum_{n=1}^{N} \exp(-j\varphi_n(\Delta h, \Delta v)) \right| \cdot P(\Delta h + h_r,\ \Delta v + v_r) \right\}, \tag{16}$$

and the MAP estimation of the remaining DSs can be calculated directly by

$$v_{MAP} = \Delta v_{MAP} + v_r, \tag{17}$$

$$h_{MAP} = \Delta h_{MAP} + h_r. \tag{18}$$

The processing workflow of the proposed Bayesian estimation of land deformation combining PSs and DSs is shown in Figure 3.

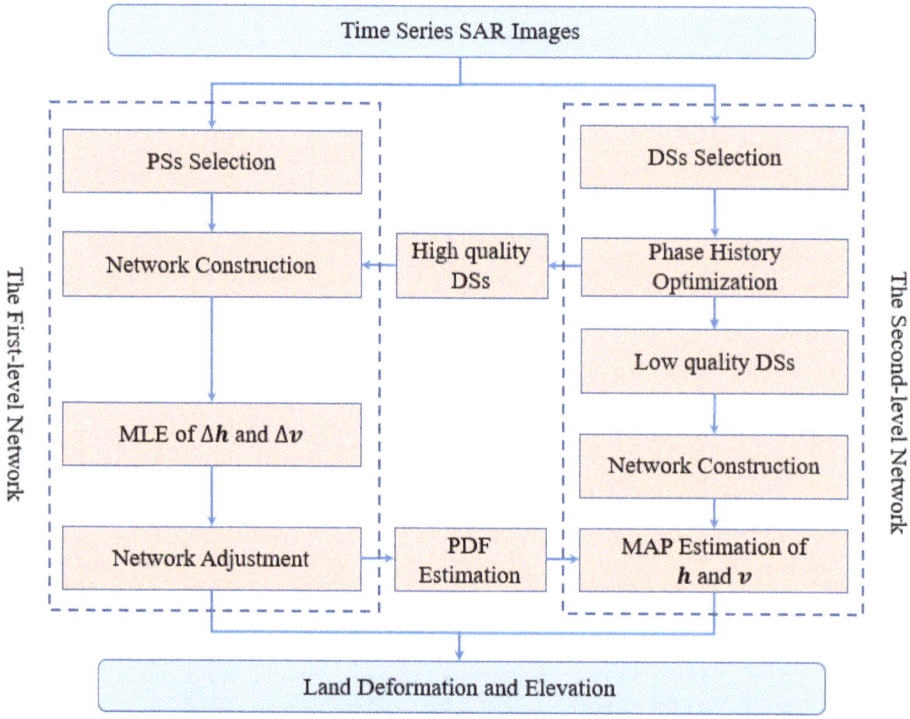

Figure 3. The processing workflow of the proposed method.

3. Analysis and Results

To validate the feasibility of the proposed method and assess its performance, two experiments are conducted based on simulated data and the Sentinel-1 real data set. Moreover, to show the reliability of the proposed method, the processing results are compared with the traditional PS–DS processing method using the MLE.

3.1. Simulated Data

A stack of 20 interferograms (with a size of 500 × 500) is generated in an existing single-master baseline configuration of the Sentinel-1A data with a time span of one year. The simulated phase consists of the deformation signals, residual elevation errors, and atmospheric phases, as shown in Figure 4a–c, respectively. Then, 450 PS pixels and 3000 DS pixels were simulated randomly in the study area, as shown in Figure 5. As mentioned above, the PSs and DSs with different scattering mechanisms have different noise levels. Therefore, the random noises with a standard deviation of 15 degrees and 45 degrees are added to the PS pixels and DS pixels (after PTA), respectively.

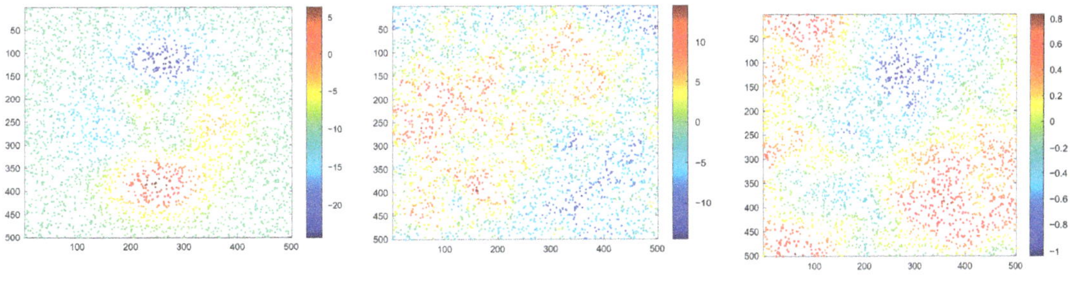

(a) (b) (c)

Figure 4. The simulated phase consists of (**a**) the deformation signals (mm/year); (**b**) residual elevation errors (m); and (**c**) atmospheric phases (rad).

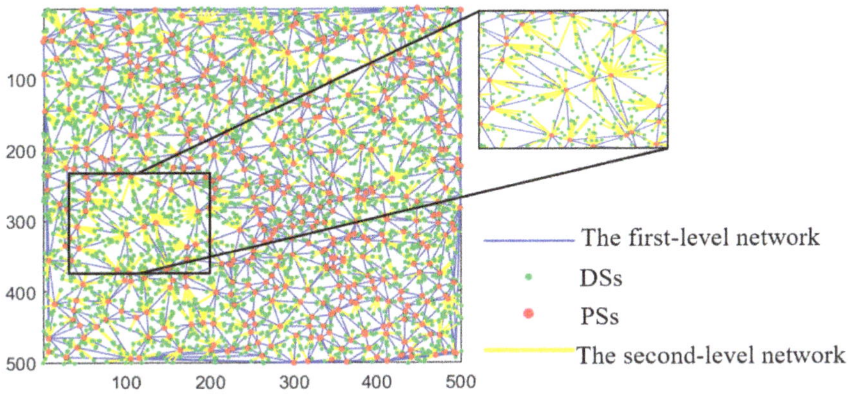

Figure 5. The proposed two-level network combining PSs and DSs.

The deformation rates estimated using PSI, MLE combining PSs and DSs, and the proposed method are shown in Figure 6a–c, respectively. It can be seen that the deformation field in the PSI is in good agreement with the simulated deformation rates. However, due to the low density of PSs, many details are missing. The MLE combining PSs and DSs can increase the density of the measuring points. However, due to the poor phase quality of DSs in the simulation and the diffusion of arc errors in the adjustment network, the inversion results in some areas cannot accurately reflect the real deformation field. The proposed method not only greatly increases the density of the measuring points, but also ensures measurement accuracy and contains more detailed information.

Figure 7 shows the performance of the MLE combining PSs and DSs and the proposed method under different noise levels of DSs. In the MLE combining PSs and DSs, the arc measurements with poor estimation are discarded by setting the threshold of the MTC to 0.75 [31]. However, this operation cannot completely eliminate the influence of overwhelming outliers or decorrelation noise. When the phase quality of DSs deteriorates by more than 35 degrees, the measurement accuracy and measuring points of the MLE combining PSs and DSs will decrease sharply. As a comparison, the proposed method can still ensure the measurement accuracy and measuring points in the case of serious noise, which confirms the robustness of the proposed method. This shows that the proposed method is more robust, especially in low coherent areas such as rural areas, vegetation coverage areas, or deserts.

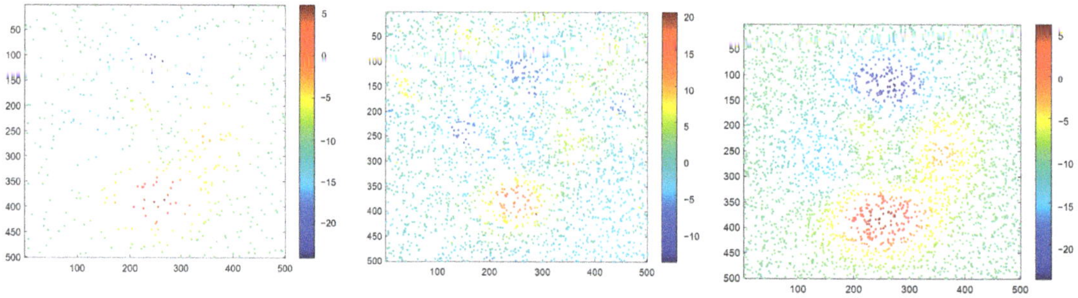

(**a**) (**b**) (**c**)

Figure 6. Deformation rates estimated using (**a**) PSI, (**b**) MLE combining PSs and DSs, and (**c**) the proposed method.

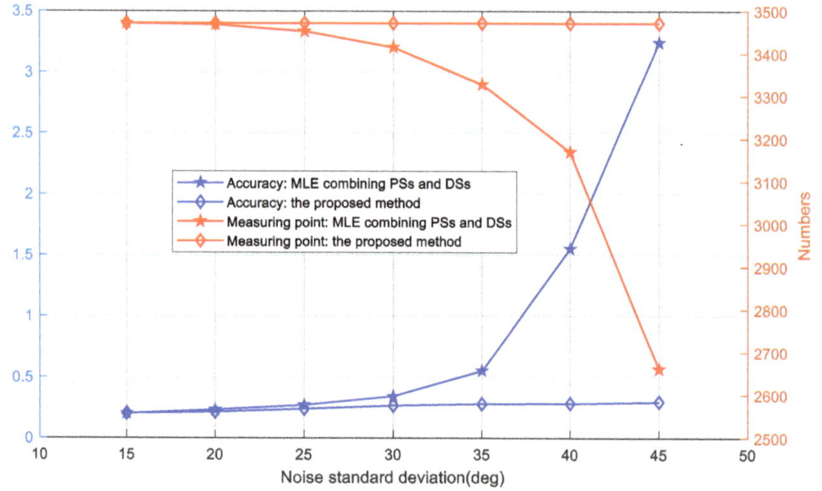

Figure 7. The performance comparison of the proposed method and MLE combining PSs and DSs.

3.2. *Real Data Set: Sentinel-1*

Further validation of the proposed method was carried out by the experiments using a stack of 31 C-band Sentinel-1A SAR images in VV polarization from 1 April 2017 to 22 March 2019. The spatial-temporal baseline configuration is shown in Figure 8.

The test site is located at Remah in the United Arab Emirates (UAE), as shown in Figure 9, a place in an arid desert where rainfall is scarce and groundwater is the only source of natural freshwater. There are many farms and plantation forests in the study area. Groundwater abstraction for agricultural or human use has a significant impact on groundwater resources. As water is pumped from a well, it causes a drawdown of the aquifer water level nearby. According to the survey [40], there are about 4175 operational wells in Al Khaznah and about 232 mio m^3 groundwater is abstracted every year. Agriculture and forests account for about 94% of the total annual groundwater use.

Abstracting groundwater leads to a depression of the groundwater surface around the well (the cone of depression). The map of groundwater level changes shows the groundwater level change from 2005 to 2017, which coincides with the agricultural areas [40,41]. At the test site, the groundwater level drops more than 14 m by monitoring the wells. In 2019, a cone-shaped water level was formed and the maximum drawdown at its center is approximately 40 to 50 m.

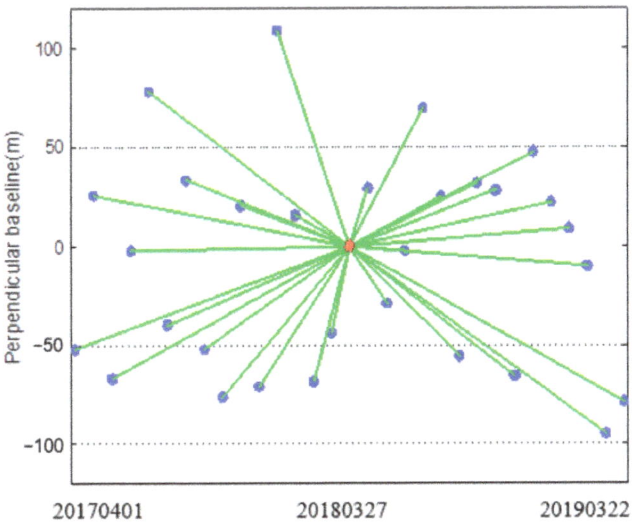

Figure 8. The baseline configuration of the used data set.

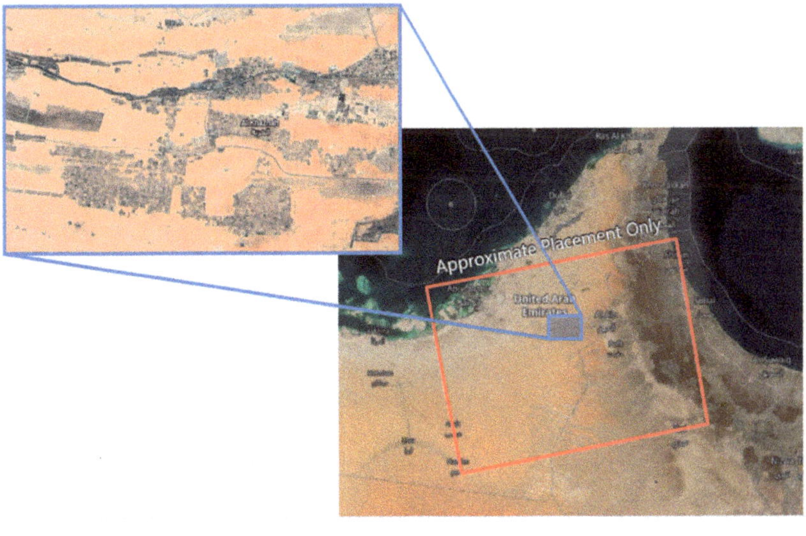

Figure 9. The test site located at Remah.

To explore the land deformation due to groundwater level drops, PSI technology is first adopted. As shown in Figure 10, the deformation rates estimated using PSI indicate that a subsidence bowl with a diameter of about 30 km is formed and the maximum deformation rate is about 55 mm/year, which is caused by the drawdown of the groundwater piezometric level. The PS pixels were selected using amplitude dispersion with a threshold of 0.25, and about 84,753 PSs are selected over the whole area of about 700 km^2. Therefore, the PS density is about 120 PSs/km^2. The arc measurements with an MTC less than 0.75 are discarded. The MTC of all available arcs has a mean value as high as 0.925. Therefore, the deformation rates using PSI are convincible.

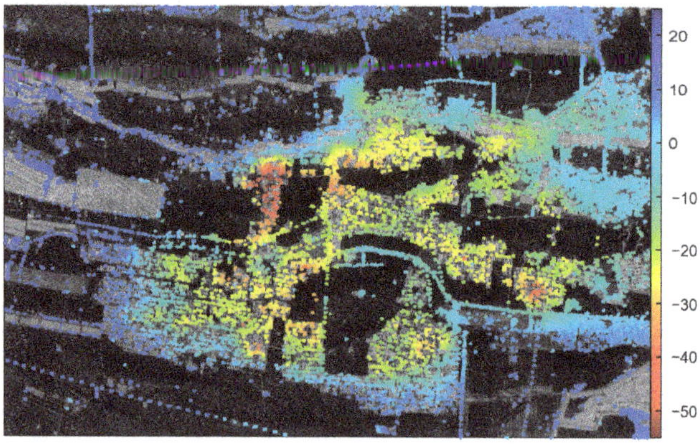

Figure 10. The land deformation estimated using PSI.

Due to the common phenomenon of spatial-temporal decorrelation, it was difficult to obtain sufficient and effective PSs. To extract information from DSs and increase the spatial density of the measured points, PTA is performed to reduce the stochastic noise in DSs [15]. Based on the intensity image set and backscattering coefficient, the SHP is firstly selected, as shown in Figure 11. The selection window size is 11 × 21 and the SHP number threshold is set to 25. Based on the SHP family, the optimum phase values of DSs in time series are estimated using the MLE on all interferograms [15]. As shown in Figure 12, compared with the original phase, the signal-to-noise ratio (SNR) of the optimum phase values has been greatly improved, which proves the effectiveness of the PTA operation. The quality of the estimated phase of DSs can be indicated by the temporal coherence γ_{PTA}, which can be used effectively for the final selection of DSs with reliable phase estimation, as shown in Figure 13. In the traditional PS–DS algorithm (MLE combining PSs and DSs), the threshold is set to 0.6. However, unlike the traditional PS–DS algorithm, the proposed method divided the DSs into high-quality DSs and low-quality DSs according to the posterior coherence factor $\rho_{PTA\_H} = 0.75$ and $\rho_{PTA\_L} = 0.6$, respectively.

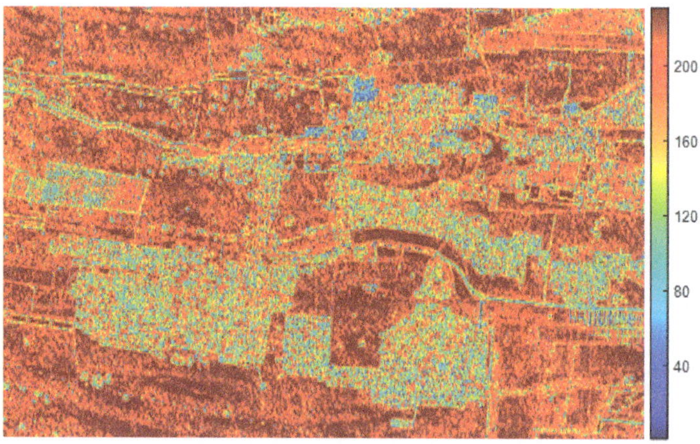

Figure 11. The SHP numbers.

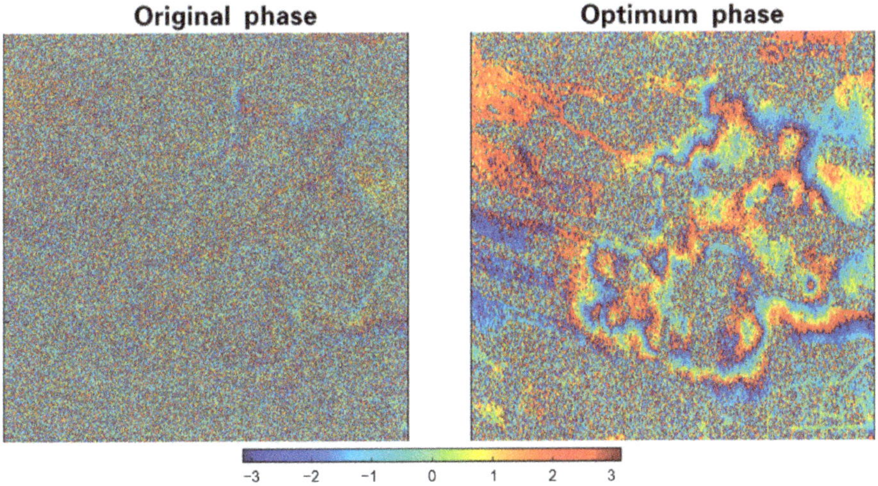

Figure 12. The original and optimum phase of DSs.

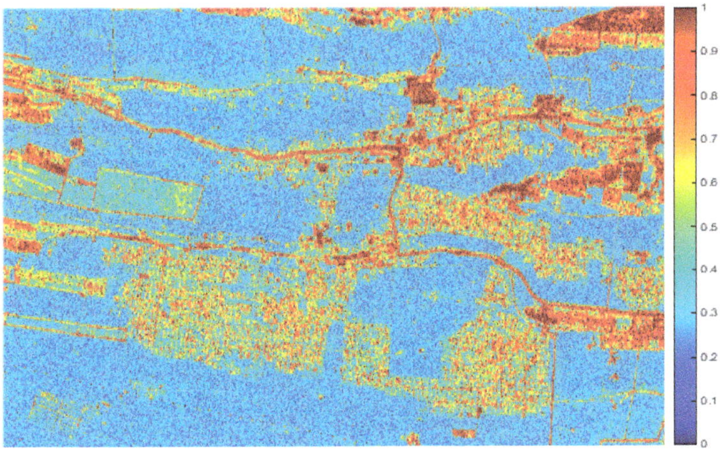

Figure 13. Temporal coherence map γ_{PTA}.

Then, the deformation rates were also estimated using the MLE combining PSs and DSs and the proposed method, as shown in Figures 14 and 15, respectively. It can be seen that the PS–DS results contain much more detailed information in comparison with the counterpart outcomes of PSI. Both the MLE combining PSs and DSs and the proposed method can reflect the depression cone correlated with the groundwater level. However, in analogy to the simulated experiment, the deformation rates estimated using the MLE combining PSs and DSs are not consistent with the results estimated using PSI in some PS pixels, as shown in the bottom row of Figure 14. There are some unexpected variations from pixel to pixel due to the poor phase quality of DSs and the diffusion of arc errors in the adjustment network, as shown in the red oval box in Figure 14. Moving on to the results estimated using the proposed method, the deformation rates are highly consistent with the results using PSI and the outliers are significantly suppressed by the employment of prior information derived from the high-quality results using PSI.

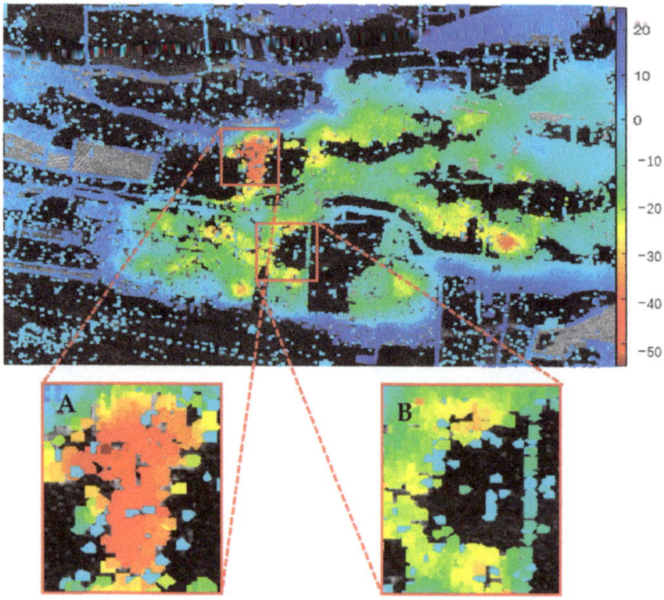

Figure 14. The land deformation estimated using MLE combining PSs and DSs. (**A**) Site A, (**B**) Site B.

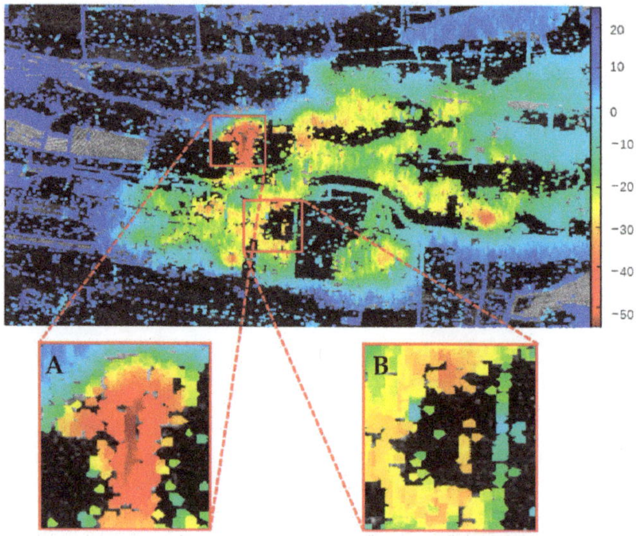

Figure 15. The land deformation estimated using the proposed method. (**A**) Site A, (**B**) Site B.

According to the survey in Refs. [40,41], the spatial distribution of the land deformation fits with the area affected by the overexploitation of the aquifers. Moreover, there is a strong positive relationship between the groundwater level drawdown in the wells and the land subsidence in the time series. When the water level in the well drops by about 10 m, the surface settlement is about 50 mm [41]. From 1 April 2017 to 22 March 2019, the groundwater level time series in region B can be monitored by the water level of the well in region B. The mean deformation rates are calculated in region B of Figures 14 and 15 and they are 12 mm/year and 25 mm/year, respectively.

As shown in Figure 16, the red circle is the groundwater level date of the GWP-060 monitoring well [41]. As mentioned earlier, due to the overexploitation of agricultural water, the groundwater level decreased significantly from May 2017 to January 2019, by up to about 8 m. According to the relationship between groundwater level and land subsidence, the land subsidence in the time series can be deduced, as shown by the solid blue line in Figure 16. The dotted blue line and dashed blue line are the deformation rates estimated using the MLE combining PSs and DSs and the proposed method, respectively. It is obvious that the deformation rate estimated using the proposed method is in good agreement with that of the groundwater level. However, the deformation rate estimated using the MLE combining PSs and DSs is too small, which is inconsistent with the groundwater level. The causes of the phenomenon are the poor phase quality of DSs in region B, which is reflected in the low temporal coherence value in Figure 13 and the error propagation in the subsequent adjustment network. In contrast, regularized by the spatial deformation field derived by the high-quality PSs and DSs, the proposed method achieves robust results even in the low coherence area.

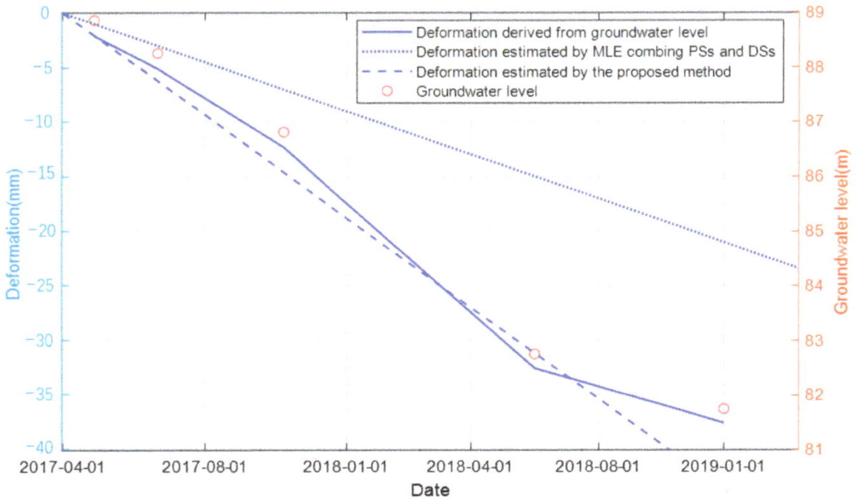

Figure 16. The correlation between the groundwater level and land deformation.

4. Discussion

The results demonstrated the potential of the proposed method for measuring land deformation velocity in low coherence areas. Especially in rural areas, vegetation coverage areas, or deserts, the temporal decorrelation is particularly serious. Even after PTA processing, the phase quality of a large number of DSs is far inferior to that of PSs, which deteriorates the deformation measurement accuracy. The proposed method regularized by the spatial deformation field has better robustness in low coherence areas, and while ensuring the measurement accuracy, the number of measurement points can be greatly increased.

The spatial deformation field is derived from PSs and high-quality DSs in the first-level network. Thus, the proposed method is a model-free implementation and does not require the surface deformation pattern to be known before starting the data analysis. The land deformation measurement in the first-level network is crucial, which not only determines the accuracy of the measurement points in the primary network but also has a direct impact on the measurement in the secondary network. Therefore, when building the first-level network, we prefer reliable points. In some areas, there are few reliable points and the proposed method will also risk failure. It would be a good choice to obtain the information on the deformation field from other instruments as a priori input.

For our test site in the UAE, the land deformation presents as a subsidence bowl caused by the overexploitation of groundwater. This type of land deformation has a strong spatial correlation and is easier to derive accurately from the first-level network. Therefore, the proposed method is expected to perform well in strong spatial correlations of land deformation, such as mining subsidence, landslides, etc.

5. Conclusions

This paper proposes a Bayesian estimation method of land deformation combining PSs and DSs to solve the contradiction between measurement accuracy and coverage. A two-level network is introduced into the traditional PSI to deal with PSs and DSs. In the first-level network, the MLE of the deformation parameters at PSs and high-quality DSs are obtained accurately. In the secondary-level network, the remaining DSs are connected to the nearest PSs or high-quality DSs and the deformation parameters are estimated using MAP. Due to the poor phase quality of the remaining DSs, MAP can achieve better estimation results than the MLE based on the spatial correlation of the deformation field. The main advantage of the proposed method lies in its robustness despite the PTA phase estimation bias and the inconsistent noise levels due to the diffserent decorrelation mechanisms. Simulation and Sentinel-1A real data results verified the feasibility and reliability of the proposed method. A total of 31 Sentinel-1A SAR data acquired between 2017 and 2019 were exploited to detect the land deformation at Remah in the United Arab Emirates (UAE) caused by the overexploitation of the aquifers. The proposed method not only greatly increased the density of the measuring points, but also ensured measurement accuracy and contained more detailed information, which reflected the depression cone correlating with the groundwater level. Even in low coherence areas, the deformation rate estimated using the proposed method is in good agreement with that of the groundwater level. Moreover, the proposed method is a model-free implementation and does not require the surface deformation pattern to be known before starting the data analysis but is affected by the spatial correlation of land deformation. Therefore, in a strong spatial correlation of land deformation, such as groundwater-related subsidence, mining subsidence, landslides, etc., the proposed method is expected to perform well.

Author Contributions: Conceptualization, G.L., Z.D. and T.Z.; methodology, G.L., Z.D. and M.L.; software, G.L.; validation, Z.H., M.L. and X.J.; formal analysis, G.L.; investigation, G.L. and H.L.; resources, Z.D. and T.Z.; data curation, G.L.; writing—original draft preparation, G.L.; writing—review and editing, G.L.; visualization, Z.H.; supervision, Z.D. and T.Z.; project administration, Z.D. and T.Z.; funding acquisition, Z.D. and T.Z. All authors have read and agreed to the published version of the manuscript.

Funding: This work was supported by the Key Program of the National Natural Science Foundation of China (No. 61931002, No. 11833001), the Natural Science Foundation of Chongqing, China, under Grant cstc2020jcyj-msxmX0663, and in part by the Beijing Natural Science Foundation (4202067).

Data Availability Statement: The Sentinel-1A data used in this study were provided by the European Space Agency (ESA), https://search.asf.alaska.edu/#/ (accessed on 5 July 2022).

Acknowledgments: Sentinel-1A data used in this study were provided by the European Space Agency (ESA) through the Sentinel-1 Scientific Data Hub. The measured groundwater level data is provided by the Environment Agency-Abu Dhabi. We are very grateful for the above support. In addition, we would like to thank the anonymous referees and the Associate Editor for useful comments that have helped to improve the presentation of this paper.

Conflicts of Interest: The authors declare no conflict of interest.

References

1. Bamler, R.; Hartl, P. Synthetic Aperture Radar interferometry. *Inverse Probl.* **1998**, *14*, R1. [CrossRef]
2. Gray, A.L.; Farris-Manning, P.J. Repeat-pass interferometry with Airborne Synthetic Aperture Radar. *IEEE Trans. Geosci. Remote Sens.* **1993**, *31*, 180–191. [CrossRef]
3. Goldstein, R. Atmospheric limitations to repeat-track radar interferometry. *Geophys. Res. Lett.* **1995**, *22*, 2517–2520. [CrossRef]

4. Zebker, H.A.; Villasenor, J. Decorrelation in interferometric radar echoes. *IEEE Trans. Geosci. Remote Sens.* **1992**, *30*, 950–959. [CrossRef]
5. Lavalle, M.; Simard, M.; Hensley, S. A temporal decorrelation model for polarimetric radar interferometers. *IEEE Trans. Geosci. Remote Sens.* **2012**, *50*, 2880–2888. [CrossRef]
6. Li, G.; Ding, Z.; Li, M.; Zhang, T.; Zeng, T.; Long, T. Earth-based repeat-pass SAR interferometry of the moon: Spatial–temporal baseline analysis. *IEEE Trans. Geosci. Remote Sens.* **2022**, *60*, 5227714. [CrossRef]
7. Ferretti, A.; Prati, C.; Rocca, F. Permanent scatterers in sar interferometry. *IEEE Trans. Geosci. Remote Sens.* **2001**, *39*, 8–20. [CrossRef]
8. Ferretti, A.; Prati, C.; Rocca, F. Nonlinear subsidence rate estimation using permanent scatterers in differential SAR interferometry. *IEEE Trans. Geosci. Remote Sens.* **2000**, *38*, 2202–2212. [CrossRef]
9. Hooper, A.; Segall, P.; Zebker, H. Persistent scatterer interferometric synthetic aperture radar for crustal deformation analysis, with application to Volcán Alcedo, Galápagos. *J. Geophys. Res.* **2007**, *112*, B07407. [CrossRef]
10. Crosetto, M.; Monserrat, O.; Cuevas-González, M.; Devanthéry, N.; Crippa, B. Persistent scatterer interferometry: A Review. *ISPRS J. Photogramm. Remote Sens.* **2016**, *115*, 78–89. [CrossRef]
11. Zhu, X.X.; Bamler, R. Very high resolution spaceborne sar tomography in urban environment. *IEEE Trans. Geosci. Remote Sens.* **2010**, *48*, 4296–4308. [CrossRef]
12. Guarnieri, A.M.; Tebaldini, S. On the exploitation of target statistics for SAR interferometry applications. *IEEE Trans. Geosci. Remote Sens.* **2008**, *46*, 3436–3443. [CrossRef]
13. Wang, Y.; Zhu, X.X.; Bamler, R. Retrieval of phase history parameters from distributed scatterers in urban areas using very high resolution SAR Data. *ISPRS J. Photogramm. Remote Sens.* **2012**, *73*, 89–99. [CrossRef]
14. Berardino, P.; Fornaro, G.; Lanari, R.; Sansosti, E. A new algorithm for surface deformation monitoring based on small baseline differential SAR interferograms. *IEEE Trans. Geosci. Remote Sens.* **2002**, *40*, 2375–2383. [CrossRef]
15. Ferretti, A.; Fumagalli, A.; Novali, F.; Prati, C.; Rocca, F.; Rucci, A. A new algorithm for processing interferometric data-stacks: Squeesar. *IEEE Trans. Geosci. Remote Sens.* **2011**, *49*, 3460–3470. [CrossRef]
16. Ho tong minh, D.; Hanssen, R.; Rocca, F. Radar Interferometry: 20 Years of development in time series techniques and future perspectives. *Remote Sens.* **2020**, *12*, 1364. [CrossRef]
17. Fan, H.; Lu, L.; Yao, Y. Method combining probability integration model and a small baseline subset for time series monitoring of mining subsidence. *Remote Sens.* **2018**, *10*, 1444. [CrossRef]
18. Hu, B.; Wang, H.-S.; Sun, Y.-L.; Hou, J.-G.; Liang, J. Long-term land subsidence monitoring of Beijing (China) using the small baseline subset (SBAS) technique. *Remote Sens.* **2014**, *6*, 3648–3661. [CrossRef]
19. Zhang, L.; Dai, K.; Deng, J.; Ge, D.; Liang, R.; Li, W.; Xu, Q. Identifying potential landslides by stacking-insar in southwestern China and its performance comparison with SBAS-Insar. *Remote Sens.* **2021**, *13*, 3662. [CrossRef]
20. Touzi, R. A review of speckle filtering in the context of estimation theory. *IEEE Trans. Geosci. Remote Sens.* **2002**, *40*, 2392–2404. [CrossRef]
21. Samiei-Esfahany, S.; Martins, J.E.; van Leijen, F.; Hanssen, R.F. Phase estimation for distributed scatterers in Insar stacks using integer least squares estimation. *IEEE Trans. Geosci. Remote Sens.* **2016**, *54*, 5671–5687. [CrossRef]
22. Shamshiri, R.; Nahavandchi, H.; Motagh, M.; Hooper, A. Efficient ground surface displacement monitoring using sentinel-1 data: Integrating Distributed Scatterers (DS) identified using two-sample t-test with persistent scatterers (PS). *Remote Sens.* **2018**, *10*, 794. [CrossRef]
23. Jiang, M.; Guarnieri, A.M. Distributed scatterer interferometry with the refinement of spatiotemporal coherence. *IEEE Trans. Geosci. Remote Sens.* **2020**, *58*, 3977–3987. [CrossRef]
24. Tebaldini, S.; Monti, A. Methods and performances for multi-pass SAR interferometry. In *Geoscience and Remote Sensing New Achievements*; InTech: London, UK, 2010; pp. 329–357. [CrossRef]
25. Wang, Y.; Zhu, X.X. Robust estimators for multipass SAR interferometry. *IEEE Trans. Geosci. Remote Sens.* **2016**, *54*, 968–980. [CrossRef]
26. Ansari, H.; De Zan, F.; Bamler, R. Efficient phase estimation for Interferogram Stacks. *IEEE Trans. Geosci. Remote Sens.* **2018**, *56*, 4109–4125. [CrossRef]
27. Jiang, M.; Ding, X.; Li, Z. Hybrid approach for unbiased coherence estimation for multitemporal insar. *IEEE Trans. Geosci. Remote Sens.* **2014**, *52*, 2459–2473. [CrossRef]
28. Wang, C.; Wang, X.S.; Xu, Y.; Zhang, B.; Jiang, M.; Xiong, S.; Zhang, Q.; Li, W.; Li, Q. A new likelihood function for consistent phase series estimation in distributed scatterer interferometry. *IEEE Trans. Geosci. Remote Sens.* **2022**, *60*, 5227314. [CrossRef]
29. Even, M.; Schulz, K. Insar deformation analysis with distributed scatterers: A review complemented by new advances. *Remote Sens.* **2018**, *10*, 744. [CrossRef]
30. Shi, G.; Lin, H.; Ma, P. A hybrid method for stability monitoring in low-coherence urban regions using persistent and distributed scatterers. *IEEE J. Sel. Top. Appl. Earth Obs. Remote Sens.* **2018**, *11*, 3811–3821. [CrossRef]
31. Du, Z.; Ge, L.; Ng, A.H.-M.; Zhang, Q.; Alamdari, M.M. Assessment of the accuracy among the common persistent scatterer and distributed scatterer based on Squeesar Method. *IEEE Geosci. Remote Sens. Lett.* **2018**, *15*, 1877–1881. [CrossRef]
32. Liang, H.; Zhang, L.; Ding, X.; Lu, Z.; Li, X.; Hu, J.; Wu, S. Suppression of coherence matrix bias for phase linking and ambiguity detection in mtinsar. *IEEE Trans. Geosci. Remote Sens.* **2021**, *59*, 1263–1274. [CrossRef]

33. Guarnieri, A.M.; Tebaldini, S. Hybrid cramér–Rao bounds for crustal displacement field estimators in sar interferometry. *IEEE Signal Process. Lett.* **2007**, *14*, 1012–1015. [CrossRef]
34. Agram, P.S.; Simons, M. A noise model for Insar Time Series. *J. Geophys. Res. Solid Earth* **2015**, *120*, 2752–2771. [CrossRef]
35. Zhang, L.; Ding, X.; Lu, Z. Modeling PSInSAR time series without phase unwrapping. *IEEE Trans. Geosci. Remote Sens.* **2011**, *49*, 547–556. [CrossRef]
36. Morishita, Y.; Hanssen, R.F. Temporal decorrelation in L-, C-, and X-band satellite radar interferometry for pasture on drained peat soils. *IEEE Trans. Geosci. Remote Sens.* **2015**, *53*, 1096–1104. [CrossRef]
37. Kellndorfer, J.; Cartus, O.; Lavalle, M.; Magnard, C.; Milillo, P.; Oveisgharan, S.; Osmanoglu, B.; Rosen, P.A.; Wegmüller, U. Global Seasonal sentinel-1 interferometric coherence and backscatter data set. *Sci. Data* **2022**, *9*, 73. [CrossRef]
38. Gauvain, J.-L.; Lee, C.-H. Maximum a posteriori estimation for multivariate gaussian mixture observations of Markov chains. *IEEE Trans. Speech Audio Process.* **1994**, *2*, 291–298. [CrossRef]
39. Oliver, M.A.; WEBSTER, R. Kriging: A method of interpolation for Geographical Information Systems. *Int. J. Geogr. Inf. Syst.* **1990**, *4*, 313–332. [CrossRef]
40. Groundwater Atlas of Abu Dhabi Emirate. Available online: https://www.researchgate.net/publication/337936386_GROUNDWATER_ATLAS_OF_ABU_DHABI_EMIRATE (accessed on 5 July 2022).
41. El Kamali, M.; Papoutsis, I.; Loupasakis, C.; Abuelgasim, A.; Omari, K.; Kontoes, C. Monitoring of land surface subsidence using persistent scatterer interferometry techniques and ground truth data in arid and semi-arid regions, the case of Remah, UAE. *Sci. Total Environ.* **2021**, *776*, 145946. [CrossRef]

remote sensing

Article

Analysis of the Spatial and Temporal Evolution of Land Subsidence in Wuhan, China from 2017 to 2021

Yizhan Zhao [1], Lv Zhou [1,*], Cheng Wang [1,2], Jiahao Li [1], Jie Qin [3], Haiquan Sheng [1], Liangke Huang [1] and Xin Li [4,5]

1 College of Geomatics and Geoinformation, Guilin University of Technology, Guilin 541004, China; zhaoyizhan@glut.edu.cn (Y.Z.); bx20211013@glut.edu.cn (C.W.); lijiahao@glut.edu.cn (J.L.); 3202080807117@glut.edu.cn (H.S.); lkhuang@whu.edu.cn (L.H.)
2 Guangxi Institute of Natural Resources Survey and Monitoring, Nanning 530023, China
3 Natural Resources and Real Estate Registration Center of Guangxi Zhuang Autonomous Region, Nanning 530029, China; 3182042041826@glut.edu.cn
4 Guangxi Key Laboratory of Spatial Information and Geomatics, Guilin University of Technology, Guilin 541004, China; lixin2017@chd.edu.cn
5 College of Geology Engineering and Geomantic, Chang'an University, Xi'an 710054, China
* Correspondence: zhoulv@glut.edu.cn

Abstract: Land subsidence is a common geological hazard. Rapid urban expansion has led to different degrees of ground subsidence within Wuhan in the past few years. The novel coronavirus outbreak in 2020 has seriously impacted urban construction and people's lives in Wuhan. Land subsidence in Wuhan has changed greatly with the resumption of work and production. We used 80 Sentinel-1A Synthetic Aperture Radar (SAR) images covering Wuhan to obtain the land subsidence change information of Wuhan from July 2017 to September 2021 by using the small baseline subset interferometric SAR technique. Results show that the subsidence in Wuhan is uneven and concentrated in a few areas, and the maximum subsidence rate reached 57 mm/yr during the study period. Compared with land deformation before 2017, the land subsidence in Wuhan is more obvious after 2020. The most severe area of subsidence is located near Qingling in Hongshan District, with a maximum accumulated subsidence of 90 mm, and obvious subsidence funnels are observed in Qiaokou, Jiangan, Wuchang and Qingshan Districts. The location of subsidence centers in Wuhan is associated with building intensity, and most of the subsidence funnels are formed in connection with urban subway construction and building construction. Carbonate belt and soft ground cover areas are more likely to lead to karst collapse and land subsidence phenomena. Seasonal changes are observed in the land subsidence in Wuhan. A large amount of rainfall can replenish groundwater resources and reduce the rate of land subsidence. The change in water level in the Yangtze River has a certain impact on the land subsidence along the rivers in Wuhan, but the overall impact is small. An obvious uplift is observed in Caidian District in the south of Wuhan, and the reason may be related to the physical and chemical expansion effects of the expansive clay.

Keywords: land subsidence; SBAS-InSAR; Wuhan city; urban construction; land uplift

Citation: Zhao, Y.; Zhou, L.; Wang, C.; Li, J.; Qin, J.; Sheng, H.; Huang, L.; Li, X. Analysis of the Spatial and Temporal Evolution of Land Subsidence in Wuhan, China from 2017 to 2021. *Remote Sens.* **2022**, *14*, 3142. https://doi.org/10.3390/rs14133142

Academic Editors: Guohong Zhang, Chisheng Wang, Wu Zhu, Daqing Ge and Siting Xiong

Received: 27 May 2022
Accepted: 27 June 2022
Published: 30 June 2022

1. Introduction

Land subsidence is a geological phenomenon of regional ground elevation reduction caused by natural or human factors [1–3]. With rapid urban expansion and excessive groundwater extraction, it has become a widespread geohazard in many cities worldwide [4,5]. Many countries and regions in the world, such as Iran, Italy, Egypt, Spain, Mexico and Texas, are facing the problem of land subsidence [6–12]. Among them, Jakarta, the capital of Indonesia, had to move its capital to East Kalimantan Province due to severe land subsidence [13]. Subsidence can decrease the water storage capacity of underground aquifers, leading to land collapse, damage to buildings and civil infrastructures, and increased risk of flooding [14–16]. The United Nations has studied and predicted that the global land subsidence will affect 1.6 billion residents by 2040 [17]. Therefore, subsidence in large cities must be monitored.

As the largest developing country in the world, China is experiencing a large-scale and high-speed urbanization. Land subsidence caused by excessive exploitation of groundwater and high-intensity urban construction is becoming increasingly exacerbated. For example, land subsidence has occurred in Beijing, Tianjin, Guangdong, Foshan and other cities [18–22]. Statistics show that the area where ground subsidence of more than 10 mm/yr occurs in China is more than 120,000 km^2 by the end of 2015, involving 94 prefecture-level cities and 425 districts and counties [23]. Wuhan is a central city in the central region of China and is the political, economic and cultural center of Hubei Province. In recent years, Wuhan is on a speed way of urbanization and rail transit construction. At present, the permanent resident population of Wuhan has reached 12.33 million, and the operating mileage of rail transit has reached 409.1 km. Large-scale urban construction and the special geological structure of Wuhan often lead to land subsidence caused by karst collapse, soft soil compression and engineering construction [24,25].

The first recorded land surface subsidence in Wuhan occurred in 1931 [26]. With the increase in human activities and the overexploitation of groundwater, the rate and extent of subsidence in Wuhan are increasing [27,28]. In particular, the rapid expansion of Wuhan in recent years has led to the emergence of several subsidence funnels within the city. Land subsidence can be divided into three types: tectonic subsidence caused by crustal subsidence movement, pumping subsidence and goaf subsidence [29]. The main causes of ground subsidence in Wuhan are the consolidation of soft soil layers and karst ground collapse caused by groundwater overdraft and urban construction [30,31]. Monitoring land subsidence in Wuhan mainly depends on about 300 benchmark points distributed in the city but lacks long-term effective land subsidence monitoring [32]. Although traditional measurement techniques (e.g., level survey and global navigation satellite system) have high measurement accuracy and sensitivity, performing subsidence monitoring on a large scale is difficult [33,34]. Interferometric Synthetic Aperture Radar (InSAR) technology can compensate for the shortcomings of traditional deformation monitoring methods [35,36]. Specifically, the time series InSAR technology overcomes the problems of temporal and spatial decorrelation and atmospheric effect [37,38] and greatly improves the measurement accuracy. This method has been widely used in the fields of land subsidence and landslide monitoring [39–42]. The often-used time series InSAR techniques include Persistent Scatterer InSAR (PS-InSAR), Small Baseline Subset InSAR (SBAS-InSAR) and Multi-Temporal InSAR (MT-InSAR) [43–45].

Bai et al. [46] first monitored the Wuhan urban area using high-resolution TerraSAR-X data and found that the land deformation rate in Wuhan ranges from −64 to 18 mm/yr between 2009 and 2010, with Hankou being the largest subsidence area, and urban construction and karst geology being the major causative factors of subsidence. Costantini et al. [47] used COSMO-SkyMed SAR data to obtain the land deformation rates in Hankou from 2013 to 2014 ranging from −80 to 40 mm/yr. Urban construction in Wuhan accelerated after 2015, and approximately 10,000 sites were under construction, leading to serious land subsidence and collapse in the city [48–51]. Zhou et al. found land subsidence was the most severe in the Houhu area between 2015 and 2016, with a maximum subsidence rate of more than 80 mm/yr [49]. Zhang et al. found that subsidence had occurred in Hankou, Wuchang, Hanyang, Qingshan Industrial Park and Baishazhou between 2015 and 2018 [51].

In this study, we used the SBAS-InSAR technique to obtain the surface deformation of Wuhan between July 2017 and September 2021. Here, we discussed the relationship between land subsidence and rainfall, the total amount of groundwater and the change in river level, which can provide a reference for the prevention of land subsidence disasters in Wuhan.

2. Study Area and Datasets

2.1. Study Area

Wuhan is located in the eastern part of Hubei Province (South East China) and it covers an area of 8494.41 km^2. It is situated at the intersection of hills, mountains and the Han River Plain, as shown in Figure 1a. The overall topography is high in the south and low in the north, with a remnant mounded impact plain in the middle. Wuhan has a maximum West–East directional distance of 134 km from east to west and a maximum North–South directional distance of about 155 km (at latitude 29°58′N to 31°22′N and longitude 113°41′E to 115°05′E) with altitudes ranging from 19.2 m to 873.7 m. Wuhan has a north subtropical monsoonal (humid) climate with rainfall concentrated between June and August each year, and is characterized by rain and heat in the same season and four distinctive seasons [52,53]. Many rivers and lakes form in the city, where the Yangtze River and Han River form a huge water network. The total water area reaches 2217.6 km^2, accounting for about 26.1% of the city's total area.

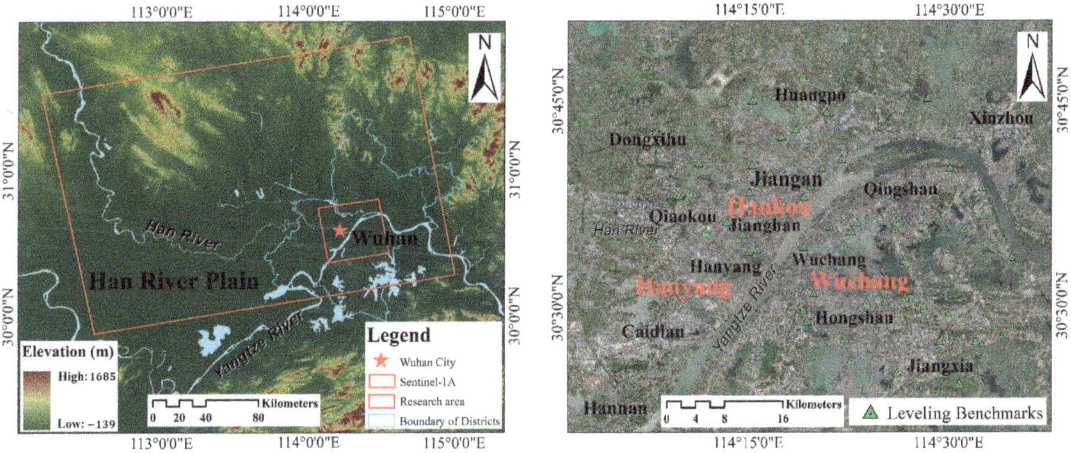

(**a**) (**b**)

Figure 1. (**a**) Geographical location of Wuhan. The orange rectangle illustrates the coverage of Sentinel-1A, and the red rectangle is the study area. (**b**) Optical satellite image of Wuhan, where green triangles indicate level benchmarks.

Wuhan has several belts of occult carbonates, consisting of the "Lower Carbonate Group" with a total thickness of more than 200 m deposited in the middle of the Late Paleozoic and the "Upper Carbonate Group" represented by marl, thin-bedded tuff, thick-bedded tuff and thick-bedded dolomite deposited in the Early Triassic. The Indochinese movement at the end of the Middle Triassic forged the basic outline of the carbonate stratigraphy in Wuhan [54]. As shown in Figure 2, the carbonate rock group is distributed in a plane from north to south, forming a total of six carbonate belts (L1–L6). Wuhan is located at the intersection of the Yangtze River and the Han River [30]. The soft soil is mainly silty soil and silt, and it has a large distribution in the sections along the rivers and around the lakes in Hankou, Wuchang, Hannan, Xinzhou and Jiangxia, with a maximum thickness of 37 m and a minimum thickness of 0.1 m. The soft soil with low strength and high compressibility covers most areas of the city and is easy to cause foundation settlement and other adverse geological problems [28]. The complex geological structure often leads to disaster events, such as sinkholes, uneven foundation settlements and landslides, in Wuhan [26,30].

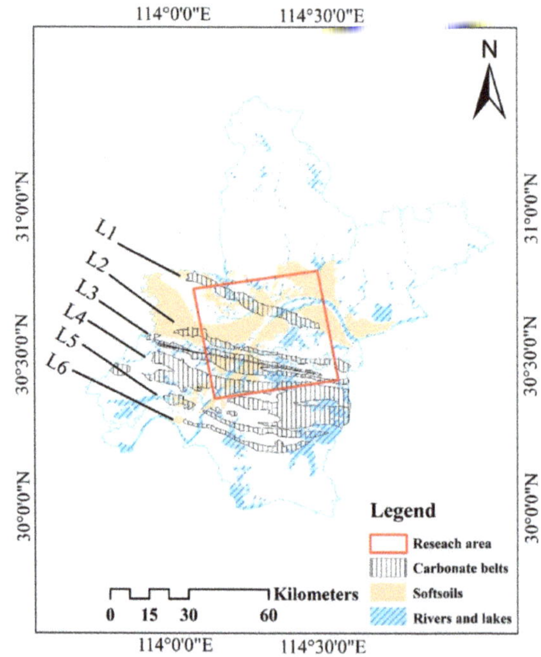

Figure 2. Geological structure map of Wuhan. Carbonate belts (L1–L6) are the Tianxingzhou karst belt, Bridge karst belt, Baishazhou karst belt, Zhuankou karst belt, Junshan karst belt and Hannan karst belt, respectively.

2.2. Datasets

We employ 80 ascending Sentinel-1A single-look complex images acquired from July 2017 to September 2021. These images cover the entire area of Wuhan, with a 1-month interval for image acquisition between July 2017 and December 2019 and a 12-day interval for images between January 2020 and September 2021. The information of Sentinel-1A images is shown in Table 1. We used the Shuttle Radar Topography Mission (SRTM) 30 m Digital Elevation Model by the National Aeronautics and Space Administration to remove topographic phases, and precise orbit ephemerides provided by the European Space Agency to eliminate orbital errors.

Table 1. Specific parameters of Sentinel-1A data.

Parameter	Value	Parameter	Value
Product type	Sentinel-1A	Incidence angle	39°
Wavelength	C-band	Path	113
Flight direction	Ascending	Resolution	2.7 m × 22 m
Polarization	VH	Number of images	80
Beam mode	IW	Time range	17 July 2017– 30 September 2021

We collected some data related to ground subsidence in Wuhan. Wuhan Metro information is obtained from Wuhan Metro Group Co., Ltd., Wuhan, China. The distribution map of soft soil and carbonate rocks is acquired from Hubei Geological Bureau. Rainfall and groundwater data are obtained from the Wuhan Water Resources Bulletin. The water level information comes from the Water Resources Department of Hubei Province (https://slt.hubei.gov.cn, accessed on 29 December 2021). To more intuitively analyze the urban changes in Wuhan, we used Google Earth historical optical images.

3. Methods

3.1. SBAS-InSAR Technique

SBAS-InSAR technology combines the data into several sets in accordance with the distribution of SAR image sequences in temporal and spatial baselines. Within the set, the interferometric image is smaller for the spatial baseline, and the interferometric image is larger for the spatial baseline between the sets, resulting in a small baseline interferogram. On the basis of the minimum norm criterion of deformation rate, the deformation rate and its time series of the coherent target are obtained by using the least square method or Singular Value Decomposition (SVD) [44].

If the target feature has $N+1$ SAR images, then time to acquire the image is $t_0 \sim t_N$. Images from the same area at different times can interfere with each other, and two images can be found in at least one subset. One of the images is selected as the master image, and the remaining images are registered with it. N view images can generate M multi-view differential interferograms, and M satisfies the following inequality [48]:

$$\frac{N+1}{2} \leq m \leq N\left(\frac{N+1}{2}\right) \tag{1}$$

The differential interferogram j is generated by interfering the SAR image maps at time t_A and time t_B ($t_B > t_A$). The extracted high coherence points are phase unwrapped after removing the flat-earth and topographic phase, and the phase of the differential interferogram at pixel coordinates (x, r) can be expressed as:

$$\begin{gathered}\delta\phi_j(x,r) = \phi(t_B,x,r) - \phi(t_A,x,r) \\ \approx \tfrac{4\pi}{\lambda}[d(t_B,x,r) - d(t_A,x,r)] + \Delta\phi^j_{topo}(x,r) + \Delta\phi^j_{APS}(t_B,t_A,x,r) \\ +\Delta\phi^j_{noise}(x,r)\end{gathered} \tag{2}$$

where $\phi(t_A, x, r)$ and $\phi(t_B, x, r)$ represent the phase values of SAR images at t_A and t_B, respectively, $d(t_B, x, r)$ and $d(t_A, x, r)$ are the cumulative form variables in the Line of Sight (LOS) direction relative to $d(t_0, x, r)$ at moments t_B and t_A. $\Delta\phi^j_{topo}(x, r)$ denotes the residual phase of the differential interferogram. $\Delta\phi^j_{APS}(t_B, t_A, x, r)$ is the phase error caused by atmospheric delay. $\Delta\phi^j_{noise}(x, r)$ denotes random noise phases.

When the effect of error is ignored, Equation (2) can be simplified as:

$$\delta\phi_j(x,r) = \phi(t_B,x,r) - \phi(t_A,x,r) \approx \frac{4\pi}{\lambda}[d(t_B,x,r) - d(t_A,x,r)] \tag{3}$$

The phase in Equation (3) is expressed as the product of the average phase rate and time between the two acquisition times to obtain the settlement time series with physical meaning:

$$v_j = \frac{\phi_j - \phi_{j-1}}{t_j - t_{j-1}} \tag{4}$$

The phase value of the interferogram j can be written as:

$$\sum_{k=t_A,j+1}^{t_B,j} (t_k - t_{k-1})v_k = \delta\phi_j \tag{5}$$

This equation indicates that the integration of the rate of each time period over the master and slave image time intervals is expressed in matrix form as:

$$Bv = \delta\phi \tag{6}$$

Rank deficiency is likely to occur because the SBAS-InSAR technique uses multiple master image spatio-temporal baselines to obtain differential interference. The generalized inverse matrix of matrix B can be obtained by using SVD to acquire the minimum norm solution of rate vector. The rate in each time period is integrated to obtain the amount of surface settlement in each time period.

3.2. SBAS-InSAR Data Processing

The main processing flow of SBAS-InSAR method is shown in Figure 3, which mainly includes the steps of generating small baseline subsets, interferogram formation and phase unwrapping, refinement and reflattening, and generation time series results. To more intuitively compare the changes in subsidence in Wuhan before and after the COVID-19, this study divides the data into two time periods (July 2017 to June 2020 and January 2020 to September 2021). The two time periods have half year data overlap to ensure the continuity of settlement results [49].

(1) Generating connection pairs: we selected the images of 13 January 2018 and 18 August 2020 as the master images and realigned the other images with the master images to avoid the effect of temporal and spatial decorrelation factors. For the first time period, the time threshold is 180 days, and the spatial baseline threshold is 45%. In the second time period, the time threshold is 60 days, and the critical value of spatial baseline is 45%.
(2) Interferogram formation and phase unwrapping: interference processing is performed on all interferometric pairs to generate differential interferograms, remove the flat-earth and topographic phase, perform phase unwrapping and generate phase diagram.
(3) Refinement and reflattening: this process mainly aims to estimate and remove the residual phases and ramp phase that still exists after phase unwrapping.
(4) Generation of time series results: the phase after unwrapping is calculated by using SVD, and the atmospheric phase and other errors are removed by two inversions to obtain the accurate displacement results on the time series. The final deformation value in the LOS direction is obtained by geocoding.

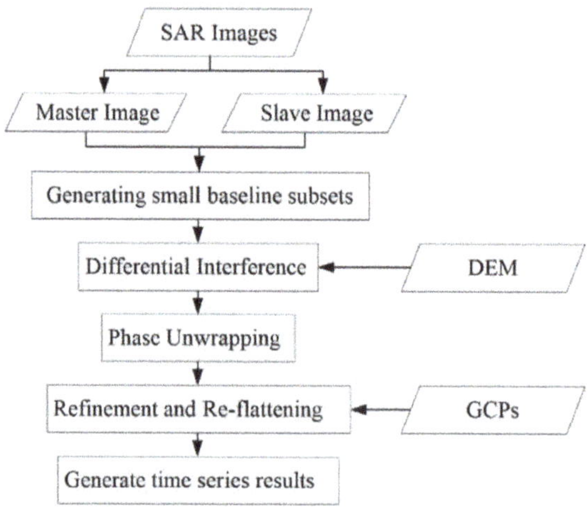

Figure 3. SBAS flow chart.

4. Results

4.1. Temporal and Spatial Distribution Characteristics of Land Subsidence in Wuhan

Figures 4 and 5 show the land subsidence rates in the LOS direction in Wuhan for the two time periods, respectively. Positive values in the graph indicate land uplift, whereas negative values indicate land subsidence. We extracted 2,851,137 PS points in the study area using SBAS InSAR technique from July 2017 to June 2020 (Figure 4), and the deformation rate in the main urban area of Wuhan ranged from −46.2 mm/yr to 24.6 mm/yr during the study period, with an average subsidence rate of 0.3 mm/yr. Four subsidence areas are found Qiaokou, Jiang'an, Qingshan and Hongshan Districts. The subsidence in Qingshan District is relatively serious, which may be related to the presence of a large number of industrial parks in the area, such as Wuhan Iron and Steel (Group) Company and Wuhan Petrochemical Plant.

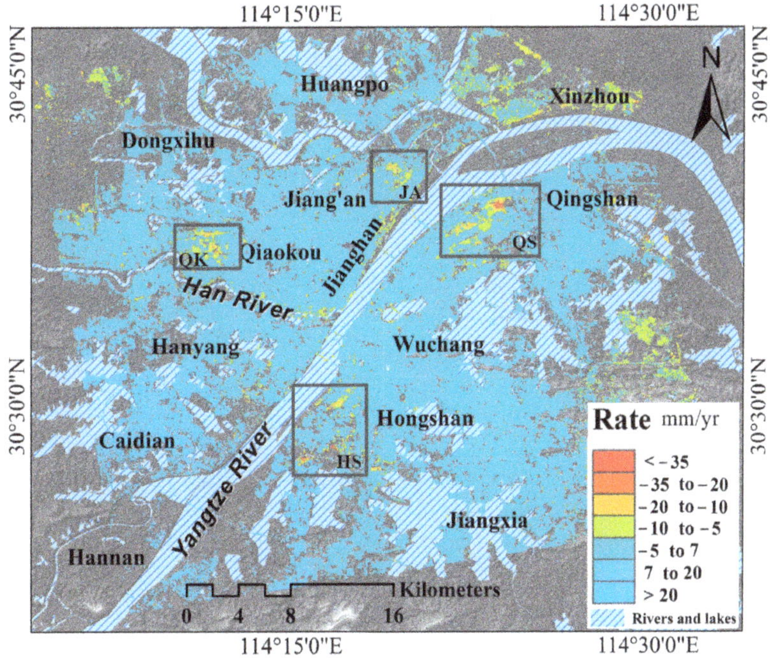

Figure 4. Land subsidence rate in the LOS direction in Wuhan from July 2017 to June 2020. The gray rectangles are the major areas of subsidence. JA, QK, QS and HS are the abbreviations of Jiangan, Qiaokou, Qingshan and Hongshan, respectively.

A total of 4,375,472 PS points were acquired from January 2020 to September 2021 (Figure 5), and the deformation rate varies from −57.1 mm/yr to 33.1 mm/yr with an average of −2.2 mm/yr. The subsidence in Wuhan has a high tendency to further expand after 2020. Five obvious subsidence areas are found in the main urban area: Jiangan, Qiaokou, Qingshan, Wuchang and Hongshan Districts. Compared with previous results, Qingshan District has a large-scale settlement phenomenon, and the settlement center is located in the Wuhan Iron and Steel (Group) Company (WISCO) ironmaking plant. The subsidence near Qingling in Hongshan District is the most serious, and the land subsidence rate in some areas exceeds 50 mm/yr under the influence of urban and subway construction. The overall subsidence range in Wuhan is small, and the deformation rate in most areas is between −5 mm/yr and 7 mm/yr. Land subsidence in Wuhan is concentrated in a few areas. Interestingly, uplift has been occurring in parts of Caidian District in the southern part of Wuhan.

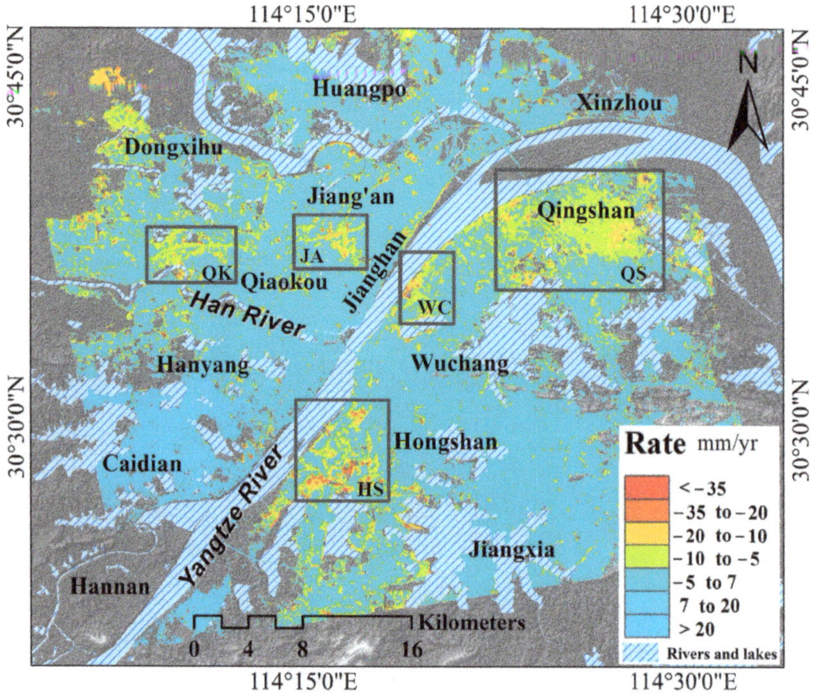

Figure 5. Land subsidence rate in the LOS direction in Wuhan from January 2020 to September 2021. The gray rectangles indicate the major area of subsidence. JA, QK, QS, WC and HS are the abbreviations of Jiangan, Qiaokou, Qingshan, Wuchang and Hongshan, respectively.

To explore the temporal evolution of land subsidence in Wuhan, we used the image acquired on 3 January 2020, as the reference image for the time series and calculated the time series of cumulative subsidence in Wuhan from 3 January 2020 to 30 September 2021. The results are shown in Figure 6. The distribution of cumulative settlement is consistent with the distribution of subsidence rate, which is relatively stable in most areas of Wuhan city, with cumulative settlement mostly between −5 and 10 mm. The maximum subsidence in Wuhan during the study period is 104 mm, located near Qingling, Hongshan District. The settlement in the main subsidence zone gradually increases with time, and the area of the subsidence funnel gradually expands. Land subsidence in Wuhan can be significantly nonlinear with seasonal changes. The area of subsidence increases rapidly between March and June during the same year. However, the change in subsidence is smaller between June and September, and a slowing trend is observed in land subsidence. Land subsidence may be influenced by rainfall. Wuhan has a considerable amount of rainfall from June to August, and a large amount of rainfall can effectively recharge groundwater, which may alleviate the land subsidence.

To check the accuracy of the Wuhan land subsidence results obtained by SBAS-InSAR technique, we evaluated the error of the average deformation rate in two time periods. The Root Mean Square Error (RMSE) is obtained by calculating the deviation of rate linear fitting. The probability density function of the RMSE of subsidence rate in the two time periods is shown in Figure 7a,b, and the average value of RMSE is 2.5 and 2.6 mm/yr, respectively. The RMSE of 97% PS points in the study area is less than 5 mm/yr, which has high accuracy for InSAR time series results.

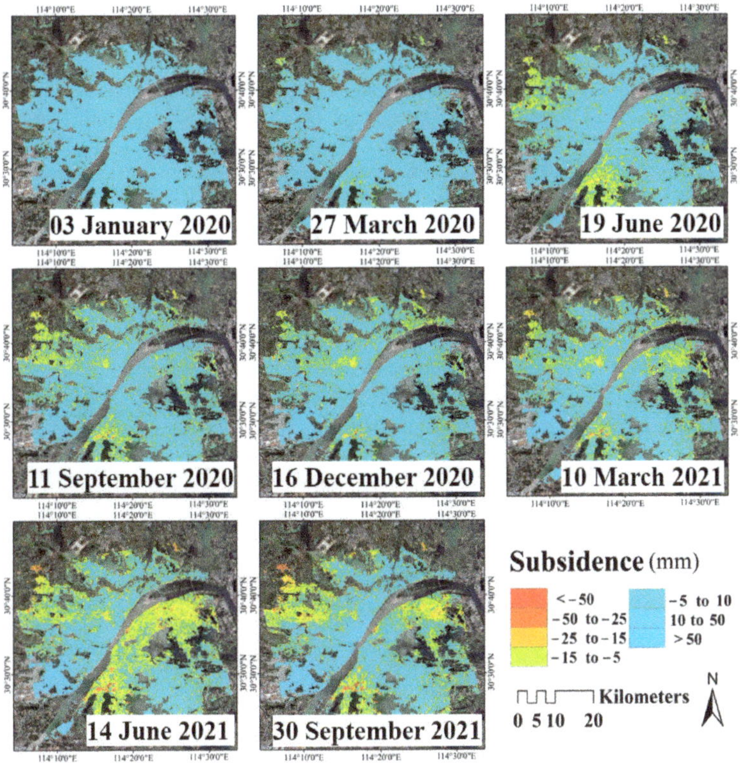

Figure 6. Time series of cumulative settlement from 3 January 2020 to 30 September 2021. The image obtained on 3 January 2020 is the reference image.

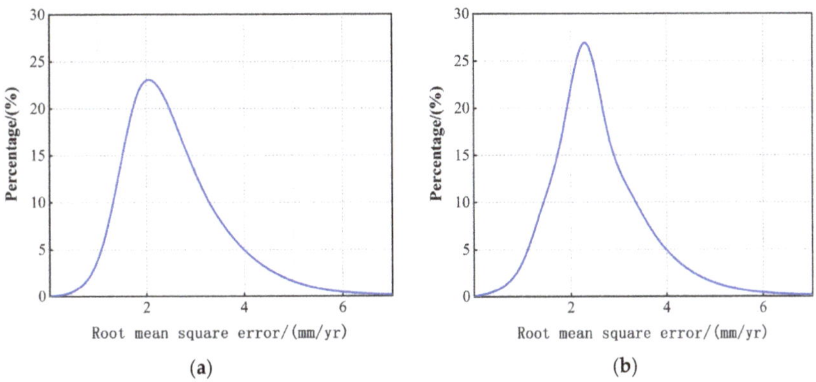

Figure 7. (**a**) RMSE of subsidence results from 17 July 2017 to 19 June 2020; (**b**) RMSE of subsidence results from 3 January 2020 to 30 September 2021.

4.2. Subsidence Changes before and after the COVID-19 in Wuhan

The changes in the main subsidence areas in Wuhan are shown in Figure 8. Five major areas of subsidence are found in Wuhan between June 2017 and September 2021, with the degree and extent of subsidence increasing in each area over time. The subsidence area in Jiangan District (JA) transfers from north to south, and the subsidence is mainly concentrated near Hankou Railway Station with the maximum subsidence rate around

30 mm/yr. The subsidence in Qiaokou (QK) District gradually transfers westward with time, and the subsidence along the Han River is more serious. A severe subsidence funnel occurred near Heping Avenue in Qingshan District (QS) between 2017 and 2020, and the subsidence of this funnel decreased after 2020. However, a widespread subsidence phenomenon occurred in the central part of Qingshan District (QS) compared with the previous period, with the deformation rate ranging from −40 mm/yr to −20 mm/yr. The Wuchang District experienced a subsidence after 2020, with severe subsidence occurring in areas close to the Yangtze River. The subsidence in Hongshan (HS) District is the most severe. Many areas have severe subsidence, and they are becoming contiguous.

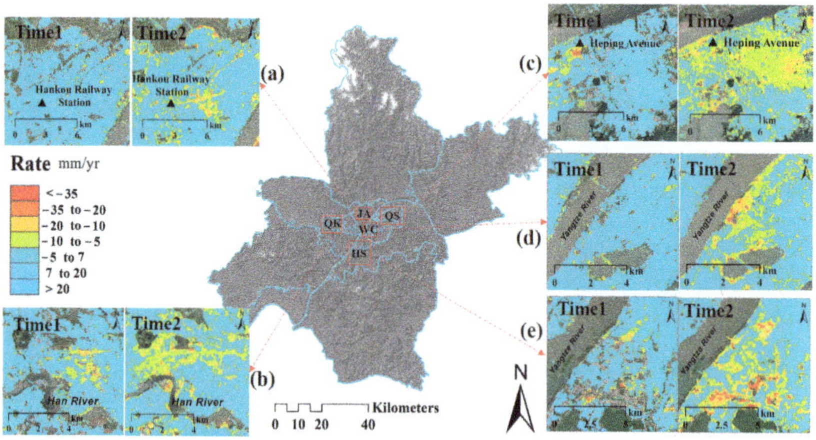

Figure 8. Changes in major subsidence areas in Wuhan from 17 June 2017 to 30 September 2021. The red rectangles are the five detected subsidence areas. Time 1 indicates June 2017 to June 2020, and time 2 denotes January 2020 to September 2021. JA, QK, QS, WC and HS are the abbreviations of Jiangan, Qiaokou, Qingshan, Wuchang, and Hongshan, respectively. The black triangle in (**a**) indicates the location of Hankou Railway Station, and the black triangle in (**b**) indicates the location of Heping Avenue. (**a**–**e**) indicate the sudsidence changes in Jiangan, Qiaokou, Qingshan, Wuchang and Hongshan, respectively.

We conducted a statistical analysis of the subsidence rate in Wuhan to analyze the subsidence changes in recent years. We calculated the percentage of PS points in different deformation rate ranges when performing the comparative analysis due to the large difference in the number of PS points in the two results. The results are shown in Figure 9a. The percentage of the PS points after the COVID-19 is higher than that before the COVID-19 in each range of deformation rate. A large change is observed in the number of PS points in the deformation rate ranging from −15 mm/yr to −5 mm/yr, and the percentage increases from 6.86 to 33.21%, which indicates a large-scale subsidence phenomenon in Wuhan after 2020. The number of PS points within the deformation rate ranging from −25 mm/yr to 15 mm/yr increases significantly, indicating that the subsidence situation in Wuhan is aggravated.

Figure 9b,c show the changes in the maximum and average deformation rates for the five major subsidence areas in Wuhan, respectively. The average deformation rate of Qingshan and Hongshan Districts changes the most due to a large-scale subsidence phenomenon in Qingshan District, and the subsidence phenomenon in Hongshan District is intensified. The maximum subsidence rate of each area exceeds 20 mm/yr between June 2017 and September 2021. After 2020, the deformation rate increases in all regions, with the maximum subsidence rate in Hongshan District exceeding 50 mm/yr.

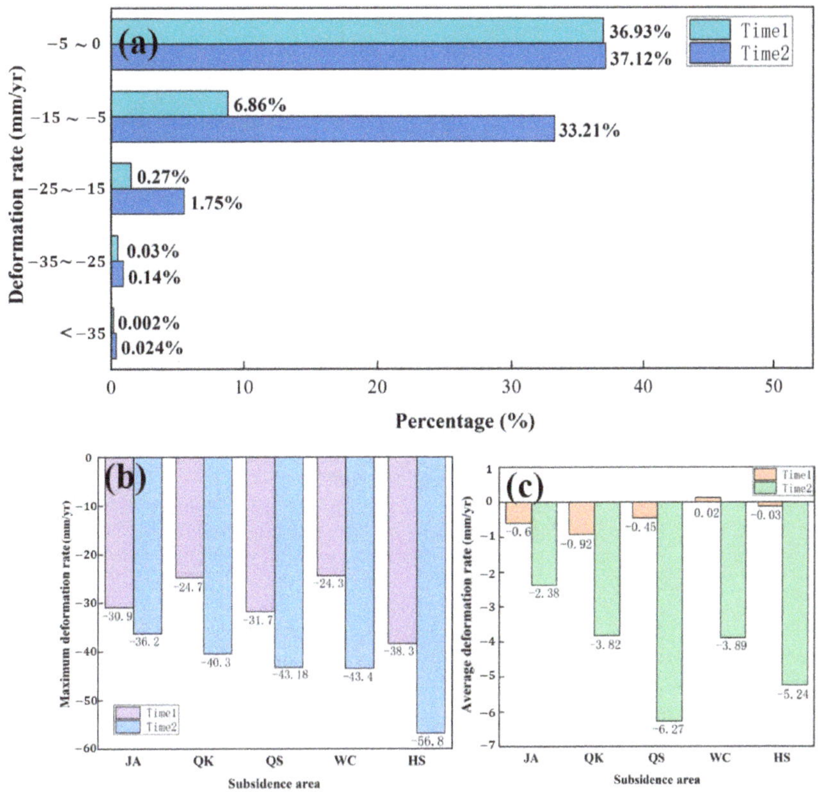

Figure 9. (**a**) Percentage of different deformation rate intervals; (**b**) changes in the maximum deformation rate of the major subsidence areas; (**c**) changes in the average deformation rate of the major subsidence areas. Time 1 denotes June 2017 to June 2020, and time 2 indicates January 2020 to September 2021.

In this study, we obtained the land subsidence results of Wuhan city in the last 4 years from 2017 to 2021. We summarized the results of previous studies on land subsidence monitoring in Wuhan to study the early subsidence, as shown in Table 2. The early subsidence funnels in Wuhan are distributed in Hankou, Qingshan, Wuchang and near Baishazhou, and the most obvious subsiding area is located in Houhu [46,47]. Historically, Houhu is a lake beach, and the lower cushion is composed of silt with a thickness of 10 m to 30 m. Urban development, especially the continuous construction of super high-rise buildings, has caused great pressure on the soft soil foundation. Previous studies showed that the Houhu area experienced severe subsidence between 2013 and 2016, with the maximum subsidence rate exceeding 80 mm/yr [48]. Qingshan District is the industrial center of Wuhan. A large amount of groundwater has been extracted because of industrial activities, resulting in land subsidence in the Qingshan District. In summary, the subsidence in Wuhan has been alleviated since 2017. The subsidence in the Houhu area has mitigated significantly, and the subsidence in other areas has also been alleviated [49]. However, the subsidence phenomenon in Wuhan intensified after 2020, with several areas experiencing subsidence [51].

Table 2. Summary of the previous studies of land subsidence in Wuhan.

Study	Method	Dataset	Deformation Rate	Main Subsidence Area
Bai et al. (2016) [46]	PS-InSAR	12 TerraSAR images (October 2009–August 2010)	−67 to 17.5 mm/yr	The largest subsidence area in Wuhan is located in Hankou
Costantini et al. (2016) [47]	PSP-InSAR approach	45 Cosmo-SkyMed images (June 2013–June 2014)	−80 to 40 mm/yr	Most areas of Hankou
Zhou et al. (2017) [48]	SBAS-InSAR	15 Sentinel-1 images (Aprirl 2015–April 2016)	−82 to 18 mm/yr	Houhu, Wuchang, Hanyang, Qingshan
Zhang et al. (2019) [49]	SBAS-InSAR	20 Radarsat-2 images (October 2015–June 2018)	−52 to 28 mm/yr	Houhu, Qingshan Industrial Park, Baishazhou
Shi et al. (2021) [51]	SBAS-InSAR	113 Sentinel-1 Images (April 2015–September 2019)	−30 to 30 mm/yr	Qingshan, Houhu, Dongxihu, Qingling

5. Discussion

Many factors are associated with the subsidence of Wuhan. Next, we will discuss the causes of land subsidence in Wuhan in terms of subway construction, urban infrastructure construction, hydrogeology, groundwater and rainfall, and river level changes. We will also briefly analyze the ground uplift phenomenon in Caidian District.

5.1. Subway Construction

Subways can effectively alleviate the pressure of urban ground traffic and improve commuting efficiency. However, subway tunneling can affect the surrounding soil during subway construction and cause ground deformation. In the area of dense buildings, subway tunneling can even lead to foundation deformation, affecting the safety of subway construction and ground buildings. Wuhan started to build subways as early as 2000. The total number of subway operating mileage in Wuhan reached 435 km with 14 subway lines by the end of 2021. The distribution of subway is shown in Figure 10a. To investigate the relationship between subway construction and land subsidence, we extracted the land subsidence within 500 m on two sides of the subway (Figure 10b). We counted the information of subway lines under construction in Wuhan during the study period, as shown in Table 3. The findings indicate that the land subsidence along Metro Line 5 is relatively severe, with the subsidence rate exceeding 35 mm/yr in some areas, and some sections of Metro Lines 6 and 12 have experienced subsidence exceeding 20 mm/yr. Two typical subsidence areas are located near Qingling Station in Hongshan District (HS) and Shiqiao Station in Jiangan District (JA).

The subsidence in the area near Qingling Station is shown in Figure 11a, with several subsiding areas along the elevated bridge section of Metro Line 5. The construction of Wuhan Metro Line 5 is difficult due to the access to the Qingling River, the elevated Third Ring Road and the karst area. A part of the line passes through the main traffic road in Baishazhou, and the surrounding communities are dense, so subsidence is extremely easy to occur during construction. Wuhan Metro Line 5 returned to construction on 24 March 2020. As a transit station for Metro Lines 5 and 12, Qingling Station has a large land subsidence. In this study, we conducted a profile analysis along the Qingling Station of Line 5 (black dashed line AA′ in the figure). Figure 11b,c indicate the profile deformation rate and cumulative settlement, respectively. The subsidence fluctuation shows an "M" shape, with large fluctuations and rapid changes in the deformation rate from 0 m to 250 m and from 850 m to 1500 m. The subsidence rate at Baisha Five Road Station and Qingling Station exceeded 35 mm/yr, with the maximum settlement reaching 70 mm, and other areas were relatively stable.

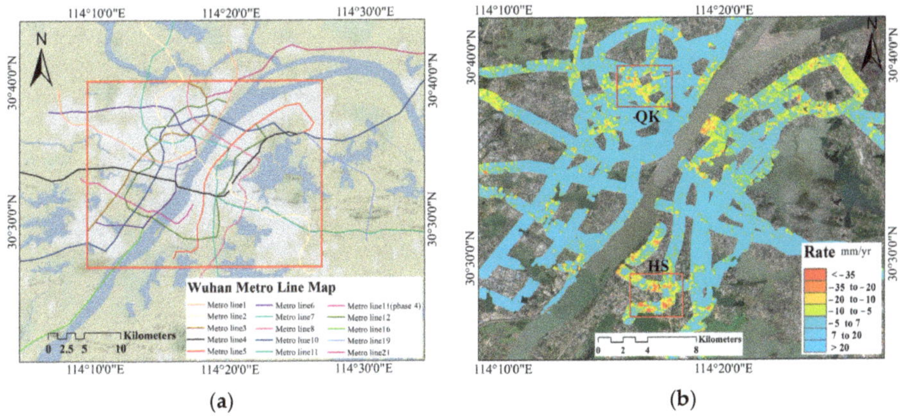

Figure 10. (**a**) Distribution of major metro lines in Wuhan; (**b**) land subsidence within 500 m on two sides of Wuhan subway. The red rectangle indicates the typical subsidence areas. JA and HS are the abbreviations of Jiangan and Hongshan, respectively. The red box in (**a**) indicates the range of optical satellite images.

Table 3. Information of Wuhan Metro under construction during the study period.

Name	Starting Point	End Point	Mileage	Construction Time
Metro Line 5	Hubei University of Chinese Medicine Station	East Square of Wuhan Railway Station	32.5 km	3 December 2015–10 December 2021
Metro Line 6	Jinyinhu Station	Xincheng 11th Road Station	7 km	28 July 2017–19 March 2021
Metro Line 12	Qingling Station	Qingling Station	59.876 km	18 December 2017–2024
Metro Line 16	South International Expo Center Station	Zhoujiahe Station	33.1 km	December 2018–September 2021
Metro Line 19	Wuhan Railway Station	Gaoxin 2nd Road Station	23.3 km	19 February 2019–2023

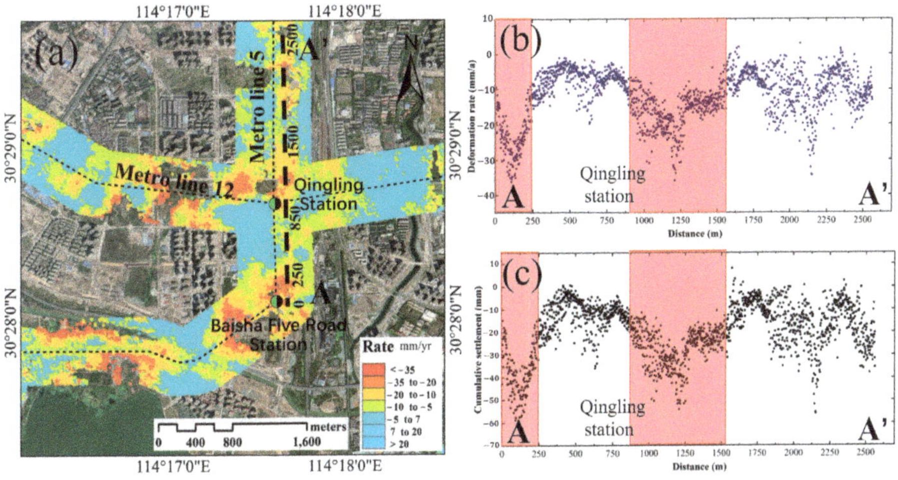

Figure 11. (**a**) Land subsidence along Qingling Station; (**b**) Deformation rate on profile AA′; (**c**) Cumulative settlement on profile AA′. The black dashed line AA′ is the profile position. The red boxes in (**b**,**c**) represent the subsidence intervals.

Shiqiao Station is also a typical area, and the subsidence is shown in Figure 12a. Two subsidence areas (S1 and S2) are found at the intersection of Metro Lines 6 and 12. Combined with the Google satellite optical images of the area, S1 is located in a residential neighborhood, and S2 is a vacant land. Residential buildings were being constructed in the S1 area during the study period. The simultaneous construction of subway and building leads to the subsidence of nearby ground. A 3D land subsidence model of S1 is shown in Figure 12c. The cumulative settlement in the study period reached 30 mm. We plotted the deformation rate and cumulative settlement profiles along BB′. The results are shown in Figure 12d,e. In a word, the subsidence fluctuation shows a "W" shape, with two obvious subsiding areas. The maximum settlement is located at S2, with a cumulative settlement of approximately 60 mm.

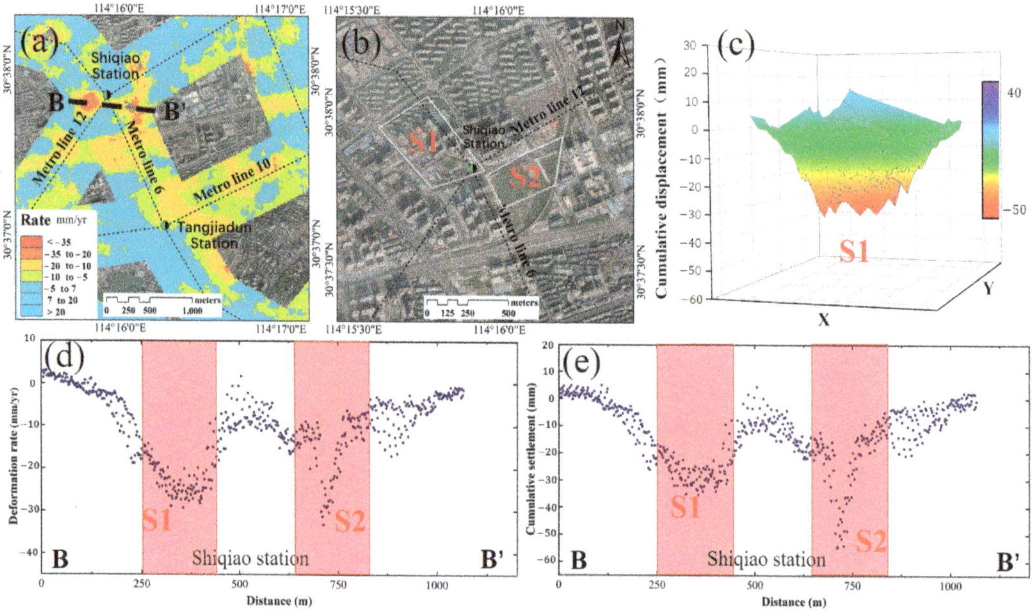

Figure 12. (**a**) Land subsidence around Shiqiao Station; (**b**) Google satellite optical images of the region; (**c**) three-dimensional land subsidence model of S1; (**d**) Deformation rate on profile BB′; (**e**) Cumulative settlement on profile BB′. The black dashed line BB′ is the profile position. The white boxes in (**b**) indicates the location of the subsidence funnel. The red boxes in (**d**,**e**) represent the subsidence intervals.

5.2. *Infrastructure Construction*

Urban construction is also an important cause of land subsidence. As the central city of central China, Wuhan has made great development in urban infrastructure construction in recent years, starting from the construction of tall buildings, city ring road, rail transit, sponge city and other projects. Various construction projects in Wuhan resumed one after another after April 2020. Groundwater needs to be pumped during the excavation of the foundation pit before the construction of the building, and the construction of the building will increase the ground load, all of which may cause subsidence. Figure 13 shows the urban surface changes in four obvious subsidence areas in Wuhan during the study period. To further investigate the settlement trend in the surrounding area during the construction of the building, we selected some reference points in the center of the settlement area to draw the cumulative settlement curve. The results are shown in Figure 14.

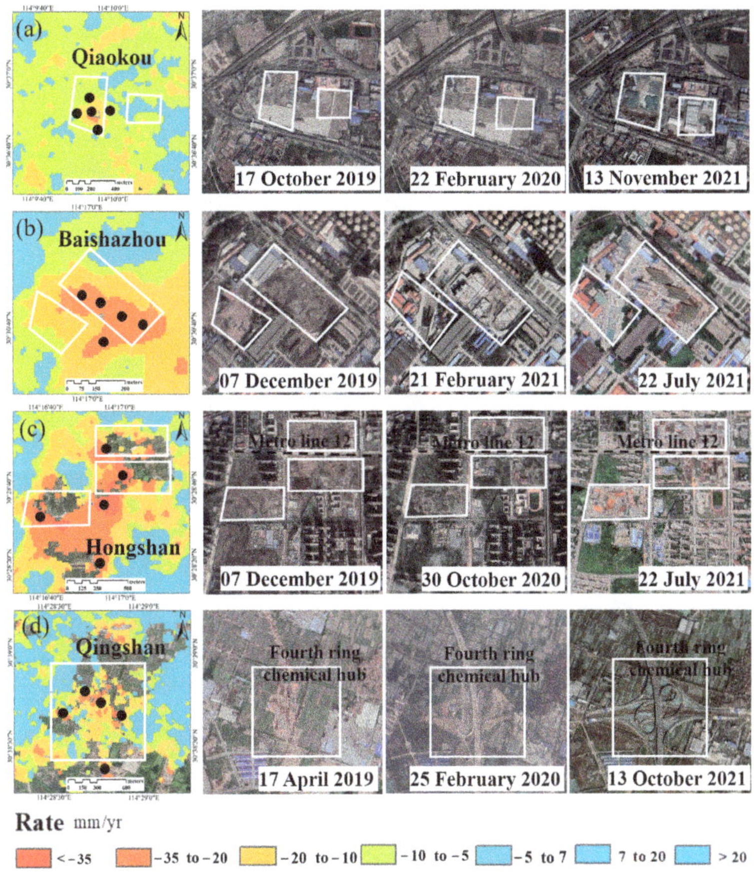

Figure 13. Urban surface changes in four obvious subsidence areas in Wuhan. The white boxes are the construction area. Black dots indicate the locations of the PS points.

Figure 13a shows an industrial park in Qiaokou District. The subsidence began to appear with the excavation of the foundation pit in September 2020, and the subsidence decreased after the construction of the foundation pit was completed in April 2021. Figure 13b shows a construction site located near Yangsigang Yangtze River Bridge in Baishazhou, where the construction of residential buildings was ongoing at the center of settlement during the study period. Figure 14b shows that subsidence continues until May 2021, with cumulative subsidence reaching about 50 mm. The rapid construction of high-rise buildings increases the ground load, which leads to the compression of soft soil layers and causes land subsidence.

In addition to a single building, multiple buildings are constructed in reality, which will cause regional large-scale subsidence. The subsidence in Hongshan District is shown in Figure 13c. Many office and residential buildings have been built near Qingling in Hongshan District in the past few years, and the area is close to Metro Lines 5 and 12 under construction. The simultaneous construction of the above ground part and the underground part has resulted in multiple subsidence funnels in this area. The optical image shows that the subsidence center is located at the people's Hospital of Wuhan University in Hongshan District under construction, and the maximum settlement exceeds 80 mm. Bridge construction can also cause land subsidence. Figure 13d shows a transportation hub under construction on the fourth ring road in Qingshan District. During the construction of the bridge, significant land subsidence occurred in the surrounding area from June 2020. The overpass has been completed, and the cumulative settlement has reached 40 mm until September 2021. The spatial distribution of land subsidence in Wuhan is affected by urban construction to a certain extent.

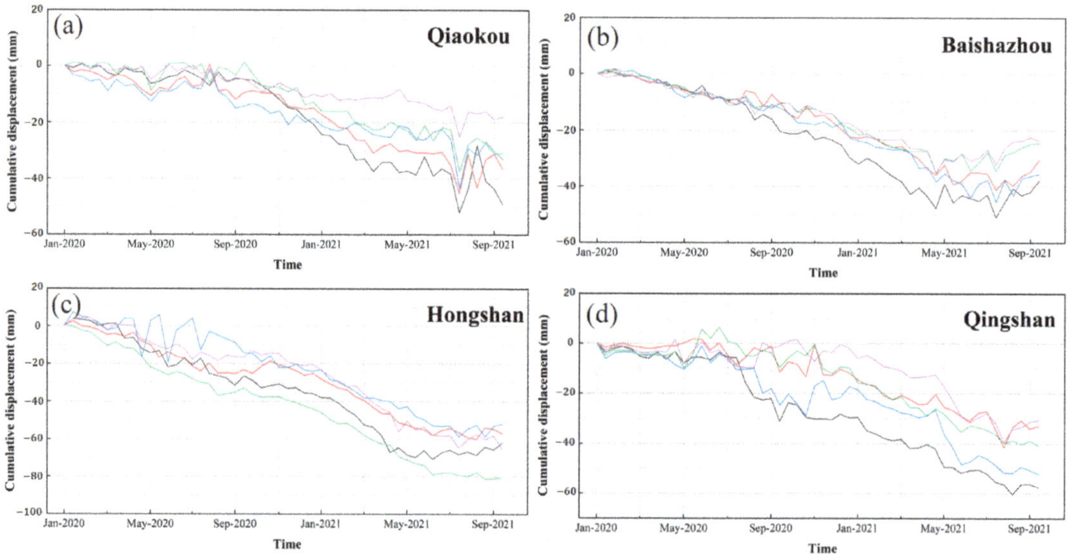

Figure 14. (**a**) Cumulative settlement curve of subsidence areas in Qiaokou District; (**b**) Cumulative settlement curve of subsidence areas in Baisahzhou; (**c**) Cumulative settlement curve of subsidence areas in Hongshan District; (**d**) Cumulative settlement curve of subsidence areas in qingshan District. Lines of different colors indicate the change in the subsidence of PS points.

5.3. *Influence of Hydrogeology on Land Subsidence*

Wuhan has complex geological conditions, carbonate rocks and soft soils are widely distributed, and special soils, such as artificial fill layers and old clay, are found. Figure 15a illustrates the spatial relationship between land subsidence and carbonate rock and soft soil. Most of the subsidence areas are located on carbonate rock and soft soil layer. For example, HH and QS are located on soft soil layer, and HS is located in carbonate rock belt and soft soil superposition area. Soft soils have high water content and low bearing capacity, and construction in areas with thick layers of soft soils can easily cause subsidence. Excessive groundwater extraction in large-scale urban construction can lead to a significant drop in the water level of fractured karst. The rapid change in karst groundwater dynamics gradually destroys the soil structure and forms soil cavities, leading to karst collapse of the ground. The special geological conditions of Wuhan make the city highly susceptible to geological hazards. We counted some of the locations of subsidence and ground collapse hazards recorded in Wuhan city, as shown in Figure 15b. Since 1977, about 30 karst collapse events have occurred in the Baishazhou carbonate belt in Hongshan District, and several subsidence disasters have occurred in the Houhu and Wuchang areas. However, the subsidence is only concentrated in a small part of the area, most of the areas located in carbonate rock or soft soil areas have relatively stable ground surface, and the land subsidence is insignificant. Areas located in carbonate rocks or soft soils do not necessarily experience land subsidence but are more likely to experience subsidence. Therefore, hydrogeology is not a decisive factor in the formation of subsidence.

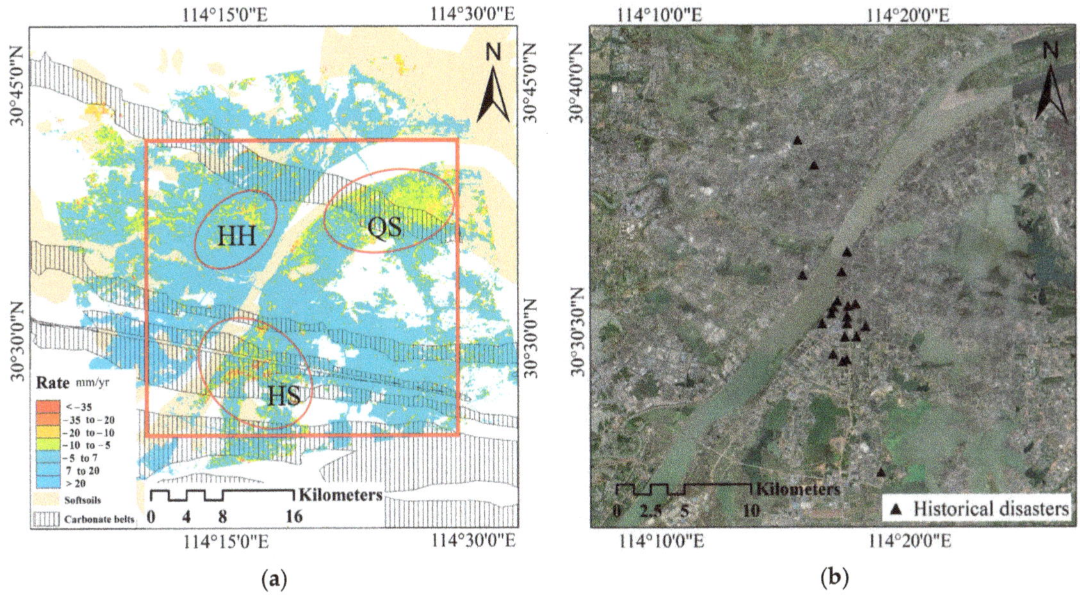

Figure 15. (**a**) Spatial relationship between land subsidence and carbonate rocks and soft soils in Wuhan; (**b**) location of subsidence and ground collapse disaster sites in Wuhan. The red ellipse in (**a**) indicates the subsidence area. HH, QS and HS are the abbreviations of Houhu, Qingshan, and Hongshan, respectively. The red rectangle in (**a**) indicates the range of optical satellite images.

5.4. Impact of Groundwater and Rainfall on Land Subsidence

An important factor in ground subsidence is the excessive extraction of groundwater. Figure 16a shows the changes in total water resources and average annual rainfall in Wuhan between 2015 and 2020. After 2016, the annual rainfall and the total amount of water resources in Wuhan decrease, and this situation is not improved until 2020. The proportional distribution of water consumption in Wuhan in 2020 is shown in Figure 16b, with industry and agriculture accounting for the largest share of water consumption. Qingshan District is a heavy industrial area and an important national steel production and chemical base. Many large industrial parks are located in this area as shown in Figure 16c. The groundwater in Qingshan District is mainly pore pressurized water in the loose accumulation layer, which is distributed in a band. One possible cause of the widespread ground subsidence phenomenon in the Qingshan district is the over exploitation of groundwater in industrial production.

To analyze the relationship between rainfall and land subsidence, we selected 15 reference points in different areas of Wuhan and compared the cumulative ground subsidence at the reference points with the monthly rainfall during the study period. The results are shown in Figure 16d. Rainfall in Wuhan is concentrated from May to September each year, with less rainfall in other months. The trend of ground deformation in Wuhan is associated with the rainfall in a certain time range. A large increase in rainfall is observed in Wuhan between May and September 2020, after which the deformation rate decreases in all subsiding areas, with some areas of slight ground uplift. After September, rainfall decreases significantly, and the rate of ground subsidence begins to increase. The results indicate a link between the magnitude of rainfall and land subsidence. In the time range of large rainfall, the land subsidence decreases, showing that a large amount of rainfall can effectively recharge groundwater resources and reduce the rate of surface subsidence.

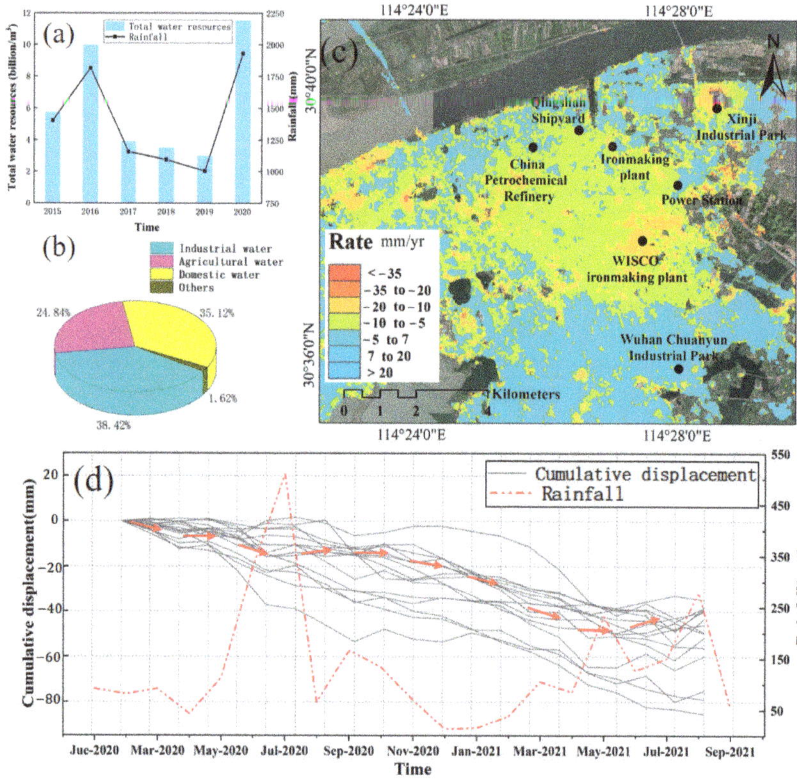

Figure 16. (**a**) Changes in total water resources and average annual rainfall in Wuhan between 2015 and 2020; (**b**) proportional distribution of water consumption in Wuhan in 2020; (**c**) distribution of industrial parks in Qingshan District; (**d**) comparison between cumulative subsidence and monthly rainfall in Wuhan. The red arrow indicates the change trend of subsidence.

5.5. Relationship between River Water Level Change and Land Subsidence

Wuhan is known as the "City of a Hundred Lakes", where the Yangtze River and Han River meet and form a water network together with many lakes. The total water area is 2117.6 km^2. The change in river water level is easy to lead to flood disaster, which affects the urban construction and people's normal lives in Wuhan. In 2020, Wuhan flood precipitation accounted for 75.8% of the annual precipitation. The Yangtze River passed through five flood peaks, and the water level of Hankou station was up to 28.77 m. Changes in water levels bring disasters and affect the ground stability along the rivers in Wuhan. The subsidence along the Han River in Qiaokou District (Figure 17b) and along the Yangtze River Road in Wuchang District (Figure 17c) may be related to water level changes. Several carbonate belts in Wuhan cross the Yangtze River, and the underground aquifers are connected to the Yangtze River. The change in water level leads to groundwater infiltration into the carbonate rock, which erodes and dissolves the carbonate rock causing more ground subsidence or collapse.

We selected two observation stations located in the Hankou area of the Yangtze River and the Hanchuan area of the Han River, respectively, and recorded the changes in water level at the stations during the study period. The locations of the stations are shown in Figure 17a. The water level changes at the two stations are the same, and the water level changes considerably during the flood season. Figure 17d,e show the relationship between water level values and land subsidence at the two observation stations of Yangtze and Han

Rivers, respectively. The peaks presented in water level changes and subsidence changes have some similarity, but the peaks in the cumulative deformation time series appear later than the water level changes. The subsidence decreases with the increase in the water level of the Yangtze River from May 2020, and the cumulative subsidence increases with the decrease in the water level of the Yangtze River after September 2020. The subsidence in areas farther away from the Han River is less influenced by the water level, and the correlation between the subsidence trend and the water level change is small. The closer to the water surface, the more similar the cumulative deformation curve is to the water level. There may be a link between land subsidence and water level changes on two sides of the Yangtze River. However, the seasonal variation of subsidence time series cannot be studied in detail due to insufficient data and short study period. More data will be used in future studies to further determine the correlation between land subsidence and water level changes.

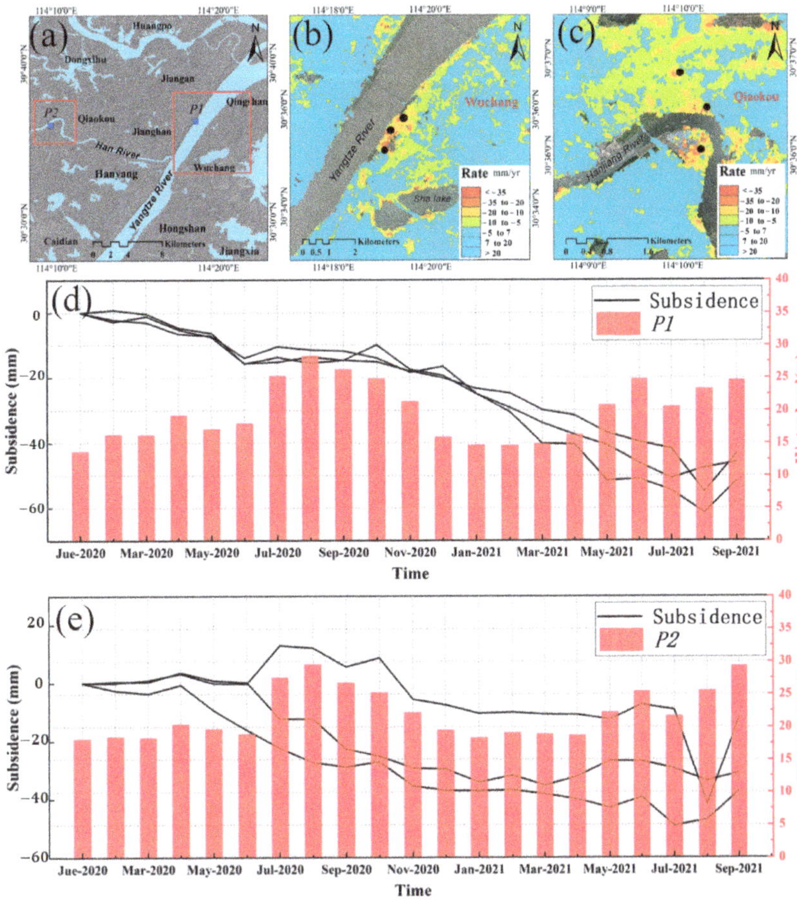

Figure 17. (**a**) Location of observation station; (**b**) subsidence along Changjiang Road in Wuchang District; (**c**) subsidence along the Han River in Qiaokou District; (**d**) comparison of water level and land subsidence in the Yangtze River; (**e**) comparison of water level and land subsidence in the Han River. P1 and P2 indicate the location of the observation point. The red boxes indicate the ranges of (**b**) and (**c**), respectively. Black dots indicate reference point locations.

5.6. Ground Uplift in Wuhan

In the result map of InSAR, the subsidence information of the ground surface can be extracted and the uplift information can be found. Although the main deformation of Wuhan is subsidence, it has significant uplift. As shown in Figure 18a, the uplift phenomenon is mainly concentrated in Caidian District in the south of Wuhan, with uplift rates ranging from 10 mm/yr to 20 mm/yr, and uplift exists in the downtown area. Dozens of lakes are located in Caidian District (Figure 18b), groundwater is abundant, and urban construction activities are low, so subsidence rarely occurs. Silty clay, muddy silty clay and old clay are widely distributed in Wuhan. The old clay is rich in expansive clay minerals, such as montmorillonite and illite. They produce physical and chemical swelling effects when they encounter water, resulting in volume expansion. Geological tectonics, flood deposition of the Yangtze River, and artificial filled soil are factors affecting the ground uplift in Wuhan [50]. At present, no clear explanation is provided for the phenomenon of ground lifting in Wuhan, and further research is still needed.

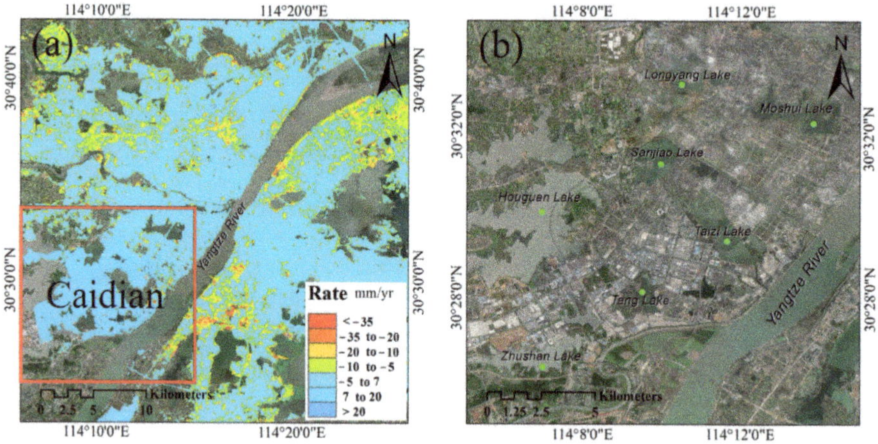

Figure 18. (**a**) Ground uplift in Wuhan city; (**b**) location distribution of major lakes in Caidian District. The green dots indicate lakes. The red box indicates the range of optical satellite images.

6. Conclusions

In this paper, we used 80 Sentinel-1A (SAR) images covering Wuhan to obtain the land subsidence information of Wuhan from July 2017 to September 2021 with the SBAS-InSAR technique. In accordance with the temporal and spatial distribution characteristics of land subsidence in Wuhan, we explored the causes of land subsidence in Wuhan in combination with subway association, infrastructure construction, hydrogeology, groundwater and rainfall, and changes in river water level. We briefly discussed the phenomenon of ground uplift in Caidian District.

The results show that the land subsidence in Wuhan is not uniformly distributed, with a maximum subsidence rate of 57.08 mm/yr. The subsidence funnel is distributed in the area covered by a soft soil layer and carbonate rock belt. Compared with that before 2017, the subsidence of Houhu area in Wuhan is obviously smaller. In contrast, the subsidence in other areas is severe. At present, the major subsidence areas in Wuhan are Jiangan, Qiaokou, Qingshan, Wuchang and Hongshan Districts. The subsidence near Qingling is the most serious, with a maximum cumulative deformation of 90 mm. The subsidence extend in this area is expanding, and multiple subsidence funnels tend to be connected.

Urban development and engineering construction are the major factors causing land subsidence in Wuhan. The construction of large buildings and subways increases the ground load, disturbs the surrounding soil and affects the ground stability, resulting in subsidence. Special geological formations are an important factor in land subsidence in Wuhan. Carbonate rock belts and large areas of soft soils make Wuhan more prone to geological problems such as karst ground collapse and uneven foundation subsidence. The amount of rainfall affects the total amount of water resources and changes in the water level of the Yangtze River. A large amount of rainfall in flood season can recharge groundwater resources and reduce the rate of land subsidence. The change in river water level has a certain impact on the land subsidence along the river in Wuhan, but the impact is small.

An obvious uplift is observed in Caidian District, and the reason for the uplift may be related to the physical expansion and chemical expansion effects of the old clay. At present, the type and accuracy of radar data continue to improve with the launch of more SAR satellites, thereby helping to better monitor urban subsidence, identify potential risks and delve into the mechanisms of subsidence.

Author Contributions: Y.Z., J.L. and L.Z. conceived and designed the experiments; C.W., J.Q. and H.S. carried out the data acquisition; Y.Z. and L.Z. performed data processing and analyses, and X.L. contributed to the manuscript of the paper; Y.Z., L.Z. and L.H. discussed and analyzed the experimental results. All authors have read and agreed to the published version of the manuscript.

Funding: This work was supported by the Guangxi Science and Technology Plan Project (Grant Nos. GUIKE AD19110107 and GUIKE AD19245060); the National Natural Science Foundation of China (Grant No. 42064002); the Natural Science Foundation of Guangxi (Grant No. 2020GXNSFBA159033); and the Guangxi Spatial Information Key Laboratory of Surveying and Mapping (Grant No. 19-185-10-05).

Data Availability Statement: Data incorporated in this research is available free through the these webpages: Sentinel-1A (https://scihub.copernicus.eu/dhus/#/home, accessed on 20 December 2021), SRTM DEM (http://srtm.csi.cgiar.org/srtmdata, accessed on 24 December 2021), satellite precise orbit data (https://qc.sentinel1.eo.esa.int/aux_poeorb, accessed on 29 December 2021).

Acknowledgments: We are grateful to the European Space Agency for providing the Sentinel-1A data and POD data for free. We would also like to thank NASA for providing the SRTM3 DEM data.

Conflicts of Interest: The authors declare no conflict of interest.

References

1. Cigna, F.; Tapete, D. Present-day land subsidence rates, surface faulting hazard and risk in Mexico City with 2014–2020 Sentinel-1 IW InSAR. *Remote Sens. Environ.* **2021**, *253*, 112161. [CrossRef]
2. Fedeski, M.; Gwilliam, J. Urban sustainability in the presence of flood and geological hazards: The development of a GIS-based vulnerability and risk assessment methodology. *Landsc. Urban Plan.* **2007**, *83*, 50–61. [CrossRef]
3. Wang, R.; Yang, T.; Yang, M.; Liao, M.; Lin, J.; Zhang, L. Attribution Analysis on Deformation Feature of the Shanghai Elevated Highway by Persistent Scatterer SAR Interferometry. *Geomat. Inf. Sci. Wuhan Univ.* **2018**, *43*, 2050–2057.
4. Agarwal, V.; Kumar, A.; Gee, D.; Grebby, S.; Gomes, R.L.; Marsh, S. Comparative Study of Groundwater-Induced Subsidence for London and Delhi Using PSInSAR. *Remote Sens.* **2021**, *13*, 4741. [CrossRef]
5. Zhang, X.; Chen, B.; Gong, H.; Lei, K.; Zhou, C.; Lu, Z.; Zhao, D. Inversion of Groundwater Storage Variations Considering Lag Effect in Beijing Plain, from RadarSat-2 with SBAS-InSAR Technology. *Remote Sens.* **2022**, *14*, 991. [CrossRef]
6. Najafabadi, R.; Ramesht, M.; Ghazi, I.; Khajedin, S.; Seif, A.; Nohegar, A.; Mahdavi, A. Identification of natural hazards and classification of urban areas by TOPSIS model (case study: Bandar Abbas city, Iran). *Geomat. Nat. Hazards Risk* **2016**, *7*, 85–100. [CrossRef]
7. Teatini, P.; Ferronato, M.; Gambolati, G.; Gonella, M. Groundwater pumping and land subsidence in the Emilia-Romagna coastland, Italy: Modeling the past occurrence and the future trend. *Water Resour. Res.* **2006**, *42*. [CrossRef]
8. El Bastawesy, M.; Cherif, O.; Sultan, M. The geomorphological evidences of subsidence in the Nile Delta: Analysis of high resolution topographic DEM and multi-temporal satellite images. *J. Afr. Earth Sci.* **2017**, *136*, 252–261. [CrossRef]
9. Bonì, R.; Herrera, G.; Meisina, C.; Notti, D.; Béjar-Pizarro, M.; Zucca, F.; González, P.; Palano, M.; Tomás, R.; Fernández, J.; et al. Twenty-year advanced DInSAR analysis of severe land subsidence: The Alto Guadalentín Basin (Spain) case study. *Eng. Geol.* **2015**, *198*, 40–52. [CrossRef]
10. Herrera, G.; Fernández, J.; Tomás, R.; Cooksley, G.; Mulas, J. Advanced interpretation of subsidence in Murcia (SE Spain) using A-DInSAR data-modelling and validation. *Nat. Hazards Earth Syst. Sci.* **2009**, *9*, 647–661. [CrossRef]
11. Chaussard, E.; Wdowinski, S.; Cabral-Cano, E.; Amelung, F. Land subsidence in central Mexico detected by ALOS InSAR time-series. *Remote Sens. Environ.* **2014**, *140*, 94–106. [CrossRef]
12. Qu, F.; Lu, Z.; Zhang, Q.; Bawden, G.; Kim, J.; Zhao, C.; Qu, W. Mapping ground deformation over Houston–Galveston, Texas using multi-temporal InSAR. *Remote Sens. Environ.* **2015**, *169*, 290–306. [CrossRef]

13. Ng, A.; Ge, L.; Li, X.; Abidin, H.; Andreas, H.; Zhang, K. Mapping land subsidence in Jakarta, Indonesia using persistent scatterer interferometry (PSI) technique with ALOS PALSAR. *Int. J. Appl. Earth Obs.* **2012**, *18*, 232–242. [CrossRef]
14. Zhu, L.; Gong, H.; Li, X.; Wang, R.; Chen, B.; Dai, Z.; Teatini, P. Land subsidence due to groundwater withdrawal in the northern Beijing plain, China. *Eng. Geol.* **2015**, *193*, 243–255. [CrossRef]
15. Qu, F.; Zhang, Q.; Lu, Z.; Zhao, C.; Yang, C.; Zhang, J. Land subsidence and ground fissures in Xi'an, China 2005–2012 revealed by multi-band InSAR time-series analysis. *Remote Sens. Environ.* **2014**, *155*, 366–376. [CrossRef]
16. Gao, M.; Gong, H.; Li, X.; Chen, B.; Zhou, C.; Shi, M.; Guo, L.; Chen, Z.; Ni, Z.; Duan, G. Land Subsidence and Ground Fissures in Beijing Capital International Airport (BCIA): Evidence from Quasi-PS InSAR Analysis. *Remote Sens.* **2019**, *11*, 1466. [CrossRef]
17. Gerardo, H.; Pablo, E.; Roberto, T.; Marta, B.; Juan, L.; Mauro, R.; Maria, M.; Dora, C.; John, L.; Pietro, T.; et al. Mapping the global threat of land subsidence. *Science* **2021**, *371*, 34–36.
18. Hu, R.; Yue, Z.; Wang, L.; Wang, S. Review on current status and challenging issues of land subsidence in china. *Eng. Geol.* **2004**, *76*, 65–77. [CrossRef]
19. Zhang, T.; Shen, W.-B.; Wu, W.; Zhang, B.; Pan, Y. Recent Surface Deformation in the Tianjin Area Revealed by Sentinel-1A Data. *Remote Sens.* **2019**, *11*, 130. [CrossRef]
20. Xiang, W.; Zhang, R.; Liu, G.; Wang, X.; Mao, W.; Zhang, B.; Cai, J.; Bao, J.; Fu, X. Extraction and analysis of saline soil deformation in the Qarhan Salt Lake region (in Qinghai, China) by the sentinel SBAS-InSAR technique. *Geod. Geodyn.* **2022**, *13*, 127–137. [CrossRef]
21. Du, Y.; Feng, G.; Liu, L.; Fu, H.; Peng, X.; Wen, D. Understanding Land Subsidence Along the Coastal Areas of Guangdong, China, by Analyzing Multi-Track MTInSAR Data. *Remote Sens.* **2020**, *12*, 299. [CrossRef]
22. Liu, Q.; Yue, G.; Ding, X.; Yang, K.; Feng, G.; Xiong, Z. Temporal and Spatial Characteristics Analysis of Deformation Along Foshan Subway Using Time Series InSAR. *Geomat. Inf. Sci. Wuhan Univ.* **2019**, *44*, 1099–1106.
23. Li, W.; Wang, L.; Guo, H.; Wang, H.; Wang, Y.; Zhu, J.; Zang, X. Effectiveness and countermeasures of land subsidence control in China. *China Water Resour.* **2021**, *7*, 32–35.
24. Huang, J.; Guo, Y.; Luo, W. Monitoring land subsidence in Wuhan city based on Sentinel-1. *Bull. Surv. Map.* **2021**, *9*, 53–58.
25. Hu, J.; Motagh, M.; Wang, J.; Qin, F.; Zhang, J.; Wu, W.; Han, Y. Karst Collapse Risk Zonation and Evaluation in Wuhan, China Based on Analytic Hierarchy Process, Logistic Regression, and InSAR Angular Distortion Approaches. *Remote Sens.* **2021**, *13*, 5063. [CrossRef]
26. Fan, S. A Discussion on Karst Collapse in Wuhan (Hubei). *Resour. Environ. Eng.* **2006**, *20*, 608–616.
27. Benattou, M.; Balz, T.; Liao, M. Measuring Surface Subsidence in Wuhan, China with Sentinel-1 Data using PSInSAR. *Int. Arch. Photogramm. Remote Sens. Spat. Inf. Sci.* **2018**, *XLII-3*, 73–77. [CrossRef]
28. Jiang, H.; Balz, T.; Cigna, F.; Tapete, D. Land Subsidence in Wuhan Revealed Using a Non-Linear PSInSAR Approach with Long Time Series of COSMO-SkyMed SAR Data. *Remote Sens.* **2021**, *13*, 1256. [CrossRef]
29. Guo, J.; Zhou, L.; Yao, C.; Hu, J. Surface Subsidence Analysis by Multi-Temporal InSAR and GRACE: A Case Study in Beijing. *Sensors* **2016**, *16*, 1495. [CrossRef]
30. Wu, C. Research and Analysis on the Subsidence of Houhu Real Estate in Wuhan. *Technol. Inno. Appl.* **2014**, *22*, 216.
31. Han, Y.; Zou, J.; Lu, Z.; Qu, F.; Kang, Y.; Li, J. Ground Deformation of Wuhan, China, Revealed by Multi-Temporal InSAR Analysis. *Remote Sens.* **2020**, *12*, 3788. [CrossRef]
32. Zhou, L.; Lu, J.; Wen, X.; Zhang, D.; Wang, C.; Yang, F. Spatial-temporal Characteristics Analysis of Ground Subsidence in Central Urban Areas of Wuhan Based on Small Baseline Subset lnterferometric Synthetic Aperture Radar. *Sci. Technol. Eng.* **2021**, *21*, 40–46.
33. Huang, B.; Zheng, F.; Bai, J.; Wang, Y. Feasibility of land surface deformation monitoring by regional CORS. *J. Geomat. Sci. Technol.* **2011**, *28*, 169–172.
34. Chen, Q.; Liu, G.; Hu, Z.; Ding, X.; Yang, Y. Mapping ground 3-D displacement with GPS and PS-InSAR networking in the Pingtung area, southwestern Taiwan. *Chin. J. Geophys.* **2012**, *55*, 3248–3258.
35. Hooper, A. A multi-temporal InSAR method incorporating both persistent scatterer and small baseline approaches. *Geophys. Res. Lett.* **2008**, *35*, 96–106. [CrossRef]
36. Chaussard, E.; Amelung, F.; Abidin, H.; Hong, S. Sinking cities in Indonesia: ALOS PALSAR detects rapid subsidence due to groundwater and gas extraction. *Remote Sens. Environ.* **2013**, *128*, 150–161. [CrossRef]
37. Zhu, J.; Yang, Z.; Li, Z. Recent progress in retrieving and predicting mining-induced 3D displacements using InSAR. *Acta Geod. Cartogr. Sin.* **2019**, *48*, 135–144. [CrossRef]
38. Karanam, V.; Motagh, M.; Garg, S.; Jain, K. Multi-sensor remote sensing analysis of coalfire induced land subsidence in Jharia Coalfields, Jharkhand, India. *Int. J. Appl. Earth Obs. Geoinf.* **2021**, *102*, 102439.
39. Liu, Z.; Xu, B.; Wang, Q.; Yu, W.; Miao, Z. Monitoring landslide associated with reservoir impoundment using synthetic aperture radar interferometry: A case study of the Yalong reservoir. *Geod. Geodyn.* **2022**, *13*, 138–150. [CrossRef]
40. Perissin, D.; Wang, Z.Y.; Lin, H. Shanghai subway tunnels and highways monitoring through cosmo-skymed persistent scatterers. *ISPRS J. Photogramm. Remote Sens.* **2012**, *73*, 58–67. [CrossRef]
41. Chen, G.; Zhang, Y.; Zeng, R.; Yang, Z.; Chen, X.; Zhao, F.; Meng, X. Detection of Land Subsidence Associated with Land Creation and Rapid Urbanization in the Chinese Loess Plateau Using Time Series InSAR: A Case Study of Lanzhou New District. *Remote Sens.* **2018**, *10*, 270. [CrossRef]

42. Pratesi, F.; Tapete, D.; Del Ventisette, C.; Moretti, S. Mapping interactions between geology, subsurface resource exploitation and urban development in transforming cities using insar persistent scatterers: Two decades of change in Florence, Italy. *Appl. Geogr.* **2016**, *77*, 20–37. [CrossRef]
43. Ferretti, A.; Prati, C.; Rocca, F. Permanent scatterers in SAR interferometry. *IEEE Trans. Geosci. Remote Sens.* **2001**, *39*, 8–20. [CrossRef]
44. Beradino, P.; Fornaro, G.; Lanari, R.; Sansosti, E. A new algorithm for Surface deformation monitoring based on small baseline differential SAR interferograms. *IEEE Trans. Geosci. Remote Sens.* **2002**, *40*, 2375–2383. [CrossRef]
45. Li, J.H.; Zhou, L.; Ren, C.; Liu, L.L.; Zhang, D.; Shi, Y.J. Spatiotemporal Inversion and Mechanism Analysis of Surface Subsidence in Shanghai Area Based on Time-Series InSAR. *Appl. Sci.* **2021**, *11*, 7460. [CrossRef]
46. Bai, L.; Jiang, L.M.; Wang, H.S.; Sun, Q.S. Spatiotemporal characterization of land subsidence and uplift (2009–2010) over wuhan in central China revealed by terrasar-x insar analysis. *Remote Sens.* **2016**, *8*, 350. [CrossRef]
47. Costantini, M.; Bai, J.; Malvarosa, F.; Minati, F.; Vecchioli, F.; Wang, R.; Hu, Q.; Xiao, J.; Li, J. Ground deformations and building stability monitoring by COSMO-SkyMed PSP SAR interferometry: Results and validation with field measurements and surveys. In Proceedings of the 2016 IEEE International Geoscience and Remote Sensing Symposium (IGARSS), Beijing, China, 10–15 July 2016; pp. 6847–6850.
48. Zhou, L.; Guo, J.; Hu, J.; Li, J.; Xu, Y.; Pan, Y.; Shi, M. Wuhan Surface Subsidence Analysis in 2015–2016 Based on Sentinel-1A Data by SBAS-InSAR. *Remote Sens.* **2017**, *9*, 982. [CrossRef]
49. Zhang, Y.; Liu, Y.; Jin, M.; Jing, Y.; Liu, Y.; Liu, Y.; Sun, W.; Wei, J.; Chen, Y. Monitoring Land Subsidence in Wuhan city (China) using the SBAS-InSAR Method with Radarsat-2 Imagery Data. *Sensors* **2019**, *19*, 743. [CrossRef]
50. Xie, P.; Zhang, S.; Zhou, L.; Li, Q.; Xiao, J.; Cai, J. Detection of the Urban Surface Deformation and New Strategy for Flood Prevention in Wuhan Central District. *Geomat. Inf. Sci. Wuhan Univ.* **2021**, *46*, 1015–1024.
51. Shi, X.; Zhang, S.; Jiang, M.; Pei, Y.; Yang, C. Spatial and temporal subsidence characteristics in Wuhan city (China) during 2015–2019 inferred from Sentinel-1 SAR Interferometry. *Nat. Hazards Earth Syst. Sci.* **2021**, *21*, 2285–2297. [CrossRef]
52. Wang, Y.; Wu, J.; Wang, H.; Zhao, D. The statistical analysis of rainfall in wuhan in the past 50 years. *J. Hubei Univ. Technol.* **2006**, *21*, 98–100.
53. Zhao, Q.; Liu, Y.; Yao, W.; Yao, Y. Hourly Rainfall Forecast Model Using Supervised Learning Algorithm. *IEEE Trans. Geosci. Remote* **2021**, *99*, 4100509. [CrossRef]
54. Luo, X. Division of "six belts and five types" of carbonate region and control of karst geological disaster in Wuhan. *J. Hydraul. Eng.* **2014**, *45*, 171–179.

remote sensing MDPI

Article

Elastic and Inelastic Ground Deformation in Shanghai Lingang Area Revealed by Sentinel-1, Leveling, and Groundwater Level Data

Yanling Chen [1,*], Minyan Liao [2], Jicang Wu [3], Xiaobo Li [1,4], Fuwen Xiong [5], Shijie Liu [3], Yongjiu Feng [3] and Xiaoya Wang [1]

1 Shanghai Astronomical Observatory, Chinese Academy Sciences, Shanghai 200030, China; lxb@ief.ac.cn (X.L.); wxy@shao.ac.cn (X.W.)
2 Chongqing Institute of Geology and Mineral Resources, Chongqing 401120, China; 1810654@tongji.edu.cn
3 College of Surveying and Geo-Informatics, Tongji University, Shanghai 200092, China; jcwu@tongji.edu.cn (J.W.); liusjtj@tongji.edu.cn (S.L.); yjfeng@tongji.edu.cn (Y.F.)
4 Institute of Earthquake Forecasting, China Earthquake Administration, Beijing 100036, China
5 Shanghai Institute of Geological Survey, Shanghai 200072, China; xiongfuwen@sigs.com.cn
* Correspondence: ylchen@shao.ac.cn

Citation: Chen, Y.; Liao, M.; Wu, J.; Li, X.; Xiong, F.; Liu, S.; Feng, Y.; Wang, X. Elastic and Inelastic Ground Deformation in Shanghai Lingang Area Revealed by Sentinel-1, Leveling, and Groundwater Level Data. *Remote Sens.* **2022**, *14*, 2693. https://doi.org/10.3390/rs14112693

Academic Editors: Guohong Zhang, Chisheng Wang, Wu Zhu, Daqing Ge and Siting Xiong

Received: 19 April 2022
Accepted: 30 May 2022
Published: 3 June 2022

Abstract: Shanghai Lingang New City, located in the southeast corner of Shanghai, was constructed by land reclamation from 2002 to 2005, in an area where the geological structure is prone to subsidence over time. Firstly, we explore the spatio-temporal pattern of ground subsidence and its mechanism using the Persistent Scatterers Interferometric Synthetic Aperture Radar (PSInSAR) technique by processing 50 scenes of Sentinel-1A images acquired from May 2016 to May 2018. In order to assess the accuracy of PSInSAR derived deformation, we collect the first-class leveling data at two benchmarks located in the study area; the comparison between the two settlement indicates that the maximum difference is 1.93 mm and 2.9 mm, respectively, which validates the PSInSAR's high accuracy. We then obtain the skeleton release coefficients by the joint analysis of PSInSAR measurements and groundwater level data. Finally, we find that this coastal area has undergone both elastic and inelastic deformation from 2016 to 2018. The outcome shows that the combination of different techniques is conductive to understand the deformation mechanism of the aquifer system in these coastal areas, which is expected to be a valuable reference for ground subsidence monitoring and groundwater extraction management.

Keywords: ground deformation; PSInSAR; leveling; groundwater; Shanghai Lingang New City

1. Introduction

Land reclamation, as an important way to use coastal areas to expand land resources, has been conducted in many regions such as the Netherlands [1,2], the Hong Kong International Airport [3–5], Macau [6], Tianjin [7,8], and Shanghai Lingang New City [9–11], which is located in the southeast corner of Shanghai, China. Due to the compaction and consolidation of the soil layer and groundwater extraction, these areas suffer from ground subsidence, which raises potential risks to urban buildings, bridges, metro lines, and other infrastructures, even threatening people's lives. The monitoring of land subsidence, the estimation of water release coefficient of the aquifer, and the correlation analysis between land subsidence and groundwater in these areas provide an important basis for the study of subsidence mechanisms and the formulation and implementation of certain protective measures.

InSAR (Interferometric Synthetic Aperture Radar) time series analysis techniques–such as PS-InSAR (Persistent Scatterers-InSAR), SBAS-InSAR (Small Baseline Line- InSAR), TCP-InSAR (Temporarily Coherent Point Interferometric Synthetic Aperture Radar) [12,13],

DS interferometry (DSI) [14], and other InSAR techniques–have been proven to be efficient in monitoring ground subsidence and the relationship between land subsidence and groundwater changes. Galloway et al. used radar data collected from ERS-1 satellite to detect and quantify land subsidence caused by aquifer system compaction in the Antelope Valley, Mojave Desert, California [15]. Hoffmann et al. used ERS-2 satellite data to monitor land subsidence in the Antelope Valley and compared these data with repeatedly surveyed benchmarks [16]. Erban et al. collected L-band ALOS PALSAR data to measure the land subsidence rate in the Mekong Delta in Vietnam [17]. Matano et.al processed three SAR datasets of ascending and descending orbits acquired over the Campania coastal sectors from June 1992 to July 2010, which provided new insights into the spatial variability of vertical ground deformation (subsidence/uplift) of the Volturno River coastal plain [18]. Wang et al. used C-band Envisat ASAR data to investigate the rate and extent of coastal land subsidence in the Pearl River Delta in China [19]. Liu et al. detected subsidence in a coastal area by implementing the ultrashort-baseline TCPInSAR algorithm with high-resolution TerraSAR-X images acquired over Tianjin (close to Bohai Bay) in China [20]. Zhao et al. used ASAR (2007–2010), CSK (2013–2016), and Sentinel-1A SAR images (2015–2017) to obtain the ground deformation rate in coastal areas in Shanghai using InSAR techniques [21–23]. There are many research results of the application of PSInSAR in Shanghai [24–33]; most scholars have analyzed the subsidence of the Shanghai area with InSAR space technology and levelling data. In this paper, we involve various technologies including not only PS-InSAR and leveling, but also groundwater level measurements taken monthly that are used to cross-validate each other, establish the relationship between settlement and groundwater in the delta area, and further distinguish the deformation types. This is helpful to further understand the evolution of land compaction and consolidation in the land reclamation area.

In this paper, we adopted 50 scenes of Sentinel-1A images acquired from May 2016 to May 2018 to detect and process persistent scatterers (PS) in Shanghai Lingang New City using the Persistent Scatterers Interferometric Synthetic Aperture Radar (PSInSAR) technique. Then, we compared the displacement time series of these high coherence points with groundwater well data and leveling data using PSInSAR technology. Finally, we estimated the aquifer parameters at groundwater wells and analyzed the deformation characteristics and their geophysical mechanism.

2. Study Area

Lingang New City is located in the southeast corner of Pudong New District, Shanghai, China, where 60% of the land was constructed by reclamation between 2002 and 2005 [9–11]. It is about 75 km from the center of Shanghai and has a total area of 152.15 km^2. Most of the Lingang New City area was built by land reclamation. Its coastal zone–a frequent landing site of typhoons and storm surges–is vulnerable to a number of natural disasters. Factors such as the natural consolidation of mucky soil and the compression effect of the underground soil layer structure often cause ground subsidence in this area.

The geographical location of Shanghai Lingang New City is shown in Figure 1. The locations of two groundwater wells (W65, W66) and two leveling points (F65, F66) are marked with asterisks in Figure 1. The names W65 and W66 under these asterisks are abbreviations for groundwater monitoring wells. The names F65 and F66 are abbreviations for leveling monitoring points. Groundwater well W65 and leveling point F65 are located at the same place, and W66 and F66 are similarly in the same location.

Figure 1. The geographical location of Shanghai Lingang New City and the location of groundwater wells and leveling monitoring points (marked with asterisk).

As shown in Table 1, the aquifers of Shanghai include: the submerged aquifer (A0), first confined aquifer (A1), second confined aquifer (A2), third confined aquifer (A3), fourth confined aquifer (A4), and the fifth confined aquifer (A5). Weakly permeable layers include: the topsoil layer (B0), first weakly permeable layer (B1), second weakly permeable layer (B2), third weakly permeable layer (B3), fourth weakly permeable layer (B4), fifth weakly permeable layer (B5), and the sixth weakly permeable layer (B6). This is according to the hydrogeological profile of the Shanghai area, as shown in Figure 2; aquifers and weakly permeable layers are missing in some areas. The second, third, fourth, and fifth confined aquifers contain abundant groundwater resources and are the main targets for emergent groundwater exploitation. To explore the relationship between groundwater and ground subsidence in Shanghai Lingang New City, we mainly consider here the thickness changes of the fourth confined aquifer, as the first, second, and third confined aquifers are in a reverse relationship between the upper and lower soft soil layers and the fifth aquifer is rare in the Lingang area. The large thickness index of the third and fourth aquifer sand layer in Lingang New City indicates that the development of the sand layer may increase ground subsidence during exploitation, especially during unreasonable groundwater extraction. Otherwise, it will not lead to ground subsidence disasters [34]; that is, the cumulative thickness indicators of the fourth confined aquifer can reflect the abundance of groundwater resources, and the thickness changes can reflect the surface subsidence. We take the changes in the water levels of the fourth confined aquifer as the research object in this paper.

Table 1. The List of Hydrogeological Section.

Hydrogeological Section	Full Name	Short Name
Aquifers	Submerged aquifer	A0
	First confined aquifer	A1
	Second confined aquifer	A2
	Third confined aquifer	A3
	Fourth confined aquifer	A4
	Fifth confined aquifer	A5
Weakly permeable layers	Topsoil layer	B0
	First weakly permeable layer	B1
	Second weakly permeable layer	B2
	Third weakly permeable layer	B3
	Fourth weakly permeable layer	B4
	Fifth weakly permeable layer	B5
	Sixth weakly permeable layer	B6

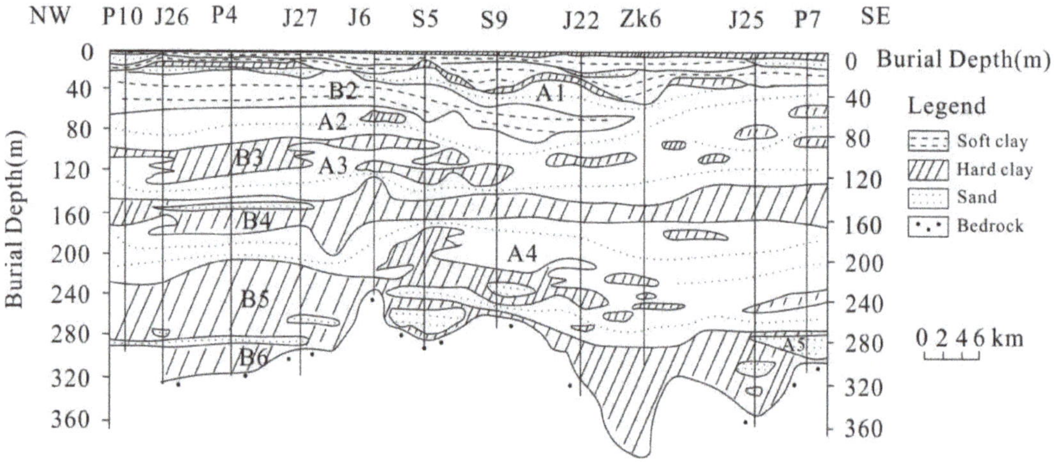

Figure 2. Hydrogeological Section Map of Shanghai City [34].

3. Materials and Methods

3.1. SAR Data

A total of 50 scenes of Sentinel-1A images acquired from 15 May 2016 to 5 May 2018 (seen in Table 2) are used to detect and process the high coherence points in Lingang New City, Shanghai. Figure 3 shows the distribution of the spatial and temporal baselines, where the image acquired on 27 June 2017 was selected as the reference image. In [35–38] the maximum temporal baseline was 408 days, and the maximum spatial baseline was 90 m. And thus the thresholds of temporal and spatial baselines were empirically set as 408 days and 90 m, respectively, to select interferograms.

Table 2. The List of SAR data and acquisition times.

No	Acquisition Time	No	Acquisition Time	No	Acquisition Time
1	20160515	18	20170311	35	20171013
2	20160608	19	20170404	36	20171106
3	20160726	20	20170416	37	20171118
4	20160819	21	20170428	38	20171130
5	20160912	22	20170510	39	20171212
6	20161006	23	20170522	40	20171224
7	20161018	24	20170603	41	20180117
8	20161030	25	20170615	42	20180129
9	20161111	26	20170627	43	20180210
10	20161123	27	20170709	44	20180222
11	20161205	28	20170721	45	20180306
12	20161217	29	20170802	46	20180318
13	20161229	30	20170814	47	20180330
14	20170122	31	20170826	48	20180411
15	20170203	32	20170907	49	20180423
16	20170215	33	20170919	50	20180505
17	20170227	34	20171001		

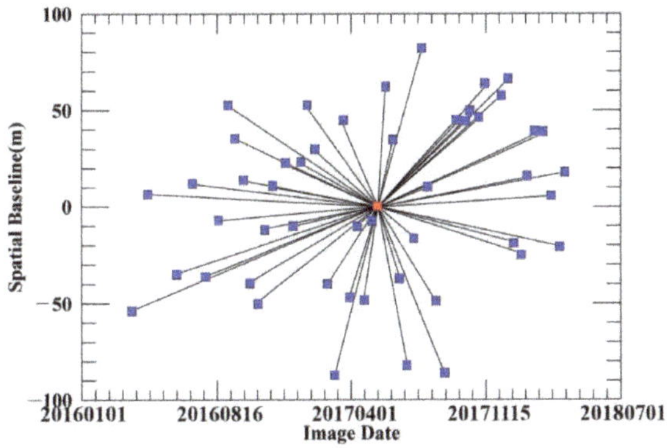

Figure 3. Distribution of the temporal and spatial baselines. The red square indicates the reference image; blue squares indicate the slave images.

3.2. Leveling Data

We collected the measurements from two first-class leveling points (i.e., F65 and F66, marked as asterisks in Figure 1) located in the study area to verify the accuracy of the InSAR measurements. The Trimble DINI03 digital level was used to obtain the leveling data with first-class leveling accuracy. The accuracy of the vertical displacement monitoring was ±0.3 mm. The F65 and F66 leveling points were observed monthly since 15 February 2011 and 25 February 2009, respectively.

3.3. Water Level Well Data

The Shanghai Institute of Geological Survey has conducted long-term monitoring of groundwater level changes in Shanghai. There are two groundwater level wells (W65, W66) in the study area, as shown in Figure 1, and the groundwater level is monitored monthly.

3.4. PSInSAR Technique

PSInSAR is an advanced InSAR technology that can accurately detect ground surface deformation from a stack of SAR images [39–43]. The PSInSAR technique has been widely used in urban areas with a high density of coherent points. This algorithm is a time series analysis method based on point targets with stable scattering characteristics (e.g., buildings, bridges, roads, and bare rocks). Then, time series analysis is performed on the interference phase to obtain high-precision surface observation information from these persistent scatterers.

When processing these SAR images, one SAR image is usually selected as the master image and the others are used as slave images. After registration and mitigation of the flattening effect, the external DEM (e.g., SRTM DEM 90 m × 90 m) is used to remove the topographic phase, and the differential interferograms are generated. Then, persistent scatterers (PS) that can keep high coherence in all the interferograms are selected. A Delaunay triangle network is generated from these PS points [44–48]. The wrap phase (φ_i^k) at the point i in the k th interferogram can be written as follows:

$$\varphi_i^k = W\left\{\varphi_{i,def}^k + \varphi_{i,hgt}^k + \varphi_{i,orb}^k + \varphi_{i,atm}^k + \varphi_{i,noise}^k\right\} \quad (1)$$

where $W\{\cdot\}$ is the wrapping operator; $\varphi_{i,def}^k$ is the phase contributed by ground deformation, $\varphi_{i,hgt}^k$ is the phase associated with height errors, $\varphi_{i,orb}^k$ is the phase associated with orbital errors, $\varphi_{i,atm}^k$ is the phase that is related to the atmospheric delay, and $\varphi_{i,noise}^k$ is the noise.

The atmospheric phase and the orbit phase can be separated by temporal and spatial filtering. Finally, the time series deformation phase and average deformation rate of the line of sight of the satellite are obtained [49–53]. The deformation phase reflects the displacement of the ground during the observation time. SAR belongs to active remote sensing. The signal is measured twice from transmission to reception. Therefore, the phase change caused by ground point deformation can be expressed as:

$$\varphi_{i,def}^k = -\frac{4\pi}{\lambda} Def_{LOS} \quad (2)$$

where λ is the wavelength of the radar signal. Def_{LOS} is the deformation of the radar line of sight, and in the data analysis and comparison, it is necessary to transfer it to the vertical direction to compare it with levelling data or other vertical measuring data.

3.5. Reduction of InSAR Measurement

In order to investigate the relationship between ground deformation monitoring by PSInSAR and groundwater well observations, we processed them to the same scale by using the inverse distance square weighted (IDW) method, which has been widely used in many fields such as meteorological research, mine reserves research, oceans research, and other fields.

In this case study, the coherent points located within 100 m of the groundwater well and their subsidence time series were extracted. The deformation value of the groundwater well was calculated by the former IDW method for the coherent points. Then, the weighted deformation was seen as the surface deformation of the groundwater well within this time.

The weight function to calculate the weight of each PS point:

$$w_i = \frac{D_i^{-P}}{\sum_{i=1}^{n} D_i^{-P}} \quad (3)$$

where w_i is the weight of each PS point, D_i is the Euclidean distance between the PS point and the groundwater level point, $P$is the power parameter, and n is the number of PS point. In this paper, we use $P = 2$, which is the commonly used method of the inverse distance square weighted method.

Then, we use the inverse distance weighted function to obtain the estimated deformation value at the coordinates of the groundwater well:

$$Disp(h_0) = \frac{D_i^{-P} \sum_{i=1}^{n} Disp(h_i)}{\sum_{i=1}^{n} D_i^{-P}} \tag{4}$$

where $Disp(h_0)$ is the estimated deformation value of the groundwater well, $Disp(h_i)$ is the deformation value of the PS point.

Based on this method, the characteristics of the unknown geographic space are predicted.

3.6. Aquifer Parameters Estimation

According to the Terzaghi-Jacob theoretical model [54], the total stress of the confined aquifer (σ_T) is equal to the sum of the pore stress (p) and the effective stress (σ_e) of the aquifer.

$$\sigma_T = p + \sigma_e \tag{5}$$

When the groundwater in the confined aquifer is extracted, the groundwater level of the confined aquifer will decrease, resulting in the decrease of the pore stress and increase of the effective stress. The sum of these two stresses will introduce compression to the aquifer, which will cause ground subsidence.

When the effective stress of the aquifer σ_e is less than the historical effective stress $\sigma_{e(max)}$, the aquifer system undergoes elastic deformation, and the surface settlement can be recovered by measures such as recharging the groundwater. If the effective stress of the aquifer $\sigma_e$is continuously greater than the historical effective stress $\sigma_{e(max)}$, the aquifer system will undergo inelastic deformation, i.e., consolidation deformation, and the ground surface will have permanent ground subsidence.

The water release capacity of the confined aquifer is expressed by the water release coefficient. According to the Terzaghi-Jacob theoretical model, the relationship between the aquifer system deformation and groundwater level change can be represented by two different skeleton water release coefficients. Such coefficients are key hydraulic parameters for evaluating the water storage capacity of groundwater aquifer systems.

$$s_{ke}^* = \frac{\Delta b^*}{\Delta h}, \sigma_e < \sigma_{e(max)} \tag{6}$$

$$s_{ki}^* = \frac{\Delta b^*}{\Delta h}, \sigma_e > \sigma_{e(max)} \tag{7}$$

where s_{ke}^* is the elastic water release coefficient of the aquifer skeleton, s_{ki}^* is the inelastic water release coefficient of the aquifer skeleton, Δb^* is the deformation of the aquifer system obtained from the deformation results of PSInSAR, Δh is the change of groundwater level, which can be obtained by the change of groundwater level at the well.

When the aquifer thickness changes due to changes in the groundwater level, the actual observed water release coefficient can be compared with the theoretical value to determine whether the water release coefficient is elastic or inelastic. Therefore, whether elastic deformation or inelastic deformation occurs in the aquifer system can be determined.

4. Results and Discussion

4.1. PSInSAR Derived Deformation

Based on PSInSAR technology, the Sentinel-1A images covering Shanghai Lingang New City acquired from 15 May 2016 to 05 May 2018 were processed. The deformation rate map generated from 21,447 PS points is shown in Figure 4. PS points are mainly distributed on the west side of Dishui Lake in the experimental area. The deformation rate ranges from −67 mm/year to 1.1 mm/year in the LOS (line of sight) direction, and the average deformation rate is −3.4 mm/year in the LOS direction.

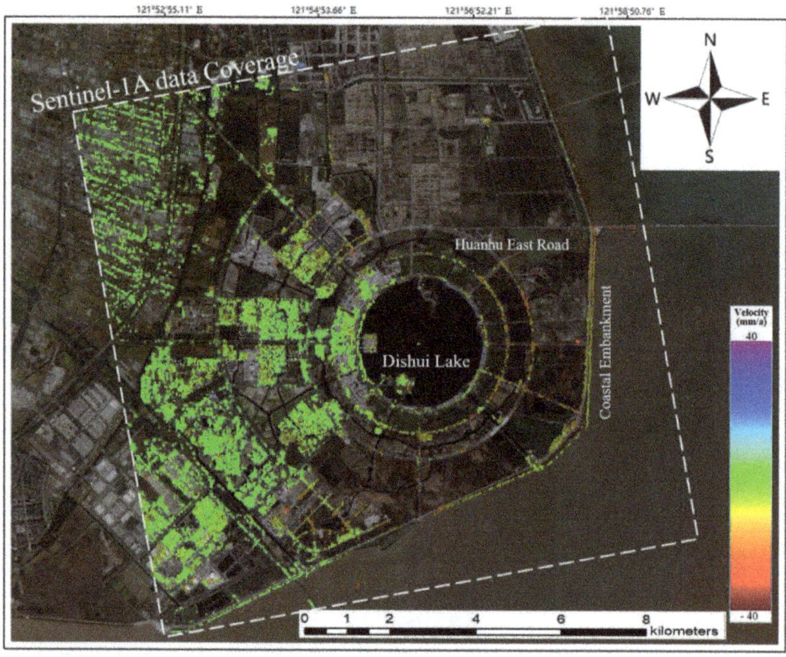

Figure 4. The map of the deformation rate map of Shanghai Lingang New City.

Dishui Lake is an artificial lake built on the beach in Lingang New City, the construction of which was started in 2002. The east and north sides of Dishui Lake are mainly farmlands and wetlands, where quite sparse PS points were identified. As shown in Figure 4, the area with large deformation is mainly concentrated on the east side of Dishui Lake. Although ground subsidence also occurred on the west side of Dishui Lake, it is relatively stable.

The coastal embankment on the east side of Dishui Lake suffered large subsidence, and notable subsidence also occurred around Huanhu East Road. Due to the late formation of land on the east side of Dishui Lake, such subsidence is mainly contributed by the compaction and consolidation of the soil layer in the area.

The coastal embankment, located on the east side of Dishui Lake in Shanghai Lingang New City, has a length of 4 km and was built on the beach with cement or bare rocks. The coastal embankment can help avoiding the erosion of the coast by external factors and is also an important guarantee for flood prevention. Figure 4 indicates the coastal embankment has larger deformation. To investigate the spatial pattern of the deformation, the PS points on the coastal embankment were extracted and their deformation rates were plotted in Figure 5. There are 334 PS points in total, and the density of PS points is about 85.6 points per kilometer. The deformation rate of the coastal embankment ranges from −18.23 mm/year to −4.69 mm/year in the LOS direction. The average deformation rate is −10.55 mm/year in the LOS direction. The statistics of the deformation rate of the coastal embankment from south to north is shown in Figure 6.

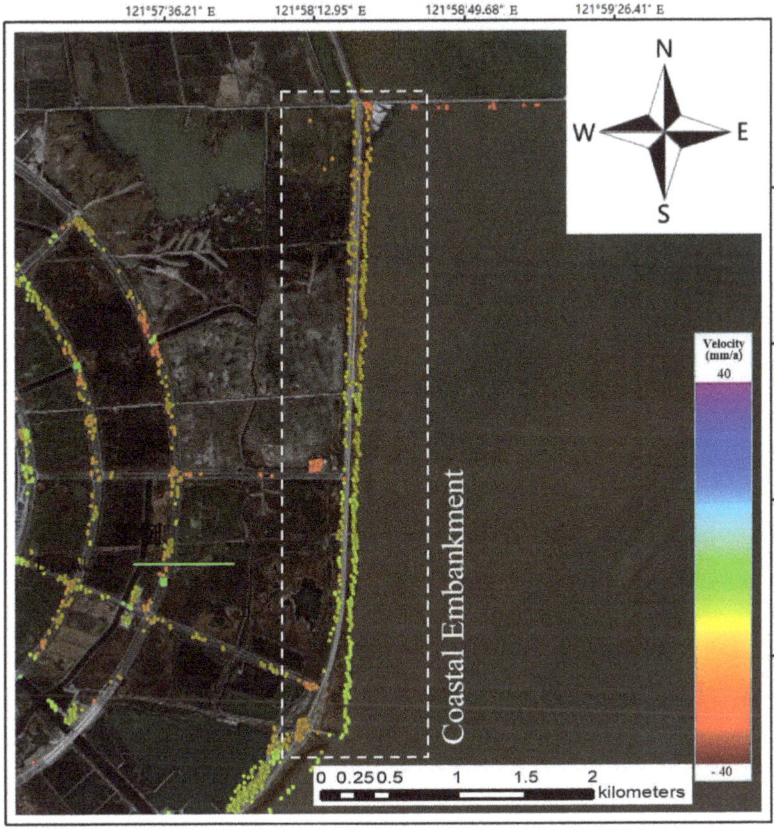

Figure 5. The deformation rate map of the coastal embankment.

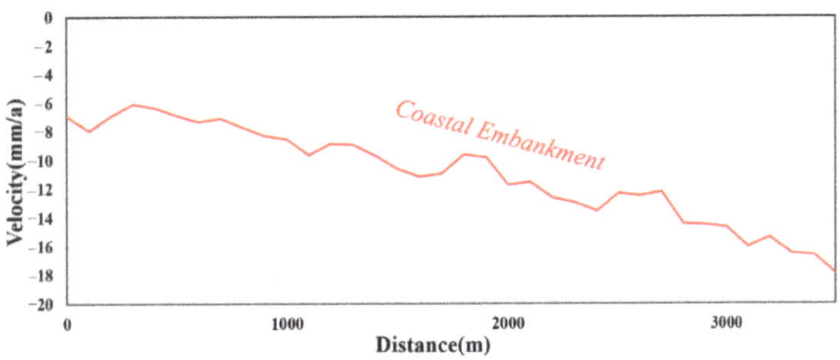

Figure 6. The statistics of the deformation rate in the LOS direction of the coastal embankment from south to north (x-axis is the distance from the coherence point to the starting point of the south).

The results show that the deformation rate on the north side of the coastal embankment is larger than that on the south side, see Figure 6. The deformation rate of the coastal embankment is continuous. Since there are no GPS measurements or leveling data on the coastal embankment section, the deformation results derived from InSAR technology provide useful data for the study of settlements in the area.

4.2. Analysis of Subsidence Characteristics and Leveling Verification

A comparison between the leveling observation data and the PSInSAR result at groundwater monitoring wells W65 and W66 is shown in Figures 7 and 8. Figure 7 shows the time series relationship between the leveling observation data F65 and the PSInSAR result at groundwater monitoring well W65. Figure 8 shows the time series relationship between the leveling observation data F66 and the PSInSAR result at groundwater monitoring well W66. InSAR results have been converted from the LOS direction to the vertical direction and then compared with the leveling data. The maximum difference of these two datasets at W65 and W66 is 1.93 mm and 2.9 mm, respectively. It indicates that the PSInSAR derived deformation time series has comparable accuracy with the leveling data. As the noise level at the selected PS points is similar, the deformation retrieved is expected to be reliable for exploring the geophysical mechanism behind the data.

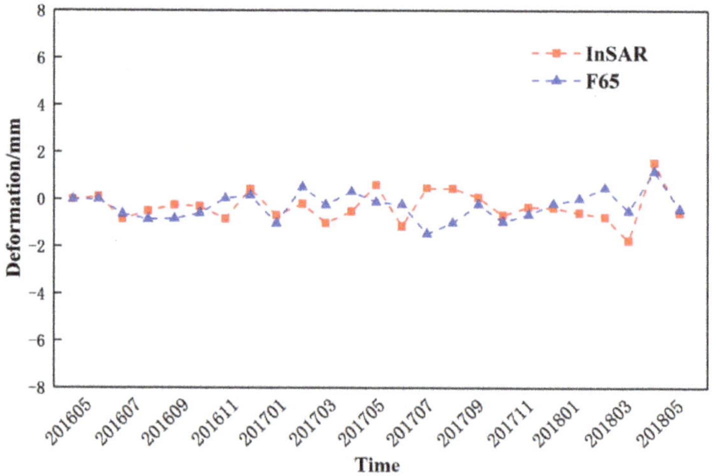

Figure 7. The time series relationship between the leveling observation data F65 and the PSInSAR result at groundwater monitoring well W65.

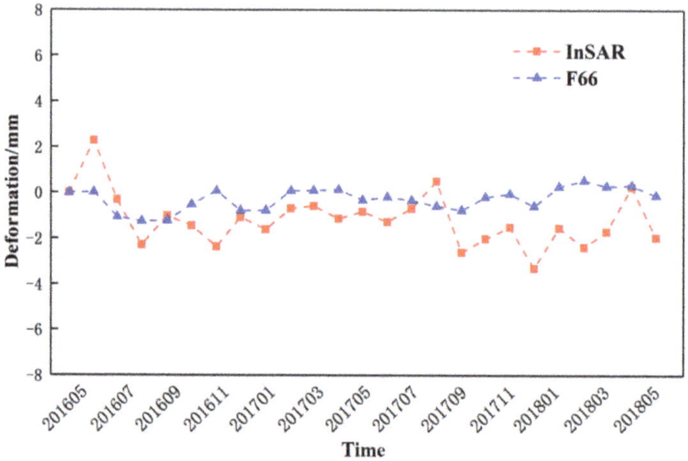

Figure 8. The time series relationship between the leveling observation data F66 and the PSInSAR result at groundwater monitoring well W66.

4.3. The Relationship between Groundwater Level Changes and Ground Subsidence

In order to study the ground subsidence of Shanghai Lingang New City and its mechanism, we used the groundwater level data at two monitoring wells (i.e., W65 and W66) to analyze the relationship between ground subsidence and groundwater level changes. The distribution map of groundwater wells is shown in Figure 1. The landscapes around these wells are shown in Figures 9 and 10. The groundwater monitoring well W65 is close to the coastline, surrounded by wasteland, houses, roads, and ponds. There was no new construction during our observation period. The W66 monitoring well is located on the west side of Hucheng Ring Road and is also near the river. This area is mostly occupied by farmlands.

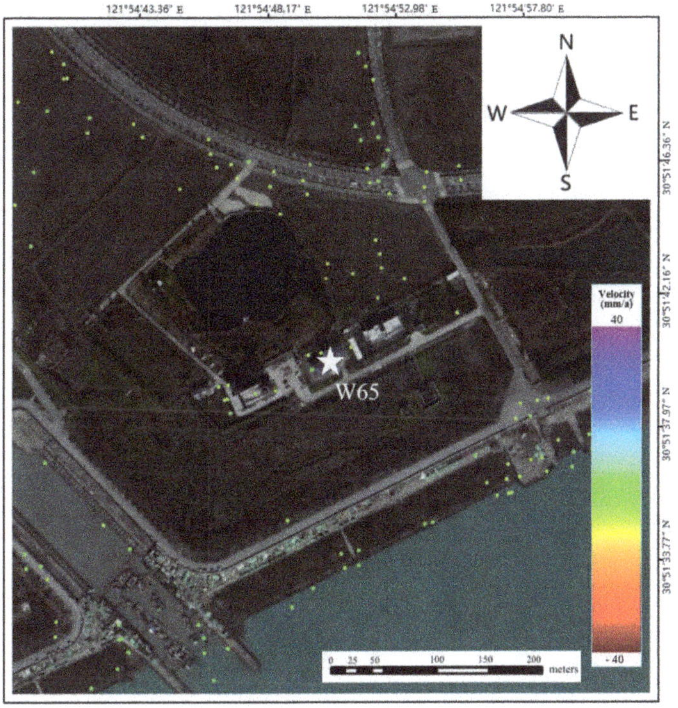

Figure 9. The distribution and deformation rate of PS points around groundwater well W65.

Figures 11 and 12 show the relationships between the groundwater level monitoring value of the fourth confined aquifer and the surface deformation obtained by the InSAR technique. During the observation period, the groundwater level of W65-4 (the fourth confined aquifer of groundwater monitoring well W65) varied within 2 m and had notable seasonal fluctuations. It is well documented that the groundwater level starts to rise in winter and spring and starts to decrease in summer and autumn. The ground deformation value obtained by the PSInSAR method at the monitoring well W65 is around 2 mm within two years, and the average deformation rate is −0.3 mm/year. The ground deformation trend is consistent with the variation of groundwater level data. The variation between InSAR results and groundwater level data had some relevance. It can be considered that it is stable relatively. On the other hand, during the observation period, the groundwater well level of W66-4 was quite steady before August 2017, and from August 2017 to May 2018 had undergone tremendous changes, but it still has the same deformation trend as the InSAR deformation results in the region.

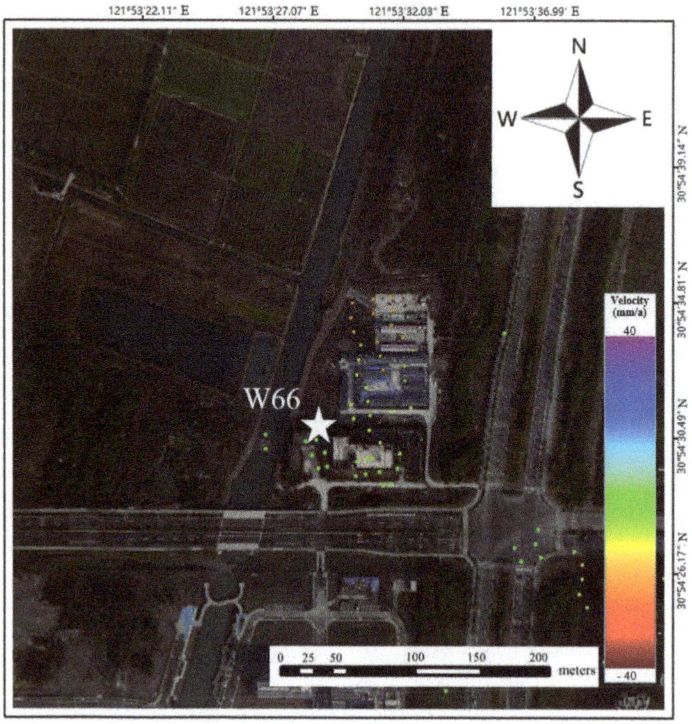

Figure 10. The distribution and deformation rate of PS points around groundwater well W66.

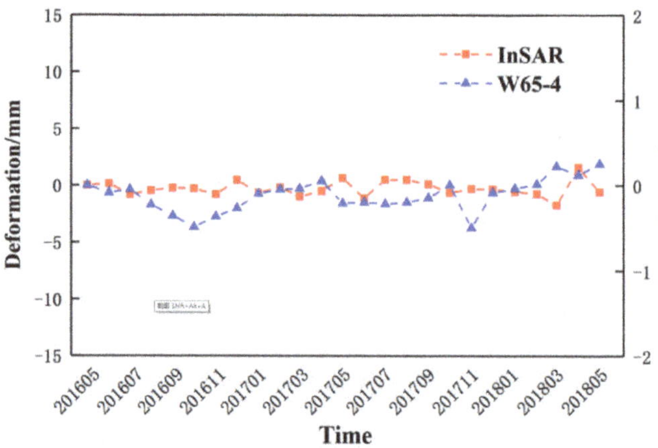

Figure 11. The relationship between PSInSAR derived deformation time series and groundwater level at well W65-4.

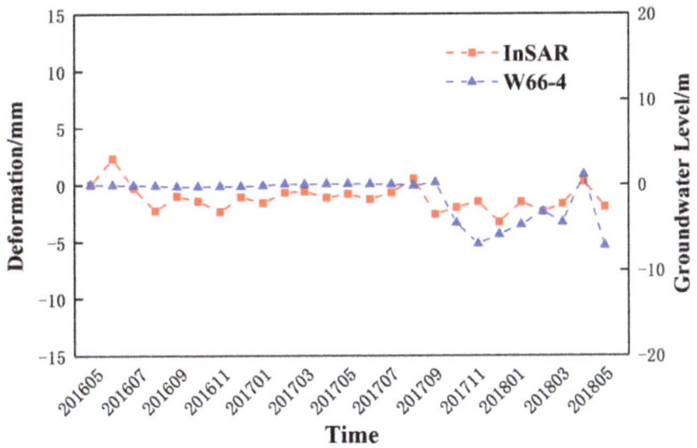

Figure 12. The relationship between PSInSAR derived deformation time series and groundwater level at well W66-4.

Table 3 shows elastic and inelastic skeleton release coefficients determined by PSInSAR deformation and groundwater level variations. W65 groundwater well skeleton release coefficient is between 0.0013–0.0087, and the correlation coefficient between PSInSAR deformation and groundwater level of W65-4 is 0.24 (positive correlation). Groundwater well skeleton release coefficient at well W66 is between 0.0024–0.0054, and the max correlation coefficient between InSAR deformation and groundwater level is 0.68 (positive correlation). According to the study conducted by Hoffmann in 2003 [55], the theoretical value of the elastic water release coefficient is generally between 10^{-5} and 10^{-3} for aquifers dominated by loose clay and silt, and the inelastic water release coefficient often reaches tens to hundreds of times the value of the elastic water release coefficient. Therefore, this indicates that the aquifer at the W65 well and W66 well had undergone elastic deformation.

Table 3. Elastic and inelastic skeleton release coefficients determined by InSAR deformation and groundwater level variations.

ID	Time Span (year/month)	Well (m)	InSAR (mm)	S^{*}_{ki}
W65-4	201605–201611	−0.37	−0.83	0.0022
	201611–201707	0.15	1.3	0.0087
	201706–201805	0.45	0.56	0.0013
W66-4	201605–201611	−0.19	−0.83	0.0044
	201611–201707	0.24	1.3	0.0054
	201706–201804	1.11	2.71	0.0024

4.4. Discussion

By using the Sentinel-1 A SAR images, leveling data, and groundwater well data, we have monitored the land subsidence in the delta area and reveal the geophysical mechanism behind this deformation.

In terms of the precision of PSInSAR vertical subsidence, the maximum difference of PSInSAR and leveling time series at two level points is 1.93 mm and 2.9 mm, respectively, which is basically consistent with Zhao et al.'s result that the mean of absolute difference values between COSMO-SkyMed and leveling measurements is 3.0 mm, and the mean difference value between Sentinel-1 A and leveling measurements is 3.6 mm; these data were obtained by studying Shanghai coastal deformation from February 2007 to April 2017 [21]. It validates that the PSInSAR derived deformation time series has millimeter-level comparable accuracy with the leveling data [32], which shows that the PSInSAR derived deformation is reliable for exploring the geophysical mechanism behind the data.

After extracting the settlement from the PSInSAR LOS deformation, we analyze the displacement of these PS points and the fluctuation of the two groundwater wells and confirm that the changes of the fourth layer confined aquifer data of Shanghai can better reflect the surface deformation in this area from 2016 to 2018. Previous research results related to the zone of the ocean-reclaimed lands of Shanghai are subject to subside due to soil consolidation and compression. Our results show that except for this reason, the settlement in our research of interest may be related to groundwater extraction. This was demonstrated by the comparative analysis of the PSInSAR settlement and groundwater level in the period from May 2016 to May 2018.

Furthermore, according to the Terzaghi-Jacob theoretical model [54], we determine the types of the ground subsidence near the coastal area in the delta by calculating skeleton water release coefficients, which shows that part of Shanghai Lingang New City suffered inelastic ground deformation; this type of deformation is difficult to recover and may result in secondary disasters [16].

The coastal embankment on the southeast side of Lingang New City suffered large subsidence, as shown through the PSInSAR from May 2016 to May 2018, which has been investigated in other period stages using ASAR, COSMO-SkyMed, and Sentinel-1 SAR images from February 2007 to April 2017 [21–23]. Its subsidence mechanism has not yet been fully studied due to lack of sufficient ground-truth validation, however this settlement should be taken seriously due to its special geographical location.

5. Conclusions

In this paper, we mainly used 50 scenes of Sentinel-1A images acquired from May 2016 to May 2018 to obtain the displacement time series of Lingang New City after 14 years' reclamation using the PSInSAR technique. By establishing the relationship between land subsidence and groundwater changes, we firstly combine ground in-situ observations such as first-class leveling data to validate the precision of PSInSAR, which indicates that the PSInSAR derived deformation time series has comparable accuracy with leveling data. Then, we integrate the groundwater well data to obtain the elastic and inelastic skeleton release coefficients, and finally obtain the type of deformation of this area.

The results show that the combination of multiple technologies is more conductive to understand the deformation mechanisms and fluctuation of aquifer systems in these coastal areas.

Therefore, we will next consider carrying out longer time series of SAR combined with ground monitoring to investigate the evolution of Shanghai Lingang New City for the management of coastal embankment safety risk assessment and adjustment, which is also extended to other case study areas.

Author Contributions: Conceptualization, Y.C., M.L. and J.W.; methodology, Y.C., M.L., J.W. and X.L.; software, Y.C., M.L. and X.L.; validation, Y.C., M.L., J.W., S.L., Y.F. and X.W.; data curation, M.L. and F.X.; writing—original draft preparation, Y.C., M.L., J.W. and F.X.; writing—review and editing, Y.C., M.L., J.W., X.L., F.X., S.L., Y.F. and X.W.; project administration, Y.C., M.L. and J.W.; funding acquisition, Y.C. and M.L. All authors have read and agreed to the published version of the manuscript.

Funding: This research is supported by the technical standard project of Shanghai "scientific and technological innovation action plan" (grant no. 21DZ2205300), and Chongqing Science and Technology Bureau (grant no. cstc2020jxjl00001).

Data Availability Statement: The sentinel-1A SAR data is available at: https://scihub.copernicus.eu/ (accessed on 19 April 2022). The leveling data and groundwater level data are supported by Shanghai Institute of Geological Survey, China.The data used to support the findings of this study are included within the article.

Acknowledgments: We want to thank European Space Agency (ESA) for providing free SAR data of Sentinel-1A.

Conflicts of Interest: The authors declare no conflict of interest.

References

1. Hoeksema, R. Three stages in the history of land reclamation in the Netherlands. *Irrig. Drain. J. Int. Comm. Irrig. Drain.* **2007**, *56*, S113–S126. [CrossRef]
2. Cuenca, M.C.; Hanssen, R.; Hooper, A.; Arikan, M. Surface deformation of the whole Netherlands after PSI analysis. In Proceedings of the Fringe 2011 Workshop, Frascati, Italy, 19–23 September 2011.
3. Zhao, Q.; Lin, H.; Gao, W.; Zebker, H.; Chen, A.; Yeung, K. InSAR detection of residual settlement of an ocean reclamation engineering project: A case study of Hong Kong International Airport. *J. Oceanogr.* **2011**, *67*, 415–426. [CrossRef]
4. Plant, G.; Covil, C.; Hughes, R. *Site Preparation for the New Hong Kong International Airport*; Thomas Telford: Telford, UK, 1998.
5. Liu, G.; Ding, X. Detecting of Surface Deformation in Coastlands by DInSAR: Accuracy and Applicability Analysis. *Bull. Surv. Mapp.* **2006**, *9*, 9–13.
6. Wan, X.; Hu, B.; Ma, Q. Monitoring ground subsidence in Macau with Persistent Scatters Interferometry. *Eng. Surv. Mapp.* **2012**, *21*, 39–43.
7. Luo, Q.; Perissin, D.; Li, Q.; Lin, H.; Duering, R. Tianjin INSAR time series analysis with L-and X-band. In Proceedings of the 2011 IEEE International Geoscience and Remote Sensing Symposium, Vancouver, BC, Canada, 24–29 July 2011; pp. 1477–1480.
8. Lv, X.; Niu, W.; Luo, L.; Zhang, Y.; Yi, N. Research on monitoring and control of land subsidence in a sea reclamation region of Tianjin. *Shanghai Land Resour.* **2016**, *37*, 55–59.
9. Zhang, J.; Hu, K.; Yue, W.; Liu, B.; Wang, J.; Gao, Q. Land Subsidence in Shanghai City and Its Response to Groundwater exploitation and reinjection. *Resour. Environ. Yangtze Basin* **2016**, *25*, 567–572.
10. Shen, S. Geological environmental character of Lin-Gang new city and its influences to the construction. *Shanghai Geol.* **2008**, *105*, 24–28.
11. Li, J.; Dai, Z.; Ying, M.; Wu, R.; Fu, G.; Xu, H. Analysis on the development and evolution of tidal flats and reclamation of land resource along shore of Shanghai city. *J. Nat. Resour.* **2007**, *22*, 361–371.
12. Zhang, L.; Ding, X.; Lu, Z.; Jung, H.; Hu, J.; Feng, G. A novel multitemporal InSAR model for joint estimation of deformation rates and orbital errors. *IEEE Trans. Geosci. Remote Sens.* **2013**, *52*, 3529–3540. [CrossRef]
13. Zhang, L.; Lu, Z.; Ding, X.; Jung, H.; Feng, G.; Lee, C. Mapping ground surface deformation using temporarily coherent point SAR interferometry: Application to Los Angeles Basin. *Remote Sens. Environ.* **2012**, *117*, 429–439. [CrossRef]
14. Wang, C.; Wang, X.S.; Xu, Y.; Zhang, B.; Jiang, M.; Xiong, S.; Zhang, Q.; Li, W.; Li, Q. A new likelihood function for consistent phase series estimation in distributed scatterer interferometry. *IEEE Trans. Geosci. Remote Sens.* **2022**, *60*, 5227314. [CrossRef]
15. Galloway, D.; Hudnut, K.; Ingebritsen, S.; Phillips, S.; Peltzer, G.; Rogez, F.; Rosen, P. Detection of aquifer system compaction and land subsidence using interferometric synthetic aperture radar, Antelope Valley, Mojave Desert, California. *Water Resour. Res.* **1998**, *34*, 2573–2585. [CrossRef]
16. Hoffmann, J.; Galloway, D.; Zebker, H. Inverse modeling of interbed storage parameters using land subsidence observations, Antelope Valley, California. *Water Resour. Res.* **2003**, *39*, 1031. [CrossRef]
17. Erban, L.; Gorelick, S.; Zebker, H.; Fendorf, S. Release of arsenic to deep groundwater in the Mekong Delta, Vietnam, linked to pumping-induced land subsidence. *Proc. Natl. Acad. Sci. USA* **2013**, *110*, 13751–13756. [CrossRef]
18. Matano, F.; Sacchi, M.; Vigliotti, M.; Ruberti, D. Subsidence trends of volturno river coastal plain (northern campania, southern italy) inferred by sar interferometry data. *Geosciences* **2018**, *8*, 1–22. [CrossRef]
19. Wang, H.; Wright, T.; Yu, Y.; Lin, H.; Jiang, L.; Li, C.; Qiu, G. InSAR reveals coastal subsidence in the Pearl River Delta, China. *Geophys. J. Int.* **2012**, *191*, 1119–1128. [CrossRef]
20. Liu, G.; Jia, H.; Nie, Y.; Li, T.; Zhang, R.; Yu, B.; Li, Z. Detecting subsidence in coastal areas by ultrashort-baseline TCPInSAR on the time series of high-resolution TerraSAR-X images. *IEEE Trans. Geosci. Remote Sens.* **2013**, *52*, 1911–1923. [CrossRef]
21. Zhao, Q.; Ma, G.; Wang, Q.; Yang, T.; Liu, M.; Gao, W.; Falabella, F.; Mastro, P.; Pepe, A. Generation of long-term InSAR ground displacement time-series through a novel multi-sensor data merging technique: The case study of the Shanghai coastal area. *ISPRS J. Photogramm. Remote Sens.* **2019**, *154*, 10–27. [CrossRef]
22. Zhao, Q.; Pepe, A.; Gao, W.; Lu, Z.; Bonano, M.; He, M.L.; Wang, J.; Tang, X. A DInSAR investigation of the ground settlement time evolution of ocean-reclaimed lands in Shanghai. *IEEE J. Sel. Top. Appl. Earth Obs. Remote Sens.* **2015**, *8*, 1763–1781. [CrossRef]
23. Zhao, Q.; Pepe, A.; Gao, W.; Li, X.; Lu, Z.; Bonano, M.; Manunta, M.; Lanari, R. Derivation of Ground Settlement Spatiotemporal Characteristics of Reclaimed Area in Nanhui New City, Shanghai, China, from Time-Series in InSAR Interpretation. *Dragon 3mid Term Results* **2014**, *724*, 39.
24. Luo, Y.; Yan, X.; Yang, T.; Ye, S.; Wu, J. Space-time characteristics of exploitation and recharge of groundwater and land subsidence if Shanghai land area. *J. Nanjing Univ. Nat. Sci. Ed.* **2019**, *3*, 449–457.
25. Yu, L.; Yang, T.; Zhao, Q.; Liu, M.; Pepe, A. The 2015–2016 ground displacements of the Shanghai coastal area inferred from a combined COSMO-SkyMed/Sentinel-1 DInSAR analysis. *Remote Sens.* **2017**, *9*, 1194. [CrossRef]
26. Zhang, K.; Yue, W.; Liu, B. Land Subsidence in Shanghai and Its Response to Groundwater Irrigation. *Yangtze River Basin Resour. Environ.* **2016**, *25*, 4.
27. Liao, M.; Pei, Y.; Wang, H.; Fang, Z.; Wei, L. Subsidence monitoring in Shanghai using the PSInSAR technique. *Shanghai Land Resour.* **2012**, *33*, 5–10.
28. Hu, S.; Wu, J.; Ban, B.; Zhang, L. PALSAR interferometry for urban subsidence monitoring: An experiment in Shanghai area. In Proceedings of the 2009 2nd Asian-Pacific Conference on Synthetic Aperture Radar, Xi'an, China, 26–30 October 2009; pp. 140–143.

29. Yang, M.; Jiang, Y.; Liao, M.; Wang, H. The analysis of the subsidence patterns in Lingang New City (Shanghai) using high-resolution SAR images. *Shanghai Land Resour.* **2013**, *34*, 12–16.
30. Shi, X.; Wu, J.; Ye, S.; Zhang, Y.; Xue, Y.; Wei, Z.; Li, Q.; Yu, J. Regional land subsidence simulation in Su-xi-Chang area and Shanghai City, China. *Eng. Geol.* **2008**, *100*, 27–42. [CrossRef]
31. Liu, G.; Luo, X.; Chen, Q.; Huang, D.; Ding, X. Detecting land subsidence in Shanghai by PS-networking SAR interferometry. *Sensors* **2008**, *8*, 4725–4741. [CrossRef]
32. Yang, M.; Yang, T.; Zhang, L.; Lin, J.; Qin, X.; Liao, M. Spatio-Temporal Characterization of a Reclamation Settlement in the Shanghai Coastal Area with Time Series Analyses of X-, C-, and L-Band SAR Datasets. *Remote Sens.* **2018**, *10*, 3291. [CrossRef]
33. Ding, J.; Zhao, Q.; Tang, M.; Calò, F.; Zamparelli, V.; Falabella, F.; Liu, M.; Pepe, A. On the Characterization and Forecasting of Ground Displacements of Ocean-Reclaimed Lands. *Remote Sens.* **2020**, *12*, 297. [CrossRef]
34. Qiu, J.; Li, X. *The Quaternary Strata and Sedimentary Environment of Shanghai City*; Shanghai Science and Technology Press: Shanghai, China, 2007.
35. Attema, E.; Davidson, M.; Snoeij, P.; Rommen, B.; Floury, N. Sentinel-1 mission overview. In Proceedings of the 2009 IEEE International Geoscience and Remote Sensing Symposium, Cape Town, South Africa, 12–17 July 2009; pp. I-36–I-39.
36. Panetti, A.; Torres, R.; Lokas, S.; Bruno, C.; Croci, R.; L'Abbate, M.; Marcozzi, M.; Pietropaolo, A.; Venditti, P. GMES Sentinel-1: Mission and satellite system overview. In Proceedings of the EUSAR 2012; 9th European Conference on Synthetic Aperture Radar, Nuremberg, Germany, 23–26 April 2012 ; pp. 162–165.
37. Rostan, F.; Riegger, S.; Pitz, W.; Torre, A.; Torres, R. The C-SAR instrument for the GMES sentinel-1 mission. In Proceedings of the 2007 IEEE International Geoscience and Remote Sensing Symposium, Barcelona, Spain, 23–28 July 2007; pp. 215–218.
38. Yague, M.; De, Z.; Prats-Iraola, P. Coregistration of interferometric stacks of Sentinel-1 TOPS data. *IEEE Geosci. Remote Sens. Lett.* **2017**, *14*, 1002–1006. [CrossRef]
39. Ferretti, A.; Prati, C.; Rocca, F. Permanent scatterers in SAR interferometry. *IEEE Trans. Geosci. Remote Sens.* **2001**, *39*, 8–20. [CrossRef]
40. Ferretti, A.; Prati, C.; Rocca, F. Nonlinear subsidence rate estimation using permanent scatterers in differential SAR interferometry. *IEEE Trans. Geosci. Remote Sens.* **2000**, *38*, 2202–2212. [CrossRef]
41. Massonnet, D.; Feigl, K. Radar interferometry and its application to changes in the Earth's surface. *Rev. Geophys.* **1998**, *36*, 441–500. [CrossRef]
42. Bert, M. *Radar Interferometry: Persistent Scatterers Technique*; Springer: Berlin/Heidelberg, Germany, 2006.
43. Hooper, A.; Segall, P.; Zebker, H. Persistent scatterer interferometric synthetic aperture radar for crustal deformation analysis, with application to Volcán Alcedo, Galápagos. *J. Geophys. Res. Solid Earth* **2007**, *112*, 1–12. [CrossRef]
44. Hooper, A.; Zebker, H.; Segall, P.; Kampes, B. A new method for measuring deformation on volcanoes and other natural terrains using InSAR persistent scatterers. *Geophys. Res. Lett.* **2004**, *31*, L23611. [CrossRef]
45. Tofani, V.; Raspini, F.; Catani, F.; Casagli, N. Persistent Scatterer Interferometry (PSI) technique for landslide characterization and monitoring. *Remote Sens.* **2013**, *5*, 1045–1065. [CrossRef]
46. Kampes, B.; Hanssen, R. Ambiguity resolution for permanent scatterer interferometry. *IEEE Trans. Geosci. Remote Sens.* **2004**, *42*, 2446–2453. [CrossRef]
47. Hanssen, R. *Radar interferometry: Data interpretation and error analysis*; Springer Science & Business Media: Berlin/Heidelberg, Germany, 2001.
48. Czikhardt, R.; Papco, J.; Bakon, M.; Liscak, P.; Ondrejka, P.; Zlocha, M. Ground stability monitoring of undermined and landslide prone areas by means of sentinel-1 multi-temporal InSAR, case study from Slovakia. *Geosciences* **2017**, *7*, 87–103. [CrossRef]
49. Crosetto, M.; Monserrat, O.; Cuevas-González, M.; Devanthéry, N.; Crippa, B. Persistent scatterer interferometry: A review. *ISPRS J. Photogramm. Remote Sens.* **2016**, *115*, 78–89. [CrossRef]
50. Colesanti, C.; Locatelli, R.; Novali, F. Ground deformation monitoring exploiting SAR permanent scatterers. In Proceedings of the IEEE International Geoscience and Remote Sensing Symposium, Toronto, ON, Canada, 24–28 June 2002; pp. 1219–1221.
51. Perissin, D.; Wang, T. Repeat-pass SAR interferometry with partially coherent targets. *IEEE Trans. Geosci. Remote Sens.* **2011**, *50*, 271–280. [CrossRef]
52. Hooper, A.; Bekaert, D.; Spaans, K.; Arıkan, M. Recent advances in SAR interferometry time series analysis for measuring crustal deformation. *Tectonophysics* **2012**, *514*, 1–13. [CrossRef]
53. Hooper, A.; Zebker, H. Phase unwrapping in three dimensions with application to InSAR time series. *JOSA A* **2007**, *24*, 2737–2747. [CrossRef]
54. Terzaghi, K. *Erdbaumechanik auf bodenphysikalischer Grundlage*; F. Deuticke: Leipzig, Vienna, 1925.
55. Hoffmann, J. The application of satellite radar interferometry to the study of land subsidence over developed aquifer systems. Ph.D. Thesis, Department of Geophysics, Stanford University, Stanford, CA, USA, 2003; pp. 1–211.

remote sensing

MDPI

Article

Interpretation of the Spatiotemporal Evolution Characteristics of Land Deformation in Beijing during 2003–2020 Using Sentinel, ENVISAT, and Landsat Data

Shuangcheng Zhang [1,2,*], Yafei Zhang [1], Jing Yu [1], Qianyou Fan [1], Jinzhao Si [1], Wu Zhu [1] and Mingxin Song [3]

1 School of Geology Engineering and Geomatics, Chang'an University, Xi'an 710054, China; 2020126037@chd.edu.cn (Y.Z.); 2019226003@chd.edu.cn (J.Y.); 2020026013@chd.edu.cn (Q.F.); 2020126029@chd.edu.cn (J.S.); zhuwu@chd.edu.cn (W.Z.)

2 State Key Laboratory of Geo-Information Engineering, Xi'an 710054, China

3 Xi'an Institute of Prospecting and Mapping, Xi'an 710054, China; 2018226008@chd.edu.cn

* Correspondence: shuangcheng369@chd.edu.cn

Citation: Zhang, S.; Zhang, Y.; Yu, J.; Fan, Q.; Si, J.; Zhu, W.; Song, M. Interpretation of the Spatiotemporal Evolution Characteristics of Land Deformation in Beijing during 2003–2020 Using Sentinel, ENVISAT, and Landsat Data. *Remote Sens.* **2022**, *14*, 2242. https://doi.org/10.3390/rs14092242

Academic Editors: Francesca Ardizzone and Michele Saroli

Received: 13 March 2022
Accepted: 5 May 2022
Published: 7 May 2022

Abstract: Since the 1930s, due to the rapid development of the city and the increase of population, the demand from Beijing residents for water resources has gradually increased. Land deformation in the Beijing Plain is a serious issue. In order to warn of, and mitigate, disasters, it is urgently necessary to obtain the latest rate, extent, and temporal evolution of land subsidence in Beijing. Firstly, the temporal and spatial distribution characteristics of land deformation in Beijing during 2003–2020 were unveiled using the time-series interferometric synthetic aperture radar (InSAR) technique and two different satellite datasets, sentinel-1a/1b and ENVISAT ASAR. By means of combining calibration of InSAR results with the global positioning system (GPS), we studied the evolutionary process of long-term land subsidence in Beijing. The precision of our InSAR annual subsidence results is less than 10 mm. Land subsidence in Beijing is unevenly distributed, and so five main land subsidence zones were monitored. The time-series results showed that the rate of land subsidence rate continued to increase from 2003 to 2015, but has gradually shown a slowing trend from 2015 to 2020. Further, we used the quadratic polynomial fitting method to interpolate the time-series deformation results from 2010 to 2015, and compared these with GPS. The results demonstrated that although the InSAR observation method is not strictly registered with GPS in time, its deformation trend is consistent. In addition, the calibrated long time-series was consistent with the three deformation stages of land subsidence evolution in Beijing. Finally, we analyzed the deformation information obtained by InSAR technology in combination with land use type data, precipitation and groundwater data. The results demonstrated that the central area is mostly stable, and land deformation in the northeast is obvious and uneven. In addition, land use type and precipitation have little influence on land subsidence. Changes in land subsidence were closely related to changes in groundwater level, and seasonal variations in deformation correlated with precipitation.

Keywords: land subsidence; dataset calibration; quadratic polynomial fitting; spatiotemporal evolution characteristics

1. Introduction

Beijing has an extensive geopolitical history and cultural diversity. It is the political, cultural, and international communication capital of China. Social, economic, and urban development have been rapidly increasing in Beijing. In addition, Beijing has been affected by land subsidence since 1935 [1]. In the 1970s, rapid, large-scale, and concentrated land subsidence was observed. Consequently, new subsidence areas, such as Changping, Shunyi, and Daxing, have gradually appeared. Land subsidence is a local descending movement caused by consolidation and compression of the underground loose strata under the influence of human engineering economic activities. It is also the main regional environmental geological disaster in China [2]. Therefore, analyzing long-term land subsidence in Beijing is necessary.

With acceleration in urbanization a number of technologies have been commonly used to monitor surface subsidence in Beijing: precise leveling surveys [3], global positioning system (GPS) [4], interferometric synthetic aperture radar (InSAR) [5] and other technologies. Among these, leveling has several advantages, including high precision, advanced technology, and reliable results. However, it can only reflect the overall situation of surface deformation in Beijing, and requires permanent observations which can be time-consuming and labor-intensive. GPS technology can provide high-precision geographical coordinates, which can be used to monitor global and regional crustal deformation and land subsidence. However, GPS technology can only provide a limited number of discrete points for land subsidence monitoring, has a small monitoring range, and cannot obtain real-time and macroscopic deformation monitoring results in most cases. In contrast, InSAR, a novel method of determining land subsidence, can provide high-precision three-dimensional terrain information and small-scale land deformation information [6–10]. It greatly reduces costs and labor, and provides risk analysis and early warning signals during rapid urban development [11–15]. In contrast to conventional measurements, it can provide land subsidence results with centimeter-level, or even millimeter-level, accuracy over a large region [16]. The conventional differential interferometry synthetic aperture radar (D-InSAR) method was based on individual interferogram to obtain land subsidence [17]. However, it does not capture changes in deformation rate within the study area. In addition, changes in atmospheric conditions, satellite orbit errors and digital elevation model (DEM) errors also significantly reduce the accuracy of D-INSAR measurements. The Small Baseline Subset (SBAS) technique, proposed by Berardino et al. [18,19], can continuously and stably monitor the land surface for a long period in a large area with a specific revisit period, thereby minimizing the limitations of conventional measurement methods [20]. Regarding land subsidence monitoring in Beijing using the InSAR method, some annual InSAR deformation results have been exhibited in recent years. Shi [21], Yuan [22], Zhou [23], Li [24] and Pan et al. [25]. obtained the deformation sequence of Beijing for a long time period using SBAS-InSAR technology. In addition, many experts and scholars have conducted relevant studies on the spatiotemporal evolution of surface deformation in Beijing and its genesis mechanism. Zhou et al. [26] used ENVISAT ASAR and RadarSAT-2 data, and found several settlement funnels in the north and east of Beijing. Further, Du et al. [27] used ALOS data, and found that the subsidence rate in northern Beijing accelerated from 2007 to 2016. Liu et al. [28] used ENVISAT ASAR and RadarSAT-2 data and SBAS-InSAR technology to obtain vertical deformation characteristics of Beijing during 2003–2013. Muhetaer et al. [29], Yang [30], and Ng [31] analyzed the spatiotemporal distribution characteristics of land deformation in Beijing using time-series InSAR and GPS technologies. Hu et al. [32] analyzed development and changes in land deformation and their relationship with geological faults in Beijing from 2015 to 2017 using an atmospheric estimation model time-series analysis method. Further, Zhou et al. [33] introduced the machine learning approach to analyze the influence of groundwater, compressible layer thickness and other influencing factors on land deformation. Previous studies either identified land subsidence stages, by the analysis of individual interferograms based on conventional InSAR method, or detected land subsidence in a limited time, by SBAS-InSAR. New data and more advanced InSAR techniques are demanded to gain the latest rate, scope of activity, and temporal evolution of ground deformation in Beijing. In addition, there are some time gaps between different SAR datasets.

To better study long-term evolution of land deformation in Beijing, we need to continuously update the time series of different sensor deformations. SAR images provided by a single radar satellite are insufficient to reflect the long-term evolution of land subsidence, thus we need to combine the Sentinel satellite with others to update the time series of different sensors' monitoring of deformation. Samsonov [34] and Pepe et al. [35] used a singular value decomposition algorithm to fuse multi-satellite time-series data with time overlap data, further fusing the vertical deformation time-series of ENVISAT ASAR and COSMO -SkyMed using the consolidation settlement model, thus realizing the deformation time series over a long time period. Based on the Sentinel-1A/1B, and ENVISAT ASAR datasets, and GPS measurements since 2003, our study aimed to acquire the spatial and temporal distribution characteristics of land subsidence in Beijing from 2003 to 2020 provided by advanced InSAR technology. First, the interferograms of two different InSAR datasets are processed by SBAS-InSAR technology respectively, and time series deformation measurements from 2003 to 2020 are generated to reveal the temporal and spatial changes of land subsidence in Beijing. Second, combined with calibration of the InSAR results by GPS, we obtained the spatial and temporal changes of land subsidence in Beijing over nearly twenty years. In addition, the results were verified by the Mann-Kendall method. Finally, we discuss the correlation between land subsidence and land use type, groundwater, and precipitation. The rest of the manuscript is organized in the following way: Section 2 introduces the study area; Section 3 introduces the SBAS-InSAR algorithm, GNSS processing strategy and the data used for processing and verification; Section 4 introduces the temporal and spatial changes of land subsidence in different SAR datasets and the settlement results of fusion time series; and Section 5 introduces the relationship between land subsidence and land use type, groundwater level, precipitation, ground fissures and fault activity, Section 6 summarizes our conclusions.

2. Study Area

Beijing is located in northern China and north of the North China Plain, ad-jacent to Tianjin in the east and Hebei in the west. It has a total area of 16,412 km^2. The elevation of Beijing is high in the northwest and low in the southeast, with an average elevation of 43.5 m. Beijing is surrounded by mountains in the west, north and northeast, whereas the southeast region comprises a plain gradually sloping towards the Bohai Sea. Figure 1 shows the geographical location of Beijing. The depressions in the Beijing plain area are loose sediments with thicknesses of more than 1000 m, which are prone to ground fractures and land subsidence [36].

According to results from the Beijing earthquake station, the faults in the northeast mainly include Huangzhuang-Gaoliying fault and Shunyi's buried active fault, and the horizontal deformation of these faults is mainly dextral. However, the Sunhe-Nankou fault is a typical fault in the northwest; its horizontal deformation is mainly characterized by sinistral deformation [37]. The faults in the northeast and northwest reflect tectonic compression to a certain extent. In addition, some earthquakes in Beijing are mainly related to characteristics of the crustal structure and physical properties of the rocks. The fault structure has a significant controlling effect on quaternary sediments in the Beijing plain and is closely related to seismic activity. The quaternary tectonic activities have both inheritance and cenozoic factors, and the distribution of depositional centers is more complex than those in the northeast. The tertiary and tectonic activities in the northwest are more obvious. Studies have shown that spatial distribution of land deformation was consistent with that of the quaternary compressible layers in Beijing [38]. Land subsidence in the plain is more severe, and the frequency of land subsidence in Chaoyang, Shunyi, Tongzhou, Changping, Haidian and other areas in Beijing is higher and the degree of harm greater. More than half of the Beijing plain area is seriously affected by land subsidence [39].

The geological environment in Beijing is complicated. Therefore, combined with groundwater, geological information, land use type data, and the result of time-series SBAS, it is necessary to continuously monitor land subsidence in Beijing for a long time.

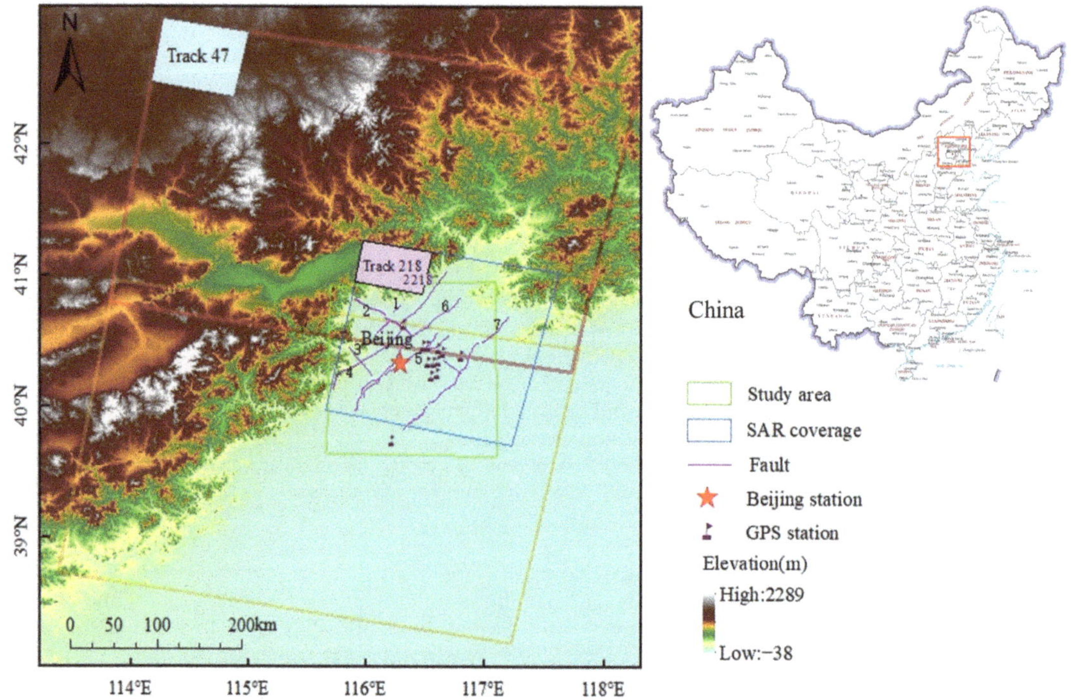

Figure 1. Study area geographical location and coverage of the Synthetic Aperture Radar (SAR) data. In the background is the shaded terrain-generated digital elevation model (SRTM DEM) of the Shuttle Radar Topography Mission. The red and yellow rectangles represent the coverage of sentinel-1 SAR data, the blue indicate coverage of ENVISAT ASAR data, and the green rectangle represents the location of the study area. The Beijing Plain is mainly controlled by northeast and northwest direction active faults. The northeast direction fault systems are composed mainly of the Xiaotang Mountain (1), Huangzhuang-Gaoliying (4), Tongxian-Nanyuan (5), Shunyi-Liangxiang (6), and Xiadian (7) faults. The northwest direction fault systems are composed mainly of the Nankou Sunhe (2) and the Yongding River (3) faults. The figure in the upper right corner shows the location of the study area in China. It can be downloaded from http://bzdt.ch.mnr.gov.cn/index.html, accessed on 12 January 2022. The part of the red rectangle in the map corresponds to the figure on the left.

3. Materials and Methods

3.1. SAR Images

ENVISAT and Sentinel-1A/1B image data of Beijing were used to monitor land subsidence, and to analyze the spatiotemporal evolution characteristics of land subsidence in Beijing. ENVISAT ASAR is a polar-orbiting Earth observation satellite launched by the European Space Agency (ESA) in 2002, while Sentinel-1A/1B is an active radar imaging satellite launched by the ESA in 2014. In total, our study used 40 SAR images from the ENVISAT ASAR satellite from 2003 to 2010, and 90 SAR images from Sentinel-1A/1B satellite from 2015 to 2020. Both datasets had a VV polarization model. The specific imaging parameters are listed in Table 1, and the time span of the SAR dataset used in this study is shown in Figure 2. The processed SAR datasets span the entire time interval between June 2003 and December 2020. Due to the absence of comprehensive SAR datasets, there is a significant temporal gap (about five years) between the available ENVISAT and Sentinel datasets. The spatiotemporal baseline diagrams of the two data types are shown in Figure 3.

Table 1. Information of the SAR dataset.

Sensor	ENVISAT ASAR	Sentinel-1A/1B
Band	C	C
Wavelength (cm)	5.5	5.56
Heading (°)	−164	−166
Track	2218	47
Polarization	VV	VV
Orbit directions	Descending	Descending
Number of images	40	90
Data range	18 June 2003–29 September 2010	24 January 2015–22 January 2021
Incidence angle (°)	22.8	33.9

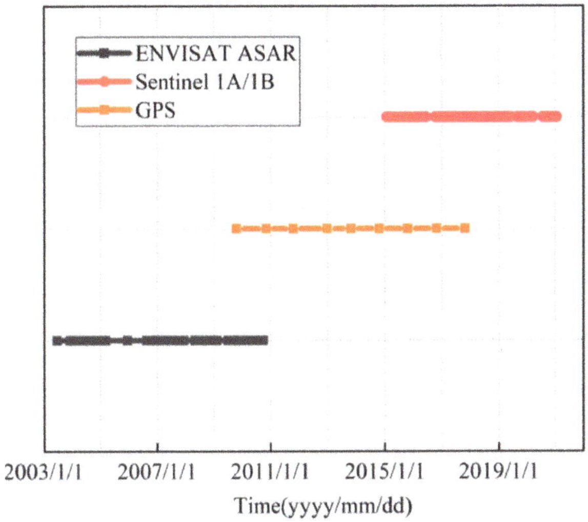

Figure 2. Time span of the SAR datasets and GPS measurement.

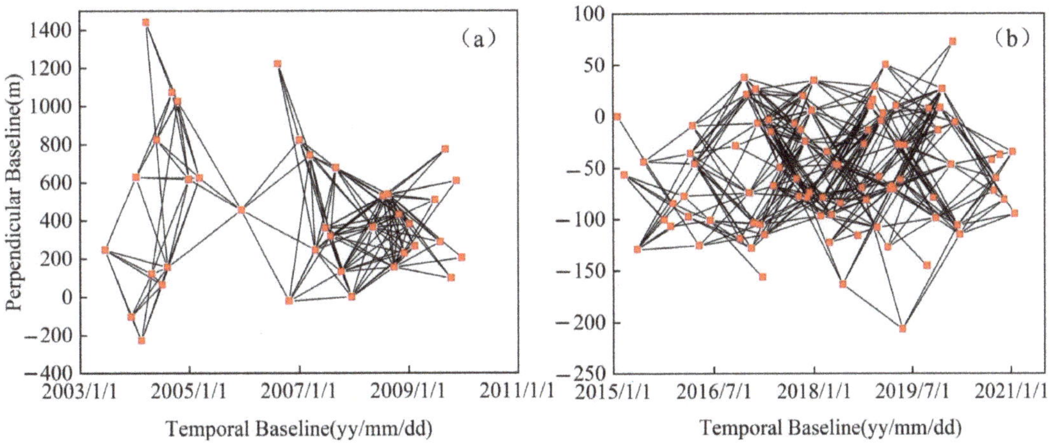

Figure 3. Temporal and perpendicular baseline maps of (**a**) ENVISAT ASAR and (**b**) Sentinel-1A/1B.

3.2. Other Datasets

Digital elevation model (DEM) data, having a spatial resolution of 30 m and provided by the National Aeronautics and Space Administration (NASA) Shuttle Radar Topography Mission (SRTM), were used in this study. In addition, the five year interval land use data (spatial resolution: 30 m) from 2005 to 2020, provided by Landsat data through the Google Earth Engine (GEE) platform [40], were used. Groundwater exploitation and depth data were provided by the Beijing Water Authority, and precipitation data of the meteorological station were provided by the National Oceanic and Atmospheric Administration (NOAA). Groundwater exploitation and depth data can be downloaded from http://swj.beijing.gov.cn/, accessed on 10 February 2022. Precipitation can be downloaded from https://www.ncei.noaa.gov/maps-and-geospatial-products, accessed on 10 February 2022. Moreover, the data from 12 global positioning system monitoring sites in Beijing, from 2009 to 2020, were acquired by the Beijing Institute of Geological & Prospecting Engineering in China. Each monitoring site had 12 data periods, and each monitoring site was a bedrock point. The distribution of monitoring points is shown in Figure 1.

3.3. Land Subsidence Monitoring Using SBAS-InSAR

The principle of SBAS-inSAR technology is to combine all SAR data into several small sets by setting certain spatiotemporal baseline thresholds, with smaller baselines within the sets and larger baselines between the sets. Finally, through the least square solution within the set and the singular value decomposition method between the sets, the joint solution result of the surface deformation information of the whole time-series is obtained [41,42]. In this way the data utilization rate is improved to overcome the incoherent problem of image information. In addition, the influence of DEM error on deformation measurement can be reduced.

Assuming that there are $N + 1$ SAR images covering the same area and the time series is $(t_0, \cdots, t_N)$, M differential interferograms can be obtained under the condition that the spatiotemporal threshold is satisfied. Let the acquisition times of the primary and secondary images of the differential interferogram i be t_B and t_A, respectively, and $t_B > t_A$. Thus, the differential interference phase generated by SAR images at pixels (x, y) of t_A and t_B can be expressed as:

$$\delta_{\varphi_i}(x,y) = \varphi(t_B, x, y) - \varphi(t_A, x, y) \approx \varphi_{def,i}(x,y) + \varphi_{topo,i}(x,y) + \varphi_{atm,i}(x,y) + \varphi_{noise,i}(x,y) \tag{1}$$

where, (x, y) are azimuth direction and distance direction coordinates, $\varphi(t_B, x, y)$ and $\varphi(t_A, x, y)$ is the image phase generated by the differential interferogram phase, $\varphi_{def,i}(x, y)$ is the deformation phase caused by interference on the land displacement between t_A and t_B, $\varphi_{topo,i}(x, y)$ is the topography phase error caused by topographic undulation, $\varphi_{atm,i}(x, y)$ is the atmospheric phase error, $\varphi_{noise,i}(x, y)$ is the random noise phase.

According to Equation (1), a set of equations including M equations and N variables can be obtained [18]:

$$\delta_{\varphi} = A\varphi \tag{2}$$

where δ_{φ} is the matrix formed by M differential interferogram phase, φ is the matrix of the deformation phase to be determined by an image on N SAR. A is an $M \times N$ matrix, each row corresponds to an interference pair. When the number of small baseline subsets is one, A is the full rank matrix, and its least square solution is:

$$\hat{\varphi} = [(A^T A)^{-1} A^T] \cdot \delta_{\varphi} \tag{3}$$

When the number of small baseline subsets is greater than 1, A is a rank-deficient matrix and the rank deficient is $N - L + 1$. $A^T A$ is a singular matrix, and there are infinite solutions to the equation. To solve this problem, the equation must be transformed into the phase average change rate in the sense of the minimum norm, and the pseudo-inverse matrix can then be obtained by the singular value decomposition method [43]. Let the parameter to be solved be expressed as:

$$v = \left[\frac{\varphi_1 - \varphi_0}{t_1 - t_0}, \cdots, \frac{\varphi_N - \varphi_{N-1}}{t_N - t_{N-1}} \right]^T \tag{4}$$

where the value of φ_0 is zero. It can be further obtained as:

$$\delta_{\varphi_i} = \sum_{i=tS_i+1}^{tM_i} v(t_i - t_{i-1}), \quad i = 1, \cdots, M \tag{5}$$

where tM_i and tS_i respectively represent the acquisition time of the master and slave image corresponding to the interferogram i. For M differential interferogram, equations can be formed:

$$B \cdot v = \delta_\varphi \tag{6}$$

where, B is a $M \times N$ matrix corresponding to the unknown vector. For each row element in B, the value of $B(j,i)$ is $t_{i+1} - t_i$. Everything else in the matrix is 0. Subsequently, singular value decomposition was performed on matrix B, the value of which is USV^T [44], where U is an $M \times N$ orthogonal matrix, S is a $M \times N$ diagonal matrix, and V is an $N \times N$ orthogonal matrix. According to least squares, the phase change rate can be expressed as:

$$v = VS^{-1}U^T\delta_\varphi \tag{7}$$

3.4. GPS Processing Strategy

In order to obtain the results of high-precision positioning of GNSS data, we used the GAMIT/GLOBK10.7 software package developed by the Massachusetts Institute of Technology (MIT) and Scripps Institution of Oceanography. GAMIT/GLOBK software has the characteristics of high precision, is a free and open source for scientific research institutes, has continuous upgrading and updating, and has been widely used in high-precision GNSS data processing at home and abroad [45]. Using precise ephemeris and high precision starting points, the relative accuracy of the baseline solution can reach about 10 mm, and the accuracy of the short baseline solution can be better than 1 mm [46]. The data processing procedure of GPS measurements include observation data preprocessing, GPS baseline calculation, GPS network adjustment, and extraction of GPS station elevation change information [47]. Firstly, prior to high-precision processing of GPS data, preprocessing the original GPS observation data was necessary. We used GPS observation data quality analysis software (TEQC) to analyze the quality of GPS original observation data. Secondly, we converted the qualified GPS data into RINEX standard format data as the input data of the high-precision GPS data processing software GAMIT/GLOBK. Thirdly, combined with the data of the IGS global tracking station, we used GAMIT/GLOBK 10.7 software to calculate the baseline of GPS survey data, and adopted the double difference relative positioning mode for baseline calculation [48]. In addition, in order to ensure the accuracy of GPS baseline calculation, we used the IGS precise ephemeris and atmospheric correction model to eliminate errors in orbit and signal propagation and ensure the accuracy of GPS baseline calculations. The adjustment principle of GLOBK software can be found in references [49,50]. Fourthly, we obtained the three-dimensional coordinates of the frame points with the high-precision GPS elevation change monitoring software package (HPGP-SADJ). In addition, the International terrestrial Reference Frame (2008) was adopted in GPS monitoring of land subsidence in Beijing from 2009 to 2020 to reduce the error caused by coordinate conversion calculation. Finally, based on the network adjustment results of GPS measurement points, and the GPS monitoring results, we obtained the settlement amount and development trend of each GPS measurement point of land subsidence in Beijing.

3.5. Dataset Calibration for Different Platforms

While processing ENVISAT ASAR and Sentinel-1A/1B data using SBAS-InSAR technology, selecting the same coordinates of unwrapped reference points of the two datasets is necessary. The specific operational process for processing involves transforming the same selected geographical coordinates into phase coordinates. Due to the deformation information captured by ENVISAT and Sentinel in different spatial regions, it was necessary to extract overlapping regions of the two datasets. To obtain a development trend of land subsidence over the period 2003 to 2020, we projected LOS direction deformation from different sensors to the vertical direction and geocoded them in the same geographic coordinate system. By using continuous GPS measurement and quadratic polynomial interpolation, we referenced two measurement data to the same location and corrected the offset in time to obtain GPS and InSAR observation data in the same temporal and spatial reference frame.

Before calibration, due to differences in the incident angle and resolution of different satellite data, land subsidence in the Beijing area was mainly vertical, and the deformations in the east–west and south–north directions were relatively small [51]. Therefore, the LOS deformation results can be projected in the vertical direction using the following formula:

$$V_{Vertical} = \frac{V_{LOS}}{cos\theta} \quad (8)$$

where θ is the SAR satellite incidence angle. Notably, before the LOS deformation results are projected, we need to resample the deformation acquired by different sensors to have the same spatial resolution.

4. Results

4.1. Spatiotemporal Characteristics of Land Subsidence and Reliability Analysis

To obtain land subsidence information in Beijing from 2003 to 2020, we processed SAR images from two different datasets employing SBAS-InSAR technology to obtain the annual average deformation rate chart and cumulative deformation chart along the line of sight (LOS) in Beijing during 2003–2010 and 2015–2020 (Figure 4). Figure 4a,b show annual average deformation rate during 2003–2010 and 2015–2020, whereas Figure 4c,d show the cumulative subsidence during 2003–2010 and 2015–2020. The deformation result was positive when the ground objects moved closer to the satellite. Conversely, the deformation result was negative when the ground object moved away from the satellite. Here, the positive value in blue indicates movement towards the satellite (representing lifting), and the negative value in red indicates movement in a direction away from the satellite (representing deformation). Due to the deformation information being captured by ENVISAT and Sentinel in different spatial regions, it was necessary to extract overlapping regions of the two datasets.

The spatial evolution process indicated that the distribution of land subsidence in Beijing was not uniform. The deformation in the central Beijing plain area was relatively small, and the land subsidence in the northeast was more severe. Subsidence was mainly observed in Changping, Haidian, Chaoyang, Shunyi and Tongzhou districts. Chaoyang, located in the northeast of Beijing, and Tongzhou showed the most severe subsidence. Furthermore, the subsidence of Chaoyang and Tongzhou districts tends to be contiguous. In addition, the deformation rate in Beijing during 2003–2010 ranged from −128 mm/year to 3 mm/year, and the maximum cumulative deformation was −808 mm, which was located in Chaoyang. The average deformation rate in Beijing during 2015–2020 ranged from −135 mm/year to 12 mm/year, with the maximum cumulative deformation being −734 mm. The maximum cumulative deformation was also observed in Chaoyang, Beijing.

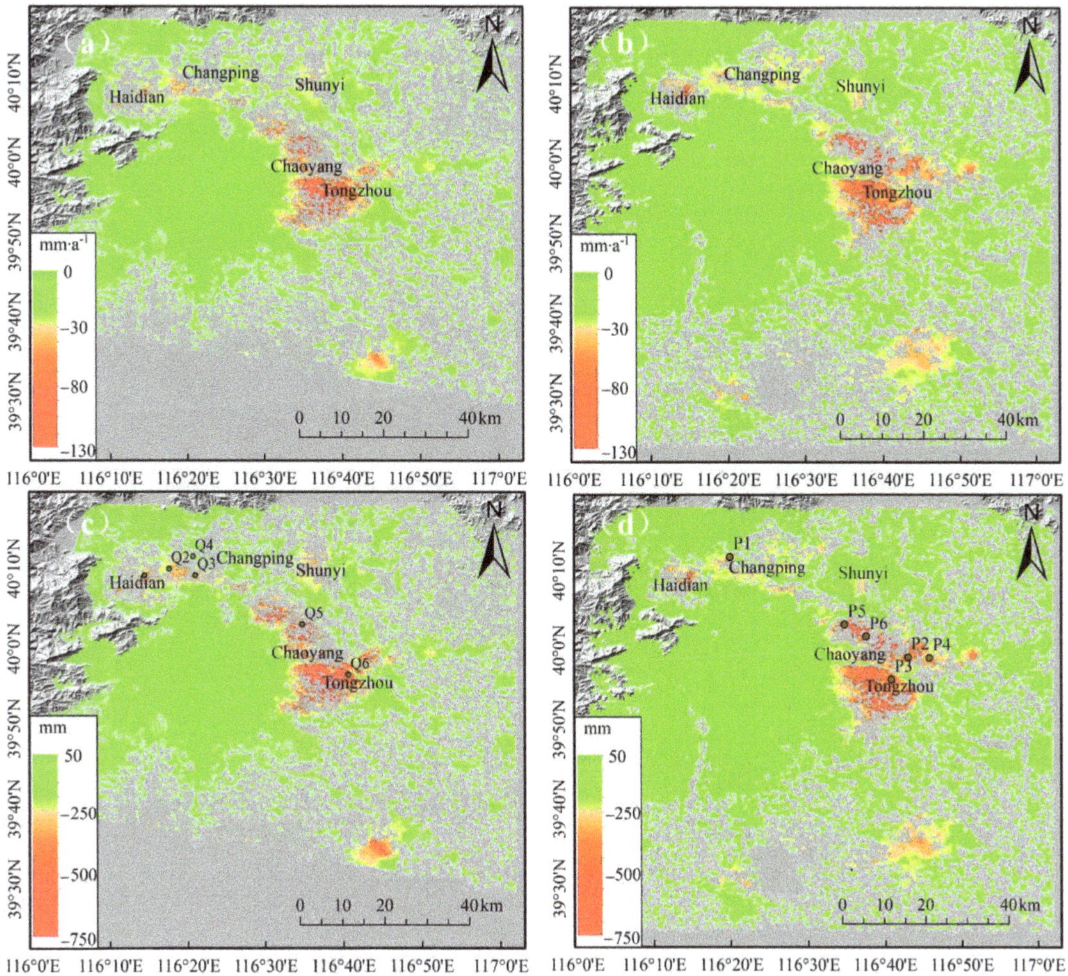

Figure 4. Deformation results of land subsidence along the LOS in Beijing during 2003–2010 and 2015–2020. Mean deformation velocity during 2003–2010 (**a**) and 2015–2020 (**b**). Cumulative deformation during 2003–2010 (**c**) and 2015–2020 (**d**). Note: Q1–Q6 (**c**) and P1–P6 (**d**) represent the extracted deformation feature points in typical subsidence area. The black marks in the figure are mainly deformation zones. The colored dots from green to red represent InSAR values.

In addition, to verify the reliability and accuracy of the deformation results, 35 GPS measurements were collected for verification. Due to the InSAR deformation result mainly reflecting deformation along the LOS direction [52], according to the heading and incident angle of the pixels corresponding to the GPS point, the GPS deformation result was uniformly projected to the LOS direction for checking of accuracy. GPS deformation in the line of sight can be expressed as

$$d_{\text{GPS}} = \sin\theta \times \sin\left(\alpha - \frac{3\pi}{2}\right) \times d_x + \sin\theta \times \cos\left(\alpha - \frac{3\pi}{2}\right) \times d_y + \cos\theta \times d_z \quad (9)$$

where, α is the satellite heading angle and θ is the radar incidence angle. d_x, d_y and d_z are the deformation of GPS in the three-dimensional direction, respectively.

There are three methods to verify the comparison between GPS measurement values and InSAR deformation results, which are point-to-point verification, point-to-surface verification, and point-to-line verification. In order to quantitatively evaluate the deformation monitoring accuracy obtained by InSAR and GPS, point-to-point verification was adopted to verify accuracy. In order to highlight the correlation of deformation between InSAR and GPS, we compared InSAR-derived deformation rates with GPS deformation results in the LOS direction (Figure 5). The difference between the deformation results of InSAR and GPS is less than 10 mm/a, and the root mean square error is 4.6 mm/a. In addition, the correlation between INSAR and GPS is high with a correlation value of about 0.9. The reliability of InSAR monitoring results is illustrated.

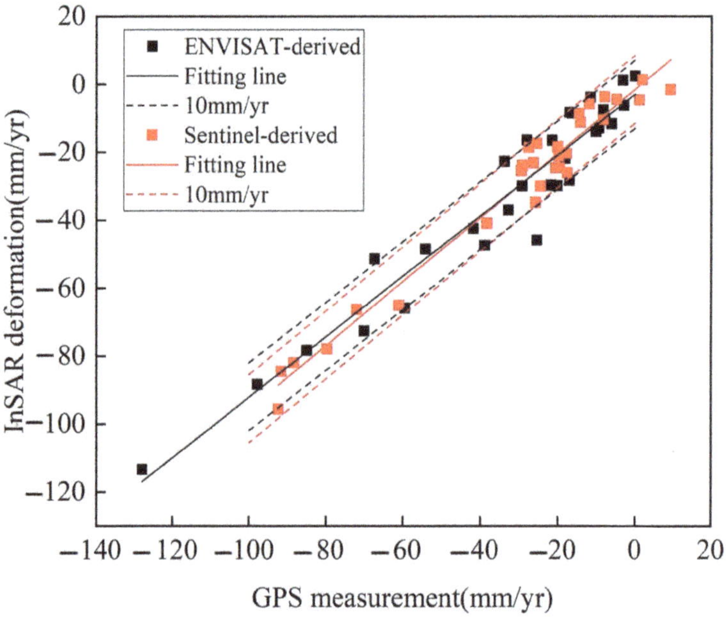

Figure 5. Correlation between GPS measurements and annual LOS deformation results of InSAR. Black scatter points are the ENVISAT-derived results, and black straight line is the line fitted by ENVISAT-derived results and GPS. Red scatter points are the sentinel-derived results, and red straight line is the line fitted by sentinel-derived results and GPS.

4.2. *Time Series Land Subsidence*

To better analyze the long-term evolution characteristics of land subsidence in Beijing, six deformation feature points at different positions were extracted from the subsidence funnel in the accumulated deformation charts for 2003–2010 and 2015–2020. The location distribution of each feature point are shown in Figure 4b,d. A time-series deformation analysis was performed on the extracted feature points and the corresponding results are shown in Figure 6. Moreover, since only Gaussian noise is permitted in deformation time series, it is necessary to correct atmospheric error, orbit error and DEM error phase.

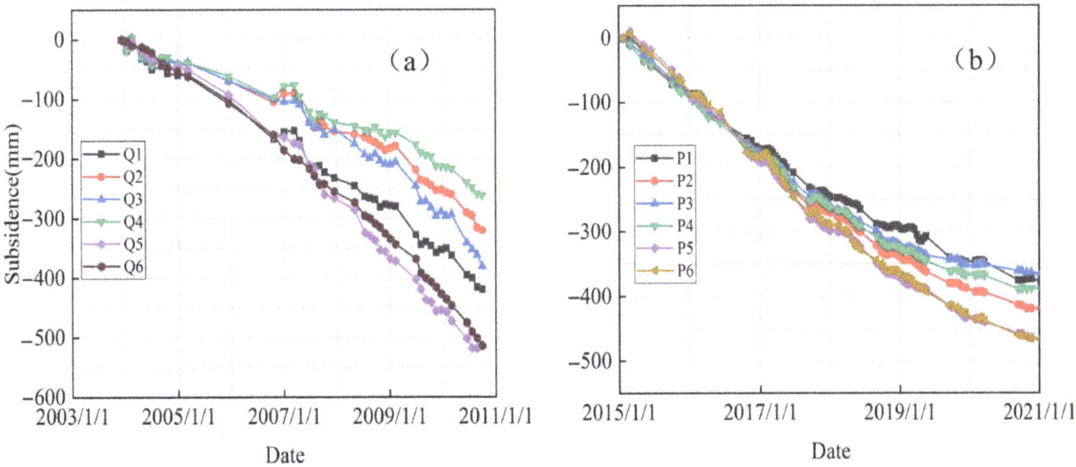

Figure 6. Time-series deformation of typical feature points from (**a**) 2003–2010 and (**b**) 2015–2020.

The time-series of the extracted feature points during 2003–2010 and 2015–2020 showed nonlinear subsidence at different velocities (Figure 6). From 2003 to 2010 (Figure 6a), land subsidence along the LOS in Beijing continued to increase, and its land deformation rate also increased, with the maximum accumulative deformation amount reaching −514 mm. The deformation was relatively small from 2003 to 2005. After 2006, the amount of land subsidence in Beijing increased gradually. Further, five observed deformation funnels expanded over approximately seven years in scale and number. The largest settlement center was located in Chaoyang. In addition, the time-series deformation diagram (Figure A1 in Appendix A) further demonstrates the formation process of settlement funnels from 2003 to 2010. From 2015 to 2020 (Figure 6b), land subsidence continued to increase. The subsidence in Chaoyang was the most severe, with the maximum accumulative deformation reaching −472 mm. The deformation changed slightly from 2015 to 2016, and then gradually increased from 2017 (Figure A2). Although the accumulated subsidence amount in Beijing continued to increase from 2005 to 2020, land subsidence showed a slowing trend from 2015 to 2020. In analyzing the time-series of each feature point, the time-series subsidence of Changping, Haidian, and Tongzhou, corresponding to the feature points Q1, Q2, Q3, Q4, P4, and P6, showed obvious seasonal characteristics. Considering the time series of the Q1 feature point as an example, land subsidence showed a rebound trend from October 2006 to October 2007, and subsequently continued to sink.

The development process of land subsidence in Beijing included three deformation stages: initial subsidence, serious subsidence and then gradual stability [53]. In the 1930s, land subsidence was found in Xidan and Dongdan of Beijing. With the expansion of industry in the city and over-exploitation of groundwater, several new settlement centers, such as Changping and Shunyi, emerged in Beijing. At the same time, problems, such as foundation settlement, buildings cracking and environmental pollution also occurred. It brought great financial loss and harm to urban construction and peoples' lives. In the case of vertical deformation, in order to analyze the development and change of land deformation in Beijing from 2003 to 2020, we projected LOS direction deformation from different sensors to the vertical direction and geocoded them in the same geographic coordinate system. In order to compare the time-series displacements of InSAR inversion with those from GPS, we generated a buffer 100 m radius around the GPS reference point. On this basis, we obtained the average displacement of all slow varying filtered phase (SFP) pixels in the buffer. After referring the two measurements to the same location, and correcting the offset in time, we obtained GPS and InSAR observations in the same temporal and spatial reference frame. Finally, long-time and high-time resolution deformation time series were generated. The specific strategies were as follows:

The time-series results of the ENVISAT ASAR, GPS, and Sentinel-1A/1B data are shown in Figure 7. ENVISAT ASAR and Sentinel-1A/1B satellites have different incident Angles, spatial resolutions, and selected reference points. To address the problem of different reference points, we selected the same coordinate reference points during the SBAS-InSAR process. The ENVISAT ASAR and Sentinel-1A/1B data had incident angles of 22.8° and 33.9°, respectively. The deformation results obtained from the ENVISAT ASAR and Sentinel-1A/1B data were mainly vertical. Therefore, directly projecting the LOS deformation results obtained from ENVISAT ASAR and Sentinel-1A/1B data in the vertical direction was appropriate. Moreover, we resampled the deformation results obtained from the two datasets to 30m to ensure consistency of their spatial resolution. In this study, the time-series of the two feature points P3 (Figure 8a) and P5 (Figure 8b) were selected for calibration. Due to the fact that the ENVISAT satellite stopped operating around 2012 and the Sentinel satellite began operating in 2014, there is no temporal overlap between the two satellites. It was impossible to directly calibrate the time series deformation results obtained from ENVISAT ASAR and sentinel-1a/1b data. Therefore, we used GPS data and a polynomial fitting algorithm to calibrate the deformation results. The time-series deformations of the ENVISAT, Sentinel-1A/1B, and GPS data were referenced to the same position, and corrected and offset temporally. For the time-series deformation results obtained from InSAR, we selected deformation points within 100 m of the GPS monitoring points. ENVISAT ASAR and Sentinel-1A/1B data acquired on 10 December 2003 and 24 January 2015, respectively, were used as temporal data. Both had different temporal datum for obtaining deformation results; therefore, the temporal datum of ENVISAT ASAR and Sentinel-1A/1B had to be unified. The GPS data used for calibration were based on the image from 17 October 2009. Thus, long time-series deformation was acquired (as shown in Figure 8a,b) after rectifying the deformation results from 2003 to 2010 and 2015 to 2020. In the same time period, we found that, although the InSAR observation method is not strictly registered in time with GPS, its deformation trend was consistent. However, the time series was not continuous. To acquire the missing data from 2010 to 2015, we adopted quadratic polynomial fitting interpolation and a mean algorithm. The resultant deformation results for the long time-series are shown in Figure 8c,d. We used continuous GPS time series results to analyze the obtained results. It was found that the deformation trend of this result was consistent with GPS. In addition, the calibrated long time-series was consistent with the three deformation stages of land subsidence evolution in Beijing. This was mainly demonstrated through rapid land subsidence after the over-exploitation of groundwater in Beijing. As the South-to-North Water Transfer Project was implemented at the end of 2014, the rate of land subsidence gradually decreased, and gradually became stable. The maximum accumulative deformation amount at P5 reached −1368 mm. In addition, on comparing GPS deformation results from 2009 to 2020 with those of ENVISAT and Sentinel-1A/1B fused data, the temporal sequence results after supplementing were found to be consistent with GPS trends through visual interpretation. By observing the data supplemented by the quadratic polynomial fitting method and the mean value algorithm, the inflection point of land subsidence in Beijing was found to be located in the corresponding time span of the supplemented data. Quantitatively, the inflection point occurred around 2015. The result obtained by the Mann-Kendall method was consistent with the analysis results (Figure 9). The extent and magnitude of land subsidence are related to geological structures [54–56]. Since geological structural changes occur over a certain time period, the magnitude of land subsidence does not change quickly. However, the rate of land subsidence can change with changes in groundwater [57,58]. The South-to-North Water Transfer Project significantly reduced the amount of groundwater exploitation. Thus, based on the analysis of the actual scenarios, the inflection point could have occurred due to the implementation of this project, which slowed the deformation rate of land subsidence to a certain extent.

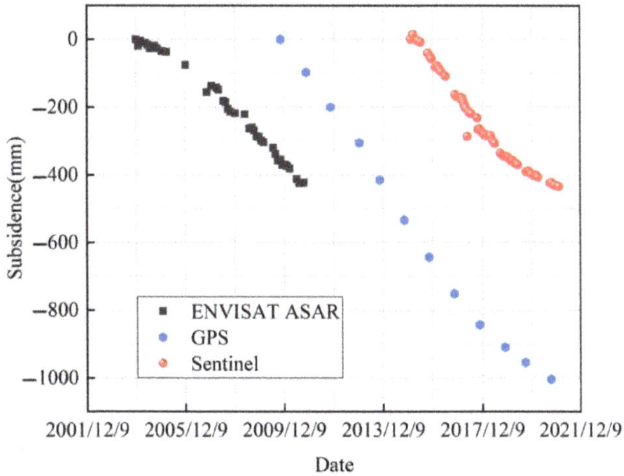

Figure 7. Time-series deformation results of typical feature points from 2003 to 2010.

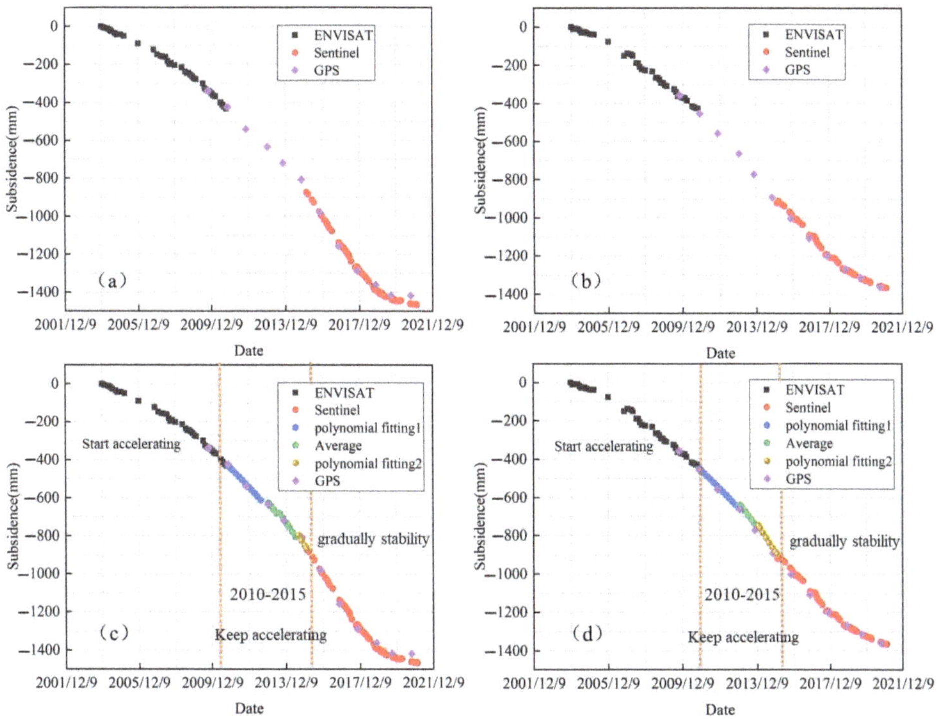

Figure 8. Comparison of the corrected time series settlement with GPS observation results: (**a**,**b**) are the corrected results of time series deformation at P3 station and P5 station, respectively. (**c**,**d**) are the supplemented results at P3 station and P5 station during 2010–2015.

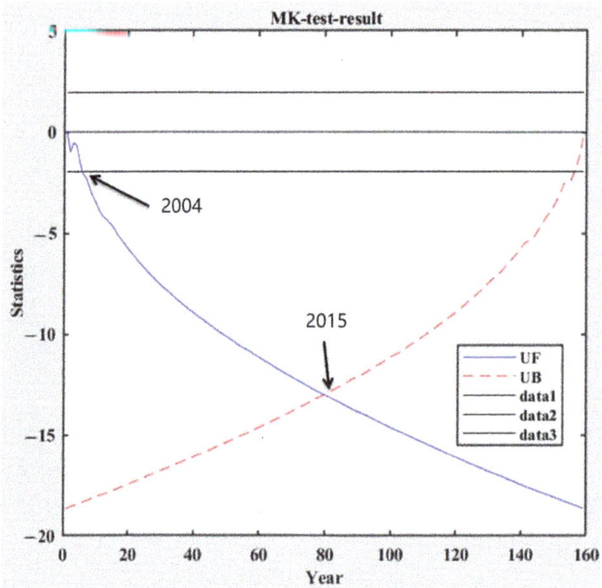

Figure 9. Mann-Kendall test result.

5. Discussion

5.1. Relationship between Land Subsidence and Land Use Type in Beijing during 2003–2020

Land subsidence is a slow time-dependent subsidence process, which may have a slow and correlated response to land use and cover [59,60]. In order to explore the relationship between surface deformation and land use types in Beijing, we used previously obtained information of surface deformation rate from 2003 to 2020 to explore the evolutionary relationship between surface deformation and land use types in Beijing, based on land cover data.

With rapid urbanization, the proportion of construction land in Beijing is gradually increasing, whereas that of cropland is decreasing. To analyze the changes in land use types in Beijing over a long period of time we acquired the distribution data of land use types in Beijing from 2005 to 2020 over a five-year interval, and used ArcGIS software to perform simple operations, such as reclassification and superposition analysis. Figure 10 shows the land use types in Beijing from 2005 to 2020.

According to the distribution maps of land use types in 2005, 2010, 2015 and 2020, change of land use type in Beijing was obvious during the study period, especially the change in construction land. This type of land use is prone to land subsidence, because the loads of construction land and human activities lead to subsidence of construction land area [61]. The specific magnitude will be explained later.

In order to display the magnitude of change of land use types in detail during the study period, we calculated the area proportion of each land use type, and the corresponding statistical results are shown in Table 2. As time goes on, we found that construction land expanded in space, while cropland continued to decrease. The proportion of construction land increased by 5%, i.e., from 16.7% in 2005 to 21.8% in 2020. Further, during 2005–2020, the proportion of cropland decreased, whereas that of other land use types remained unchanged.

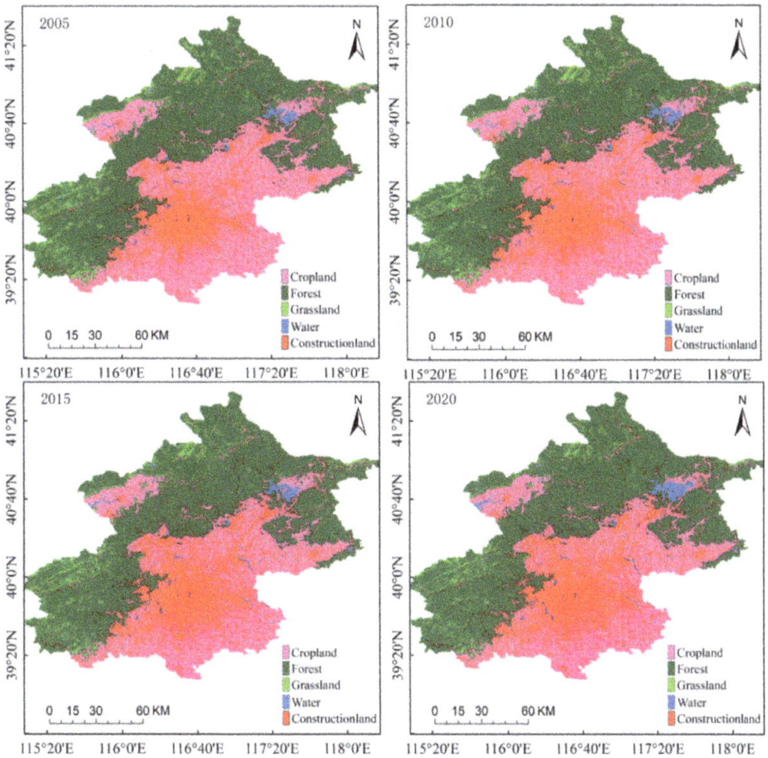

Figure 10. Land use types in Beijing from 2005 to 2020.

Table 2. Statistical results of different land types in Beijing (2005–2020).

Land Use and Cover	Proportion of the Corresponding Area			
	2005	2010	2015	2020
Constructionland	16.7%	19%	21.1%	21.8%
Cropland	31.2%	28.6%	26.5%	25.6%
Forest	47.6%	47.8%	48.4%	48.8%
Grassland	3.4%	3.6%	3%	2.4%
Water	1%	1%	1%	1.4%

Superposition analysis is a very important spatial analysis function in GIS. Under the unified spatial reference system, in order to analyze the relationship between evolution of land use type and land subsidence, we superimposed land use type data with land subsidence results from InSAR based on the 2005–2020 land use type data, having a resolution of 30 m. The land subsidence areas were mainly located in Changping, Tongzhou, Shunyi, Haidian and Chaoyang. Therefore, we analyzed the proportion of each land use type in these five land subsidence areas. The corresponding results are shown in Figure 11. The proportion of forest, grassland, and water in each of the five land subsidence areas was relatively small from 2005 to 2020, while that of cropland was the highest in 2005. However, over time, the proportion of cropland gradually decreased, and building density continued to increase, resulting in the proportion of construction land in each land subsidence area increasing, with the highest proportion observed in Tongzhou and Changping. The settlement is mainly concentrated in construction land, and is pretty small in other land use types. Therefore, land subsidence is related to land use type. Compared with other land use types, construction land is more prone to land subsidence.

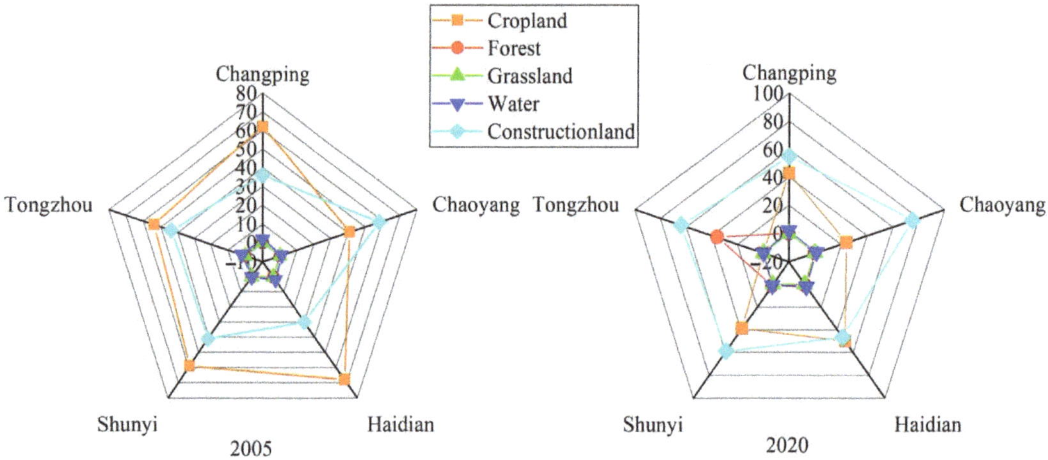

Figure 11. Proportion of land use types in the five subsidence areas of Changping, Haidian, Shunyi, Tongzhou, and Chaoyang.

Since changes in forest, grassland and water in the land subsidence area were pretty small, in order to better analyze the development and change of land subsidence from 2003 to 2020, we focused on the area of construction land and subsidence area in the study area during 2003–2020; the extent of change is shown in Figure 12. We also calculated the corresponding number of pixels. From 2005 to 2010, the proportion of construction land (the pixel count increased from 3,076,955 to 3,522,020) and land subsidence both showed an upward trend, and the change in both showed a positively correlation. It shows that construction land has an impact on land subsidence. Previous studies have proven that overexploitation of groundwater was the major cause of land subsidence in the Beijing plain. Groundwater is distributed in confined aquifers. Overexploitation of groundwater makes it difficult for the aquifer to bear the weight of its upper layer. Solari et al. [62] found that building load is a factor promoting high displacement rate. In the case of overexploitation of groundwater, the large-scale construction of buildings aggravates the rate of land subsidence. Therefore, the impact of construction land on land subsidence is closely related to massive exploitation of groundwater. After 2015, although the area of construction land continued to increase (the pixel count increased from 3,883,157 to 4,003,512), the proportion of land subsidence area decreased, which may be largely related to the South-to-North Water Transfer project, implemented at the end of 2014. In addition, Yang et al. [30] found that the use of long piles in high-rise buildings can effectively reduce the settlement, and increase of building load does not cause the same proportion of ground settlement. After effective measures were adopted in construction, the impact of construction land on land subsidence in Beijing was relatively small or even nothing. The reasons for the change of land subsidence1 rate with groundwater level will be explained in the next chapter.

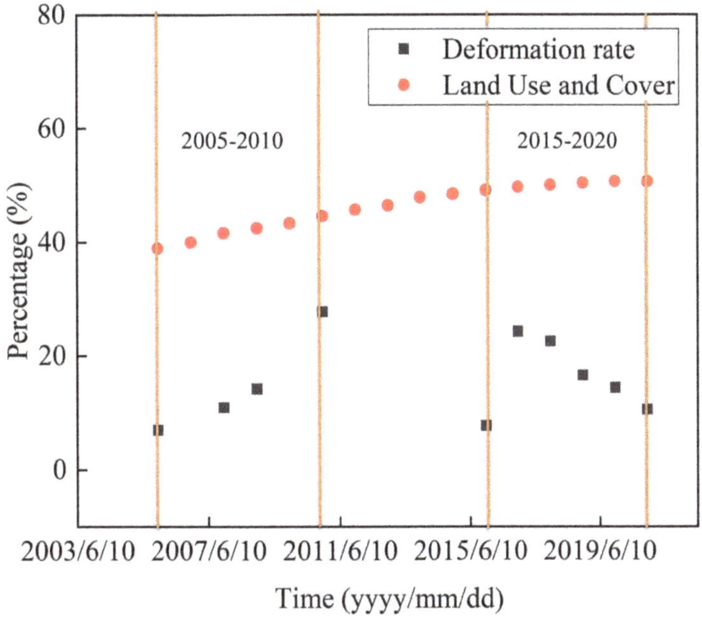

Figure 12. Change extent of construction land and deformation.

5.2. Relationship between Land Subsidence and Groundwater in Beijing during 2003–2020

According to previous studies, land subsidence and groundwater in Beijing are closely related [63]. We obtained groundwater depth and exploitation data from 2003 to 2020 from the Beijing Water Resources Bulletin [64], provided by the Beijing Water Authority. The corresponding land subsidence results were also analyzed. The results are shown in Figure 13.

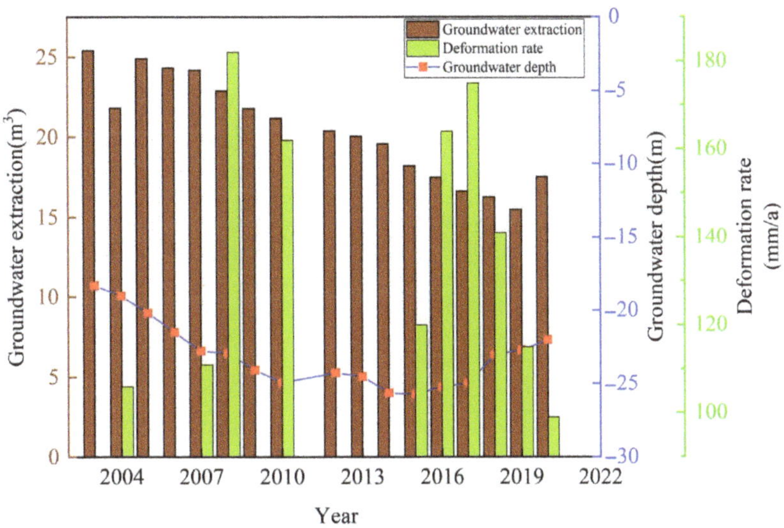

Figure 13. Evolutionary trend of groundwater data and deformation results from 2003 to 2020.

As shown in Figure 13, groundwater depth continued to increase from 2003 to 2010, and the annual average groundwater depth from 2013 to 2020 showed an important turning point around 2015; mainly reflected in slowing down of groundwater depth and a downward trend around 2015. Further, groundwater exploitation gradually decreased from 2003 to 2020, while the land subsidence rate increased between 2003 and 2010. Although groundwater exploitation decreased during this period, the decrease was not significant. Thus, the conclusion is that extreme exploitation of groundwater may have aggravated land subsidence. The changes in land subsidence rate from 2003 to 2015 in Beijing were not regular, but showed an overall increasing trend. The land subsidence rates were 106 mm/year in 2004 and 120 mm/year in 2015, and the deformation rate in 2015 was 14 mm/year, higher than that in 2004. To solve the problem of excessive groundwater exploitation caused by large-scale demand for water resources by residents, China implemented the South-to-North Water Transfer Project at the end of 2014. After the middle route of the project was implemented, the demand for residential water shifted from groundwater exploitation to south-to-north water diversion; thus, preventing land subsidence to a certain extent. During 2015–2020, the land subsidence rate in 2020 was 99 mm/year, which was 21 mm/year lower than that in 2015. Although the land subsidence rate increased from 2015 to 2017, it showed a declining trend after 2017, whereas an overall decreasing trend was observed from 2015 to 2020.

Figure 14 shows monthly precipitation deformation during the study period. In Figure 14, it can be observed that rainfall is mainly concentrated in the summer months of May-August, and the rainfall in summer is higher than that in other seasons. On this basis, we combined surface deformation with rainfall trends and analyzed the relationship between them. During the concentrated rainfall period, there is an obvious rebound deformation phenomenon along the LOS direction. We discovered that higher rainfall from May to August 2006 reduced the need to use groundwater for farm irrigation, and, thus, land subsidence rate showed a gradually slowing trend from October 2006 to March of the following year. Meanwhile, rainfall in autumn and winter of the same year decreased, so that land subsidence in the spring of the following year showed intensification. This indicates that peoples' demand for groundwater decreases after rainfall, which slows down land subsidence to a certain extent. Ground settlement is negatively correlated with precipitation.

5.3. Relationship between Land Subsidence and Geological Structures and Faults in Beijing

Beijing is located at the intersection of Yanshan, North China Plain, and Taihang Mountain fault. Due to its complex tectonic environment and active neo-tectonic movement, many active fault zones, including Huangzhuang-Gaoliying and Nankou-Sunhe in the northeast, Shunyi-Qianmen-Liangxiang, and Nanyuan-Tongxian, have been formed in the boundary area [65]. The sediments in Beijing are mainly distributed in the Beijing Plain and vary in thickness from a few meters to several hundred meters. The presence of a compressible layer is the elementary criterion of ground subsidence. The lithologic conditions and structural characteristics of Beijing provide an inner geological background for land subsidence, which mainly occurs in the quaternary strata. The sedimentary time, genetic type, lithology and thickness of the quaternary strata affect land subsidence [66]. Previous studies have shown that when the groundwater level declines in Beijing, the pore water pressure of the clayey aquifer decreases, and the effective stress in the soil increases, thus resulting in compression deformation and land subsidence [67].

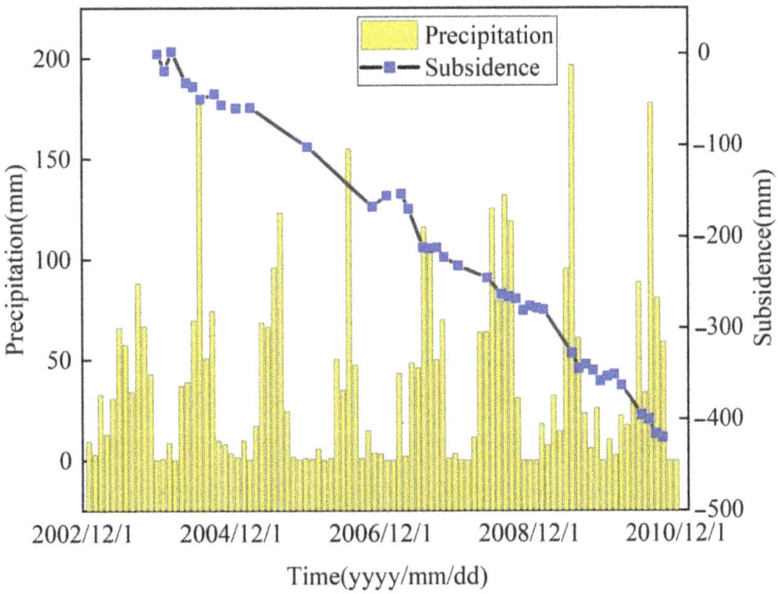

Figure 14. Changes in precipitation from 2003 to 2010.

Section 4.2 shows the feasibility of using a quadratic polynomial fitting algorithm and mean algorithm to supplement missing data required for land subsidence analysis. It not only reduces the cost of using other expensive satellite data, such as ALOS-2 and Radarsat-2, but also successfully reflects the changing trend of land subsidence in Beijing from 2010 to 2015. Moreover, the result obtained by the Mann-Kendall method was in good agreement with the analysis results. However, the quadratic polynomial fitting algorithm and mean algorithm are relatively simple. Therefore, it is necessary to combine groundwater change and the subsidence model to connect long-term time-series deformation in the future.

6. Conclusions

Using Beijing as the study area, we studied spatiotemporal evolution characteristics and their influencing factors for the time period from 2003 to 2020. First, 40 ENVISAT images and 90 Sentinel-1A/1B images were selected, and SBAS-InSAR technology was used to acquire LOS deformation rate charts and accumulated deformation charts during 2003–2010 and 2015–2020 in Beijing. We then analyzed the spatiotemporal variation characteristics of land subsidence during 2003–2010 and 2015–2020. This study provides significant deformation information for the Beijing area. Furthermore, the results of this study have important reference significance for stability analysis, large-scale engineering construction, urban planning and comprehensive utilization of land resources in Beijing. We can draw the following main conclusions from this study. Firstly, the distribution of land subsidence in Beijing was found to be uneven. The settlement is mainly located in the northeast and more stable in the middle. In the former stage, the maximum deformation rate was −128 mm/year, and the maximum cumulative deformation amount was −808 mm, which is located in Chaoyang. In the latter stage, the maximum deformation rate in Beijing was −135 mm/year, and the maximum cumulative deformation amount reached was −734 mm. Using GPS measurements, we demonstrated the reliability of our InSAR annual subsidence results. Secondly, modified surface deformation results were obtained by using the quadratic polynomial fitting method combined with GPS data to report on the evolutionary characteristics of surface subsidence during 2003–2020. The reported results are consistent with those obtained by the Mann-Kendall method. On the other hand, it showed that, although the InSAR observation method is not strictly registered with GPS in time, the deformation trend was consistent, which further demonstrated the reliability of InSAR monitoring results. In addition, the calibrated long time-series was consistent with the three deformation stages of land subsidence evolution in Beijing. Thirdly, compared with other land use types, construction land is more prone to land subsidence and its influence on land subsidence is closely related to overexploitation of groundwater. Fourthly, overexploitation of groundwater can aggravate land subsidence. The implementation of the middle line project at the end of 2014 prevented land subsidence to a certain extent. In addition, during the concentrated rainfall period, there is an obvious rebound deformation phenomenon along the LOS direction. It shows that the seasonal variation in deformation is related to precipitation. Finally, geological factors show that when groundwater level declines in Beijing, the pore water pressure of the clayey aquifer decreases, and the effective stress in the soil increases, thus resulting in compression deformation and land subsidence. Therefore, land use type and precipitation have little influence on land subsidence. The change of groundwater level is the main influence factor of land subsidence.

Author Contributions: Conceptualization, S.Z. and Y.Z.; methodology, S.Z. and Y.Z.; software, Y.Z.; validation, Y.Z., J.Y., Q.F. and J.S.; writing—original draft preparation, Y.Z. and S.Z.; writing—review and editing, Q.F., J.S. and W.Z.; supervision, M.S. All authors have read and agreed to the published version of the manuscript.

Funding: This research was funded by the National Key Research and Development Program of China (2019YFC1509802, 2020YFC1512000); the National Natural Science Foundation of China (42074041,41731066); State Key Laboratory of Geo-Information Engineering (SKLGIE2019-Z-2-1); Shaanxi Natural Science Research Program (2020JM-227).

Data Availability Statement: ENVISAR ASAR datasets provided in the article is available at http://esar-ds.eo.esa.int/socat/ASA_IMS_1P, accessed on 12 January 2022. Sentinel-1A/1B data provided in the article is available at https://vertex.daac.asf.alaska.edu/, accessed on 20 January 2022. DEM provided in the study is available at https://earthexplorer.usgs.gov/, accessed on 12 January 2022. Groundwater provided in the article is available at http://swj.beijing.gov.cn/, accessed on 10 February 2022. Precipitation provided in the article is available at https://www.ncei.noaa.gov/maps-and-geospatial-products, accessed on 10 February 2022.

Acknowledgments: The author would like to thank ESA for providing sentinel-1A/1B, ENVISAT ASAR data. Thanks to NASA for the SRTM DEM elevation data. Thanks to Beijing Water Authority for the data of groundwater exploitation and groundwater depth. Thanks to GEE platform for providing Landsat data and land use and classification data. Thanks to NOAA for providing precipitation data.

Conflicts of Interest: The authors declare no conflict of interests.

Appendix A

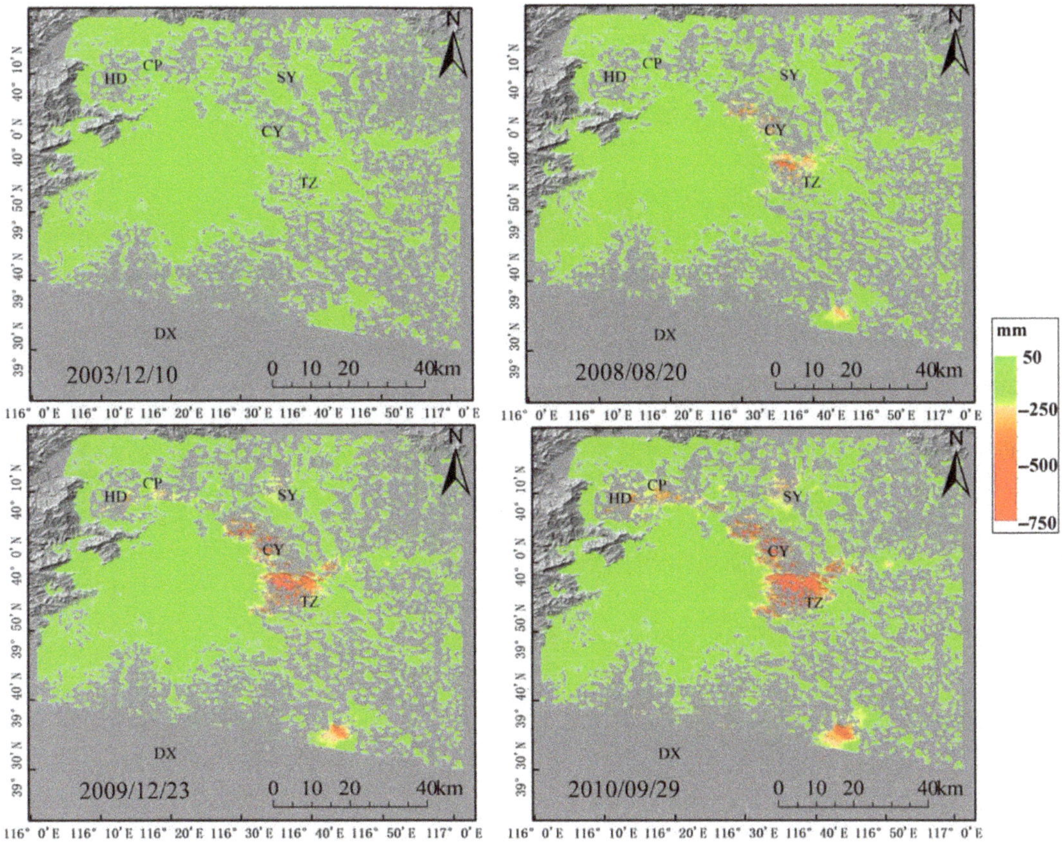

Figure A1. Cumulative time-series deformation in LOS from 2003 to 2010. CY: Chaoyang District, CP: Changping District, HD: Haidian District, SY: Shunyi District, and TZ: Tongzhou District.

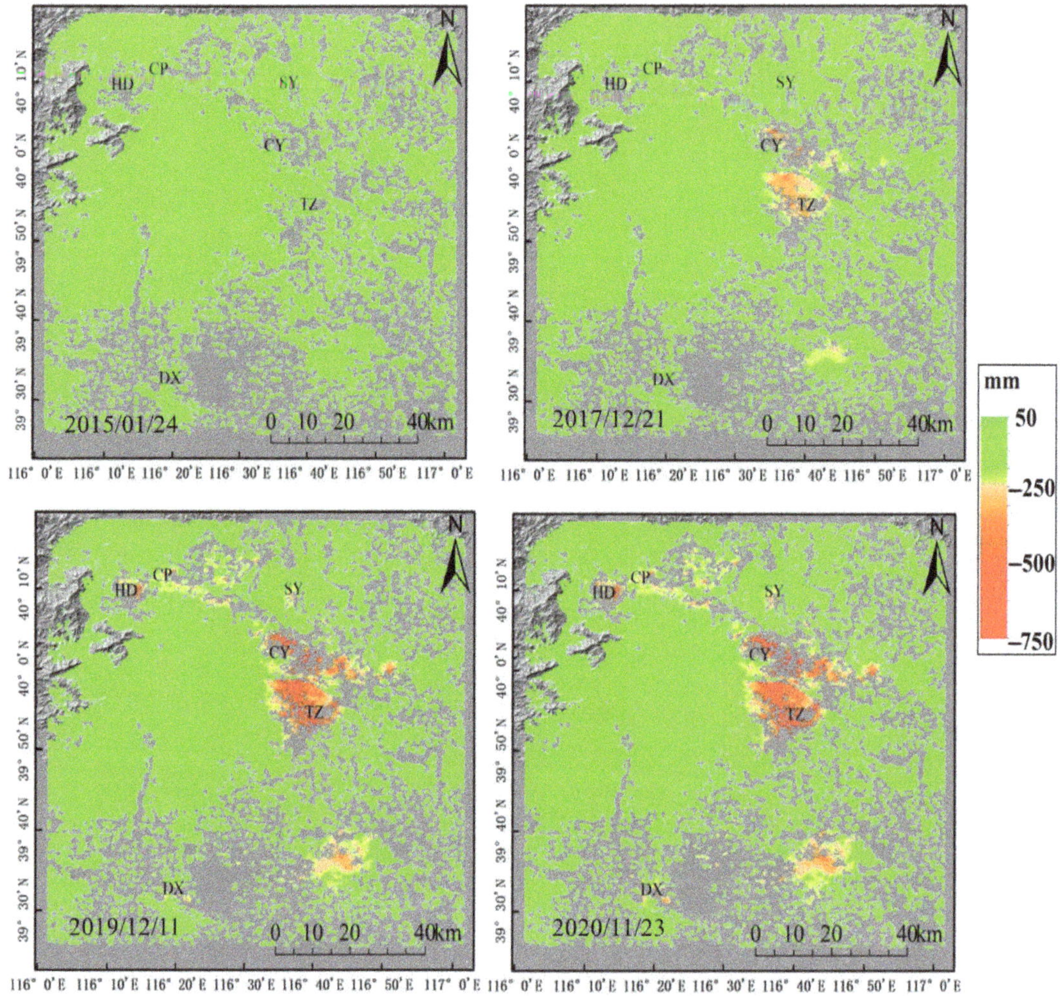

Figure A2. Cumulative time-series deformation along the LOS from 2015 to 2020. CY: Chaoyang District, CP: Changping District, HD: Haidian District, SY: Shunyi District, and TZ: Tongzhou District.

References

1. Xu, Y.S.; Shen, S.L.; Cai, Z.Y.; Zhou, G.Y. The state of land subsidence and prediction approaches due to groundwater withdrawal in China. *Nat. Hazards* **2008**, *45*, 123–135. [CrossRef]
2. Chen, B.B.; Gong, H.L.; Li, X.J.; Lei, K.C.; Gao, M.L.; Zhou, C.F.; Ke, Y.H. Spatial–temporal evolution patterns of land subsidence with different situation of space utilization. *Nat. Hazards* **2015**, *77*, 1765–1783. [CrossRef]
3. Zhang, Y.Q.; Gong, H.L.; Gu, Z.Q.; Wang, R.; Li, X.J.; Zhao, W.J. Characterization of Land Subsidence Induced by Groundwater Withdrawals in the Plain of Beijing City, China. *Hydrogeol. J.* **2014**, *22*, 397–409. [CrossRef]
4. Bürgmann, R.; Hilley, G.; Ferretti, A.; Novali, F. Resolving Vertical Tectonics in the San Francisco Bay Area from Permanent Scatterer InSAR and GPS Analysis. *Geology* **2006**, *34*, 221–224. [CrossRef]
5. Osmanoğlu, B.; Sunar, F.; Wdowinski, S.; CabralCano, E. Time series analysis of InSAR data: Methods and trends. *Int. Soc. Photogramm. Remote Sens.* **2016**, *115*, 90–112. [CrossRef]
6. Yan, Y.; Doin, M.P.; Lopez-Quiroz, P.; Tupin, F.; Fruneau, B.; Pinel, V.; Trouvé, E. Mexico City subsidence measured by InSAR time series: Joint analysis using PS and SBAS approaches. *IEEE J. Sel. Top. Appl. Earth Obs. Remote Sens.* **2012**, *5*, 1312–1326. [CrossRef]
7. Hu, B.; Wang, H.S.; Jiang, L.M. Monitoring of the reclamation-induced ground subsidence in Macao (China) using the PSInSAR technique. *J. Cent. South Univ.* **2013**, *20*, 1039–1046. [CrossRef]

8. Li, Z.H.; Liu, J.N.; Xu, C.J. Error Analysis in InSAR Data Processing. *Geomat. Inf. Sci. Wuhan Univ.* **2004**, *29*, 72–76. (In Chinese)
9. Yan, W.; Ming-Sheng, L.; De-Ren, L.; Zi-Xin, W.; Zheng, F. Subsidence velocity retrieval from long-term coherent targets in radar interferometric stacks. *Chin. J. Geophys.* **2007**, *50*, 598–604. (In Chinese)
10. He, X.F.; Gao, Z.; Xiao, R.Y.; Luo, H.B.; Feng, C. Monitoring and analysis of subsidence along Lian-Yan railway using multitemporal Sentinel-1A InSAR. *Acta Geod. Et Cartogr. Sin.* **2021**, *50*, 600–611. (In Chinese)
11. Dai, K.; Liu, G.X.; Li, Z.H.; Li, T.; Yu, B.; Wang, X.W.; Singleton, A. Extracting Vertical Displacement Rates in Shanghai (China) with Multi-Platform SAR Images. *Remote Sens.* **2015**, *7*, 9542–9562. [CrossRef]
12. Perissin, D.; Wang, T. Time-Series InSAR Applications Over Urban Areas in China. *IEEE J. Sel. Top. Appl. Earth Obs. Remote Sens.* **2010**, *4*, 92–100. [CrossRef]
13. Heleno, S.I.; Oliveira, L.G.; Henriques, M.J.; Falcão, A.P.; Lima, J.N.; Cooksley, G.; Ferretti, A.; Fonseca, A.M.; Ferreira, J.P.; Fonseca, J.F. Persistent Scatterers Interferometry detects and measures ground subsidence in Lisbon. *Remote Sens. Environ.* **2011**, *115*, 2152–2167. [CrossRef]
14. Sousa, J.J.; Hooper, A.J.; Hanssen, R.F.; Bastos, L.C.; Ruiz, A.M. Persistent Scatterer InSAR: A comparison of methodologies based on a model of temporal deformation vs. spatial correlation selection criteria. *Remote Sens.* **2011**, *115*, 2652–2663. [CrossRef]
15. Wang, C.S.; Wang, X.S.; Xu, Y.P.; Zhang, B.C.; Jiang, M.; Xiong, S.T.; Zhang, Q.; Li, W.D.; Li, Q.Q. A new likelihood function for consistent phase series estimation in distributed scatterer interferometry. *IEEE Trans. Geosci. Remote. Sens.* **2022**, 1. [CrossRef]
16. Ferretti, A.; Prati, C.; Rocca, F. Permanent scatterers in SAR interferometry. *IEEE Trans. Geosci. Remote Sens.* **2001**, *39*, 8–20. [CrossRef]
17. Lu, Z.; Danskin, W. InSAR analysis of natural recharge to define structure of a groundwater basin, San Bernardino, California. *Geophys. Res. Lett.* **2001**, *28*, 2661–2664. [CrossRef]
18. Berardino, P.; Fornaro, G.; Lanari, R.; Sansosti, E. A New Algorithm for Surface Deformation Monitoring Based on Small Baseline Differential SAR Interferograms. *IEEE Trans. Geosci. Remote Sens.* **2002**, *40*, 2375–2383. [CrossRef]
19. Lanari, R.; Mora, O.; Manunta, M.; Mallorquí, J.J.; Berardino, P.; Sansosti, E. A small-baseline approach for investigating deformations on full-resolution differential SAR interferograms. *IEEE Trans. Geosci. Remote Sens.* **2004**, *42*, 1377–1386. [CrossRef]
20. Mora, O.; Mallorqui, J.J.; Broquetas, A. Linear and nonlinear terrain deformation maps from a reduced set of interferometric SAR images. *IEEE Trans. Geosci. Remote Sens.* **2003**, *41*, 2243–2253. [CrossRef]
21. Shi, M.; Gong, H.L.; Gao, M.L.; Chen, B.B.; Zhang, S.K.; Zhou, C.F. Recent Ground Subsidence in the North China Plain, China, Revealed by Sentinel-1A Datasets. *Remote Sens.* **2020**, *12*, 3579. [CrossRef]
22. Yuan, M.Z.; Li, M.; Liu, H.; Lv, P.; Li, B.; Zheng, W. Subsidence Monitoring Base on SBAS-InSAR and Slope Stability Analysis Method for Damage Analysis in Mountainous Mining Subsidence Regions. *Remote Sens.* **2021**, *13*, 3107. [CrossRef]
23. Zhou, L.; Guo, J.M.; Li, X. Monitoring and analyzing surface subsidence based on SBAS-InSAR in Beijing region, China. In Proceedings of the International Conference on Intelligent Earth Observing and Applications, Guilin, China, 9 December 2015.
24. Li, Y.S.; Zhang, J.F.; Li, Z.H.; Luo, Y. Land Subsidence in Beijing City from InSAR Time Series Analysis with Small Baseline Subset. *Geomat. Inf. Sci. Wuhan Univ.* **2013**, *38*, 1374–1377. (In Chinese)
25. Pan, C.; Jiang, L.M.; Sun, Q.S.; Jiang, Y.N. Monitoring and Analyzing Chengdu Ground Subsid-ence Based on InSAR Technology by Using Sentinel-1 Radar Image. *J. Geod. Geodyn.* **2020**, *40*, 198–203.
26. Zhou, C.; Gong, H.; Chen, B.; Zhou, F.; Duan, G.; Gao, M.; Lu, W. Land Subsidence under Different Land Use in the Eastern Beijing Plain, China 2005–2013 revealed by InSAR timeseries analysis. *GISci. Remote Sens.* **2016**, *53*, 671–688. [CrossRef]
27. Du, Z.; Ge, L.; Ng, A.H.M.; Xiaojing, L.; Li, L. Mapping land subsidence over the eastern Beijing city using satellite radar interferometry. *Int. J. Digit. Earth* **2018**, *11*, 504–519. [CrossRef]
28. Liu, L.M.; Yu, J.; Chen, B.B.; Zhu, F.; Duan, G.Y.; Gao, M.L.; Lu, W. Urban subsidence monitoring by SBAS-InSAR technique with multi-platform SAR images: A case study of Beijing Plain, China. *Eur. J. Remote Sens.* **2020**, *53*, 141–153. [CrossRef]
29. Muhetaer, N.; Yu, J.; Wang, Y.B.; Yue, J. Temporal and Spatial Evolution Characteristics Analysis of Beijing Land Subsidence Based on InSAR. In Proceedings of the 3rd International Forum on Geoscience and Geodesy, Shenyang, China, 13–15 November 2020; Volume 658, pp. 13–15.
30. Yang, Q.; Ke, Y.H.; Zhang, D.Y.; Chen, B.B.; Gong, H.L.; Lv, M.Y.; Zhu, L.; Li, X.J. Multi-Scale Analysis of the Relationship between Land Subsidence and Buildings: A Case Study in an Eastern Beijing Urban Area Using the PS-InSAR Technique. *Remote Sens.* **2018**, *10*, 1006. [CrossRef]
31. Ng, A.H.M.; Ge, L.L.; Li, X.J.; Zhang, K. Monitoring ground deformation in Beijing, China with persistent scatterer SAR interferometry. *J. Geod.* **2011**, *86*, 375–392. [CrossRef]
32. Hu, L.Y.; Dai, K.R.; Xing, C.Q.; Li, Z.; Tomás, R.; Clark, B.; Shi, X.; Chen, M.; Zhang, R.; Qiu, Q.; et al. Land subsidence in Beijing and its relationship with geological faults revealed by Sentinel-1 InSAR observations. *Int. J. Appl. Earth Obs. Geoinf.* **2019**, *82*, 101886. [CrossRef]
33. Zhou, C.F.; Gong, H.L.; Chen, B.B.; Li, X.; Li, J.; Wang, X.; Gao, M.; Si, Y.; Guo, L.; Shi, M.; et al. Quantifying the contribution of multiple factors to land subsidence in the Beijing Plain, China with machine learning technology. *Geomorphology* **2019**, *335*, 48–61. [CrossRef]
34. Samsonov, S.V.; D'Oreye, N.; González, P.J.; Tiampo, K.F.; Ertolahti, L.; Clague, J.J. Rapidly accelerating subsidence in the Greater Vancouver region from two decades of ERS ENVISAT RADARSAT-2 DInSAR measurements. *Remote Sens. Environ.* **2014**, *143*, 180–191. [CrossRef]

35. Pepe, A.; Zhao, Q.; Bonano, M.; Lu, Z.; Zhou, Y. The study of the deformation time evolution in coastal areas of Shanghai:a joint C/X-band SBAS-DInSAR analysis. In Proceedings of the Name of the 2015 IEEE International Geoscience and Remote Sensing Symposium (IGARSS), Milan, Italy, 26–31 July 2015; pp. 306–309.
36. Gao, M.L.; Gong, H.L.; Chen, B.B.; Zhou, C.F.; Chen, W.F.; Liang, Y.; Shi, M.; Si, Y. InSAR time-series investigation of long-term ground displacement at Beijing Capital International Airport, China. *Tectonophysics* **2016**, *691*, 271–281. Available online: http://geodata.geoscience.cn/cjy/cjy.action (accessed on 14 January 2022). [CrossRef]
37. Wang, Z.; Cai, X.; Yan, J.; Wang, J.; Liu, Y.; Zhang, L. Using the integrated geophysical methods detecting active faults: A case study in Beijing, China. *J. Appl. Geophys.* **2018**, *156*, 82–91. [CrossRef]
38. Yang, Y.; Jia, S.M.; Wang, H.G. The Status and Development of Land Subsidence in Beijing Plain. *Shanghai Geol.* **2010**, *31*, 23–27.
39. Yang, Y.; Jia, S.M.; Wang, H.G.; Zhou, Y. Analysi s on impact of land subsidence on planned new cities in Beijing. *City Plan. Rev.* **2013**, *37*, 67–71. (In Chinese)
40. Yang, J.; Huang, X. The 30 m annual land cover dataset and its dynamics in China from 1990 to 2019. *Earth Syst. Sci. Data* **2021**, *13*, 3907–3925. [CrossRef]
41. Zhu, J.J.; Li, Z.W.; Hu, J. Research Progress and Methods of InSAR for Deformation Monitoring. *Acta Geod. Et Cartogr. Sinica* **2017**, *46*, 1717–1733.
42. Yang, C.S.; Zhang, Q.; Zhao, C.Y.; Ji, L.Y. Small Baseline Subset InSAR Technology Used in Datong Basin Ground Subsidence, Fissure and Fault Zone Monitoring. *Geomat. Inf. Sci. Wuhan Univ.* **2014**, *39*, 945–950.
43. Song, Y.F.; Liu, G.X.; Zhang, R. Monitoring Significant Ground Subsidence by Using New Small Baseline Subset Multiple Temporal InSAR Approach. *Remote Sens. Inf.* **2016**, *31*, 84–88.
44. Golub, G.H.; VanLoan, C.F. *Matrix Computation*; Johns Hopkins University Press: Baltimore, MD, USA, 1996.
45. Zhu, Y.X.; Feng, L.P.; Jia, X.L.; Zhang, Q.H.; Ruan, R.G. The PPP precision analysis based on BDS regional navigation system. *Acta Geod. Et Cartogr. Sin.* **2015**, *44*, 377–383.
46. Liu, Y. *GAMIT/GLOBK Installation and Application Based on Linux System*; Beijing Surveying and Mapping: Beijing, China, 2014; pp. 30–33.
47. Liu, X.W.; Pu, D.X.; Gao, X.; Zhang, S.Y.; Xia, D.H. High Precision GPS/BDS Data Processing and Precision Contrast Analysis Based on GAMIT10.61. *GNSS World China* **2018**, *43*, 77–83.
48. Zhang, S.C.; Wang, Q.Y.; Liu, Q.; Zhang, C.L.; He, Y.F.; Zhao, G.S. Analysis of precision relative positioning accuracy of BDS by GAMIT. *Sci. Surv. Mapp.* **2018**, *43*, 92–97.
49. DONG, D.; HERRING, T.A.; KING, R.W. Estimating regional deformation from a combination of space and terrestrial geodetic data. *J. Geod.* **1998**, *72*, 200–214. [CrossRef]
50. Sun, F.L.; Song, S.L.; Zhu, W.Y.; Zheng, Z.Y. Utilizing GLOBK to Achieve Weekly Comprehensive Solution of GPS. *GNSS World China* **2013**, *38*, 61–65.
51. Wang, Y.; Ge, D.Q.; Zhang, L.; Li, M.; Guo, X.F.; Wang, Y. Inter-comparison and time series fusion of ascending and descending PSInSAR data for land subsidence monitoring. *Remote Sens. Land Resour.* **2014**, *26*, 125–130.
52. Dai, K.R.; Zhuo, G.C.; Xu, Q.; LI, Z.; LI, W.; Guan, W. Tracing the Pre-failure Two-dimensional Surface Displacements of Nanyu Landslide, Gansu Province with Radar Interferometry. *Geomat. Inf. Sci. Wuhan Univ.* **2019**, *44*, 1778–1786.
53. Zhang, J.J.; Zhang, H.X.; Zhang, M.S.; Tao, H.; Dong, Y.; Xu, C.M. Study on the seepage deformation induce ground fissures caused and subsidence in Xi'an. *J. Eng. Geol.* **2018**, *26*, 301–309.
54. Zhang, Y.; Xue, Y.Q.; Wu, J.C.; Wang, H.; He, J. Mechanical modeling of aquifer sands under long-term groundwater withdrawal. *Eng. Geol.* **2012**, *125*, 74–80. [CrossRef]
55. Tosi, L.; Teatini, P.; Carbognin, L.; Brancolini, G. Using high resolution data to reveal depth-dependent mechanisms that drive land subsidence: The venice coast, Italy. *Tectonophysics* **2009**, *474*, 271–284. [CrossRef]
56. Zhu, L.; Gong, H.L.; Li, X.J.; Wang, R.; Chen, B.; Dai, Z.; Teatini, P. Land subsidence due to groundwater withdrawal in the northern Beijing plain, China. *Eng. Geol.* **2015**, *193*, 243–255. [CrossRef]
57. Chen, M.; Tomas, R.; Li, Z.H.; Motagh, A.; Li, T.; Hu, L.Y.; Gong, H.L.; Li, X.J.; Yu, J.; Gong, X.L. Imaging Land Subsidence Induced by Groundwater Extraction in Beijing (China) Using Satellite Radar Interferometry. *Remote Sens.* **2016**, *8*, 468. [CrossRef]
58. Chen, B.B.; Gong, H.L.; Li, X.J.; Lei, K.C.; Zhang, Y.Q.; Li, J.W.; Gu, Z.Q.; Dang, Y.N. Spatial-temporal Characteristics of Land Subsidence Corresponding to Dynamic Groundwater Funnel in Beijing Municipality, China. *Chin. Geogr. Science* **2011**, *21*, 753–764. [CrossRef]
59. Zhou, C.D.; Gong, H.L.; Chen, B.B.; Li, J.W.; Gao, M.L.; Zhu, F.; Chen, W.F.; Liang, Y. InSAR time-series analysis of land subsidence under different landuse types in the eastern Beijing plain, China. *Remote Sens.* **2017**, *9*, 380. [CrossRef]
60. Zhou, C.F.; Gong, H.L.; Chen, B.B.; Gao, M.L.; Cao, Q.; Cao, J.; Duan, L.; Zuo, J.J.; Shi, M. Land Subsidence Response to Different Land Use Types and Water Resource Utilization in Beijing-Tianjin-Hebei, China. *Remote Sens.* **2020**, *12*, 457. [CrossRef]
61. Zhou, C.D.; Gong, H.L.; Zhang, Y.Q.; Duan, G.Y. The Influence of Building Load to Land Subsidence in Beijing Plain based on PS-InSAR and GIS. *J. Geo-Inf. Sci.* **2016**, *18*, 1551–1562.
62. Solari, L.; Ciampalini, A.; Raspini, F.; Bianchini, S.; Moretti, S. PSInSAR Analysis in the Pisa Urban Area (Italy): A Case Study of Subsidence Related to Stratigraphical Fac-tors and Urbanization. *Remote Sens.* **2016**, *8*, 120. [CrossRef]
63. Guo, H.P.; Li, W.P.; Wang, L.Y.; Chen, Y.; Zang, X.S.; Wang, Y.L.; Zhu, J.Y.; Bian, Y.Y. Present situation and research prospects of the land subsidence driven by groundwater levels in the North China Plain. *Hydrogeol. Eng. Geol.* **2021**, *48*, 162–171.

64. Long, D.; Yang, W.T.; Scanlon, B.R.; Zhao, J.S.; Liu, D.G.; Burek, P.; Pan, Y.; You, L.Z.; Wada, Y. South-to-North Water Diversion stabilizing Beijing's groundwater levels. *Nat. Commun.* **2020**, *11*, 3665. [CrossRef]
65. Chen, B.; Gong, H.; Li, X.; Lei, K.; Zhu, L.; Gao, M.; Zhou, C. Characterization and causes of land subsidence in Beijing, China. *Int. J. Remote Sens.* **2017**, *38*, 808–826. [CrossRef]
66. Zheng, J.R.; Gong, H.L.; Li, Q.Y.; Cui, Y.Z.; Chen, B.B. The Control Factors on Subsidence of Beijing Plain Area in 2003–2009 Based on PS-InSAR Technology. *Bull. Surv. Mapp.* **2014**, *12*, 40–43.
67. Karen, M.W.; Yehuda, B.; David, T.S. Satellite Interferometric Observations of Displacements Associated with Seasonal Groundwater in the Los Angeles Basin. *J. Geophys. Res. Atmos.* **2002**, *107*, 265–296.

remote sensing

Article

Asymmetric Interseismic Strain across the Western Altyn Tagh Fault from InSAR

Yunhua Liu *, Dezheng Zhao and Xinjian Shan

State Key Laboratory of Earthquake Dynamics, Institute of Geology, China Earthquake Administration, Beijing 100029, China; dezhengzhao@ies.ac.cn (D.Z.); xjshan@ies.ac.cn (X.S.)
* Correspondence: liuyunhua@ies.ac.cn

Abstract: As the northern boundary of the Tibetan Plateau, the long Altyn Tagh fault (ATF) controls the regional tectonic environment, and the study of its long-term fault slip rate is key to understanding the tectonic evolution and deformation of the northern Tibetan Plateau. In this paper, we measure the fault slip rate of the western segment of the ATF using InSAR observations between 2015 to 2020. The Multi-Temporal Interferometric InSAR analysis is applied to obtain the two-dimensional fault-parallel and vertical displacement fields. The spatially dense InSAR observations clearly illustrate the asymmetrical pattern of displacement fields across the fault. Constrained by our InSAR observations, the fault slip rate and locking depth of the western segment of the ATF are inverted using four different models in a Bayesian framework. The two-layer viscoelastic model incorporating lateral heterogeneity of rheology in the lower crust indicates that the fault slip rate of the western ATF is estimated to be 9.8 ± 1.1 mm/yr (at 83.8°E across the ATF) and 8.6 ± 1.1 mm/yr (at 85.1°E), respectively, and the locking depth is 15.8 ± 4.3 km and 14.8 ± 4.9 km. Our new estimates generally agree with the previous estimates of fault slip rate constrained by GPS observations. We conclude that the contrast between the thickness of the elastic layer and the shear modulus of the Tibetan plateau and the Tarim basin jointly contribute to the asymmetric interseismic strain accumulation on the ATF.

Keywords: Altyn Tagh Fault; asymmetry interseismic strain; viscoelastic; InSAR

Citation: Liu, Y.; Zhao, D.; Shan, X. Asymmetric Interseismic Strain across the Western Altyn Tagh Fault from InSAR. *Remote Sens.* **2022**, *14*, 2112. https://doi.org/10.3390/rs14092112

Academic Editor: José Fernández

Received: 27 February 2022
Accepted: 20 April 2022
Published: 28 April 2022

Publisher's Note: MDPI stays neutral with regard to jurisdictional claims in published maps and institutional affiliations.

1. Introduction

The 1600-km-long Altyn Tagh fault (ATF) is the northern boundary of the Tibetan Plateau (TP). It separates the rigid Tarim basin from the relatively weak TP [1,2] and connects almost all orogenic belts and the thrust systems on the northern margin of the plateau. Investigating the earthquake cycle behavior of this large left-lateral strike-slip fault system informs us of the uplift process, stress transfer, and crustal deformation mechanism of the TP [3–5]. For instance, the slip rate of the ATF has been a matter of debate to discriminate among continental deformation models [6–8]. The large fault slip rate (20–30 mm/yr) of the ATF supports the 'lithosphere extrusion model,' which proposes that the India–Eurasia collision be accommodated laterally along large-scale narrow bounding fault zones by lithosphere extrusion [2,9,10]. Whereas slow slip rate along the ATF supports the 'continuum model,' which stresses that the India–Eurasia collision is partitioned by numerous, isolated fault structures [9–11]. Determining the fault slip rate of ATF helps to understand the interrelationship between the large-scale plateau boundary faults and continental dynamics [12–14].

Due to the harsh environment and inconvenient transportation in most areas along with the ATF, conventional geodetic measurements such as Global Positioning System (GPS) and leveling are limited. Interferometric Synthetic Aperture Radar (InSAR) substantially improves the temporal and spatial resolution of surface deformation via stacking individual interferograms, thus providing crustal deformation maps at a decadal scale [15–18].

Wright et al. [19] processed the European Remote Satellite (ERS) images (1992–1999) across the western ATF (79°E) and obtained the interseismic fault slip rates of the Karakoram fault and the western ATF. Their work pioneered the use of InSAR to obtain interseismic crustal deformation. Elliott et al. [20] found that the ATF (85°E) slips at 11 ± 5 (1σ) mm/yr, and their derived line-of-sight (LOS) displacement field features an asymmetric pattern across the fault. However, due to the phase decorrelation in interferograms around the Altyn mountains, the LOS velocities show a sense of asymmetry to the north of the ATF, opposite to that expected for the rigid Tarim basin. Zhu et al. [21] processed the Environmental Satellite (ENVISAT) data spanning 2003–2010 at almost the same location as Elliott et al. [20] but identified an asymmetric velocity pattern to the south of the ATF. Jolivet et al. [22] processed InSAR images across the eastern ATF (94°E) and observed apparent velocity asymmetry on both sides of the fault. They linked the asymmetric strain pattern to rigidity differences across the ATF or the geometry of a south-dipping fault. More recently, using a combination of GPS and InSAR measurements at 86°E, Li et al. [23] favored no asymmetric strain pattern across the ATF. Several factors might contribute to the discrepancies among previous studies. For example, vertical crustal deformation is neglected in the InSAR observations, which are more sensitive to vertical deformation than horizontal deformation. In addition, simplified two-dimensional elastic models might not be reasonable for modeling the interseismic strain.

Compared with previous ERS and ENVISAT satellites, the Sentinel-1 satellite has the advantages such as wide coverage, a short revisit period, and free public access to the data policy. Besides, Terrain Observation with Progressive Scans (TOPs) imaging technology is adopted, ensuring a large area of ground monitoring and reducing the revisit period [24–26]. The Sentinel SAR data have been widely used in geological disaster monitoring [27–29].

In this study, we processed three ascending and three descending Interferometric Wide swath (IW) mode images across the western ATF. We decomposed the ascending and descending LOS maps to obtain the horizontal and vertical deformation fields. We modeled the fault-parallel velocity profiles using different models to investigate possible asymmetric interseismic strain across the ATF. Our results provide new insights into the kinematic characters along with the ATF.

2. Data and Methods

2.1. GPS Data

He et al. [30] established a 400-km-long transect consisting of 17 GPS stations across the ATF (86°E) in 2009 (indicated by the red triangle symbol in Figure 1). Those campaign GPS stations was routinely observed for 2–3 periods in 2009–2011. In December 2017, Li et al. [31] re-measured the same GPS stations and processed those campaign-mode GPS data together with data at seven regional Crustal Movement Observation Network of China (CMONOC, https://iag.dgfi.tum.de/media/archives/prchina03/CRUSTAL%20MOVEMENT%20OBSERVATION.htm (accessed on 8 February 2022)) continuous sites near this region. We combined the GPS velocity field of Li et al. [31] with that of Wang and Shen [32] by minimizing the velocity residuals within common sites. The combined GPS velocity is referenced to a Eurasian frame. To highlight the relative motion of the crust on both sides of the Altyn Tagh fault, we transformed the GPS site velocities into a Tarim-fixed reference frame. The Euler pole in Wang et al. [33] is selected, then the GPS velocity field relative to Tarim Basin is obtained by deducting the rotational motion and global motion of the Euler pole from the whole velocity field data. A more detailed description of the GPS data process can be found in Li et al. [31].

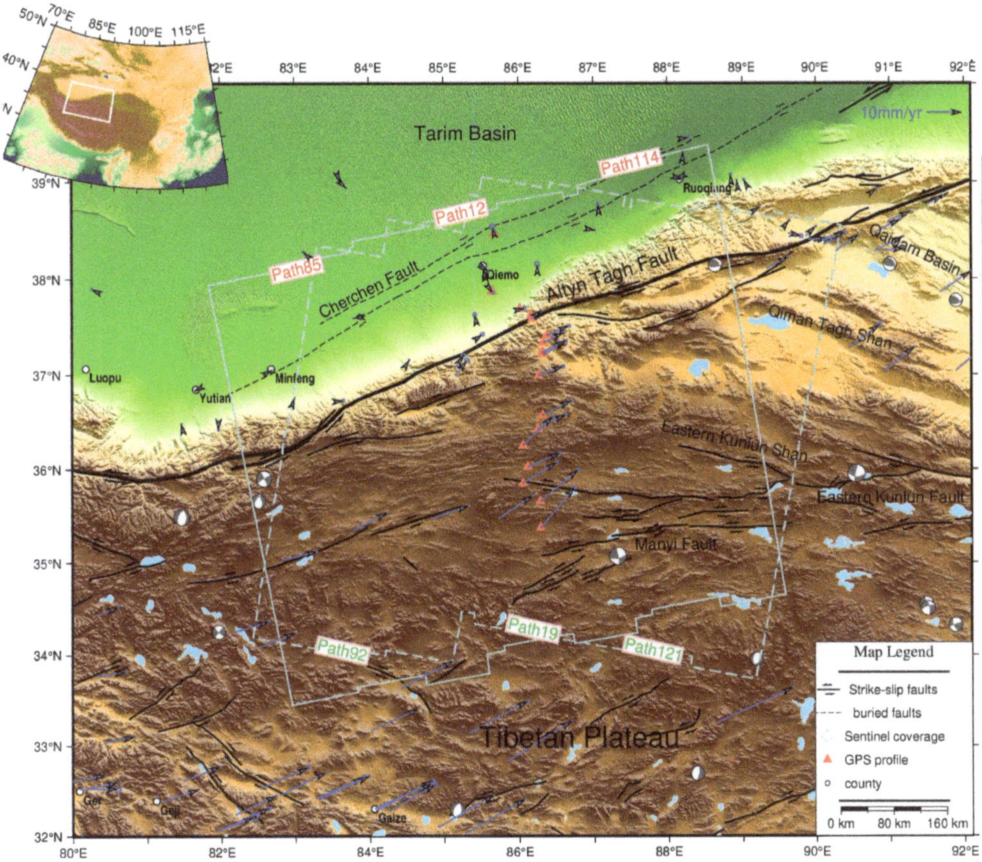

Figure 1. Tectonic map of northern Tibet showing major faults, sutures, and InSAR data coverage. It shows major faults (dashed line denoting buried faults, strike-slip faults in black). Cyan and dashed cyan rectangles show the spatial extent of the ascending and descending Sentinel-1 images, respectively. Beach balls in gray displaying the focal mechanisms are from the global CMT catalog [34]).

2.2. InSAR Data and Processing

Three ascending and three descending Interferometric Wide (IW) swath mode imagery are selected, and three frames on each orbit are concentrated into a long strip. The coverage of each strip is about 250 km × 700 km, and the observation period is from 2014 to 2020. See Figure 1 for InSAR data coverage.

GAMMA software performs the InSAR processing [35], starting with Single Look Complex (SLC) level data. For each track, we take the middle acquisition as the single prime, then use the imaging geometry as the reference for the images of other dates, and perform registration resampling after stitching until the registration accuracy of all images relative to the reference is higher than 1/1000 pixels in the azimuth direction [36,37]. In this way, interferograms can be generated with all combinations among those scenes without re-registration. The selection criterion of the interferogram pairs according to the spatial baseline is less than 30 m, and the time baseline is more than 1 year and less than 3 years, considering that the study area in the summer by the influence of the thickness of active layers change with the melting permafrost [38–42], to eliminate the interferograms between June to September [43,44], the final baseline network is shown in Figure A1. In the interferometry process, a multilooking operation with 20 looks in the range direction and 4 looks in the azimuth direction is applied to suppress noises. The simulation and removal of the topography phase are performed using the 30 m ALOS Global Digital Surface Model ALOS World 3d-30 m (AW3D30) [45,46], and the two-pass InSAR approach is used to produce differential interferograms, then the interferograms are filtered using a power spectrum filter [47]. The rock area with a stable phase is selected as a reference point and the region with coherence smaller than 0.2 is masked. The unwrapped phase is derived from the filtered phase using a minimum cost flow (MCF) algorithm [48]. Finally, the unwrapped phase is geocoded in the geographic coordinate system.

For ATF, an area with significant relief, spatial variations in hydrostatic delays are strong, it is necessary to reduce the tropospheric effects carefully [49–51]. The traditional atmospheric correction method based on the empirical linear correlation formula of phase elevation may remove the concerned deformation signal [52–55], especially in this region where long-wavelength deformation signals correlate with topography. For this purpose, the stratified tropospheric delay is corrected by Generic Atmospheric Correction Online Service (GACOS) [56], which is based on reanalysis data from the European Centre for Medium-Range Weather Forecasts (ECMWF) with a spatial resolution of 0.125° and a temporal resolution of 6 h. The Iterative Tropospheric Decomposition (ITD) model is used to separate the elevation-related signals and turbulence signals from the Zenith Tropospheric Delay (ZTD) and interpolate this to generate high-resolution tropospheric delay maps, and the high-resolution tropospheric delay is obtained by interpolation [56–58]. Compared with other atmospheric delay correction methods, its effectiveness has been verified [59–61], and it has been applied in many studies recently and achieved good results [62–64].

Detailed atmospheric phase correction steps of GACOS can be found in reference [65]. The interferograms corrected by the GACOS atmospheric correction are obtained by further differencing between the unwrapped interferograms and the atmospheric correction maps. A quadratic surface model and iterative least-squares method are used to remove residual orbit errors. Here, the correction process is briefly described. Figure 2 shows the 12-day ascending track interferogram from 15 to 27 December 2019 (20191215–20191227). Since the interval is only 12 days, it can be considered that the interferogram does not contain a crustal deformation signal, while atmospheric behavior and orbital error play a dominant role. Figure 2b shows the phase difference between two GACOS ZTD maps obtained at the date of Sentinel-1 acquisitions and project the result into line-of-sight (LOS) using the incidence angle of the SAR image. By only applying the GACOS corrections, the corrected phase still shows an insignificant correlation with the topography, we observed long-wavelength residual signals, and we also used the spatial information contained in the cumulative displacement to determine the empirical correction on each interferogram [66]. Figure 2 shows an example where we have removed a quadratic ramp in latitude and a linear ramp in longitude, a latitudinal cross dependency of the phase-elevation relationship, and then apply the procedure to all the interferograms. Figure 2d shows the phase value after correction. From the histogram, the original interference phase diagram and GACOS simulation have two peaks (Figure 2g,h). After correction, the phase histogram has only one peak value and is near 0 value, showing obvious atmospheric delays have been most effectively corrected. However, we admit that there is still a left side heavy tail distribution in the GACOS corrected phases, as shown in Figure 2i. As can be seen from the original interferogram, this signal has always existed in the upper left corner of the graph, and we interpret it as atmospheric turbulence signal, so neither GACOS nor the phase-elevation empirical formula can remove it, but it can be removed by filtering in the subsequent time-series processing.

The corrected interferograms are then imported into the poly-interferogram rate and time-series estimator algorithm (Π-RATE) software package for time-series analysis [67–69]. Unlike other Small Baseline Subset (SBAS) methods, Π-RATE uses a minimum distance spanning tree algorithm to pick out independent observations for time series analysis. In this method, pixels that are not always coherent in different interferograms are employed to calculate deformation rates to provide as much constraint on the rate as possible [70,71]. The observed values also include a random Atmospheric Phase Screen (APS), a temporal high-pass filter followed by a spatial low-pass filter applied to mitigate atmospheric disturbances. LOS velocity and its uncertainty are obtained using an iterative weighted least squares approach on a pixel-wise basis.

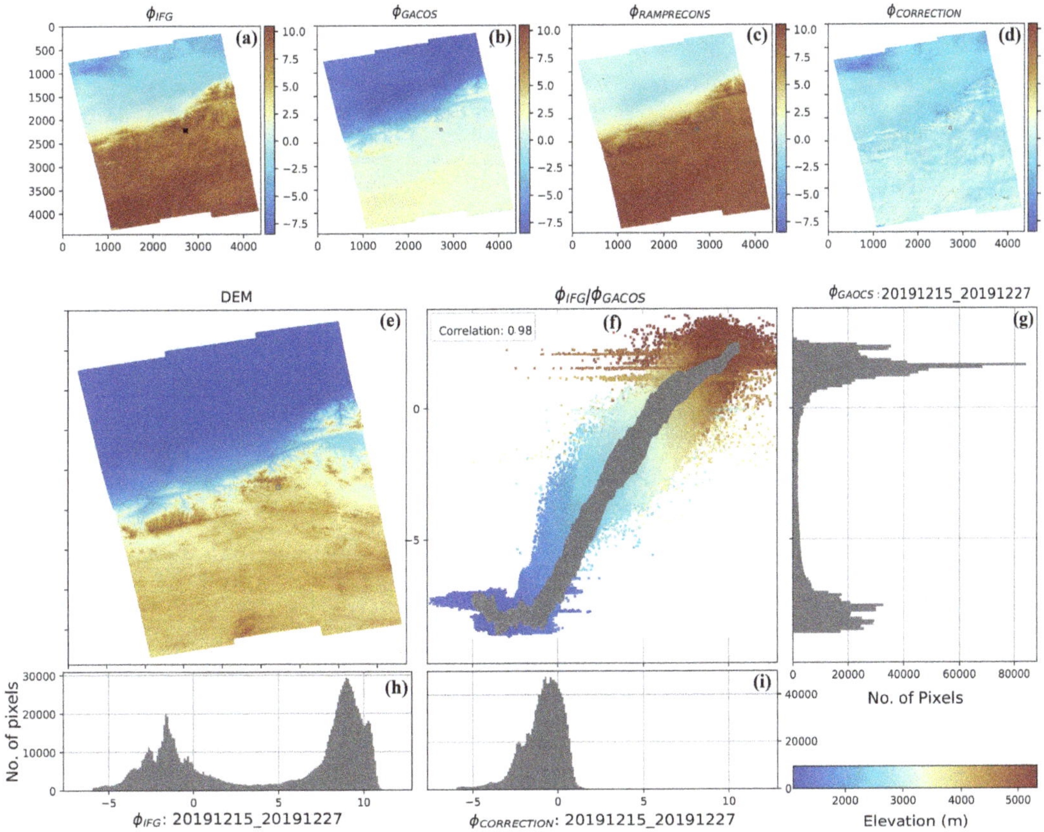

Figure 2. Example of an empirical correction of the stratified delays after the GACOS correction. (**a**) Interferogram for date pair 15–27 November 2019 (format: YYYYMMDD). (**b**) Modeled tropospheric delays using GACOS product. (**c**) Reconstruction of the stratified delays considering an empirical displacement-elevation relationship. (**d**) Corrected interferogram. (**e**) Digital Elevation Model from AW3D30. (**f**) Phase/GACOS correlation. (**g**) Histogram for reconstructed GACOS. (**h**) Histogram for original interferogram. (**i**) Histogram for corrected interferogram.

3. Results

3.1. InSAR LOS Velocity Field

To obtain a large-scale velocity field including several tracks, it is a common procedure that ties the InSAR to the GNSS velocities like Weiss et al. [72]. However, due to the scarity of GPS stations (most of the research area is uninhabited), the final result of tie to GPS may be biased by removing the best-fit plane. Here we adopted a simple method; first, we calculate the datum offset using the overlapped zone of the adjacent track, then the offset is subtracted from the adjacent track to unify the datum. After removing the average offset in each overlapping area between every two adjacent tracks in ascending and descending, respectively, we obtain the merged LOS velocity fields. We project the GPS velocities into the local satellite line-of-sight and calculate the difference from the InSAR velocities. Three

GPS stations (AT13, AT14, and AT15) are chosen as reference sites to align those datasets, as shown in Figure 3a,c. all the velocity maps were computed with respect to an average value at the location of the three GPS stations. There is an obvious cross-fault deformation gradient on both sides of the fault, and the color is relatively uniform on the south or north side of the fault, but there is an obvious difference near the fault zone. The blue movement is toward the satellite, and the red movement is away from the satellite, which is consistent with a left-lateral motion across the fault.

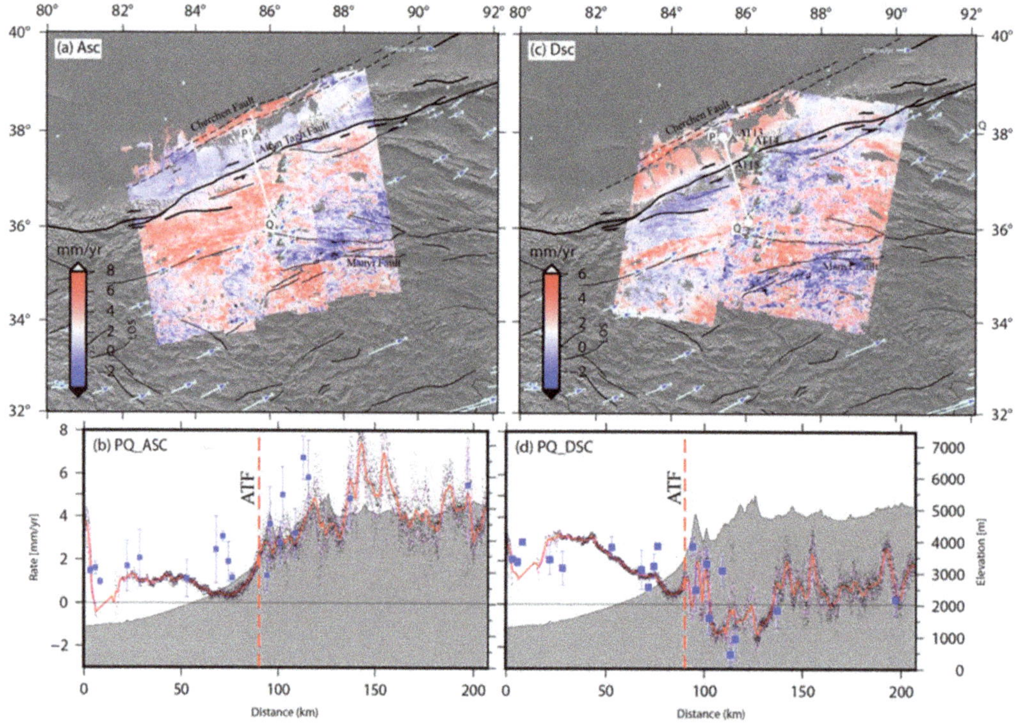

Figure 3. Line-of-sight (LOS) velocities on (**a**) ascending (mosaic of tracks 85, 12, and 114) and (**c**) descending (mosaic of tracks 92, 19, and 121) tracks. The solid white line indicates the position of the velocity profile. Negative (blue) and positive (red) values indicate relative motion towards and away from the satellite, respectively. (**b**,**d**) Ascending and descending LOS velocity along with PQ profiles across the ATF. Red dashed lines represent the ATF. The blue squares with error bars are the velocities projected into LOS within 100 km of the fault crossing profiles. The scattered points show InSAR LOS velocities within a 10 km perpendicular distance from the profiles. The solid red lines mark binned mean along with profiles. Grey shaded areas are topographic profiles along the PQ line from DEM.

Figure 3b,d shows the cross-fault rate profile (PQ) in the ascending and descending InSAR rate maps. From the profile of PQ, there are obvious velocity gradients across the fault, and the LOS velocity difference is about 4–5 mm/yr. In the study area, the ATF is the major fault, whereas the velocity profiles also reflect the deformation of the Altyn North Margin fault (NATF) and Cherchen fault (CHEF). Such an effect may lead to an asymmetrical pattern in the velocity profiles. The ascending and descending velocity fields exhibit a clear transition, reflecting the left-lateral strike-slip fault motion.

We compare the LOS-converted GPS (no vertical velocities) profile (Figure 1) in He et al. [30] at longitude 86°E (Figure 3c,d) with our InSAR observations. The result shows that the difference between them is distributed within ±1 mm/yr, comparable to InSAR measurement accuracy (about 1 mm/yr). The good agreement indicates the reliability of our InSAR rate map.

3.2. ATF-Parallel and Vertical Surface Velocities

According to the previous tectonic results, considering the geometry, structural continuity of ATF, and the earthquakes, the whole Altyn Tagh fault system can be divided into eight segments [73]. We are concerned about the Nierkule and Kurukuole segments in this paper. The continuous Nierkule section from 82.75°E to 84.5°E has a length of 190 km, characterized by the development of a fault depression valley. The Kurukuole segment from 84.5°E to 87.2°E has a length of about 230 km and an overall trend of NE70°, characterized by obvious linear characteristics and a large fault slip rate.

When InSAR observations are made in the northeastern margin of Qinghai-Tibet Plateau, LOS deformation inherently contains a contribution from vertical deformation, which may lead to deviation of crustal deformation interpretation based on InSAR data. Therefore, after obtaining the velocity fields from different perspectives of the ascending and descending orbits, we derive maps of horizontal and vertical average velocities by combining different orbits. By combining two independent observation directions and assuming that the third component is zero, we can project the information in the direction of any two orthogonal basis vectors [74]. Here, we use the method in Lindsey et al. [75]:

$$\begin{bmatrix} E_a sin\varphi + N_a cos\varphi & U_a \\ E_d sin\varphi + N_d cos\varphi & U_d \end{bmatrix} \begin{bmatrix} D_H \\ D_V \end{bmatrix} = \begin{bmatrix} LOS_a \\ LOS_d \end{bmatrix} \quad (1)$$

where E, N, and U are east, north, and up components (subscript character a for ascending and d for descending) of the unit look vectors for the satellite tracks considering the radar incidence angle; D_H and D_V are horizontal and vertical surface displacement components; φ is the local fault strike, and LOS_a and LOS_d are InSAR LOS displacements from the ascending and descending tracks, respectively. Assuming that the horizontal movement is N70°E along the average strike of ATF, Figure 4 shows the fault-parallel and vertical displacement results.

To compare the decomposed components with GPS results, GPS measurements are projected in parallel and vertical fault directions and compared with InSAR velocity profiles. Therefore, in addition to the PQ profile, an MN profile is added in the west, as shown in Figure 4. Within 100 km on both sides of the MN profile, there are only a few GPS stations on one side of the Tarim Basin.

The most prominent feature in the deformation field is the large-scale tectonic motion across the fault boundary with sharp transitions. Superimposed on this tectonic pattern are numerous nontectonic deformations mostly related to permafrost. Firstly, in addition to the strike-slip rate of the ATF, the Cherchen fault zone is also reflected in the rate map. Cherchen fault zone is a boundary fault in the northwestern part of the southeastern Tarim Basin, which is approximately parallel to the ATF. The fault has been active for a long time since the end of the early Paleozoic and has controlled the sedimentation and affected the tectonic deformation in the southeastern Tarim Basin. The SAR data mainly cover the western segment, with strong Cenozoic activity intensity and developed branch faults, which can be divided into Cherchen deep fault and Cherchen main fault. The western segment of the Cherchen fault bears not only the compressive stress caused by the west Kunlun uplift but also the compressive stress caused by the Altyn uplift, so the fault activity intensity is high, and branch faults are developed. At the same time, because of the multiple compressive stress directions in the western section, the Cherchen fault plane is complex and has the characteristics of strike-slip thrust activity. This characteristic is shown in both vertical and horizontal decomposition rate fields, and similar characteristics are also shown in Daout et al. [40]. In Figure 4b, there is a vertical component, which shows the thrust characteristic.

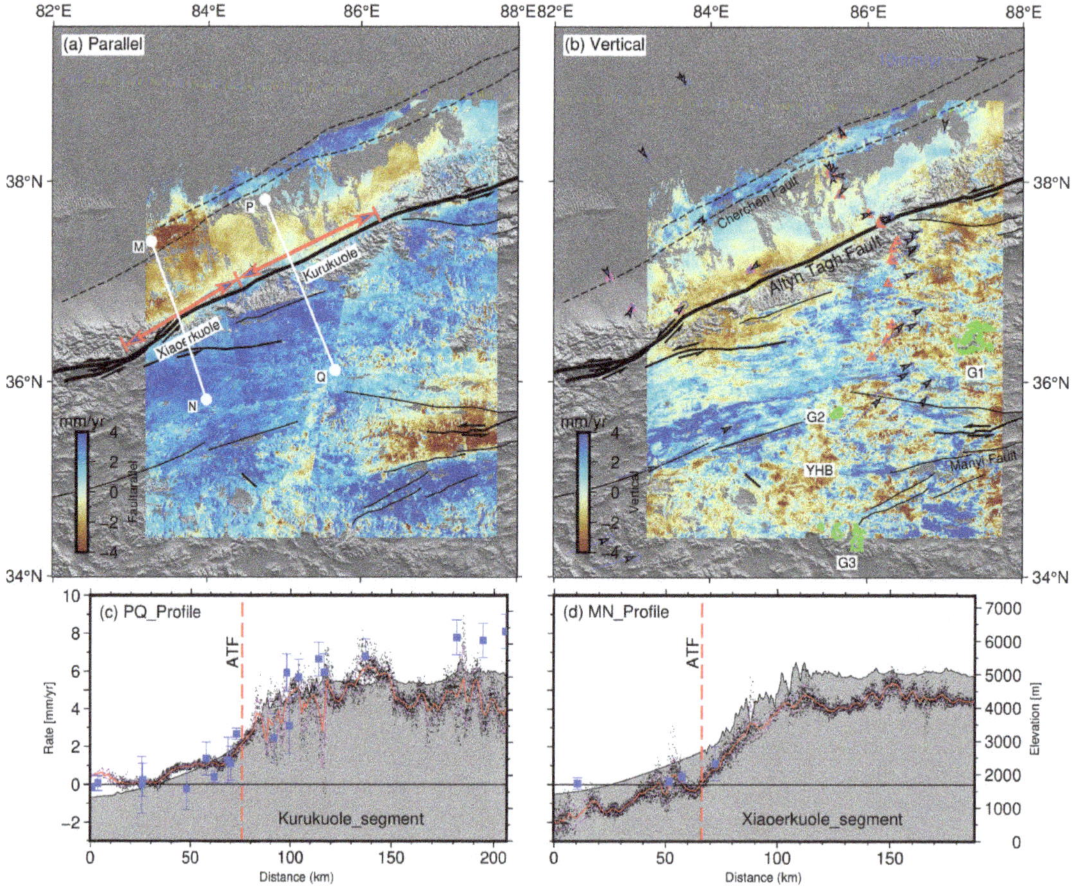

Figure 4. Horizontal and vertical surface velocities for the Altyn west segment region. (**a**) Fault-parallel and (**b**) Vertical components of InSAR observations derived from a combination of ascending and descending LOS directions (Figure 3), assuming a constant horizontal direction of motion of N70°E. The red line segment marks the beginning of the fault segment. (**c**) InSAR fault-parallel velocities within 10 km of the Kulukuole segment crossing profile PQ are shown in (**a**). The blue squares with error bars are the fault-parallel Global Positioning System (GPS) velocities, which location is shown as a red triangle in (**b**). The solid red lines and purple dashed line mark binned mean and standard deviations along with profiles, respectively. (**d**) Same as above but for a profile MN that crosses Xiaoerkule segment of ATF, but GPS stations used are marked with a purple triangle in (**b**). The solid white line indicates the position of the velocity profile, and vertical dashed red lines represent the locations of the ATF. The green line represents the outline of the glacier, and the symbol G1 represents Songzhiling, G2 for Muztag Feng, and G3 for Zangser Kangri. YHB-Yanghu Basin.

Secondly, obvious and steep displacement gradient changes are observed on the Manyi fault trace, and the post-earthquake effect can still be observed. The post-earthquake deformation seems to have a local pattern, with the deformation mainly concentrated at about 50 km from the fault. The seismogenic fault of the Manyi earthquake in 1997 is a strike-slip fault with a relatively simple structure. Using InSAR, the surface coseismic rupture length is about 170 km, the maximum displacement is 7 m, and the overall strike is N76°E [76]. Some scholars have studied the post-earthquake mechanism of the Manyi earthquake. Ryder et al. [77] found that both the standard linear body model and the afterslip model can be used to describe the post-earthquake deformation. It has been 24 years since the earthquakes occurred, and further research is needed to determine whether it has gradually entered the interseismic stage from post-earthquake [78,79].

Finally, there are large areas with sporadic permafrost in the vertical rate map. The annual average temperature of the Qinghai–Tibet Plateau is very low, forming a large area of permafrost, which is the highest elevation and largest permafrost area in the world [80–83]. Permafrost in the Tibetan Plateau is very sensitive to climate change and unstable in nature [84]. In the cold environment, the soil is frozen, the water in the soil is redistributed during the freezing process, and the structure and density of soil particles change with the generation of pressure, resulting in frost heave. In a warm environment, the frozen soil melts, the soil drains and consolidates under the action of gravity and external load, and the soil layer compresses and deforms, resulting in settlement. These are the characteristics of frost heaving and thawing settlement, and such changes will be repeated and reversible with the alternation of seasons [85–87]. Thawing subsidence and freezing uplift are reflected in the InSAR rate map [40]. Therefore, when selecting data for time series processing, we try to avoid the summer with serious frozen soil melting to minimize the impact [44]. Nevertheless, there are still large vertical deformation variables near major glaciers and the Yanghu basin.

3.3. Locking Depth and Slip Rate

The screw dislocation model of Savage & Burford [88] in homogeneous elastic half-space is used to retrieve the far-field velocity and the locking depth of ATF firstly. The model assumes that the slip on the fault is zero (locked) to depth D but is an amount v_0 everywhere below that depth. The part above locking depth cannot slide freely, the shear strain will accumulate near the fault, and the strain value will increase with the distance closer to the fault. For an infinitely long vertical strike-slip fault, the relationship between slip rate and distance can be expressed by an arctangent function as follows [70]:

$$V_x = \frac{v_0}{\pi} tan^{-1}\left(\frac{x}{D}\right) + V_{offset} \quad (2)$$

where V_x is observed fault-parallel velocity, v_0 is the slip rate, x is the perpendicular distance to the fault trace, D is the locking depth, and V_{offset} is the rate offset constant related to the reference frame [89].

To solve the locking depth and slip rate along with other parameters given in Equation (2) with the uncertainty in consideration, we use the Bayesian probability density method to find the best-fitting fault parameters as well as their uncertainties [89–92]. Referring to previous research results, a prior constraint is carried out on the range of parameter values. The slip rate is constrained from 1 mm/yr to 30 mm/yr, the locking depth is constrained from 1 km to 65 km, and all parameters are a uniform prior probability distribution over each range. Based on the Markov chain Monte Carlo (MCMC) sampler [93,94], 600 initial walkers are selected to explore the parameter space. For each InSAR slip rate profile across the fault, the model ran over 1 million iterations and produced over 20,000 independent random samples from which we estimate both the maximum a posteriori probability (MAP) solution and corresponding parameter uncertainties. The MAP solution is taken as the optimal solution for the model parameters.

The results of our analysis are shown in Figure 5. The best-fitting results (blue curve in the figure) show that the slip rate of the Nierkule segment is 7 ± 0.1 mm/yr and the locking depth is 18.1 ± 0.6 km. At the Kurukuole segment, the fault slip rate decreases to 6.1 ± 0.3 mm/yr, and the locking depth becomes 11.7 ± 0.4 km. The results show that the slip rate and locking depth change obviously from west to east and decrease gradually from west to east.

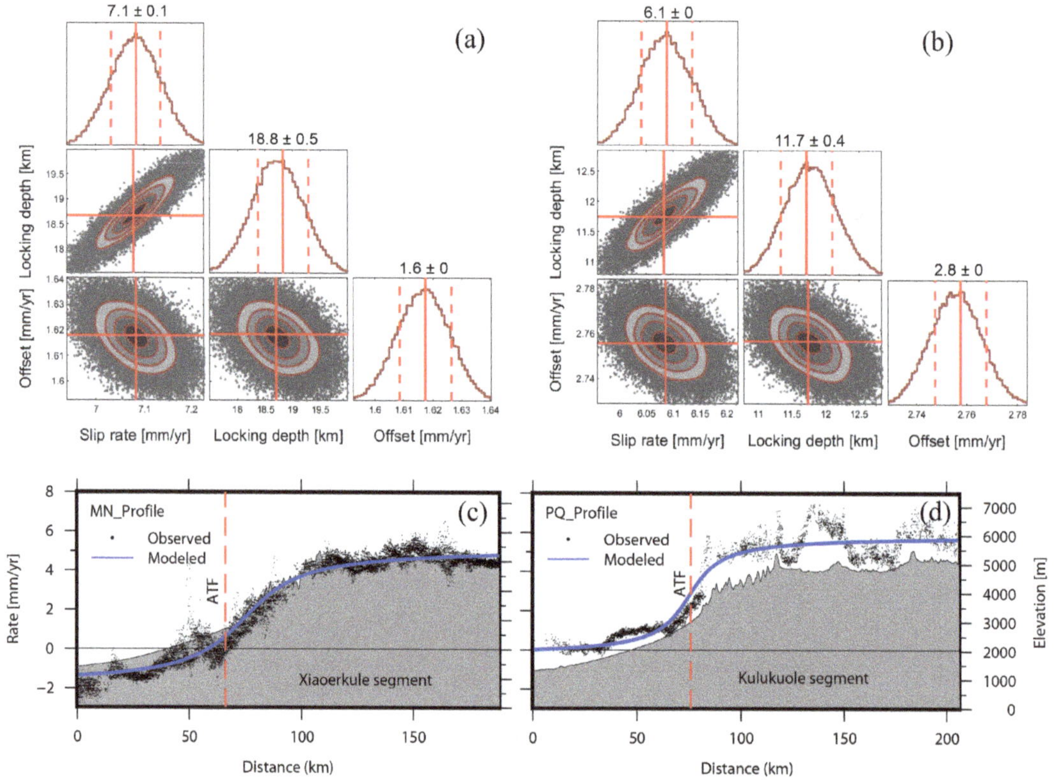

Figure 5. Probability distributions for profiles MN (**a**) and PQ (**b**). The red line and dot indicate the maximum a posteriori probability (MAP) solution from the Markov chain Monte Carlo (MCMC) analysis. We modeled fault-parallel velocity profiles along ATF, (**c**) for profile MN, and (**d**) for profile PQ. The profile's location is shown in Figure 4. The red dashed lines are the fault location of ATF.

4. Discussion

4.1. Strain Asymmetry

Over the last two decades, the proliferation of geodetic measurements, including GPS and InSAR, has helped to document asymmetric patterns of cross-fault velocity on major continental strike-slip faults [95–97], such as the San Andreas fault [98–100] and the Anatolian fault [101,102]. Elliott et al. [20] and He et al. [30] observed asymmetric patterns in the Altyn fault zone and excluded the possibility of offsetting the location of the fault trace, meaning that the deep fault plane shifted southward for more than 10 km. Therefore, the southward dip of the fault is nearly 45°, which is obviously unreasonable for the strike-slip ATF with a vertical fault plane. Jolivet et al. [22] found that the rigidity contrast coefficient across the fault is 0.85 and concluded that such asymmetry is the joint effect of a rigidity decrease from Tarim to Qaidam and a southward fault trace deviation.

The possible factors leading to the asymmetric velocities across the strike-slip faults are summarized as follows: (1) the difference in medium properties on both sides of the fault [103]; (2) the varying thickness of the elastic layer (hereafter *Te*) on both sides of the fault [104]; (3) the rheological contrast in the lower crust and upper mantle across faults [105]; (4) the deviation of the fault positions relative to fault trace below the brittle layer [106].

The reason in part accounting for the above disputes is the scarcity of geodetic measurements in some regions and the limited resolution of the diverse models. Constrained by the high-resolution rate map, we establish different models to fit and explain the asymmetry of the velocity profile across the ATF. In Model A, we used the Savage–Burford model in an elastic half-space (Figure 6a); Models B and C have two layers, and they are both the Savage–Prescott coupling model but different in thickness of the elastic layer (Figure 6b,c). Model D is the three-layer channel viscous model (Figure 6d), which is discussed in Section 4.1.3.

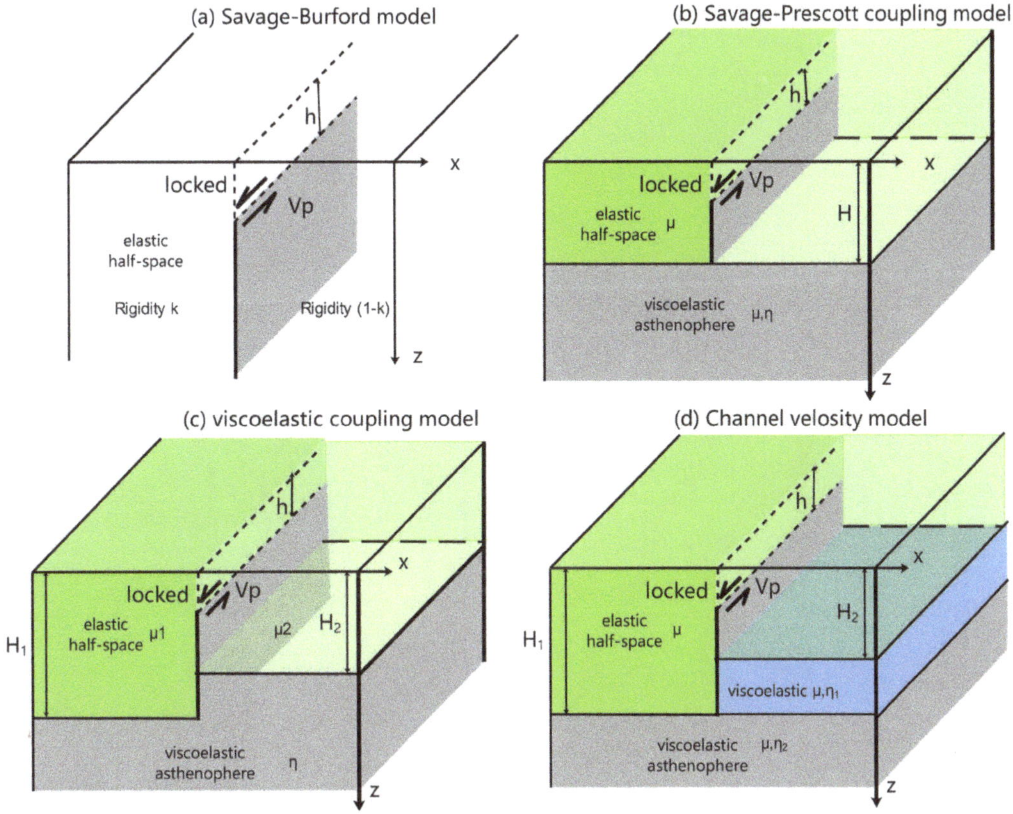

Figure 6. Schematic diagrams of the (**a**) modified Savage–Burford model in elastic half-space [88]. Interseismic deformation is modeled as slip on a buried dislocation that slides at the plating rate, Vp. (**b**) Savage–Prescott coupling model. Cyclic motion down to depth H and steady sliding below H on a fault in an elastic layer over viscoelastic half-space. The slip rate on the fault is equal to the plate velocity, Vp. (**c**) Same as model b, but considering the different thickness of the elastic layer across the fault. (**d**) Three-layer channel velocity model geometries. The thicknesses of the elastic upper crust and viscoelastic mid-crustal layers are H and h, respectively. Figure adapted from Figure 3 in Ref. [107].

4.1.1. Modified Half-Space Model

For the Model A, we calculate an asymmetric surface velocity field on each side of the fault accounting for the medium difference on both sides of the fault [22]:

$$V_x \begin{cases} \frac{2Kv_0}{\pi} tan^{-1}\left(\frac{x}{D}\right) + V_{offset} & \text{if } x > 0 \\ \frac{2(1-K)v_0}{\pi} tan^{-1}\left(\frac{x}{D}\right) + V_{offset} & \text{if } x > 0 \end{cases} \quad (3)$$

where K is the rigidity contrast, representing the rigidity ratio of the upper crust on both sides of the fault, ranging from 0 to 1. If $K = 0.5$, it means that the rigidity of the upper crust on both sides is the same. Inversion results are shown in Figures 7a and 8a. The ratio for rigidity of the upper crust in the Nierkule segment is 0.6/0.4, and in the Kurukuole segment is 0.7/0.3, which means that the crustal rigidity on the north side of the fault is about 0.5–1.3 times larger than that on the south side. At the same time, the resolved slip rate of 6.8 ± 0.1 mm/yr is similar to the inversion result of the model without considering the rigidity parameters. The locking depth of 17.2 ± 0.4 km is about 1 km shallower in the Nierkule segment and 1 km deeper in the Kurukuole segment. Generally, the inversion result with the addition of the asymmetry coefficient does not change much for the elastic half-space model.

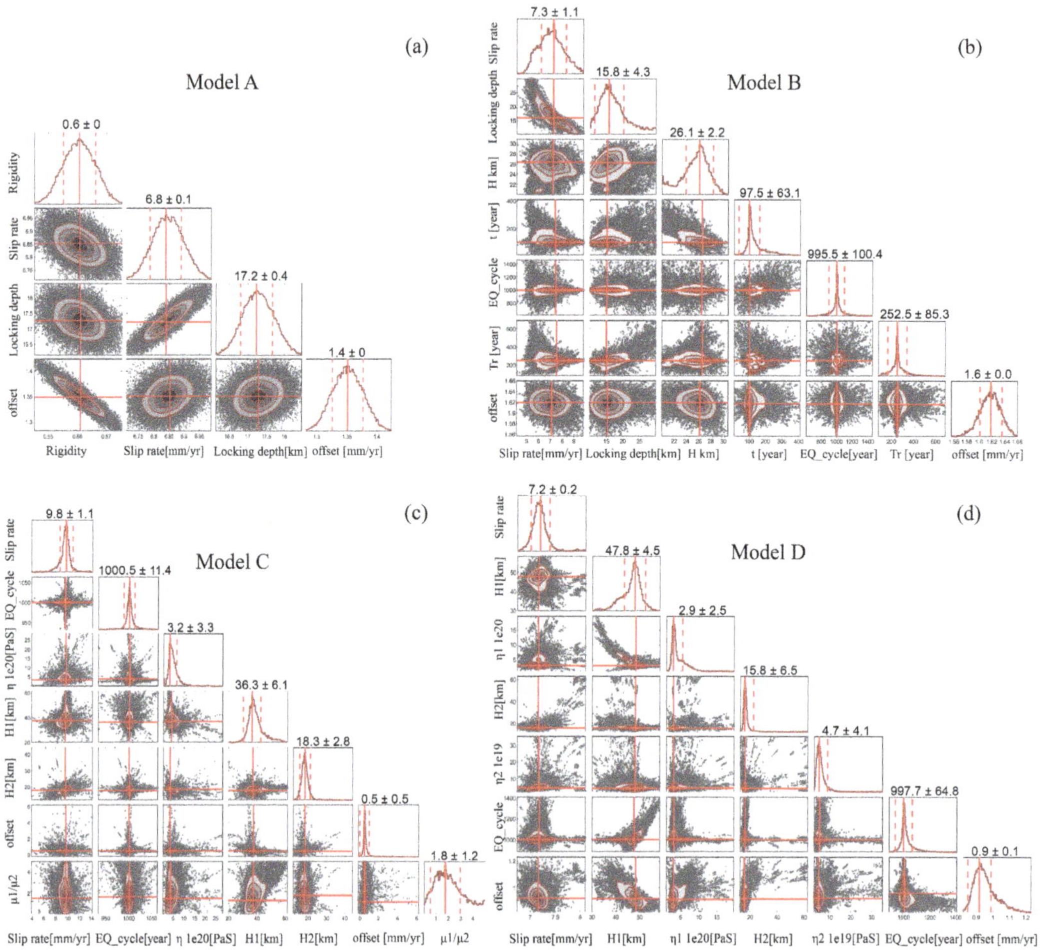

Figure 7. Inversion results of profile MN using the MCMC method. Results of (**a**) Model A, (**b**) Model B, (**c**) Model C, (**d**) Model D.

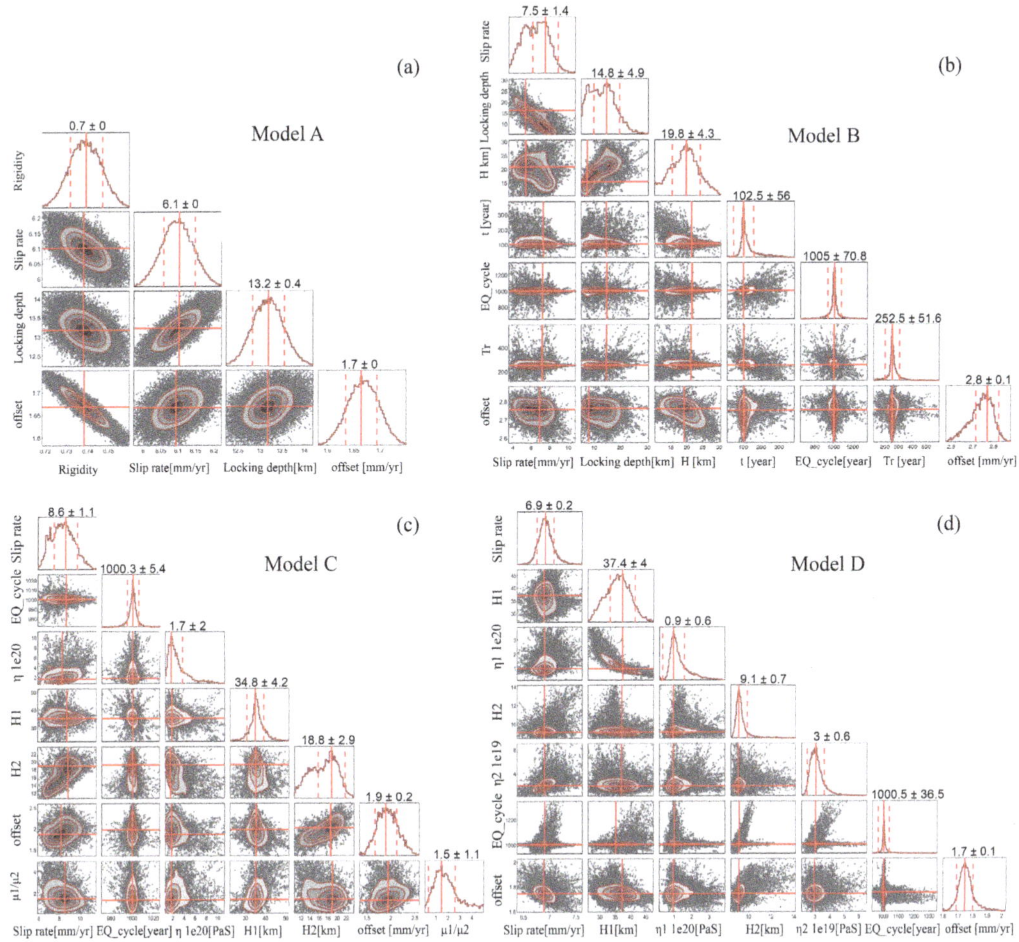

Figure 8. Inversion results of profile PQ using the MCMC method. Results of (**a**) Model A, (**b**) Model B, (**c**) Model C, and (**d**) Model D.

4.1.2. Viscoelastic-Coupling Models

Both Model B and C are viscoelastic-coupling models [108]. It is assumed that an elastic upper crust is overlying a viscoelastic (the Maxwell body) lower crust or upper mantle, and there is a linear coupling relationship between the two layers.

$$v(x,z) = \frac{\Delta u}{\pi t_R} e^{-t/t_R} \sum_{n=1}^{\infty} \frac{(t/t_R)^{n-1}}{(n-1)!} F_n(x,z,D,H) \quad (4)$$

The surface velocity of the viscoelastic model is related to the elastic layer thickness (H), the earthquake recurrence period T, the relaxation time τ_M, and other parameters [109]. A prior constraint of the unknown parameters is employed according to Table A1.

For a ground point with $z = 0$, the F_n is simplified as [107]:

$$F_n(x, z=0, D, H) = tan^{-1}\left(\frac{x}{2nH - D}\right) - tan^{-1}\left(\frac{x}{2nH + D}\right) \quad (5)$$

In the case of model B, the best-fitting model parameters are shown in Figures 7b and 8b. For the Nierkule segment (MN in Figure 4), the inferred long-term slip rate is 7.3 ± 1.1 mm/yr, the locking depth is 15.8 ± 4.3 km, and the elastic layer thickness is 26.1 ± 2.2 km. For the Kurukuole section (PQ in Figure 4), the optimal values are estimated to be 7.5 ± 0.3 mm/yr slip rate, 14.8 ± 4.9 km locking depth, and 19.8 ± 4.3 km elastic layer thickness. The slip rate obtained by this model is slightly larger than that obtained by the buried screw dislocation model in elastic half-space, but it is still smaller than the long-term geological rate.

Model C is still a viscoelastic model adopted in [107]. The difference is that different effective elastic layer thicknesses on both sides of the fault are considered. To reduce the number of parameters in the inversion, the locking depth is fixed as the results from Model B and no longer participates in inversion because of the trade-off between fault slip rate and locking depth. We assume that the Tarim Basin and the TP have different effective elastic thickness and shear modulus, respectively, and other prior conditions are the same as Model B to inverse the difference in elastic layer thickness and shear modulus ratio on both sides of the fault.

The inversion results are shown in Figures 7c and 8c. The long-term slip rate of the Nierkule segment (MN profile inversion) estimated by the model is 9.8 ± 1.1 mm/yr, the viscosity coefficient of the viscoelastic layer in the lower crust is $3.2 \pm 3.3 \times 10^{20}$ Pa s, and the thickness of the elastic layer on the side of the Tarim Basin is 36.3 ± 6.1 km. The thickness of the elastic layer on the side of the Tibetan Plateau is 18.3 ± 2.3 km, and the ratio of the shear modulus on both sides is 1.8 ± 1.2. For the Kurukuole segment (PQ profile), the inferred slip rate is 8.6 ± 1.1 mm/yr, and the viscosity coefficient in the lower crust is $1.7 \pm 2 \times 10^{20}$ Pa s. The thickness of the elastic layer on the side of the Tarim Basin is 34.8 ± 4.2 km, and that on the other side of TP is 18.8 ± 2.9 km. The ratio of shear modulus on both sides is 1.5 ± 1.1, and the inversion results of fault slip rate by this model are close to the long-term geological rate. The best-fitting model is shown in Figure 9.

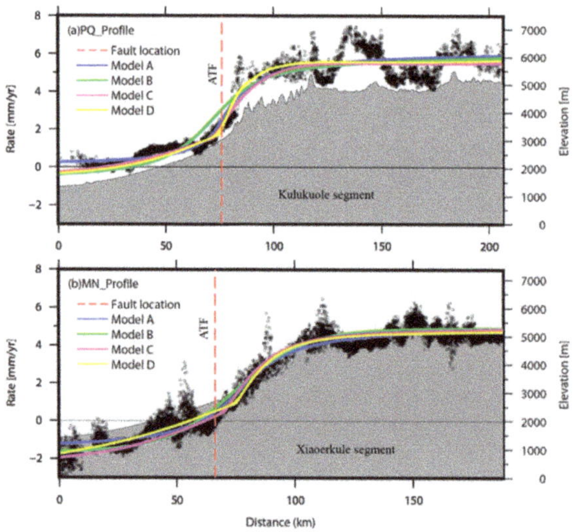

Figure 9. Model prediction from four types of models constrained by the horizontal velocity profiles. (**a**) for profile PQ and (**b**) for profile MN. The location of the profile is shown in Figure 4a. Blue, green, pink, and yellow curves are model predictions from Model A, B, C, and D, respectively.

In addition, two other cases are also tested. One is that the thickness of the elastic layer on both sides of the fault is the same, but the shear modulus is different (that is, the ratio parameter of shear modulus on both sides of the fault is added based on Model B). The test results showed that the long-term slip rate increased to 9.5 ± 0.5 mm/yr and

9.3 ± 0.5 mm/yr, respectively. The ratio of the shear modulus of the upper crust on the side of the Tarim Basin to that on the other side of TP is 0.5 ± 0.2 (see Figure A2), which means that the rigidity of the upper crust on the side of Tarim Basin is weaker than that on the other side of TP, which is not consistent with general understanding. In the other case, the shear modulus on both sides of the fault is the same, but the elastic layer thickness is different (the shear modulus ratio parameter is removed based on Model C). The test results are shown in Figure A3. The inferred long-term slip rate estimated is close to model C, which is 9.3 ± 0.2 mm/yr and 9.4 ± 0.2 mm/yr, respectively. However, the obtained elastic layer thickness on the side of Tarim Basin is large, the inversion result for the MN profile is 41.5 ± 2.5 km, and the inversion result for the PQ profile is 50.6 ± 3.4 km. The elastic layer thickness on the side of TP does not change much with Model C. That is to say, the big difference in elastic layer thickness on both sides of the fault can also cause such asymmetry of deformation field in the Altyn fault zone. The study of *Te* based on topographic and gravity data shows that the Tarim Basin has a moderate elastic layer thickness, with *Te* values ranging from 20 to 60 km. The closer it is to the TP, the lower the value is, and the higher value, the near center of the basin is [110–112]. Intermediate *Te* values appear in the Qaidam Basin, filled with thick sediments. The Altyn and East Kunlun regions obviously have low elastic layer thickness, with *Te* values ranging from 10 to 30 km. Generally, large deformation also occurs in these weak zones, which are also the main areas absorbing the shortening of the India-Euroasia continental collision [112]. It can be considered that the addition of the shear modulus ratio parameter does not affect model inversion. As a characterization of lithospheric strength, the spatial variation of effective elastic thickness is of great significance to lithospheric deformation. In addition, because elastic thickness mainly depends on crustal thickness, temperature, composition, etc., it can be used to reflect the lateral changes of deep lithosphere structure, and the elastic layer thickness on both sides of the fault has a great influence on the inversion results. Results based on the four combined test results of models B and C and the previous research results on the thickness of the elastic layer around the TP, we prefer the results of Model C. The differences in elastic layer thickness and shear modulus on both sides of the fault jointly cause the asymmetry in deformation across the ATF.

4.1.3. Three-Layer Model

The large-scale tectonic evolution of the Qinghai Tibet Plateau and its marginal areas show that the weak ductile lower crust plays an important role. Therefore, Model D, following the practice of DeVries and Meade [113], establishes a simple three-layer viscoelastic model on the side of the TP to restrict the viscosity and thickness of the lower crust of the ATF, a low viscosity viscoelastic channel (H1) with a thickness of 20 km is embedded on the right side between the elastic block and the bottom substrate. The two-layer elastic, viscoelastic model (as shown in Figure 6d) is still used on the side of the Tarim Basin. The inversion results are shown in Figures 7d and 8d. The inferred long-term slip rate of the Nierkule segment (MN profile) is 7.2 ± 0.2 mm/yr, and the viscosity coefficient of the lower crust is $2.9 \pm 2.5 \times 10^{20}$ Pa s, the viscosity coefficient of the embedded layer is $4.7 \pm 4.1 \times 10^{19}$ Pa s, the thickness of the elastic layer on the side of the Tarim Basin is 47.6 ± 4.5 km, and on TP side is 15.8 ± 6.5 km. For the Kulukuole section (PQ profile), the inferred long-term slip rate is 6.9 ± 0.2 mm/yr, and the viscosity coefficient of the lower crust is $0.9 \pm 0.6 \times 10^{20}$ Pa s, the viscosity coefficient of the embedded layer is $3.0 \pm 0.6 \times 10^{19}$ Pa s, the thickness of the elastic layer on the Tarim Basin side is 37.4 ± 4 km, and the thickness of the elastic layer on the TP side is 9.1 ± 0.7 km [114]. The geomagnetic data revealed the existence of two huge middle and lower crustal low-resistance anomalous zones, which are considered as two weak material flows in the middle and lower crust, where thermal fluids are abundant [115]. The purpose of Model D is to test whether there is a low viscosity viscoelastic channel on one side of the TP. Compared with the Altyn collision orogenic belt where the Model D section is located, the model may be more suitable for the Lhasa block and Qiangtang terrane suture belt.

4.1.4. Summary

In short, Model A is a half-space model with different rigidity on both sides of the fault, highlighting the asymmetric pattern is only due to a lateral variation in rigidity. Models B and C are viscoelastic-coupling models, where both the elastic thickness and shear modulus are free parameters, highlighting the difference between elastic layer thickness and shear modulus on both sides of the fault might contribute to the asymmetric strain pattern across the western ATF. The Model D is a three-layer model in which a viscoelastic mid-crustal layer is sandwiched between an elastic upper crust and an elastic upper mantle, aiming to test whether the middle to lower crust may be weak.

Integrating the above comparisons and results, we conclude that the difference in elastic layer thickness and shear modulus might lead to the asymmetric strain across the ATF. Nevertheless, our current study could not distinguish which factor plays a dominant role. Further studies are needed in the future.

In conclusion, there are differences in the physical properties of the media on the north and south sides of the ATF, and the above results are verified by inversion. This difference may be related to the collision orogeny between the Tarim block and the Altyn block. In addition, there are some shortcomings in this paper. On the north side of the ATF, there is a parallel Cherchen fault, which should be classified into the Altyn fault zone System from the perspective of its distribution characteristics and kinematic characteristics. On the south side, there are also several large faults whose deformation characteristics are shown in the InSAR deformation field. The large-scale surface deformation signal observed by satellites should be the interaction of these faults. Here, it is only simplified to be controlled by a single fault, and the research of multi fault combination model can be strengthened in the future.

4.2. Interseismic Slip Rate

As a boundary strike-slip fault, the fault slip rate of the ATF is useful to constrain the current continental dynamic process [116], as well as to further explore the deformation mechanism of the TP and quantitatively evaluate seismic risk. Therefore, a great deal of research work has been carried out to obtain the slip rate of each segment of the ATF by using Geodesy (GPS, InSAR) and geological methods.

The geological method is mainly based on accumulated displacements of the offset landforms and the corresponding dating data to estimate the average slip rate of fault movement. Xu et al. [117] and Meriaux et al. [6,118] obtained a slip rate of 17–20 mm/yr through a detailed study of each segment of the ATF. Zhang et al. [119] and Cowgill et al. [120] resummarized and analyzed previous research data and believed that the ATF slip rate should be ~10 mm/yr. Cowgill et al. [120] assumed that the abandonment age for the upper terrace was the offset time of the alluvial terrace scarp, and the calculated geological slip rate was 9.4 ± 2.3 mm/yr. Zhang et al. [119] also recalculated the fault slip rate from the abandoned age of upper and lower terraces and the activity of fault scarps and yielded a result consistent with the slip rate observed by GPS and InSAR, which could solve the mismatch between the geological slip rate and the slip rate obtained by GPS. Later, Gold et al. [121] and Cowgill et al. [122] obtained geological slip rates of different sites along with the ATF, which were basically at low slip rates. In the past 20 years, with the advance in dating of the geomorphological markers, it seems that both the slip rate obtained by geodesy [20,22,123–125] and the slip rate obtained by geology in recent years [123,125,126] have been well consistent. That is, along most of the ATF, there is a uniform slip rate of ~10 mm/yr. Our results show that after using the layered model, considering the vertical variation of viscosity coefficient, the long-term slip rate is 9.8 ± 1.1 mm/yr, and the estimated slip rate of ATF is consistent with the geological data. If the viscous effect of the lower crust and upper mantle is ignored, it will lead to a misestimation of the slip rate.

5. Conclusions

The fault slip rate and locking status of the boundary faults are important for understanding regional tectonic deformation and seismogenic potential. The rapid development of spatial geodesy techniques (GPS and InSAR) provides technical support for obtaining high-resolution and high-precision deformation maps, and studying fault slip rate and locking depth during the earthquake cycle. Our key conclusions are summarized as follows:

1. Different models are used to invert the fault slip rate and locking depth of the western segment of ATF. In the viscoelastic model, the media on both sides of the fault are divided into the elastic upper crust, viscoelastic lower crust, and upper mantle, and stratified lateral inhomogeneity parameters are considered. From the west to east section, the inferred sinistral strike-slip rate of the ATF is 9.8 ± 1.1 mm/yr and 8.6 ± 1.1 mm/yr, respectively, and the locking depth is 15.8 ± 4.3 km and 14.8 ± 4.9 km respectively. Based on the elastic model fitting, the left-lateral strike-slip rates of the ATF are 7.1 ± 0.1 mm/yr and 6.1 ± 0 mm/yr, respectively, and the locking depths are 18.1 ± 0.6 km and 11.7 ± 0.4 km, respectively. The results show that ignoring the viscoelastic effect will significantly affect the estimation of fault slip rate and locking degree;
2. The decomposition of the ascending and descending LOS velocities into fault parallel and vertical components in this paper shows that although the ATF is mainly strike-slip, it is accompanied by a small amount of thrust, and the thrust component gradually increases from west to east, reaching 1 mm/yr in the Altyn Mountains. If the vertical deformation of InSAR data is ignored, it may bias the interpretation of crustal deformation. Constrained by the fault parallel velocity field from west to east, the sinistral strike-slip rates of the ATF obtained in this paper are 9.8 ± 1.1 mm/yr and 8.6 ± 1.1 mm/yr, respectively, in line with the characteristics that the ATF is in low slip rates and decreases gradually from west to east which is obtained by GPS inversion;
3. The velocity profile across the ATF shows that there is asymmetry across the fault. The results integrate the combined test results of different parameters of Models B and C and the previous research results on the thickness of the elastic layer around the TP; we believe that the difference between elastic layer thickness and shear modulus on both sides of the fault jointly causes the asymmetry of interseismic velocity.

Author Contributions: Conceptualization, InSAR data processing, writing—original draft preparation, Y.L.; software, model and editing D.Z.; validation and supervision, X.S. All authors have read and agreed to the published version of the manuscript.

Funding: This research was funded by the National Natural Science Foundation of China (41874027), the National Key Technologies R&D Program (2018YFC1503605), and the Basic Scientific Funding of the Institute of Geology, China Earthquake Administration (IGCEA1804).

Data Availability Statement: Copernicus Sentinel-1 SAR data for the years 2014–2020 was retrieved from ASF DAAC in September 2020, processed by ESA (https://search.asf.alaska.edu, accessed on 8 February 2022). The glacier outlines dataset is provided by National Cryosphere Desert Data Center. (http://www.ncdc.ac.cn, accessed on 8 February 2022). The AW3D30 DEM is provided by the Japan Aerospace Exploration Agency (JAXA) (https://www.eorc.jaxa.jp/ALOS/en/aw3d30/data/index.htm, accessed on 8 February 2022). The GACOS data can be downloaded from the website (http://www.gacos.net/, accessed on 8 February 2022).

Acknowledgments: The conceptualization originates from Yunhua's collaborating supervisor, Tim Wright, at the University of Leeds. The author highly appreciated his help, advice, support, and for his constructive suggestions. We are also grateful to Richard Walters (now at Durham University) and Ekbal Hussain for their help in InSAR data processing when Yunhua visited the University of Leeds. We thank Yanchuan Li for providing the GPS data. The figures in this study were generated using the public domain Generic Mapping Tools (GMT) software [127] and the Matplotlib Python libraries.

Conflicts of Interest: The authors declare no conflict of interest.

Appendix A

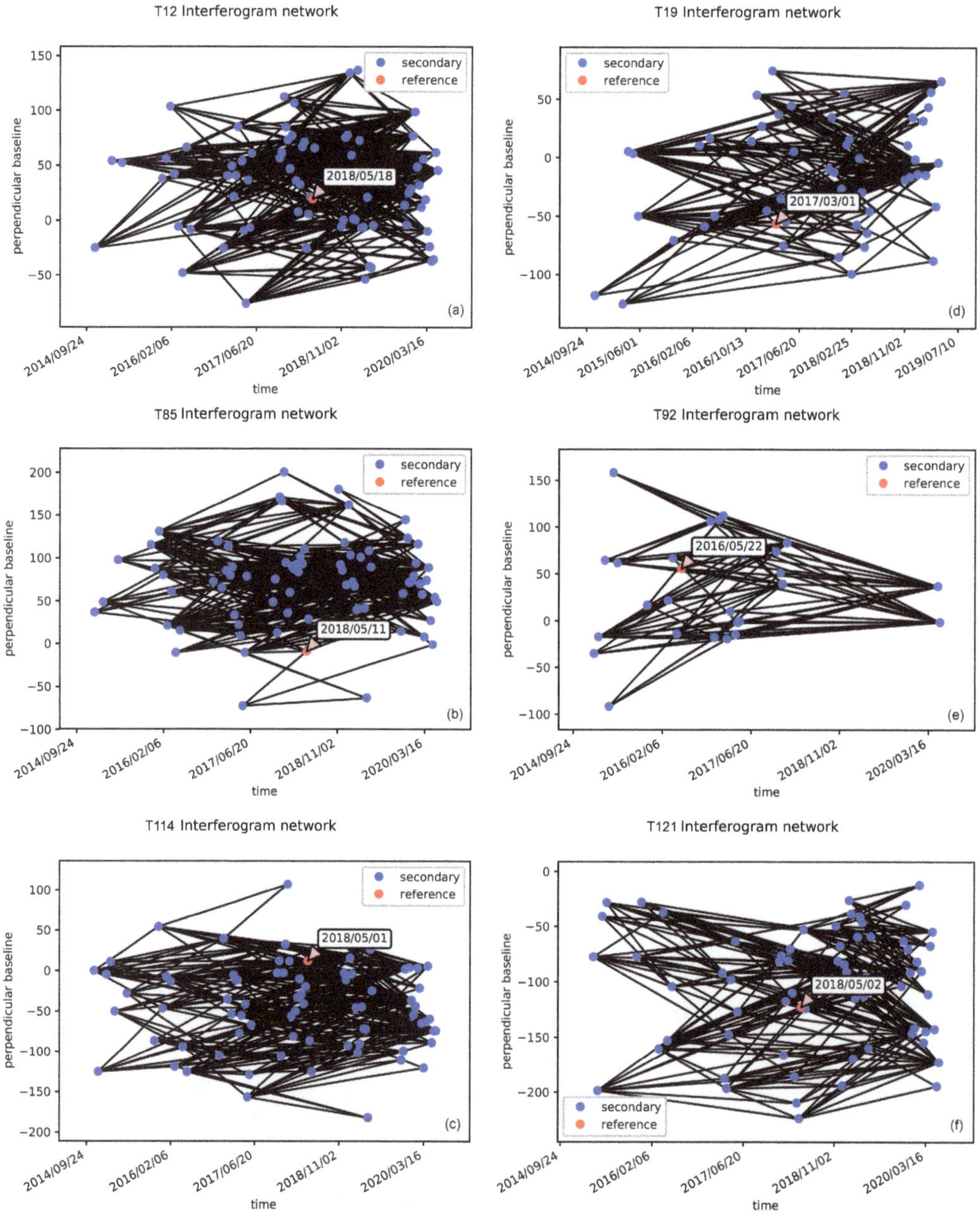

Figure A1. Computed interferograms for the six Sentinel-1 tracks. The reference images are shown with red dots.

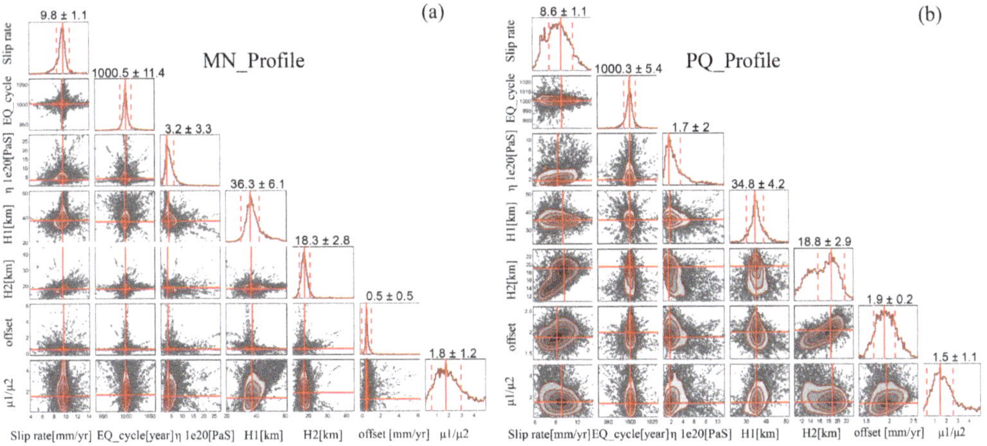

Figure A2. Maximum a posteriori probability (MAP) solution for profiles MN (**a**) and PQ (**b**). With the model, it has the same elastic thickness but different elastic shear modulus on both sides of the fault.

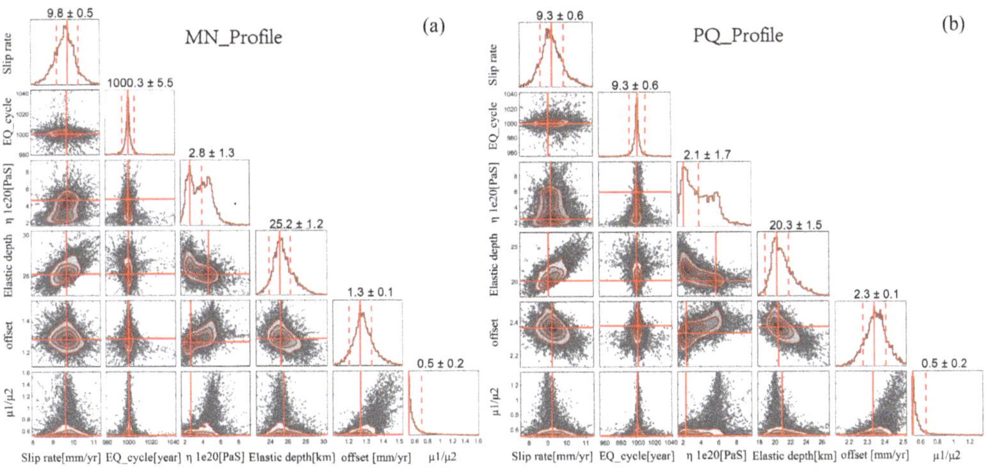

Figure A3. Maximum a posteriori probability (MAP) solution for profiles MN (**a**) and PQ (**b**). With the model, it has the same elastic thickness on both sides of the fault.

Table A1. A Priori Bounds in inversion for ATF.

Parameter	Minimum	Maximum	Priori Value
Fault slip rate (mm/yr)	−20	20	10
Locking depth (km)	0	35	18
Velocity offset (mm/yr)	0	15	3
Recurrence interval (years)	800	1500	1000
Time since last earthquake (years)	100	100	100
Elastic thickness of F_{NE}, H1 (km)	10	65	35
Elastic thickness of F_{SW}, H2 (km)	10	65	20
Viscosity η	10^{16}	10^{22}	10^{19}
Viscoelastic lower crust $\eta 2$	10^{16}	10^{22}	10^{19}
Rigidity ratio, $\mu 1/\mu 2$	0	5	2

55. Tymofyeyeva, E.; Fialko, Y. Mitigation of Atmospheric Phase Delays in InSAR Data, with Application to the Eastern California Shear Zone. *J. Geophys. Res. Solid Earth* **2015**, *120*, 5952–5963. [CrossRef]
56. Yu, C.; Li, Z.H.; Penna, N.T.; Crippa, P. Generic Atmospheric Correction Model for Interferometric Synthetic Aperture Radar Observations. *J. Geophys. Res. Solid Earth* **2018**, *123*, 9202–9222. [CrossRef]
57. Yu, C.; Penna, N.T.; Li, Z. Generation of Real-time Mode High-resolution Water Vapor Fields from GPS Observations. *J. Geophys. Res. Atmos.* **2017**, *122*, 2008–2025. [CrossRef]
58. Yu, C.; Li, Z.; Penna, N.T. Interferometric Synthetic Aperture Radar Atmospheric Correction Using a GPS-Based Iterative Tropospheric Decomposition Model. *Remote Sens. Environ.* **2018**, *204*, 109–121. [CrossRef]
59. Murray, K.D.; Bekaert, D.P.S.; Lohman, R.B. Tropospheric Corrections for InSAR: Statistical Assessments and Applications to the Central United States and Mexico. *Remote Sens. Environ.* **2019**, *232*, 111326. [CrossRef]
60. Wang, Q.; Yu, W.; Xu, B.; Wei, G. Assessing the Use of GACOS Products for SBAS-InSAR Deformation Monitoring: A Case in Southern California. *Sensors* **2019**, *19*, 3894. [CrossRef]
61. Wang, Y.Q.; Chang, L.; Feng, W.P.; Samsonov, S.; Zheng, W. Topography-Correlated Atmospheric Signal Mitigation for InSAR Applications in the Tibetan Plateau Based on Global Atmospheric Models. *Int. J. Remote Sens.* **2021**, *42*, 4364–4382. [CrossRef]
62. Jiang, G.Y.; Wang, Y.B.; Wen, Y.M.; Liu, Y.J.; Xu, C.; Xu, C.J. Afterslip Evolution on the Crustal Ramp of the Main Himalayan Thrust Fault Following the 2015 Mw 7.8 Gorkha (Nepal) Earthquake. *Tectonophysics* **2019**, *758*, 29–43. [CrossRef]
63. Feng, W.P.; Samsonov, S.; Liang, C.R.; Li, J.H.; Charbonneau, F.; Yu, C.; Li, Z.H. Source Parameters of the 2017 Mw 6.2 Yukon Earthquake Doublet Inferred from Coseismic GPS and ALOS-2 Deformation Measurements. *Geophys. J. Int.* **2019**, *216*, 1517–1528. [CrossRef]
64. Dogru, F. The Importance of Atmospheric Corrections on InSAR Surveys Over Turkey: Case Study of Tectonic Deformation of Bodrum-Kos Earthquake. *Pure Appl. Geophys.* **2020**, *177*, 5761–5780. [CrossRef]
65. Yu, C.; Li, Z.H.; Penna, N.T. Triggered Afterslip on the Southern Hikurangi Subduction Interface Following the 2016 Kaikōura Earthquake from InSAR Time Series with Atmospheric Corrections. *Remote Sens. Environ.* **2020**, *251*, 112097. [CrossRef]
66. Cavalié, O.; Doin, M.P.; Lasserre, C.; Briole, P. Ground Motion Measurement in the Lake Mead Area, Nevada, by Differential Synthetic Aperture Radar Interferometry Time Series Analysis: Probing the Lithosphere Rheological Structure. *J. Geophys. Res. Solid Earth* **2007**, *112*, 1–18. [CrossRef]
67. Biggs, J.; Wright, T.; Lu, Z.; Parsons, B. Multi-Interferogram Method for Measuring Interseismic Deformation: Denali Fault, Alaska. *Geophys. J. Int.* **2007**, *170*, 1165–1179. [CrossRef]
68. Wang, H.; Wright, T.J. Satellite Geodetic Imaging Reveals Internal Deformation of Western Tibet. *Geophys. Res. Lett.* **2012**, *39*, L07303. [CrossRef]
69. Wang, H.; Wright, T.J.; Liu-Zeng, J.; Peng, L. Strain Rate Distribution in South-Central Tibet From Two Decades of InSAR and GPS. *Geophys. Res. Lett.* **2019**, *46*, 5170–5179. [CrossRef]
70. Hussain, E.; Hooper, A.; Wright, T.J.; Walters, R.J.; Bekaert, D.P.S. Interseismic Strain Accumulation across the Central North Anatolian Fault from Iteratively Unwrapped InSAR Measurements. *J. Geophys. Res. Solid Earth* **2016**, *121*, 9000–9019. [CrossRef]
71. Walters, R.J.; Parsons, B.; Wright, T.J. Constraining Crustal Velocity Fields with InSAR for Eastern Turkey: Limits to the Block-like Behavior of Eastern Anatolia. *J. Geophys. Res. Solid Earth* **2014**, *119*, 5215–5234. [CrossRef]
72. Weiss, J.R.; Walters, R.J.; Morishita, Y.; Wright, T.J.; Lazecky, M.; Wang, H.; Hussain, E.; Hooper, A.J.; Elliott, J.R.; Rollins, C.; et al. High-Resolution Surface Velocities and Strain for Anatolia From Sentinel-1 InSAR and GNSS Data. *Geophys. Res. Lett.* **2020**, *47*. [CrossRef]
73. "Altyn Active Fault Zone" Team. *The Altyn Tagh Active Fault System*; Seismology Publishing House: Beijing, China, 1992. (In Chinese)
74. Fialko, Y.; Sandwell, D.; Agnew, D.; Simons, M.; Shearer, P.; Minster, B. Deformation on Nearby Faults Induced by the 1999 Hector Mine Earthquake. *Science* **2002**, *297*, 1858–1862. [CrossRef]
75. Lindsey, E.O.; Fialko, Y.; Bock, Y.; Sandwell, D.T.; Bilham, R. Localized and Distributed Creep along the Southern San Andreas Fault. *J. Geophys. Res. Solid Earth* **2014**, *119*, 7909–7922. [CrossRef]
76. Peltzer, G.; Crampé, F.; King, G. Evidence of Nonlinear Elasticity of the Crust from the Mw7.6 Manyi (Tibet) Earthquake. *Science* **1999**, *286*, 272–276. [CrossRef]
77. Ryder, I.; Parsons, B.; Wright, T.J.; Funning, G.J. Post-Seismic Motion Following the 1997 Manyi (Tibet) Earthquake: InSAR Observations and Modelling. *Geophys. J. Int.* **2007**, *169*, 1009–1027. [CrossRef]
78. Feng, M.X.; Bie, L.D.; Rietbrock, A. Probing the Rheology of Continental Faults: Decade of Post-Seismic InSAR Time-Series Following the 1997 Manyi (Tibet) Earthquake. *Geophys. J. Int.* **2018**, *215*, 600–613. [CrossRef]
79. Zhao, D.Z.; Qu, C.Y.; Bürgmann, R.; Gong, W.Y.; Shan, X.J. Relaxation of Tibetan Lower Crust and Afterslip Driven by the 2001 Mw7.8 Kokoxili, China, Earthquake Constrained by a Decade of Geodetic Measurements. *J. Geophys. Res. Solid Earth* **2021**, *126*, 1–33. [CrossRef]
80. Ran, Y.H.; Li, X.; Cheng, G.D.; Zhang, T.J.; Wu, Q.B.; Jin, H.J.; Jin, R. Distribution of Permafrost in China: An Overview of Existing Permafrost Maps. *Permafr. Periglac. Process.* **2012**, *23*, 322–333. [CrossRef]
81. Guo, D.L.; Wang, H.J.; Li, D. A Projection of Permafrost Degradation on the Tibetan Plateau during the 21st Century. *J. Geophys. Res. Atmos.* **2012**, *117*, 1–15. [CrossRef]
82. Fort, M.; Vliet-lanoe, B. Van Permafrost and Periglacial Environment of Western Tibet. *Landf. Anal.* **2014**, *5*, 25–29.

83. Yang, M.; Nelson, F.E.; Shiklomanov, N.I.; Guo, D.; Wan, G. Permafrost Degradation and Its Environmental Effects on the Tibetan Plateau: A Review of Recent Research. *Earth-Sci. Rev.* **2010**, *103*, 31–44. [CrossRef]
84. Cheng, G.D.; Wu, T.H. Responses of Permafrost to Climate Change and Their Environmental Significance, Qinghai-Tibet Plateau. *J. Geophys. Res. Earth Surf.* **2007**, *112*, 1–10. [CrossRef]
85. Rouyet, L.; Lauknes, T.R.; Christiansen, H.H.; Strand, S.M.; Larsen, Y. Seasonal Dynamics of a Permafrost Landscape, Adventdalen, Svalbard, Investigated by InSAR. *Remote Sens. Environ.* **2019**, *231*, 111236. [CrossRef]
86. Li, Z.W.; Zhao, R.; Hu, J.; Wen, L.X.; Feng, G.C.; Zhang, Z.Y.; Wang, Q.J. InSAR Analysis of Surface Deformation over Permafrost to Estimate Active Layer Thickness Based on One-Dimensional Heat Transfer Model of Soils. *Sci. Rep.* **2015**, *5*, 15542. [CrossRef]
87. Zhao, R.; Li, Z.W.; Feng, G.C.; Wang, Q.J.; Hu, J. Monitoring Surface Deformation over Permafrost with an Improved SBAS-InSAR Algorithm: With Emphasis on Climatic Factors Modeling. *Remote Sens. Environ.* **2016**, *184*, 276–287. [CrossRef]
88. Savage, J.C.; Burford, R.O. Geodetic Determination of Relative Plate Motion in Central California. *J. Geophys. Res.* **1973**, *78*, 832–845. [CrossRef]
89. Lindsey, E.O.; Fialko, Y. Geodetic Constraints on Frictional Properties and Earthquake Hazard in the Imperial Valley, Southern California. *J. Geophys. Res. Solid Earth* **2016**, *121*, 1097–1113. [CrossRef]
90. Johnson, K.M.; Fukuda, J. New Methods for Estimating the Spatial Distribution of Locked Asperities and Stress-Driven Interseismic Creep on Faults with Application to the San Francisco Bay Area, California. *J. Geophys. Res. Solid Earth* **2010**, *115*, 1–28. [CrossRef]
91. Fukuda, J.; Johnson, K.M. Mixed Linear-Non-Linear Inversion of Crustal Deformation Data: Bayesian Inference of Model, Weighting and Regularization Parameters. *Geophys. J. Int.* **2010**, *181*, 1441–1458. [CrossRef]
92. Fukuda, J.; Johnson, K.M. A Fully Bayesian Inversion for Spatial Distribution of Fault Slip with Objective Smoothing. *Bull. Seismol. Soc. Am.* **2008**, *98*, 1128–1146. [CrossRef]
93. Foreman-Mackey, D.; Hogg, D.W.; Lang, D.; Goodman, J. Emcee: The MCMC Hammer. *Publ. Astron. Soc. Pac.* **2013**, *125*, 306–312. [CrossRef]
94. Goodman, J.; Weare, J. Ensemble Samplers with Affine invariance. *Commun. Appl. Math. Comput. Sci.* **2010**, *5*, 65–80. [CrossRef]
95. Le Pichon, X.; Kreemer, C.; Chamot-Rooke, N. Asymmetry in Elastic Properties and the Evolution of Large Continental Strike-Slip Faults. *J. Geophys. Res. Solid Earth* **2005**, *110*, 1–11. [CrossRef]
96. Vaghri, A.; Hearn, E.H. Can Lateral Viscosity Contrasts Explain Asymmetric Interseismic Deformation around Strike-Slip Faults? *Bull. Seismol. Soc. Am.* **2012**, *102*, 490–503. [CrossRef]
97. Huang, W.J.; Johnson, K.M. Strain Accumulation across Strike-Slip Faults: Investigation of the Influence of Laterally Varying Lithospheric Properties. *J. Geophys. Res. Solid Earth* **2012**, *117*, 1–16. [CrossRef]
98. Fialko, Y. Interseismic Strain Accumulation and the Earthquake Potential on the Southern San Andreas Fault System. *Nature* **2006**, *441*, 968–971. [CrossRef] [PubMed]
99. Lindsey, E.O.; Fialko, Y. Geodetic Slip Rates in the Southern San Andreas Fault System: Effects of Elastic Heterogeneity and Fault Geometry. *J. Geophys. Res. Solid Earth* **2013**, *118*, 689–697. [CrossRef]
100. Jolivet, R.; Bürgmann, R.; Houlié, N. Geodetic Exploration of the Elastic Properties across and within the Northern San Andreas Fault Zone. *Earth Planet. Sci. Lett.* **2009**, *288*, 126–131. [CrossRef]
101. Meade, B.J. Estimates of Seismic Potential in the Marmara Sea Region from Block Models of Secular Deformation Constrained by Global Positioning System Measurements. *Bull. Seismol. Soc. Am.* **2002**, *92*, 208–215. [CrossRef]
102. Aktuğ, B.; Doğru, A.; Özener, H.; Peyret, M. Slip Rates and Locking Depth Variation along Central and Easternmost Segments of North Anatolian Fault. *Geophys. J. Int.* **2015**, *202*, 2133–2149. [CrossRef]
103. Fay, N.P.; Humphreys, E.D. Fault Slip Rates, Effects of Elastic Heterogeneity on Geodetic Data, and the Strength of the Lower Crust in the Salton Trough Region, Southern California. *J. Geophys. Res. B Solid Earth* **2005**, *110*, 1–14. [CrossRef]
104. Chéry, J. Geodetic Strain across the San Andreas Fault Reflects Elastic Plate Thickness Variations (Rather than Fault Slip Rate). *Earth Planet. Sci. Lett.* **2008**, *269*, 352–365. [CrossRef]
105. Lundgren, P.; Hetland, E.A.; Liu, Z.; Fielding, E.J. Southern San Andreas-San Jacinto Fault System Slip Rates Estimated from Earthquake Cycle Models Constrained by GPS and Interferometric Synthetic Aperture Radar Observations. *J. Geophys. Res. Solid Earth* **2009**, *114*, 1–18. [CrossRef]
106. Le Pichon, X.; Chamot-Rooke, N.; Rangin, C.; Sengör, A.M.C. The North Anatolian Fault in the Sea of Marmara. *J. Geophys. Res. Solid Earth* **2003**, *108*, 1–20. [CrossRef]
107. Johnson, K.M.; Segall, P. Viscoelastic Earthquake Cycle Models with Deep Stress-Driven Creep along the San Andreas Fault System. *J. Geophys. Res. Solid Earth* **2004**, *109*, 1–19. [CrossRef]
108. Savage, J.C.; Prescott, W.H. Asthenosphere Readjustment and the Earthquake Cycle. *J. Geophys. Res.* **1978**, *83*, 3369–3376. [CrossRef]
109. Meade, B.J.; Hager, B.H. Block Models of Crustal Motion in Southern California Constrained by GPS Measurements. *J. Geophys. Res. B Solid Earth* **2005**, *110*, 1–19. [CrossRef]
110. Ravikumar, M.; Singh, B.; Pavan Kumar, V.; Satyakumar, A.V.; Ramesh, D.S.; Tiwari, V.M. Lithospheric Density Structure and Effective Elastic Thickness Beneath Himalaya and Tibetan Plateau: Inference from the Integrated Analysis of Gravity, Geoid, and Topographic Data Incorporating Seismic Constraints. *Tectonics* **2020**, *39*, 1–26. [CrossRef]

111. Chen, B.; Liu, J.X.; Chen, C.; Du, J.S.; Sun, Y. Elastic Thickness of the Himalayan-Tibetan Orogen Estimated from the Fan Wavelet Coherence Method, and Its Implications for Lithospheric Structure. *Earth Planet. Sci. Lett.* **2015**, *409*, 1–14. [CrossRef]
112. Chen, B.; Chen, C.; Kaban, M.K.; Du, J.S.; Liang, Q.; Thomas, M. Variations of the Effective Elastic Thickness over China and Surroundings and Their Relation to the Lithosphere Dynamics. *Earth Planet. Sci. Lett.* **2013**, *363*, 61–72. [CrossRef]
113. Devries, P.M.R.; Meade, B.J. Earthquake Cycle Deformation in the Tibetan Plateau with a Weak Mid-Crustal Layer. *J. Geophys. Res. Solid Earth* **2013**, *118*, 3101–3111. [CrossRef]
114. Bai, D.H.; Unsworth, M.J.; Meju, M.A.; Ma, X.B.; Teng, J.W.; Kong, X.R.; Sun, Y.; Sun, J.; Wang, L.F.; Jiang, C.S.; et al. Crustal Deformation of the Eastern Tibetan Plateau Revealed by Magnetotelluric Imaging. *Nat. Geosci.* **2010**, *3*, 358–362. [CrossRef]
115. Sun, Y.J.; Dong, S.W.; Zhang, H.; Li, H.; Shi, Y.L. 3D Thermal Structure of the Continental Lithosphere beneath China and Adjacent Regions. *J. Asian Earth Sci.* **2013**, *62*, 697–704. [CrossRef]
116. Zhang, P.Z.; Deng, Q.D.; Zhang, G.M.; Ma, J.; Gan, W.J.; Min, W.; Mao, F.Y.; Wang, Q. Active Tectonic Blocks and Strong Earthquakes in the Continent of China. *Sci. China Ser. D Earth Sci.* **2003**, *46*, 13–24. [CrossRef]
117. Xu, X.; Wang, F.; Zheng, R.; Chen, W.; Ma, W.; Yu, G.; Chen, G.; Tapponnier, P.; Van Der Woerd, J.; Meriaux, A.-S.; et al. Late Quaternary Sinistral Slip Rate along the Altyn Tagh Fault and Its Structural Transformation Model. *Sci. China Ser. D Earth Sci.* **2005**, *48*, 384–397. [CrossRef]
118. Mériaux, A.S.; Tapponnier, P.; Ryerson, F.J.; Xu, X.W.; King, G.; Van der Woerd, J.; Finkel, R.C.; Li, H.B.; Caffee, M.W.; Xu, Z.Q.; et al. The Aksay Segment of the Northern Altyn Tagh Fault: Tectonic Geomorphology, Landscape Evolution, and Holocene Slip Rate. *J. Geophys. Res. Solid Earth* **2005**, *110*, 1–32. [CrossRef]
119. Zhang, P.Z.; Molnar, P.; Xu, X.W. Late Quaternary and Present-Day Rates of Slip along the Altyn Tagh Fault, Northern Margin of the Tibetan Plateau. *Tectonics* **2007**, *26*, 1–24. [CrossRef]
120. Cowgill, E. Impact of Riser Reconstructions on Estimation of Secular Variation in Rates of Strike-Slip Faulting: Revisiting the Cherchen River Site along the Altyn Tagh Fault, NW China. *Earth Planet. Sci. Lett.* **2007**, *254*, 239–255. [CrossRef]
121. Gold, R.D.; Cowgill, E.; Arrowsmith, J.R.; Gosse, J.; Chen, X.; Wang, X.F. Riser Diachroneity, Lateral Erosion, and Uncertainty in Rates of Strikeslip Faulting: A Case Study from Tuzidun along the Altyn Tagh Fault, NW China. *J. Geophys. Res. Solid Earth* **2009**, *114*, 1–24. [CrossRef]
122. Cowgill, E.; Gold, R.D.; Chen, X.; Wang, X.-F.; Arrowsmith, J.R.; Southon, J. Low Quaternary Slip Rate Reconciles Geodetic and Geologic Rates along the Altyn Tagh Fault Northwestern Tibet. *Geology* **2009**, *37*, 647–650. [CrossRef]
123. Bendick, R.; Bilham, R.; Freymueller, J.; Larson, K.; Yin, G. Geodetic Evidence for a Low Slip Rate in the Altyn Tagh Fault System. *Nature* **2000**, *404*, 69–72. [CrossRef]
124. Zhang, P.Z.; Shen, Z.K.; Wang, M.; Gan, W.J.; Bürgmann, R.; Molnar, P.; Wang, Q.; Niu, Z.J.; Sun, J.Z.; Wu, J.C.; et al. Continuous Deformation of the Tibetan Plateau from Global Positioning System Data. *Geology* **2004**, *32*, 809–812. [CrossRef]
125. Wallace, K.; Yin, G.; Bilham, R. Inescapable Slow Slip on the Altyn Tagh Fault. *Geophys. Res. Lett.* **2004**, *31*, 6–9. [CrossRef]
126. Mériaux, A.S.; Van Der Woerd, J.; Tapponnier, P.; Ryerson, F.J.; Finkel, R.C.; Lasserre, C.; Xu, X. The Pingding Segment of the Altyn Tagh Fault (91°E): Holocene Slip-Rate Determination from Cosmogenic Radionuclide Dating of Offset Fluvial Terraces. *J. Geophys. Res. Solid Earth* **2012**, *117*, 1–26. [CrossRef]
127. Wessel, P.; Smith, W.H.F.; Scharroo, R.; Luis, J.; Wobbe, F. Generic Mapping Tools: Improved Version Released. *EOS Trans. Am. Geophys. Union* **2013**, *94*, 409–410. [CrossRef]

remote sensing

MDPI

Article

A New Method for InSAR Stratified Tropospheric Delay Correction Facilitating Refinement of Coseismic Displacement Fields of Small-to-Moderate Earthquakes

Wenyu Gong [1,*], Dezheng Zhao [1], Chuanhua Zhu [2], Yingfeng Zhang [1], Chenglong Li [1], Guifang Zhang [1] and Xinjian Shan [1]

1 State Key Laboratory of Earthquake Dynamics, Institute of Geology, China Earthquake Administration, Beijing 100029, China; dezhengzhao@ies.ac.cn (D.Z.); jingqing129@ies.ac.cn (Y.Z.); chenglong-li@ies.ac.cn (C.L.); gfzhang@ies.ac.cn (G.Z.); xjshan@ies.ac.cn (X.S.)

2 Ministry of Natural Resources (MNR) Key Laboratory for Geo-Environmental Monitoring of Great Bay Area & Guangdong Key Laboratory of Urban Informatics & Shenzhen Key Laboratory of Spatial Smart Sensing and Services, School of Architecture & Urban Planning, Shenzhen University, Shenzhen 518060, China; czhu@szu.edu.cn

* Correspondence: gwenyu@ies.ac.cn

Abstract: Focusing on stratified tropospheric delay correction in the small-amplitude coseismic displacement field of small-to-moderate earthquakes (<Mw 6.5), we develop a Simple-Stratification-Correction (SSC) approach based on the empirical phase-elevation relationship and spatial properties of the troposphere, via an equal-size window segmentation. We validate our SSC method using 23 real earthquakes that occurred from January 2016 to May 2021 with a moment magnitude (Mw) ranging from 4.5 to 6.5. We conclude that SSC performs well according to the amount of reduction in semi-variance and the root-mean-square value. This method primarily focuses on stratification delay correction; thus, it is especially useful in regions with complex terrain, while it can mitigate partial large-scale turbulence signals. We investigate three parameters that are empirically setup in the correction working flow and inspect their optimal settings, when implementing SSC for quick response after earthquake. Our method is ready to be integrated into an operational InSAR processing chain to produce a reliable atmospheric phase screen map, which can also serve as an auxiliary product to quickly and timely quantify stratification delays in coseismic interferograms. Through improved accuracy of the coseismic displacement field, the focal mechanism could be better constrained to facilitate the building and expansion of the geodesy-based earthquake catalogue.

Keywords: Interferometric Synthetic Aperture Radar (InSAR); coseismic displacement; tropospheric delay correction; small-to-moderate earthquakes; geodetic earthquake catalogue

Citation: Gong, W.; Zhao, D.; Zhu, C.; Zhang, Y.; Li, C.; Zhang, G.; Shan, X. A New Method for InSAR Stratified Tropospheric Delay Correction Facilitating Refinement of Coseismic Displacement Fields of Small-to-Moderate Earthquakes. *Remote Sens.* **2022**, *14*, 1425. https://doi.org/10.3390/rs14061425

Academic Editors: Masashi Matsuoka and Claudia Meisina

Received: 17 January 2022
Accepted: 9 March 2022
Published: 15 March 2022

Publisher's Note: MDPI stays neutral with regard to jurisdictional claims in published maps and institutional affiliations.

1. Introduction

Interferometric Synthetic Aperture Radar (InSAR) observations are sensitive to the magnitude and location of earthquakes and is widely applied to recognize earthquakes with small moment magnitude (Mw < 6.5) but still having observable surface deformation [1]. Many previous studies have found that InSAR-derived coseismic displacement can be used to determine the location, depth, and coseismic slip distribution, providing more accurate results than those constrained by traditional seismic data [2–7]. It is especially useful for monitoring earthquakes that occurred in remote areas, where the density of the in situ observation network is limited. Constrained by InSAR-derived coseismic displacements, various algorithms and community tools have been developed to retrieve the kinematic and dynamic source parameters of earthquakes [8–12], greatly extending our knowledge of earthquake mechanisms. Additionally, the expansion of SAR satellite missions and the development of robust processing techniques in recent years led to increasing interest in building a geodetic earthquake catalogue using InSAR datasets [13] that complement earthquake relocation using traditional seismic observations.

However, the applicability of InSAR for deformation monitoring is limited by the influence of temporal decorrelation and electromagnetic path delay variations, both of which reduce the sensitivity and accuracy of the InSAR measurements [14,15]. This is especially severe with respect to small-to-moderate earthquakes, whose coseismic displacement could be contaminated with interferometric noises, such as atmosphere noise.

The uncertainties in InSAR observations have a significant impact on the accuracy of the fault slip models [2,16]. For example, InSAR-derived coseismic slip models often contain uncertainties in geometry determination, particularly in constraining fault dip orientation in cases of blind thrust faults [17]. Observations of such events can expand our knowledge of the local fault systems, given that they often occur on unmapped buried faults (e.g., [18]). The small-to-moderate earthquakes occur far more frequently than larger events. Many of them occur in remote areas, where GPS observations are often unavailable. With a simulated dataset Dawson et al. [2] demonstrated that InSAR is insensitive to Mw 6.2 earthquakes with a focal depth deeper than 10 km, and Mw 5.5 earthquakes occurred deeper than 6 km. It provides a rough bound of magnitude and depth of the InSAR-detectable coseismic deformation.

Our research will focus on those events with Mw $\leq$ 6.5. As a major factor impeding accurate displacement measurement, atmospheric delay is caused by the refractivity changes from a neutral atmosphere (mainly from the troposphere) and the ionosphere [19–22]. The artifacts due to ionosphere effects are correlated to radar sensor wavelength [15]. As suggested by Funning et al. [23], tropospheric noise is the major factor that causes ambiguities in identifying deformation signals in Sentinel-1 interferograms. In this study, we are concerned with the mitigation of tropospheric signals, which can be further classified into turbulent mixing and vertical stratification [20], informed by the physical origin in a neutral atmosphere.

Many previous studies that relied on various auxiliary tropospheric information sources have achieved success in mitigating InSAR tropospheric delays. The atmospheric phase screen can be obtained from dense Global Navigation Satellite System (GNSS) network (e.g., [24,25]), multispectral remote sensing products [26], and numerical weather prediction [27], or from reanalysis products (e.g., [28–30]), or combined outputs from above sources (e.g., GACOS [31]). However, the external tropospheric correction schemes are either limited by spatially coarse input (e.g., of the absence of a dense GNSS network) or only effective for specific satellite missions (e.g., MERIS products or an ENVISAT mission) or significantly dependent on the quality of input reanalysis products (e.g., European Centre for Medium-Range Weather Forecasts (ECMWF) products).

Researchers have also made progress in separating atmospheric components from interferometric phase-self, based on the spatial/temporal characteristics of tropospheric signals. Focusing on coseismic displacement refinement, several studies adopted advanced multi-temporal InSAR strategies (e.g., [32–34]) to correct atmospheric delays, considering the temporally stochastic properties of the troposphere's components. However, the requirement of a large SAR dataset in these approaches implies that it takes several years to calculate a long-term time-series, which in turn causes incoherence due to a large temporal baseline. More importantly, stratification delays are seasonally correlated [35], contradicting the underlying assumption in multi-temporal InSAR analysis.

Other previous efforts explored the separation of atmospheric delays from a single interferogram. These studies included an investigation of the modeling of co-variance due to atmospheric delays, to determine a reasonable weighting in the inversion as well as to improve accuracy when estimating geophysical parameters using InSAR displacement products [36,37]. Other algorithms focused on stratified delays and estimated corresponding contributions from terrain dependence [38–42]. For those methods, some studies applied a phase-elevation model directly to an entire interferogram (e.g., [39]), while other studies applied segmentation (e.g., segmenting an interferogram into multiple windows) or spatial spectrum decomposition, and then performed the phase-elevation correction in each segment. For example, windows-based algorithms may split images into multiple windows with constant size [41,43], or adjust the window size based on local elevation gradients [44]. Lin et al. [38] developed a multi-scale approach via band-pass filtering to estimate and remove the tropospheric delay components. Bekaert et al. [42] developed a power-law-based phase-elevation model and applied a band filter to reduce the displacement signal before modeling tropospheric phases. Murray et al. [40] segmented each interferogram using a K-means clustering algorithm and constructed a tropospheric delay screen from each cluster.

To improve the accuracy of the observed coseismic displacements and better serve various geophysical applications, our study aims to develop a new tropospheric artifact mitigation method, drawing on the empirical phase-elevation relationship and spatial variant atmosphere. In the following discussion, a brief introduction to 23 small-to-moderate earthquakes in our test, InSAR datasets and the details of our proposed method will be described. The quality of our correction algorithm will be subsequently validated by applying it to multiple earthquakes and comparing the results to GACOS products. Finally, we discuss various parameter settings in our method and related them to atmospheric physical properties. The discussion will help users to better apply our method and to understand the composition of the modeled tropospheric phases.

2. Data and Methodology

2.1. Atmospheric Stratification Delay Estimation

The changes of tropospheric reflectivity between two SAR image acquisitions causes a phase shift that would interfere with surface deformation signals. According to the physical origin of tropospheric signal, interferometric phases include contributions from the turbulent mixing component and the vertical stratification:

1. The turbulent delay component typically results from turbulent processes in the troposphere and leads to three-dimensional spatial heterogeneities in refractivity [20]. The stochastic model of tropospheric spatial variability shows a power-law dependence on frequency [36]. The contribution from this type of component could be reduced by averaging independent interferograms [19].
2. Vertical stratified delay is caused by the different vertical refractivity profiles of two acquisitions, assuming no heterogeneity within the horizontal layers [20]. In contrast to the turbulent component, the vertical stratified delay is correlated with topography and affects regions with mountainous terrain conditions [20,21]. Additionally, as revealed by the ERA5 weather model dataset, the stratified delay features a seasonal fluctuation [35]. Many methods rely on a suite of assumptions about the spatial-temporal characteristics of InSAR signals (e.g., the linear progression of deformation over time and the zero-mean Gaussian nature of atmospheric phases) that are often not valid in cases of the stratification delay. This is of great significance for many earthquake studies using InSAR. The coupling in the temporal domain further complicates the correction of stratification delays. Thus, the estimated surface deformation may be ambiguous, and inappropriate correction may lead to an erroneous interpretation of the earthquake deformation.

Mitigation of the stratification delay can be performed based on the empirical relationship between the interferometric phase and elevation [31,38–40,45,46]. Cavalié et al. [39] approximated the relationship as a linear function between height and phase:

$$\varphi_{stra} = K \times H + C \quad (1)$$

where K denotes the slope relates to topography H and phase φ_{stra}, and C denotes the constant parameter. In order to better approximate the phase spatial variation at a higher part of the atmosphere, Bekaert et al. [41] described phase-elevation dependence using a power-law decaying model:

$$\phi_{stra} = A \times (h_0 - h)^{\alpha} \tag{2}$$

where A is a scale factor that relates to the elevation, α is a constant factor describing the power-law decaying, h_0 is the reference height where the tropospheric phase converges to zero (h_0 suggested to be ~7–13 km [41]), and thus $h_0 - h$ denotes a height difference relative to the reference height at each pixel.

2.2. Correction Working Flow

Focusing on the applications to earthquake studies, especially with respect to small-to-moderate earthquakes, we introduced an empirical phase-elevation-model-based tropospheric delay correction method, Simple-Stratification-Correction (termed SSC hereinafter). Two challenging factors should be considered are (1) avoiding partial or complete removal of real deformation signals, because the deformation may be correlated to elevation [38,40], and (2) preparing for significant spatial variation of the tropospheric properties at a several-kilometer scale [41,43,44]. Here, we masked the main deforming zone for each earthquake and segmented the images into small windows to deal with these two problems. The detailed steps are described below:

1. We masked the coseismic displacement area in individual interferograms. The mask was generated based on the reported epicenter. We aimed to avoid participation of pixels dominated by coseismic displacements in estimating phase-elevation model parameters.
2. We then segmented each interferogram into multiple small patches. We cropped the interferogram into $M \times 2^N$ by $M \times 2^N$ dimensions and the coseismic zone had to be included in this step. We aimed to roll-change the window numbers by a factor of 2 in a more automatic manner. The choice of the number of segmented windows was related to the scale of the estimated stratifications and the size of the earthquakes. We found that a patch number of 8 or 16 in both range and azimuth was appropriate for most of the earthquakes tested here. More scenarios on properly splitting windows will be discussed in Section 4.
3. Each window contained a cluster of pixels, which were subsequently utilized to estimate the phase-elevation model parameters via the empirical linear model (Figure 1). For a window partially impacted by coseismic deformation, the percentage of deforming pixels was computed in conjunction a with previously determined mask of the coseismic zone. The empirical threshold was 60%, meaning that if >60% pixels were not masked, they were used to estimate the phase-elevation model parameters in this window. However, if the percentage was smaller than 60%, this window was recognized as a masked one and the corresponding parameters estimation were skipped here.
4. The next step was to fill the masked windows that failed in the direct estimation in the previous step. We calculated the semi-variogram structure from the input phase over the non-deforming zone. This step aided the kriging solution in predicting the model parameters at those windows dominated by coseismic deformation zones.
5. Based on the obtained sparse C (constant parameter) and K (scale parameter) grid, we applied the kriging interpolation to calculate the values at all pixels. For each pixel, we obtained a pair of K and C. Given the known height, we applied these resolved values back to the empirical phase-elevation model to calculate ϕ_{stra}. Note that the estimated parameters in the last step were assigned to be at the center of each window. Therefore, after the spatial interpolation, it left an empty area that was not computable. This appears in Figure 1b as the region outside the dashed box. The width of this zone is the half of a segmented window.
6. Finally, we subtracted the modeled tropospheric delay from each unwrapped interferogram to obtain the corrected coseismic displacement map.

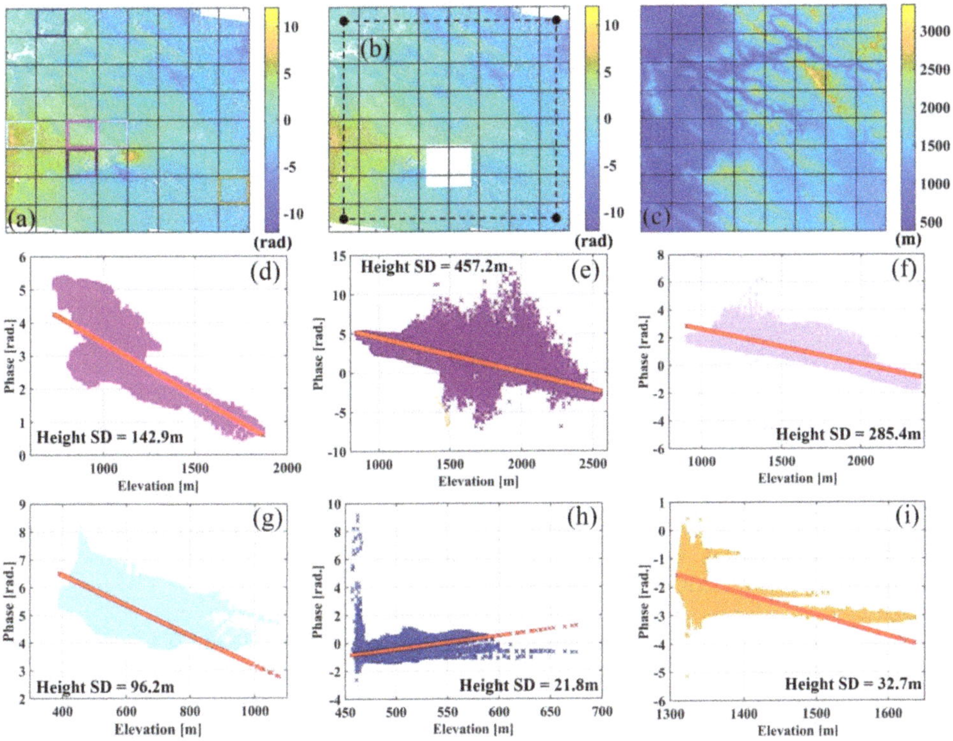

Figure 1. Demonstration of the primary ideal of the developed stratification correction scheme with the 25 August 2018, Mw 5.98 Javanrud, Iran, earthquake as an example: (**a**) is the interferogram subset with the original unwrapped phase and segmented with an 8 × 8 window; (**b**) is the unwrapped phase with coseismic zone masked; the dashed box denotes zones computed during interpolation; (**c**) is the elevation of the corresponding area; (**d**–**i**) are phase-elevation model examples for a single window. The red markers show the computed regression model; other colored markers denote original phase-elevation pixel pairs and they are associated with the box in (**a**) by the same color.

To better understand the overall procedure of SSC, Figure 1 takes the M 6.0 earthquake that occurred on 25 August 2018, near Javanrud, Iran, as an example and provides a visual demonstration. The unwrapped interferogram are segmented into 8 × 8 windows. Six phase-elevation model examples are shown in Figure 1d–i. For each plot, the color of the marks corresponds to the box with the same color in Figure 1a. They have different terrain conditions. The one in Figure 1f is partially masked, due to the nearby coseismic zone.

We calculated the standard deviation (SD) of the regional height inside each window and the coefficient of determination (R^2) for each linear regression. Both calculations were annotated in the corresponding plots. Of the unmasked windows, 60% had height SDs larger than 200 m, and 50% achieved R^2 values larger than 0.5. This indicated that this coseismic example had certain topography relief and the regression reached an ordinary quality. Figure 1h,i shows two windows as examples over a flat area. According to the distribution of these phase-elevation samples, the stratification relationship was weak here and the main atmospheric impact of this local zone was a likely turbulence component. Thus, the absence of terrain relief was a main factor that reduced the quality in the phase-elevation modeling.

In summary, the key objectives of our SSC solution are: (1) estimating atmospheric phase from the non-deforming pixels, implemented by introducing a proper mask for each earthquake; and (2) combining the phase-elevation dependence model with spatial property of the troposphere.

Note that we have also tested with the power-low model and integrated that test into SSC scheme. However, after the interpolation step (step 4), it failed in producing a physically reasonable atmospheric delay map. We suspect that the parameters of the power-law model are not likely correlated at the spatial scale we used in SSC. Therefore, our subsequent real earthquake test was performed using the linear phase-elevation model.

2.3. Earthquake Catalog and Coseismic InSAR Observations

We tested our SSC method with a total of 23 earthquakes that occurred from January 2016 to May 2021 (Table 1). The moment magnitude ranged from Mw 4.8 to 6.5. We excluded earthquakes that occurred on flat areas as we are focusing on the stratification correction. We also discarded those earthquakes that occurred at the seashore, because there were not enough coherent pixels surrounding the coseismic zones, thereby causing difficulties in determining the phase-elevation dependence. All of these earthquakes were observed by Sentinel–1 SAR images, and three of them had both descending and ascending interferograms. The majority of the coseismic interferometric pairs were downloaded from SARVIEW projects (https://sarviews-hazards.alaska.edu/, accessed on 4 March 2022). SARVIEW is a fully automatic D-InSAR processing system for Sentinel-1SAR acquisitions; it includes interferometric pair selection, phase filtering, unwrapping, geocoding, and other post-processing steps [47]. The focal mechanism parameters of the 22 May 2016, Dingjie earthquake and the 8 August 2017, Jinghe earthquake were obtained from Hou et al. [33] and Gong et al. [48], respectively, while the others were obtained from Zhu et al. [13]. We limited the temporal baseline of each interferogram to the shortest revisit period, generally 6–12 days, to maximize the coherence. It was also important to make sure that there were enough pixels surrounding the coseismic zones to ensure the sufficiency of non-deforming pixels in the atmospheric delay modeling.

Table 1. Information of earthquakes (Mw $\leq$ 6.5) tested in this study.

Index	Date	Location	Lat. (deg)	Lon. (deg)	Depth (km)	Mw	Strike (deg)	Dip (deg)	Rake (deg)	Patch Number
1	20 January 2016	Menyuan, China	37.72	101.68	12.50	5.87	129	47	77	8
2	8 February 2016	San Juan Xiutetelco, Mexico	19.67	−97.45	1.50	4.83	160	80	96	16
3	22 May 2016	Dingjie, China	28.48	87.61	2.40	5.58	188	43	−78	8
4	24 August 2016	Norcia, Italy	42.74	13.29	5.00	6.16	168	50	−83	8
5	17 October 2016	Zaduo, China	32.88	94.82	10.50	5.85	68	69	−104	8
6	26 October 2016	Visso, Italy	42.97	13.20	5.00	6.28	155	40	−91	8
7	1 December 2016	Huarichancara, Peru	−15.29	−70.84	5.50	6.18	150	44	−92	8
8	8 December 2016	Shihezi, China	43.79	86.33	19.00	5.82	280	61	80	16
9	5 April 2017	Torbat-e Jam, Iran	35.83	60.44	8.92	6.14	124	42	62	8

Table 1. *Cont.*

Index	Date	Location	Lat. (deg)	Lon. (deg)	Depth (km)	Mw	Strike (deg)	Dip (deg)	Rake (deg)	Patch Number
10 #	27 May 2017	Golmarmara, Turkey	38.73	27.79	5.00	5.58	304	60	−89	16
11	8 August 2017	Jinghe, China	44.26	82.72	14.05	6.19	90	42	89.56	8
12	1 December 2017	Kerman, Iran	30.78	57.34	7.45	6.08	119	51	80	8
13	25 August 2018	Javanrud, Iran	34.63	46.24	5.50	5.98	264	78	4	8
14	25 November 2018	Sarpol-e Zahab, Iran	34.39	45.60	14.50	6.50	25	63	−173	8
15	25 February 2019	Yanling, China	29.47	104.50	1.67	4.83	174	64	80	8
16	22 March 2020	Kasina, Croatia	45.84	16.04	9.15	5.55	46	16	9	8
17	15 May 2020	Monte Cristo Range, America	38.18	−117.93	9.84	6.44	76	73	−10	4
18	25 June 2020	Hotan, China	35.61	82.47	9.78	6.28	186	68	−89	8
19 #	22 July 2020	Western Xizang, China	33.20	86.82	10.00	6.36	31	52	−80	8
20	29 December 2020	Petrinja, Croatia	45.43	16.22	5.00	6.48	120	78	174	8
21 #	3 March 2021	Tyrnavos, Greece	39.63	22.15	6.00	6.39	292	35	−98	8 (Asc.) 6 (Des.)
22	18 April 2021	Bandar, Iran	29.72	50.60	6.50	6.10	306	60	81	8
23	21 May 2021	Dali, China	25.64	99.94	7.00	6.09	134	90	179	16

These three earthquakes had coseismic observations on both the ascending (Asc.) and descending (Des.) tracks.

3. Results

Our proposed atmospheric delay correction approach, SSC, was evaluated using 23 earthquakes in nature. Three of them had coseismic observations on both the ascending and descending tracks. Therefore, a total of 26 coseismic interferograms were used in the real data validation. We calculated the statistics before and after the corrections, in order to better quantify the effectiveness of such corrections. In order to compare the correction quality, we also applied correction with the Generic Atmospheric Correction Online Service (GACOS; http://www.gacos.net/, accessed on 4 March 2022). The GACOS product has global availability and provides high spatial resolution and easy-to-implement zenith total delay maps [31].

We took the epicenter shown in Table 1 as a priori information to mask the coseismic deformation zone. We solved the phase-elevation dependence, as described previously. We then reconstructed the atmospheric stratification delay for each coseismic interferogram and removed that factor from the original unwrapped phase. The resulting corrected coseismic displacement maps could be further used in slip distribution inversion in geophysical studies.

As shown in Figures 2 and 3, we took two earthquakes, the Mw 4.8 Mexico earthquake and the Mw 6.5 Iran earthquake, as examples for our demonstration. These two earthquakes had the minimum and maximum magnitudes in all our tested earthquakes, respectively. These two figures demonstrated interferograms before and after the atmospheric delay correction using our SSC method and the GACOS products. For these two examples, both the GACOS and SSC reduced the noises, but at different levels. The residual signals after SSC in the corrected interferograms (Figures 2c and 3c) were likely the localized phase noises (e.g., the small scale turbulence or localized stratification). The plots (d,e) of Figures 2 and 3 are the scale and offset parameter maps after kriging interpolation. The unit of digital elevation applied in the modeling was meter, therefore, Figure 2d shows a relatively small value range. Both Figures 2f and 3f show the displacement differences along a profile across the coseismic zone after the correction. Figures 2g and 3g are the histograms calculated from non-deforming pixels that were identified using the same masks in SSC implementation. They show an apparent change in the shape of displacement distribution, where the SSC largely reduced the tropospheric delays. The correction results for all other earthquakes are demonstrated in the Supplementary Materials (Figures S1–S21).

Additionally, we calculated the Root-Mean-Square (RMS) value before and after the atmospheric correction over the non-deforming zone as a statistic factor to evaluate the overall correction quality. This value was used to indicate an overall level of phase noise in each interferogram. The result is summarized in Figure 4. After implementing our SSC, most interferograms had a reduced RMS by 45%.

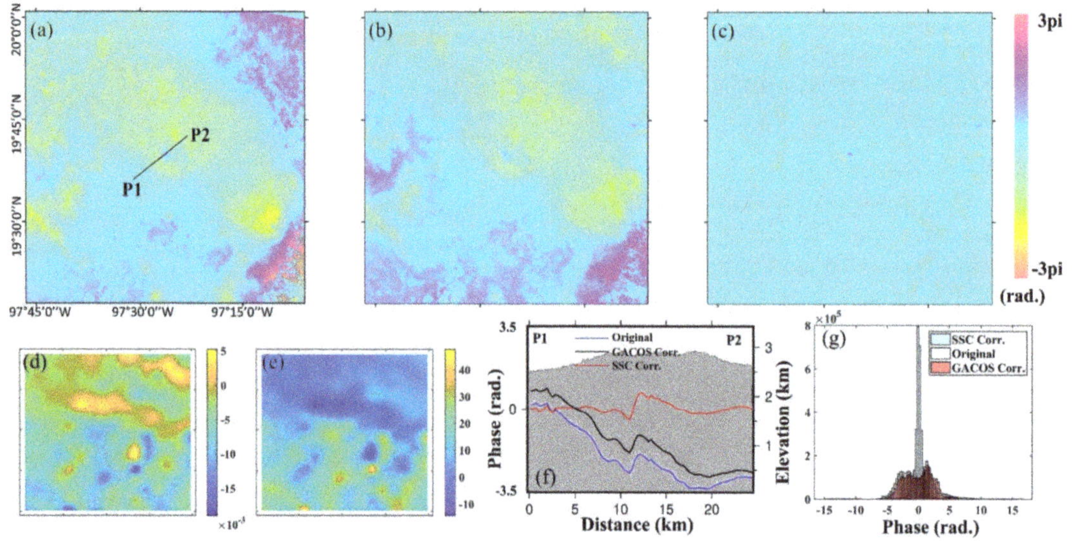

Figure 2. Examples of atmospheric correction: (**a**–**c**) are original, GACOS-corrected-, and SSC-corrected interferograms of the Mw 4.83 Mexico earthquake;(**d**) is the scale factor map; (**e**) is the offset factor map; (**f**) shows the phase changes along cross section P1–P2 in (**a**); (**g**) is the histogram of (**a**–**c**) after masking the main coseismic zone.

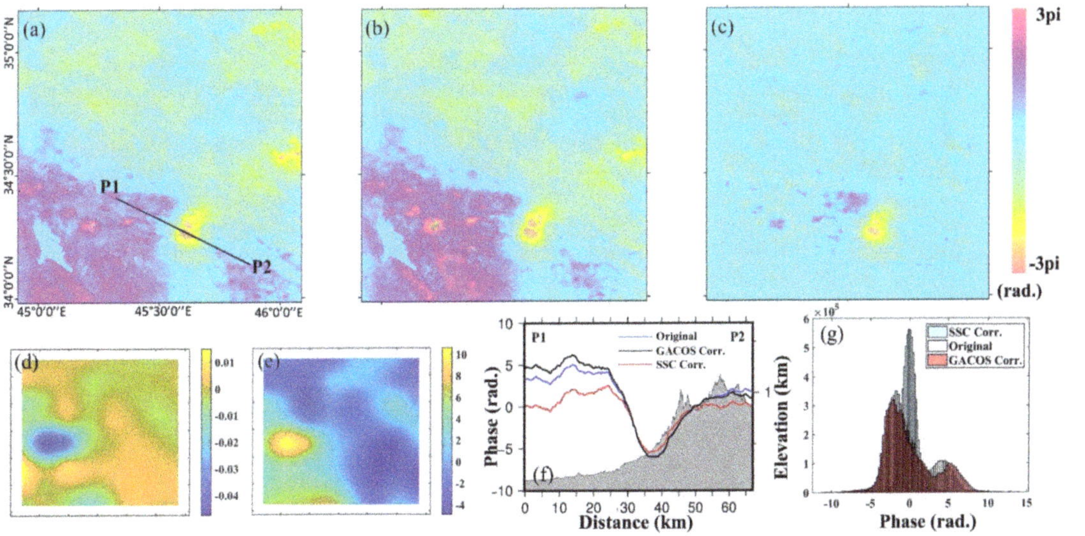

Figure 3. Examples of atmospheric correction: (**a**–**c**) are original, GACOS-corrected, and SSC-corrected interferograms of the Mw 6.5 Sarpol-e Zahab, Iran, earthquake; (**d**) is the scale factor map; (**e**) is the offset factor map; (**f**) shows the phase changes along cross section P1–P2 in (**a**); (**g**) is the histogram of (**a**–**c**) after masking the main coseismic zone.

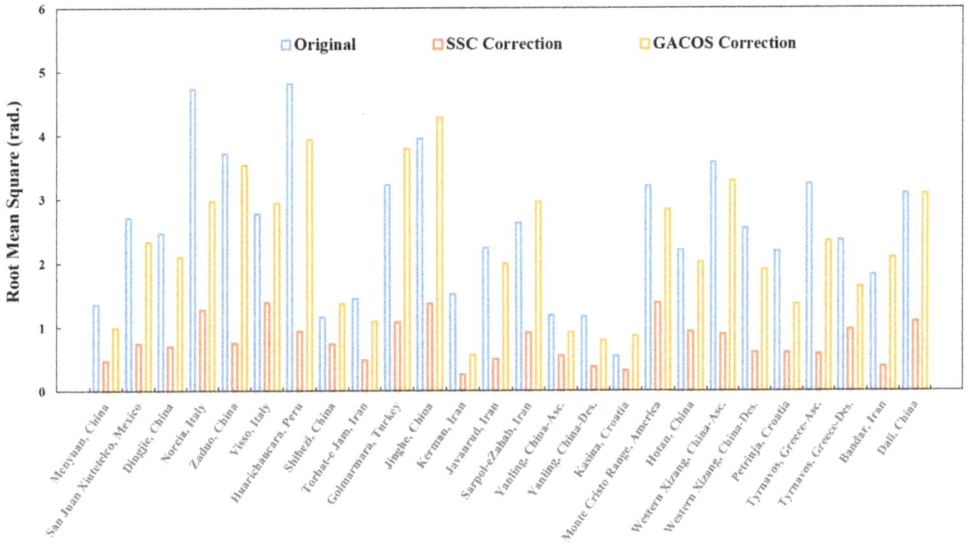

Figure 4. Statistics before and after correction. The *Y*-axis denotes the RMS value of each case with a unit of radiance. The blue bars denote the image RMS before correction. The red bars denote the image RMS after applying the method developed here. The yellow bars denote the image RMS after correction by GACOS products.

We also calculated the semivarigoram to quantify the reduction in the noise magnitude at various spatial scales. The semivariogram depicts the spatial autocorrelation of the sampled measurements and was used to quantify the atmospheric impacts in many previous studies (e.g., [20,40,49]). Note that we excluded pixels at deforming zones in the calculation.

As shown in Figure 5, SSC performs well for all tested coseismic interferograms, with a significant reduction in semi-variance. In this study, the SSC-corrected interferograms showed a reduction of 5 to 20 km at an overall spatial scale. This was likely related to (1) the spatial size of the coseismic zone; (2) the parameters used in the SSC correction, especially the number of windows; and (3) the spatial scale of the regional terrain. Here, we took the Mw 6.5 Sarpol-e Zahab, Iran, earthquake (Figure 3) as an example. After cropping the interferogram into a 2^N by 2^N dimension, the subset was then split into 8×8 windows and each window corresponded to a spatial size of ~17 km $\times$ 17 km. Note that we did not expect to accurately reconstruct the real phase-elevation parameters below this spatial scale (e.g., smaller than a single window size). Hence, the atmospheric signals related to local terrain or turbulence signal with small spatial scale remained as residuals in the corrected interferograms.

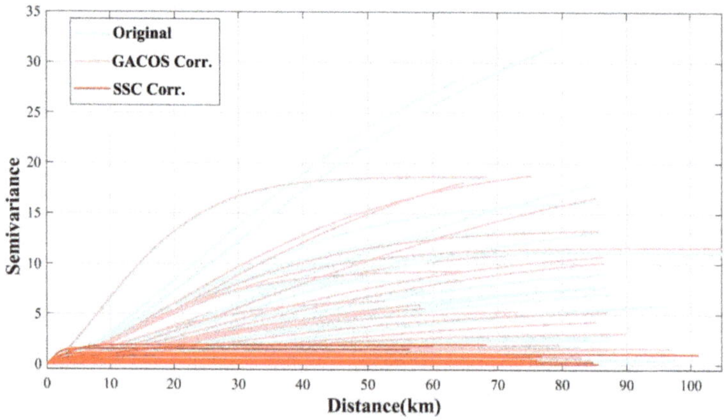

Figure 5. Semivariance of tested earthquake dataset before and after atmosphere correction.

Overall, the SSC method provided a satisfactory performance, validated by both semi-variance and RMS. Another advantage of the SSC method was that it only used a single interferogram. Compared to the multi-temporal-analysis-based tropospheric correction methods (e.g., [32–34]), our SSC could perform as long as a single interferogram was available. In this study, the GACOS products could also reduce partial atmospheric delays. The inconsistent correction quality of GACOS may be caused by the input atmospheric reanalysis products or by the GNSS meteorological products in each single case.

4. Discussion

In this section, we inspect the optimal parameter settings in SSC and its performance dependence to regional terrain. We also discuss the link of the parameters of the empirical phase-elevation model to the tropospheric contributions. This analysis help us in understanding which part of tropospheric artifacts could be modeled.

4.1. Parameter Setting

To implement the SSC, there were three parameters that had to be empirically set up. The first one was the size of image subset covering the coseismic zone, which was cropped from the uncorrected unwrapped interferogram. This was not a critical step, and the SSC could be implemented into an entire interferogram. This step was conducted in order to improve computation efficiency and to exclude large water bodies (e.g., oceans). To generate this subset, we required (1) enough land area surrounding a coseismic zone, ensuring sufficient pixels for phase-elevation parameter estimation, and (2) an image subset having a dimension of $M \times 2^N$ in width and length, ensuring that the number of windows in both directions was $\times 2^i$ ($i < 16$).

The second empirically determined input was the mask of the coseismic zone. The mask, or an equivalent method of reducing the estimation bias caused by coseismic deformation, could be realized via a spatiotemporal filter. For example, Murray et al. [40] built a temporal low-pass filter to identify and mask deforming pixels. Bekaert et al. [42] applied a band filter, which was constructed based on an initial analysis with InSAR and GNSS, to avoid contamination of the tectonic slow slip deformation signal. Given that the rupture length of small-to-moderate earthquakes with Mw $\leq$ 6.5 may be limited to 10–20 km in length or less (e.g., an Mw 6.5 event may produce a rupture of about 20 km [50]), and concentrating on a known tectonic area from a seismically determined epicenter, we built a rectangle mask for each earthquake. The spatial extent of the mask was generally controlled by the earthquake's magnitude and focal depth. In this study, the mask zone was manually set up mainly based on USGS reported epicenters. The overall length of the masks for all of the events here ranged from 3.3 to 33 km. Here, we also proposed a potential treatment to generate a proper mask in an automatic way, to allow for possible use in future applications. First, one can predict a spatial size of the coseismic displacement zone with a facility's (e.g., via U.S. Geological Survey Earthquake Hazards Program) reported focal mechanism parameters. In other words, the size of the mask could be considered as a known. Thereafter, one could search for an optimal masking location by minimizing the standard deviation of un-masked pixel phases. As suggested by a previous study [51], the average distance between an InSAR-reported epicenter and one from USGS was 3.7 km for Iranian earthquakes, and 2.7 km for Japanese earthquakes. Therefore, the search could be started from USGS-reported epicenters and bounded in a limited zone within a several-kilometers radius.

The third consideration is the number of segmented windows. As shown in Table 1, we preferentially segmented the interferogram into 8 × 8 window subsets for most of the events. For a few earthquakes with smaller magnitudes or deeper depths, we applied finer window segmentation (16 × 16). For earthquakes with similar moment magnitudes (Mw 4.5–6.5), an 8 × 8 window segmentation was sufficiently effective. To guarantee the correction quality, we suggest not splitting an image into very fine equal-size windows. Because the number of pixels of a smaller window would be reduced, the chance of them capturing stratification delays presented in areas with a sufficient signal-to-noise ratio became smaller. This would reduce the robustness of the phase-elevation model's parameter regression. In addition, it would leave more "blind" windows sitting inside the coseismic displacement zone. As mentioned above, for those windows containing insufficient number of pixels to capture the atmospheric delay, calculation could only be done by kriging interpolation instead of by direct phase-elevation regression. However, the derived windows could not be too coarse; otherwise, a single window could be controlled by a multiple trend of phase-elevation dependence.

4.2. Relationship with Regional Terrain

We compared the percentage of the RMS reduction value with the regional terrain conditions in order to quantify the terrain's dependence of correction quality. This helped us to better understand its future application scenario. We took the RMS in Figure 4 and calculated the corresponding reduction percentage in each case. Figure 6 visually demonstrates the comparison of the mean and standard deviation (SD) of the regional elevation. It shows that a reduction was >60% for most of the tested cases. For those with an average elevation <2.5 km, the reduction ranged from 30% to 90%. For the other interferograms with a higher average elevation, the correction percentage was >50%. However, we did not see a correlation between the quality of correction and the regional terrain complexity according to the SD of the height.

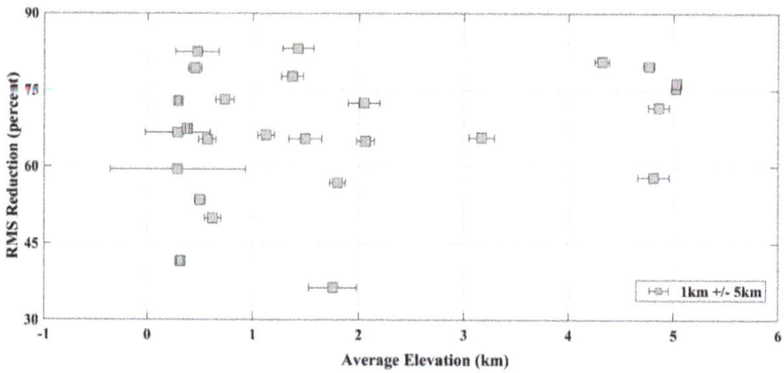

Figure 6. The RMS reduction after SSC vs. average & standard deviation of the local elevation. The X-axis has a unit of kilometer; the Y-axis is the percentage of RMS reduction; the error bar of each marker is scaled by 5 km.

4.3. Regression Models and Parameters

As presented above, the linear approximation [39] between atmospheric delays and local topography can be described by a scale factor and a constant. If we apply the phase-elevation model to an entire interferogram, the constant parameter is only a shift of the interferogram [45]. The estimation of phase-elevation model parameters is implemented for each individual window in the SSC method. Thus, the estimated constant parameter is an overall phase shift of each window. This shift describes the regional long wavelength signals and should be smooth in space. Therefore, after the spatial kriging interpolation, the whole map of the constant parameter approximates the long wavelength signal, which is likely composed of both turbulence and stratification components, while the estimation of the scale parameter (K in Equation (1)) describes the relationship between the phase and the regional elevation. Both the de-correlated pixels and those pixels within the coseismic zones should be pre-masked before the estimation.

After spatial interpolation of the K field, we obtained a pixel-by-pixel map that could be used to model stratification within the interferogram subset. Therefore, according to meaning of the parameter pair in the phase-elevation model and also the choice of spatial interpolation in the SSC procedure, this method estimated not only stratification but also some amount of large-scale troposphere phases.

As suggested by Bekaert et al. [41], the linearization assumption is only valid for the lower part of the atmosphere. Bekaert et al. suggested a power-law model that predicted that the phase delay would converge to zero at larger elevations. To integrate the power-law model into our SSC method, we modified the power-law model by adding an offset parameter ε (see Equation (3)).

$$\phi_{stra} = A \times (h_0 - h)^{\alpha} + \varepsilon \quad (3)$$

This is because the offset term can be used to represent the overall shift within each window that is caused by the atmospheric delays with a large spatial scale, as discussed above. Figure 7a–c shows the SSC resolved three parameters (scale parameter A, power parameter α, and offset parameter ε) at each window. Figure 7d–f shows the results after applying kriging interpolation. Figure 8 demonstrates the resulting stratification delay maps based on the window-wise parameters and interpolated parameter maps. Figure 8a indicates that the power-law model works well when each window is only associated with a single set of estimated parameters. However, we could not produce a meaningful phase-delay result (Figure 8b) when using interpolated parameter maps. Even the overall spatial pattern after kriging remained similar to the original window-wise parameters (see Figure 7). The failure here was likely because these variables in the power-law model remain invariant across a certain area; thus, they do not satisfy the spatially smooth assumption in the SSC strategy.

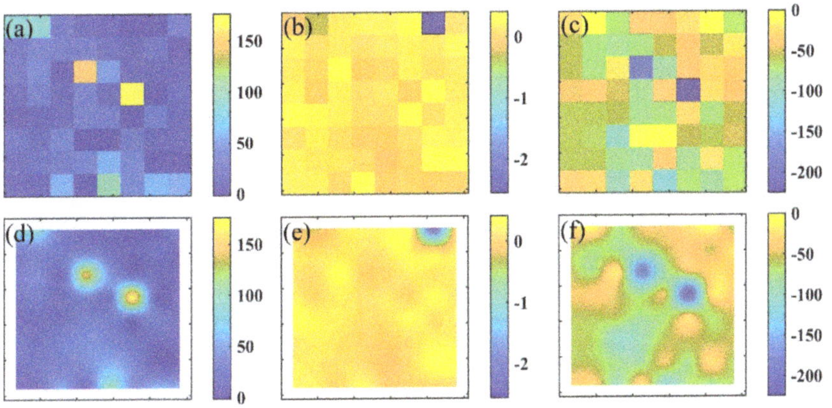

Figure 7. Demonstration of power-law-based SSC method: (**a**–**c**) show the estimated window-wise scale factor, the power factor, and the offset factor in Equation (3), respectively, and each window is described by a single set of variables; (**d**–**f**) are parameter maps after spatial interpolation, corresponding to the scale factor, the power factor and the offset factor, from left to right.

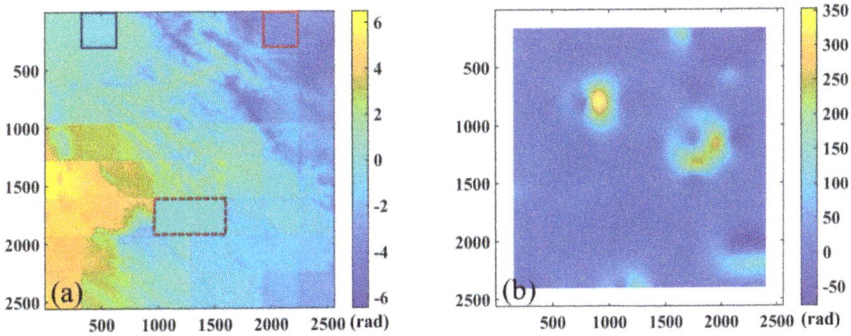

Figure 8. Resulting stratification delay maps based on the estimated power–law model parameters at each window and the interpolated parameter maps. (**a**) reconstructed atmospheric delay map base on power-law model with a single set of estimated parameters of each window; (**b**) reconstructed atmospheric delays base on power-law model with interpolated parameters maps. Dashed box denotes the masked zone. Solid boxes denote where power-law parameter estimation failed.

The dashed box in Figure 8 denotes the masked coseismic zone, while the solid box denotes windows where the nonlinear inversion failed in finding optimal values. The failure was mainly because the phase-elevation relationship was weak at these windows (e.g., a flat zone). For example, the dark blue window in Figure 8 bounds a region identical to that outline by the box with the same color in Figure 1a. Both the linear model and the power-law model show unsatisfied performance in this area. The region in the red solid box has some terrain relief (with a height SD of 180.0 m), while also failed in modeling power-law model parameters. The difficulty was caused by the mixing of multiple phase-elevation dependence in this window. Overall, the linear-model-based SSC method is preferred and performed well on the tested coseismic cases.

5. Conclusions

InSAR observations provide rich products for building a geodetic earthquake catalogue. Nevertheless, the accuracy and the capability of earthquake deformation observations are largely dependent on the atmospheric noisy level of the derived displacement field. Focusing on the coseismic displacement field reconstruction for small-to-moderate earthquakes (<Mw 6.5), we developed a Simple-Stratification-Correction (SSC) method based on the phase-elevation relationship and spatial smooth property of the troposphere. SSC employs a windows-based segmentation strategy, given its simplicity and efficiency. By applying the spatial-correlation property of atmospheric delays, SSC allows for the phase-elevation dependence variant in space. The developed SSC method is especially useful for zones with complex terrain, where the stratification delay is most severe. Our method is also helpful in mitigating some large-scale turbulence signals, according to the property of the offset term in the phase-elevation model and its usage in the SSC approach.

In SSC method, a proper mask is necessary to exclude impacts from the coseismic signal in the tropospheric phase modeling, which can reduce the bias from coseismic displacement signals. We also determined the optimal parameter setting to implement SSC for earthquake events with moment magnitude (Mw) 4.5 to 6.5. SSC only needs a single interferometic pair, rather than an interferogram stack, and thus can provide a quicker response after an earthquake. A single interferogram can maintain better coherence with the shortest available temporal baseline. This method can also be integrated in time series analysis that would have a broad implication for other tectonic studies [52,53], e.g., assessing interseismic displacement [54,55]. The key to initiate such applications with SSC is to provide a proper mask or other analysis schemes (e.g., integrating it in a small baseline approach [56]) that can reduce the bias in atmospheric signal modeling.

Overall, by improving the accuracy of the co-seismic displacement field, one could retrieve the seismogenic parameters with better quality that facilitate the building and expansion of the geodetic earthquake catalogue. Thus, it allows us to extend the capacity of InSAR in detecting and monitoring small earthquakes, increasing the number of the geodetic-based earthquake studies, and improving the quality of geodetic earthquake catalogues. Our method is ready to be integrated into an operational InSAR processing chain, e.g., one can produce an additional atmospheric phase screen map via SSC as an auxiliary product to help users understand the amount of stratification delays in each coseismic interferogram.

Supplementary Materials: The following supporting information can be downloaded at: https://www.mdpi.com/article/10.3390/rs14061425/s1. This file contains Figure S1–S21: Interferograms before and after tropospheric corrections of the rest 21 earthquake events tested in this study are listed here. Figure S1: Correction result for the Mw 5.87 earthquake in Menyuan, China, on 20 January 2016. Figure S2: Correction result for the Mw 5.58 earthquake in Dingjie, China, on 22 May 2016. Figure S3: Correction result for the Mw 6.16 earthquake in Norcia, Italy, on 24 August 2016. Figure S4: Correction result for the Mw 5.58 earthquake in Zaduo, China, on 17 October 2016. Figure S5: Correction result for the Mw 6.28 earthquake in Visso, Italy, on 26 October 2016. Figure S6: Correction result for the Mw 6.18 earthquake in Huarichancara, Peru, on 1 December 2016. Figure S7: Correction result for the Mw 5.82 earthquake in Shihezi, China, on 8 December 2016. Figure S8: Correction result for the Mw 6.14 earthquake in Torbat-e Jam, Iran, on 5 April 2017. Figure S9: Correction result for the Mw5.58 earthquake in Golmarmara, Turkey, on 27 May 2017. Figure S10: Correction result for the Mw6.19 earthquake in Jinghe, China, on 8 August 2017. Figure S11: Correction result for the Mw6.08 earthquake in Kerman, Iran, on 1 December 2017. Figure S12: Correction result for the Mw5.98 earthquake in Javanrud, Iran, on 25 August 2018. Figure S13: Correction result for the Mw 4.83 earthquake in Yanling, China, on 5 February 2019. Figure S14: Correction result for the Mw5.55 earthquake in Kasina, Croatia, on 22 March 2020. Figure S15: Correction result for the Mw6.44 earthquake in Monte Cristo Range, America, on 15 May 2020. Figure S16: Correction result for the Mw6.28 earthquake in Hotan, China, on 25 June 2020. Figure S17: Correction result for the Mw 6.36 earthquake in Western Xizang, China, on 22 July 2020. Figure S18: Correction result for the Mw6.48 earthquake in Petrinja, Croatia, on 29 December 2020. Figure S19: Correction result for the Mw 6.39 earthquake in Tyrnavos, Greece, on 3 March 2021. Figure S20: Correction result for the Mw6.10 earthquake in Bandar, Iran, on 18 April 2021. Figure S21: Correction result for the Mw6.09 earthquake in Dali, China, on 21 May 2021.

Author Contributions: Conceptualization, W.G.; methodology, W.G.; validation, W.G., C.Z. and C.L.; investigation, W.G, D.Z. and Y.Z.; writing—original draft preparation, W.G.; writing—review and editing, W.G., D.Z., Y.Z., C.L., G.Z. and X.S.; supervision, X.S.; funding acquisition, X.S. and W.G. All authors have read and agreed to the published version of the manuscript.

Funding: This research was supported by the National Key Research and Development Program of China (Grant No. 2019YFC1509201), Basic Scientific Funding of the Institute of Geology, China Earthquake Administration (No. IGCEA1719, Institute of Geology), the National Natural Science Foundation of China (Grant No. 42174015), and the China Seismic Experimental Site (CSES) program (China Earthquake Administration).

Institutional Review Board Statement: Not applicable.

Informed Consent Statement: Not applicable.

Data Availability Statement: The Sentinel-1 dataset was provided by the European Space Agency (ESA), through the Alaska Satellite Facility (https://vertex.daac.asf.alaska.edu/, accessed on 4 March 2022). Some Sentinel-1 interferograms were produced via the SARVIEWS project (https://sarviews-hazards.alaska.edu/, accessed on 4 March 2022), through download from Alaska Satellite Facility Data Center.

Acknowledgments: We would like to thank all of the anonymous reviewers for their insightful comments on the paper.

Conflicts of Interest: The authors declare no conflict of interest.

References

1. Lohman, R.B.; Simons, M. Locations of selected small earthquakes in the Zagros mountains. *Geochem. Geophys. Geosyst.* **2005**, *6*, Q03001. [CrossRef]
2. Dawson, J.; Tregoning, P. Uncertainty analysis of earthquake source parameters determined from InSAR: A simulation study. *J. Geophys Res.-Sol. Earth* **2007**, *112*, B09406. [CrossRef]
3. Funning, G.J.; Parsons, B.; Wright, T.J.; Jackson, J.A.; Fielding, E.J. Surface displacements and source parameters of the 2003 Bam (Iran) earthquake from Envisat advanced synthetic aperture radar imagery. *J. Geophys. Res. Solid Earth* **2005**, *110*, B09406. [CrossRef]
4. Mellors, R.J.; Magistrale, H.; Earle, P.; Cogbill, A.H. Comparison of four moderate-size earthquakes in southern California using seismology and InSAR. *Seismol. Soc. Am.* **2004**, *94*, 2004–2014. [CrossRef]
5. Kwong, K.B.; DeShon, H.R.; Kim, J.W.; Lu, Z. Resolving Teleseismic Earthquake Catalog and InSAR Data Discrepancies in Absolute Space to Explore Rupture Complexity Along the Ecuadorian Megathrust Fault. *J. Geophys. Res. Solid Earth* **2019**, *124*, 6703–6719. [CrossRef]
6. Weston, J.; Ferreira, A.M.G.; Funning, G.J. Global compilation of interferometric synthetic aperture radar earthquake source models: 1. Comparisons with seismic catalogs. *J. Geophys. Res. Solid Earth* **2011**, *116*, B08408. [CrossRef]
7. Ferreira, A.M.G.; Weston, J.; Funning, G.J. Global compilation of interferometric synthetic aperture radar earthquake source models: 2. Effects of 3-D Earth structure. *J. Geophys. Res. Solid Earth* **2011**, *116*, B08409. [CrossRef]
8. Wang, R.; Diao, F.; Hoechner, A. SDM—A geodetic inversion code incorporating with layered crust structure and curved fault geometry. In Proceedings of the General Assembly European Geosciences Union, Vienna, Austria, 7–12 April 2013.
9. Tong, X.; Sandwell, D.T.; Fialko, Y. Coseismic slip model of the 2008 Wenchuan earthquake derived from joint inversion of interferometric synthetic aperture radar, GPS, and field data. *J. Geophys. Res. Solid Earth* **2010**, *115*, B04314. [CrossRef]
10. Isken, M.; Sudhaus, H.; Heimann, S.; Steinberg, A.; Daout, S.; Vasyura-Bathke, H. *Kite—Software for Rapid Earthquake Source Optimisation from InSAR Surface Displacement*; V.0.1; GFZ Data Services: Potsdam, Germany, 2017.
11. Lindsey, E.O.; Natsuaki, R.; Xu, X.; Shimada, M.; Hashimoto, M.; Melgar, D.; Sandwell, D.T. Line-of-sight displacement from ALOS-2 interferometry: Mw 7.8 Gorkha Earthquake and Mw 7.3 aftershock. *Geophys. Res. Lett.* **2015**, *42*, 6655–6661. [CrossRef]
12. Elliott, J.R.; Walters, R.J.; Wright, T.J. The role of space-based observation in understanding and responding to active tectonics and earthquakes. *Nat. Commun.* **2016**, *7*, 13844. [CrossRef]
13. Zhu, C.; Wang, C.; Zhang, B.; Qin, X.; Shan, X. Differential Interferometric Synthetic Aperture Radar data for more accurate earthquake catalogs. *Remote Sens. Environ.* **2021**, *266*, 112690. [CrossRef]
14. Pepe, A.; Calò, F. A Review of Interferometric Synthetic Aperture RADAR (InSAR) Multi-Track Approaches for the Retrieval of Earth's Surface Displacements. *Appl. Sci.* **2017**, *7*, 1264. [CrossRef]

15. Hooper, A.; Bekaert, D.; Spaans, K.; Arıkan, M. Recent advances in SAR interferometry time series analysis for measuring crustal deformation. *Tectonophysics* **2012**, *514–517*, 1–13. [CrossRef]
16. Lohman, R.B.; Simons, M. Some thoughts on the use of InSAR data to constrain models of surface deformation: Noise structure and data downsampling. *Geochem. Geophys. Geosystems* **2005**, *6*, Q01007. [CrossRef]
17. Zhang, Y.; Shan, X.; Gong, W.; Zhang, G. The ambiguous fault geometry derived from InSAR measurements of buried thrust earthquakes: A synthetic data based study. *Geophys. J. Int.* **2021**, *225*, 1799–1811. [CrossRef]
18. Xu, W.; Dutta, R.; Jónsson, S. Identifying Active Faults by Improving Earthquake Locations with InSAR Data and Bayesian Estimation: The 2004 Tabuk (Saudi Arabia) Earthquake Sequence. *Seismol. Soc. Am.* **2015**, *105*, 765–775. [CrossRef]
19. Williams, S.; Bock, Y.; Fang, P. Integrated satellite interferometry: Tropospheric noise, GPS estimates and implications for interferometric synthetic aperture radar products. *J. Geophys. Res. Solid Earth* **1998**, *103*, 27051–27067. [CrossRef]
20. Hanssen, R. *Radar Interferometry: Data Interpretation and Error Analysis*, 1st ed.; Kluwer Academic Publishers: Dordrecht, The Netherlands, 2001; Volume 2.
21. Massonnet, D.; Feigl, K.L. Radar interferometry and its application to changes in the earth's surface. *Rev. Geophys.* **1998**, *36*, 441–500. [CrossRef]
22. Emardson, T.R.; Simons, M.; Webb, F.H. Neutral atmospheric delay in interferometric synthetic aperture radar applications: Statistical description and mitigation. *J. Geophys. Res.-Sol. Earth* **2003**, *108*, 2231. [CrossRef]
23. Funning, G.J.; Garcia, A. A systematic study of earthquake detectability using Sentinel-1 Interferometric Wide-Swath data. *Geophys. J. Int.* **2018**, *216*, 332–349. [CrossRef]
24. Webley, P.W.; Bingley, R.M.; Dodson, A.H.; Wadge, G.; Waugh, S.J.; James, I.N. Atmospheric water vapour correction to InSAR surface motion measurements on mountains: Results from a dense GPS network on Mount Etna. *Phys. Chem. Earth* **2002**, *27*, 363–370. [CrossRef]
25. Onn, F.; Zebker, H.A. Correction for interferometric synthetic aperture radar atmospheric phase artifacts using time series of zenith wet delay observations from a GPS network. *J. Geophys. Res. Solid Earth* **2006**, *111*, B09102. [CrossRef]
26. Li, Z.H.; Muller, J.P.; Cross, P.; Fielding, E.J. Interferometric synthetic aperture radar (InSAR) atmospheric correction: GPS, moderate resolution Imaging spectroradiometer (MODIS), and InSAR integration. *J. Geophys. Res.-Sol. Earth* **2005**, *110*, B03410. [CrossRef]
27. Gong, W.; Meyer, F.J.; Lee, C.W.; Lu, Z.; Freymueller, J. Measurement and interpretation of subtle deformation signals at Unimak Island from 2003 to 2010 using weather model-assisted time series InSAR. *J. Geophys. Res. Solid Earth* **2015**, *120*, 1175–1194. [CrossRef]
28. Webley, P.W.; Wadge, G.; James, I.N. Determining radio wave delay by non-hydrostatic atmospheric modelling of water vapour over mountains. *Phys. Chem. Earth* **2004**, *29*, 139–148. [CrossRef]
29. Jolivet, R.; Grandin, R.; Lasserre, C.; Doin, M.P.; Peltzer, G. Systematic InSAR tropospheric phase delay corrections from global meteorological reanalysis data. *Geophys. Res. Lett.* **2011**, *38*, L17311. [CrossRef]
30. Jolivet, R.; Agram, P.S.; Lin, N.N.Y.; Simons, M.; Doin, M.P.; Peltzer, G.; Li, Z.H. Improving InSAR geodesy using Global Atmospheric Models. *J. Geophys. Res.-Sol. Earth* **2014**, *119*, 2324–2341. [CrossRef]
31. Yu, C.; Li, Z.; Penna, N.T.; Crippa, P. Generic Atmospheric Correction Model for Interferometric Synthetic Aperture Radar Observations. *J. Geophys. Res. Solid Earth* **2018**, *123*, 9202–9222. [CrossRef]
32. Luo, H.; Wang, T.; Wei, S.; Liao, M.; Gong, J. Deriving Centimeter-Level Coseismic Deformation and Fault Geometries of Small-To-Moderate Earthquakes From Time-Series Sentinel-1 SAR Images. *Front. Earth Sci.* **2021**, *9*, 32. [CrossRef]
33. Hou, L.; Shan, X.; Gong, W.; Zhang, Y.; Luo, H.; Zhao, D. Characterizing seismogenic fault of 2016 Dingjie earthquake based on multitemporal DInSAR. *Chin. J. Geophys.* **2020**, *63*, 1357–1369. [CrossRef]
34. Daout, S.; Sudhaus, H.; Kausch, T.; Steinberg, A.; Dini, B. Interseismic and Postseismic Shallow Creep of the North Qaidam Thrust Faults Detected with a Multitemporal InSAR Analysis. *J. Geophys. Res. Solid Earth* **2019**, *124*, 7259–7279. [CrossRef]
35. Dong, J.; Zhang, L.; Liao, M.; Gong, J. Improved correction of seasonal tropospheric delay in InSAR observations for landslide deformation monitoring. *Remote Sens. Environ.* **2019**, *233*, 111370. [CrossRef]
36. Knospe, S.H.G.; Jonsson, S. Covariance Estimation for dInSAR Surface Deformation Measurements in the Presence of Anisotropic Atmospheric Noise. *IEEE Trans. Geosci. Remote Sens.* **2010**, *48*, 2057–2065. [CrossRef]
37. Agram, P.S.; Simons, M. A noise model for InSAR time series. *J. Geophys. Res.-Sol. Earth* **2015**, *120*, 2752–2771. [CrossRef]
38. Lin, Y.N.N.; Simons, M.; Hetland, E.A.; Muse, P.; DiCaprio, C. A multiscale approach to estimating topographically correlated propagation delays in radar interferograms. *Geochem. Geophys. Geosystems* **2010**, *11*, Q09002. [CrossRef]
39. Cavalié, O.; Doin, M.-P.; Lasserre, C.; Briole, P. Ground motion measurement in the Lake Mead area, Nevada, by differential synthetic aperture radar interferometry time series analysis: Probing the lithosphere rheological structure. *J. Geophys. Res. Solid Earth* **2007**, *112*, B03403. [CrossRef]
40. Murray, K.D.; Lohman, R.B.; Bekaert, D.P.S. Cluster-Based Empirical Tropospheric Corrections Applied to InSAR Time Series Analysis. *IEEE Trans. Geosci. Remote Sens.* **2021**, *59*, 2204–2212. [CrossRef]
41. Bekaert, D.P.S.; Hooper, A.; Wright, T.J. A spatially-variable power-law tropospheric correction technique for InSAR data. *J. Geophys. Res. Solid Earth* **2015**, *120*, 1345–1356. [CrossRef]
42. Bekaert, D.P.S.; Hooper, A.; Wright, T.J. Reassessing the 2006 Guerrero slow slip event, Mexico: Implications for large earthquakes in the Guerrero Gap. *J. Geophys. Res. Solid Earth* **2015**, *120*, 1357–1375. [CrossRef]

43. Bejar-Pizarro, M.; Socquet, A.; Armijo, R.; Carrizo, D.; Genrich, J.; Simons, M. Andean structural control on interseismic coupling in the North Chile subduction zone. *Nat. Geosci.* **2013**, *6*, 462–467. [CrossRef]
44. Liang, H.; Zhang, L.; Ding, X.; Lu, Z.; Li, X. Toward Mitigating Stratified Tropospheric Delays in Multitemporal InSAR: A Quadtree Aided Joint Model. *IEEE Trans. Geosci. Remote Sens.* **2019**, *57*, 291–303. [CrossRef]
45. Bekaert, D.P.S. InSAR Time Series Analysis of the 2006 Slow Slip Event on the Guerrero Subduction Zone, Mexico. Master's Thesis, Delft University of Technology, Delft, The Netherlands, 2010.
46. Doin, M.P.; Lasserre, C.; Peltzer, G.; Cavalie, O.; Doubre, C. Corrections of stratified tropospheric delays in SAR interferometry: Validation with global atmospheric models. *J. Appl. Geophys.* **2009**, *69*, 35–50. [CrossRef]
47. Meyer, F.J.; Whitley, M.; Logan, T.; McAlpin, D.B.; Hogenson, K.; Nicoll, J.B. The sarviews project: Automated processing of sentinel-1 sar data for geoscience and hazard response. In Proceedings of the IGARSS 2019—2019 IEEE International Geoscience and Remote Sensing Symposium, Yokohama, Japan, 28 July–2 August 2019; pp. 5468–5471.
48. Gong, W.; Zhang, Y.; Li, T.; Wen, S.; Zhao, D.; Hou, L.; Shan, X. Multi-Sensor Geodetic Observations and Modeling of the 2017 Mw 6.3 Jinghe Earthquake. *Remote Sens.* **2019**, *11*, 2157. [CrossRef]
49. Murray, K.D.; Bekaert, D.P.S.; Lohman, R.B. Tropospheric corrections for InSAR: Statistical assessments and applications to the Central United States and Mexico. *Remote Sens. Environ.* **2019**, *232*, 111326. [CrossRef]
50. Wells, D.L.; Coppersmith, K.J. New empirical relationships among magnitude, rupture length, rupture width, rupture area, and surface displacement. *Seismol. Soc. Am.* **1994**, *84*, 974–1002. [CrossRef]
51. Ghayournajarkar, N.; Fukushima, Y. Using InSAR for evaluating the accuracy of locations and focal mechanism solutions of local earthquake catalogues. *Geophys. J. Int.* **2022**. [CrossRef]
52. Dal Zilio, L.; Hetényi, G.; Hubbard, J.; Bollinger, L. Building the Himalaya from tectonic to earthquake scales. *Nat. Rev. Earth Environ.* **2021**, *2*, 251–268. [CrossRef]
53. Michel, S.; Jolivet, R.; Rollins, C.; Jara, J.; dal Zilio, L. Seismogenic Potential of the Main Himalayan Thrust Constrained by Coupling Segmentation and Earthquake Scaling. *Geophys. Res. Lett.* **2021**, *48*, e2021GL093106. [CrossRef]
54. Dal Zilio, L.; Jolivet, R.; van Dinther, Y. Segmentation of the Main Himalayan Thrust Illuminated by Bayesian Inference of Interseismic Coupling. *Geophys. Res. Lett.* **2020**, *47*, e2019GL086424. [CrossRef]
55. Grandin, R.; Doin, M.-P.; Bollinger, L.; Pinel-Puysségur, B.; Ducret, G.; Jolivet, R.; Sapkota, S.N. Long-term growth of the Himalaya inferred from interseismic InSAR measurement. *Geology* **2012**, *40*, 1059–1062. [CrossRef]
56. Lauknes, T.R. InSAR Tropospheric Stratification Delays: Correction Using a Small Baseline Approach. *IEEE Geosci. Remote Sens. Lett.* **2011**, *8*, 1070–1074. [CrossRef]

remote sensing

MDPI

Article

Coseismic Deformation Field Extraction and Fault Slip Inversion of the 2021 Yangbi M_W 6.1 Earthquake, Yunnan Province, Based on Time-Series InSAR

Xue Li [1,2], Chisheng Wang [1,2,*], Chuanhua Zhu [1,2,3], Shuying Wang [1,2], Weidong Li [4], Leyang Wang [5] and Wu Zhu [6]

1 Ministry of Natural Resources (MNR) Key Laboratory for Geo-Environmental Monitoring of Great Bay Area, Guangdong Key Laboratory of Urban Informatics, Shenzhen Key Laboratory of Spatial Smart Sensing and Services, Shenzhen University, Shenzhen 518000, China; 2060325019@email.szu.edu.cn (X.L.); czhu@szu.edu.cn (C.Z.); wangshuying2019@email.szu.edu.cn (S.W.)

2 School of Architecture and Urban Planning, Shenzhen University, Shenzhen 518000, China

3 State Key Laboratory of Earthquake Dynamics, Institute of Geology, China Earthquake Administration, Beijing 100029, China

4 School of Information Science and Engineering, Henan University of Technology, Zhengzhou 450001, China; wdli@haut.edu.cn

5 School of Surveying and Mapping Engineering, East China University of Technology, Nanchang 330013, China; wleyang@ecut.edu.cn

6 School of Geological Engineering and Surveying and Mapping, Chang'an University, Xi'an 710054, China; zhuwu@chd.edu.cn

* Correspondence: wangchisheng@szu.edu.cn

Citation: Li, X.; Wang, C.; Zhu, C.; Wang, S.; Li, W.; Wang, L.; Zhu, W. Coseismic Deformation Field Extraction and Fault Slip Inversion of the 2021 Yangbi M_W 6.1 Earthquake, Yunnan Province, Based on Time-Series InSAR. *Remote Sens.* **2022**, *14*, 1017. https://doi.org/10.3390/rs14041017

Academic Editor: Cristiano Tolomei

Received: 15 January 2022
Accepted: 15 February 2022
Published: 19 February 2022

Abstract: An earthquake of moderate magnitude (M_W 6.1) occurred in Yangbi County, Dali, Yunnan Province, China, on 21 May 2021. Compared to strong earthquakes, the measurement of the deformation fields of moderate earthquakes is more susceptible to errors associated with atmospheric, orbital, and topographic features. We adopted a new time-series InSAR method to process preseismic and postseismic Sentinel-1A SAR time-series images and separated the coseismic deformation signals from various error signals. This method uses preseismic time-series interferograms to estimate the spatially correlated look angle error induced by the digital elevation model and the atmospheric and orbital errors in the master image. The preseismic and postseismic time-series interferograms were then segmented for spatio-temporal filtering to provide a precise estimate of the atmospheric and orbital errors in slave images. Such time-series processing accurately separates various errors from the coseismic deformation signal and prevents the coseismic deformation signal from being included as noise in the error estimation during filtering. Based on this approach, we effectively eliminated the masking of the deformation signal by the errors and extracted coseismic deformation field of the Yangbi M_W 6.1 earthquake with high precision. The maximum LOS displacement in the ascending and descending tracks were determined to be −74 and −62 mm, respectively. Subsequently, we used the Geodetic Bayesian Inversion Software to invert the fault geometric parameters of this earthquake, and based on this inverted the rupture slip distribution using the least-squares method. The results showed that the fault orientation is 133.43°, dip angle is 76.98°, source depth is 5.5 km, fault sliding mode is right-lateral strike-slip. The moment magnitude (M_W) was calculated to be 6.07.

Keywords: Yangbi M_W 6.1 earthquake; time-series InSAR; error elimination; high-precision coseismic deformation; fault inversion

1. Introduction

According to the China Earthquake Networks Center (CENC; https://news.ceic.ac.cn/, accessed on 1 July 2021), an earthquake of a magnitude of M_S 6.4 occurred in Yangbi County, Dali Prefecture, Yunnan Province, China, at 21:48:34 Beijing time on 21 May 2021. The epicenter of this earthquake was located at 25.67° N and 99.87° E with a focal depth of 8 km. The location of the epicenter is shown in Figure 1. The United States Geological Survey (USGS) and Global Centroid Moment Tensor (GCMT) determined the moment magnitude of this earthquake as M_W 6.1. Multiple aftershocks followed the mainshock. The earthquake most seriously impacted Yangbi County; however, Kunming, Baoshan, and Lijiang were also significantly affected. The earthquake caused 3 deaths, 32 injuries, and 192 collapsed houses. Earthquakes of moderate magnitude not only pose a threat to human life but can also initiate a series of disasters such as tsunamis, landslides, and collapses. The acquisition of accurate and reliable coseismic deformation fields can allow the construction of accurate earthquake inversion models and the study of seismogenic mechanisms. Moreover, it can provide more objective and high-resolution earthquake characteristics, supporting scientific study and facilitating disaster assessment and emergency rescue planning.

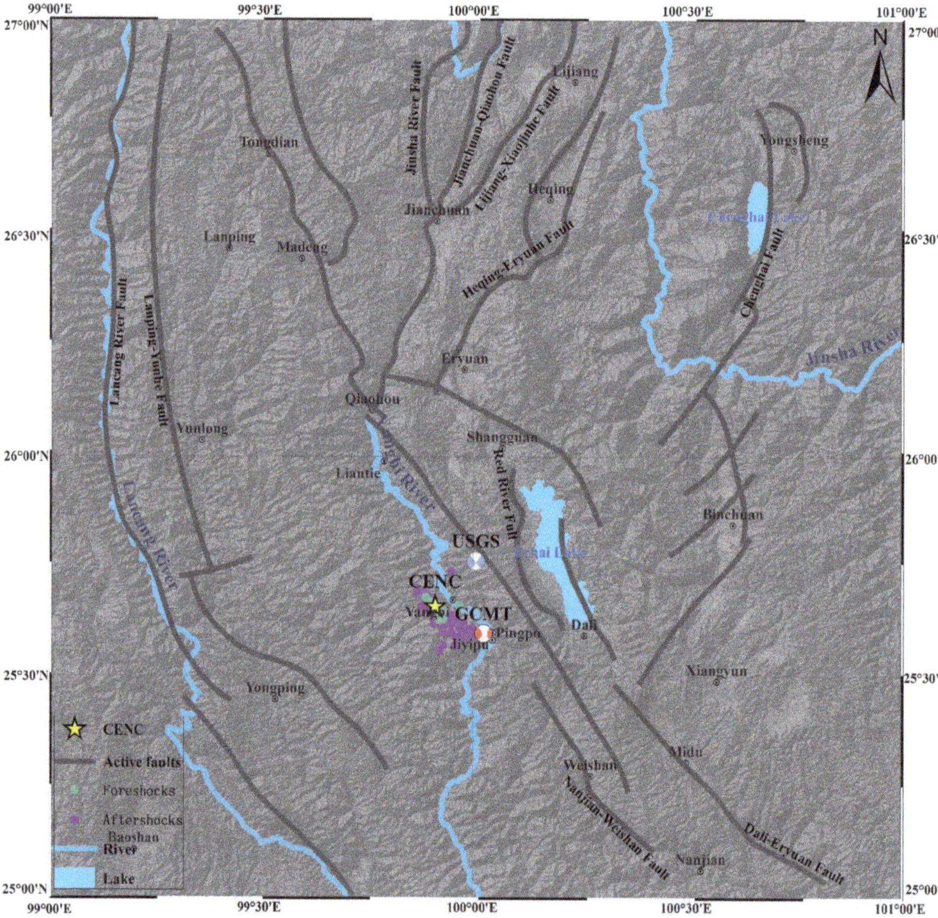

Figure 1. Regional tectonic background of Yangbi and the surrounding areas. The red and blue beach balls represent the focal mechanisms of the 2021 Yangbi M_W 6.1 earthquake provided by the Global Centroid Moment Tensor (GCMT) and the United States Geological Survey (USGS), respectively; the gray lines denote the major active faults in this area; the yellow star is the epicenter provided by the China Earthquake Networks Center (CENC); the green dots and the purple dots indicate foreshocks and aftershocks, respectively.

Coseismic deformation is routinely measured using differential radar interferometry (D-InSAR). In 1993, Massonnet successfully extracted the coseismic deformation of the 1992 Landers earthquake using differential interferometry with ESR1/2 data [1]. The use of D-InSAR has gradually increased and can facilitate large-scale and dense measurement of crustal deformation [2,3]. When extracting the coseismic deformation field of strong earthquakes that have large magnitudes and a wide range of deformation, the D-InSAR technique can obtain satisfactory results, allowing the analysis of the seismogenic mechanism [4]. However, the conventional InSAR technique has inherent error sources, including orbital error, topographic error, and atmospheric delay, which have a significant impact on the extraction accuracy of coseismic deformation signals [5]. Although the D-InSAR technique has been improved over time, it still cannot accurately eliminate certain residual errors which affect the accuracy of the deformation signals, fault slip models, and subsequent analyses.

The time-series InSAR was introduced in the late 20th century [6]. By analyzing the time and space variations of different phase components using time-series SAR data, several main phase components, such as the terrain residual phase, atmospheric effect phase, and deformation phase, can be identified. Although these phases cannot be distinguished on a single interferogram, the use of time-series InSAR can enable this and improve the accuracy of the deformation signals [7,8]. Ferretti proposed permanent scatterer interferometry (PS-InSAR) in 1999. PS-InSAR interpolates a feature target (PS candidate) that can maintain a high-level coherence over a long period of time, thereby obtaining phase information for the entire interferogram and precisely separating the atmospheric phase delays from other components to obtain fine deformation information [9,10]. The Stanford Method for Persistent Scatter (StaMPS), developed at Stanford University, is a mature algorithm for the PS-InSAR technique that directly uses a three-dimensional (3D) spatio-temporal algorithm for phase unwrapping and separates deformation signals and other noise signals by spatio-temporal filtering [11,12].

However, it is not common to apply time-series InSAR methods for the extraction of sudden deformation fields. Using time-series SAR images for coseismic deformation extraction has yielded better results than conventional D-InSAR. In 2020, Lei et al. proposed an MT-InSAR method to distinguish coseismic deformation from satellite orbit error using a set of pre-seismic SAR images and a post-seismic image; they also verified the effectiveness of the method using the Dangxiong M_W 6.3 earthquake on 6 October 2008 as a case study [13]. They concluded that the inaccurate elimination of topographic and orbital errors reduced the accuracy of coseismic deformation measurements and affected subsequent fault-slip inversions. In 2021, Luo Heng et al. proposed a stacking method using time-series SAR images to extract the coseismic deformations of three small-to-medium earthquakes on the Qinghai–Tibet Plateau and in the Tienshan region; the method effectively suppressed atmospheric phase screens and extracted weak coseismic deformation data at the sub-centimeter level [14]. To accurately extract coseismic deformation fields, long time-series interferograms are used to eliminate some disturbance terms that cannot be removed using single interferograms. Therefore, time-series InSAR methods can enable high-precision coseismic deformation field extraction.

Some studies have already studied the coseismic deformation field of the Yangbi M_W 6.1 earthquake [15,16] with conventional D-InSAR or GNSS. The main distinguishing feature of our study is that we used a time-series InSAR processing method to extract more accurate deformation data for this event. A traditional time-series InSAR procedure cannot effectively extract the coseismic deformation field because of the existence of a sudden deformation signal. Here, we improved it by processing the pre-seismic and post-seismic time series interferograms separately. Such processing improved the accuracy of error estimation and prevented the coseismic deformation signal being filtered out and excluded as an error. We successfully extracted the coseismic deformation signal of the Yangbi M_W 6.1 earthquake from long time-series Sentinel-1A SAR images. With the line-of-sight (LOS) coseismic deformation displacement of the ascending and descending tracks as the constraint, we inverted the fault geometry parameters and slip distribution of the 2021 Yangbi M_W 6.1 earthquake to analyze the underlying seismogenic mechanism of this earthquake.

2. Tectonic Background

The Yangbi M_W 6.1 earthquake occurred in Yangbi County, Dali, Yunnan Province, China. This area is located on the western edge of the Sichuan–Yunnan rhombic block. which in turn is situated near the southeast of the Tibetan Plateau. The strong crustal movement of the Tibetan Plateau has a considerable influence on the region's geological and tectonic activities, with intricate fault systems and occasional earthquakes [17]. Owing to the long-term collision and compression of the Eurasian and Indian plates, destructive geological disasters often occur in Yunnan Province [18]. Multiple active faults are present within the Sichuan–Yunnan block and most are strike-slip fault [19]. The Yangbi M_W 6.1 earthquake and its seismic sequence occurred along a secondary fault in the western part of the Weixi–Qiaohou–Weishan fault zone located at the western edge of the Sichuan–Yunnan rhombic block. The slip modes of the northern and middle sections of this fault were mainly right-lateral strike-slip. As shown in Figure 1, various fault zones are staggered near the epicenter of the earthquake, including the highly active Jinsha River Fault and Lancang River Fault. The Jinsha River Fault Zone is complex and consists of three major faults: the Jinsha River Main Fault, Jinsha River West Boundary Fault, and Jinsha River West Branch Fault, which are surrounded by several secondary faults [20]. The overall trend of the fault zone is in the north–south direction, and the main sliding mode was a right-sliding type with a high sliding rate. The Red River Fault Zone has undergone long-term tectonic deformation and evolution, and its movement is currently mainly right-lateral strike-slip [21]. The Weixi–Qiaohou–Weishan Fault is a pivotal part of the Jinsha River Fault and the Red River Fault, and no significant earthquakes have been recorded in its history. However, in recent years, small- and medium-sized earthquakes have occurred frequently near this fault zone, and the Yangbi M_W 6.1 earthquake was the strongest earthquake in this fault zone in recent decades.

According to historical earthquake information from the CENC, eight earthquakes with magnitudes greater than M_S 5.0 occurred in the area shown in Figure 1, five of which occurred within Yangbi County, and the remaining three were included an M_S 5.0 earthquake which occurred at the junction of Yangbi County and Eryun County in 2013, an M_S 5.0 earthquake in Yunlong County in 2016, and an M_S 5.1 earthquake in Changning County in 2015. The Yangbi M_W 6.1 earthquake presented a typical foreshock–primary–aftershock sequence. According to the CENC records, as of 1 July 2021, there were 54 earthquakes with magnitudes larger than Ms 2.0 during the 2021 Yangbi earthquake sequence. Before the mainshock, five tremors had magnitudes greater than M_S 4.0, the strongest of which was M_S 5.6. The mainshock was followed by multiple aftershocks, two of which were greater than M_S 5.0, and nine of these were between M_S 4.0 and 4.9. Table 1 provides a list of the focal mechanism parameters of the Yangbi earthquake provided by the USGS and GCMT to aid further investigation into the Yangbi earthquake.

Table 1. Focal mechanism solutions of the Yangbi M_W 6.1 earthquake provided by different agencies.

Source	Epicenter		M_W	Depth	Nodal Plane I			Nodal Plane II		
	Lon	Lat			Strike	Dip	Rake	Strike	Dip	Rake
USGS	100.008° E	25.727° N	6.1	17.5	45°	89°	−5°	135°	85°	−179°
GCMT	100.02° E	25.61° N	6.1	15	46°	78°	4°	315°	86°	168°

3. Materials and Methods

3.1. Time-Series InSAR Processing Method for Coseismic Deformation Field Extraction

In this study, we used a single master image interferometric configuration, specifically conventional PS-InSAR, and applied a spatio-temporal algorithm for phase unwrapping.

Because the extracted deformation signals were abrupt, the main improvement of the method suggested in this study was the segmented time-series process for estimating various errors. The phase of the coseismic deformation contained in the post-seismic time-series interferogram can affect error estimates, resulting in incorrect estimates that affect the extraction of the final coseismic deformation signal. Therefore, this method devides unwrapped interferograms into pre-seismic and post-seismic sequences. To estimate the spatially correlated look angle (SCLA) error and the atmospheric and orbit error (AOE) due to the master image, we only used the pre-seismic time-series interferogram for spatio-temporal filtering. This prevented the phase of the coseismic deformation in the post-seismic interferograms from affecting the error estimation. Finally, AOE due to the slave images were separated from the interferogram by performing temporal filtering on the pre-seismic and post-seismic interferograms, respectively. In a general analysis, filtering was performed uniformly for all time-series interferograms and weak coseismic deformation signals may be erroneously filtered out as noise, which must be accounted for in any processing flow. The complete process flow is illustrated in Figure 2.

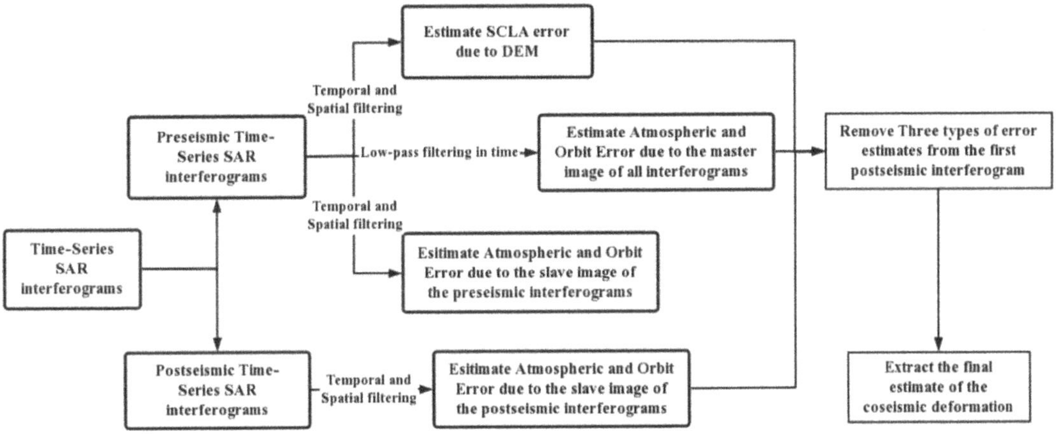

Figure 2. Time-series InSAR processing flow for coseismic deformation extraction.

An SAR image acquired just before the earthquake was used as the master image, and co-registered with other pre-seismic and post-seismic images that were used as the slave images. The master and slave images were co-registered individually to perform differential interferometry. In this case, the coseismic deformation signal is not present in the pre-seismic interferogram but rather in the post-seismic interferogram. In StaMPS, the threshold is first set using the amplitude deviation method combined with phase stability estimates to select PS candidates, and then the wrapped phase is corrected to filter out spatially uncorrelated look angle errors. Finally, the phase of the PS candidates were unwrapped to recover the absolute deformation phase using the 3D spatio-temporal method. The phases in the unwrapped interferograms were divided into five terms:

$$\varphi_{\text{int}} = \varphi_D + \varphi_A + \varphi_O + \varphi_\theta^c + \varphi_n \tag{1}$$

where φ_{int} indicates the unwrapped phase, φ_D indicates the surface deformation phase, φ_A indicates the error due to atmospheric effect, φ_O is the residual orbital error, φ_θ^c is the residual phase due to the SCLA error, and φ_n is the term for noise due to other factors, for example, co-registration and phase unwrapping.

Because the pre-seismic and post-seismic time-series interferograms needed to be processed separately, two different expressions were used for the unwrapped phases before and after the earthquake. The spatially uncorrelated look angle errors were screened out before the interferogram was unwrapped, so that the remaining residual errors in the unwrapped interferometric phase of the digital elevation model (DEM) only included the SCLA errors. The other spatially correlated errors were divided into AOEs due to the master and slave images, where the AOEs due to the master image are present in each interferogram. The deformation phase and other error terms in the pre-seismic and post-seismic interferograms can be expressed by the following equations:

$$\varphi_{x,i} = \varphi^{non}_{D,x,i} + \varphi^{c}_{\theta,x,i} + \varphi^{s}_{A,x,i} + \varphi^{m}_{A,x,i} + \varphi^{s}_{O,x,i} + \varphi^{m}_{O,x,i} + \varphi_{n,x,i} \tag{2}$$

$$\varphi_{x,j} = \varphi^{non}_{D,x,j} + \varphi_{\cos eismic,x,j} + \varphi^{c}_{\theta,x,j} + \varphi^{s}_{A,x,j} + \varphi^{m}_{A,x,j} + \varphi^{s}_{O,x,j} + \varphi^{m}_{O,x,j} + \varphi_{n,x,j} \tag{3}$$

where x denotes the master image, $\varphi_{x,i}$ represents the unwrapped phase of the pre-seismic interferogram, $\varphi_{x,j}$ represents the unwrapped phase of the post-seismic interferogram, and the superscripts m and s denote the contributions of the master and slave images to each spatially correlated error, respectively. φ^{non}_{D} denotes the surface deformation phase that does not include coseismic deformation signals and is present in all interferograms, and $\varphi_{\cos eismic,x,j}$ represents the coseismic deformation phases included in the post-seismic interferogram.

When using a DEM for terrain phase estimation, DEM errors can lead to other terrain-related residual phases in the interferometric phase. DEM errors tend to be partially spatially correlated and mapped to look angle errors; therefore, the SCLA error accounts for most DEM errors [22]. An SCLA error is present in each pair of interferograms. Because the terrain undergoes substantial deformation after an earthquake, the post-seismic time-series interferograms containing the coseismic deformation signals were not involved in the estimation of the SCLA error in this study. The following equation represents the SCLA error due to the DEM:

$$\Delta\varphi_{\theta} \approx \frac{4\pi}{\lambda} B\cos(\theta - \omega)\Delta\theta = \frac{4\pi}{\lambda} B_{\perp}(\theta)\Delta\theta \tag{4}$$

where λ denotes the radar wavelength; B denotes the baseline distance between the slave image and master image sensor position; θ denotes the look angle in the master image geometry; ω denotes the angle between the baseline and horizontal vectors; and $B_{\perp}(\theta)$ denotes the vertical component of the baseline.

Atmospheric conditions are unlikely to be identical when satellites make repeated observations of the ground; therefore, the path of the electromagnetic waves changes during propagation and produces phase delays during radar imaging, resulting in substantial errors in the measurement of surface deformation [23,24]. Both the positioning accuracy of the satellite orbit and atmospheric effects during operation can lead to biases when a satellite revisits the same position. Even with orbital parameters, it is not possible to completely remove orbital error streaks, and residual errors propagate with subsequent processing [25]. Especially in coseismic deformation measurements, orbital errors propagate into the interferometric phase and affect the final estimation of the deformation. The orbital and atmospheric errors are similar in their spatial and temporal properties; therefore, we adopted spatiotemporal filtering to separate the coseismic deformation signal from the AOEs. The AOE due to the master image in the interferometric phase was estimated first. Considering the effect of post-earthquake coseismic deformation on the estimation, only pre-seismic interferograms were selected to estimate this error. Low frequencies characterize the nonlinear deformation signal in both the time and space domains, whereas high frequencies characterize the AOE in the time domain [12]. Therefore, we performed low-pass filtering in the time dimension of the unwrapped phase for the pre-seismic interferogram to separate the surface deformation signal from the AOE due to the master image. The AOE due to the master image was removed from the interferogram, and an estimate of the master image AOE was obtained:

$$\Phi^{L}\{\varphi_{x,i}\} \approx \varphi_{D,x,i} - \left(\varphi^{m}_{A,x,i} + \varphi^{m}_{O,x,i}\right) \tag{5}$$

where Φ^L is the temporal low-pass filter operator.

To estimate the AOE due to slave images, separate spatio-temporal filtering was applyied to pre-seismic and post-seismic time-series interferograms. Given the high-frequency characteristics of the AOE in the time dimension, high-pass filtering was performed for the remaining phases in the pre-seismic and post-seismic time-series interferograms. While the AOE has low-frequency characteristics in the spatial dimension, the high-pass filter allows spatial low-pass filtering for the signal output, and the spatial filter outputs the final estimate of the AOE due to the slave image:

$$\Psi^L\left\{\Phi^H\{\varphi_{x,i}\}\right\} \approx \varphi^s_{A,x,i} + \varphi^s_{O,x,i} + \varphi_{n,x,i} \tag{6}$$

$$\Psi^L\left\{\Phi^H\{\varphi_{x,j}\}\right\} \approx \varphi^s_{A,x,j} + \varphi^s_{O,x,j} + \varphi_{n,x,j} \tag{7}$$

where Φ^H is the temporal high-pass filtering operator and Ψ^L is the spatial low-pass filtering operator.

The phase of the interferogram formed by the SAR images on the nearest date before and after the earthquake date contains the most accurate coseismic deformation signal; therefore, the final coseismic deformation estimate is only based on the interferometric phase in this interferogram. The DEM-induced SCLA error, AOE due to the master image, and estimated AOE due to the slave image were first removed from the interferometric phase, and other elevation-related errors were then removed from the remaining phases. Finally, the extracted coseismic deformation phases were spatially filtered to remove other spatial noises and obtain a final coseismic deformation phase estimate with high precision. The coseismic deformation estimate was given by the following equation:

$$\begin{aligned}\varphi_{\cos\ eismic} &\approx \varphi_{x,x+1} - \varphi^c_{\theta,x,x+1} - \left(\varphi^S_{A,x,x+1} + \varphi^S_{O,x,x+1}\right) - \left(\varphi^m_{A,x,x+1} + \varphi^m_{O,x,x+1}\right) \\ &\approx \varphi_{\cos\ eismic,x,x+1}\end{aligned} \tag{8}$$

where $x+1$ indicates the first post-seismic image to the date of the earthquake. As other surface deformation $\varphi^{non}_{D,x,x+1}$ is negligible, $\varphi_{\cos eismic}$ represents the final estimate of coseismic deformation.

3.2. Inversion of Fault Geometry Parameters and Slip Distribution

Taking the coseismic deformation fields extracted by the above time-series InSAR processing method as constraints, we performed an inversion of the 2021 Yangbi M_W 6.1 earthquake to obtain the fault geometry parameters and fault slip distribution. Based on this, we studied the fault activity and rupture characteristics of this earthquake in detail. In this study, we used the Geodetic Bayesian Inversion Software (GBIS) developed by Bagnardi and Hooper in 2018 to perform nonlinear homogeneous and slip inversion of the best fault geometry parameters for this earthquake [26]. GBIS uses a Bayesian probabilistic inversion algorithm for InSAR observations derived from the original datasets. In the Bayesian framework, an optimal set of source model parameters is extracted from the posterior probability density function (PDF) by quickly estimating the best model parameters and associated uncertainties through efficient sampling the posterior PDF. In the process of seismic inversion using this method, the Markov chain Monte Carlo (MCMC) method was used to perform sampling [27], which was controlled in combination with the Metropolis Hastings algorithm [28,29]. Automatic step-size selection can control the random walk step of each model parameter to ensure sampling efficiency. After multiple iterations, the set of model parameters with the maximum a posterior probability after pre-assigned iterations was selected.

We inverted the fault based on the elastic half-space dislocation model proposed by Okada in 1985, which provides the relationship between the surface deformation and fault slip at depth [30]. This elastic dislocation model provides nine source model parameters, including the length, width, depth, X and Y coordinates of the lower edge midpoint, dip and strike of the fault, and dip slip and strike slip. First, the seismic region was modeled as a uniform rectangular fault plane and the geometric parameters of the fault were accurately estimated using the GBIS algorithm. To obtain the dislocation distribution across a fault, it is necessary to invert the slip distribution across the fault plane. Because the slip distribution is nonuniform, the inversion of the fault slip distribution is equivalent to solving the optimal value of a linear problem. Based on the parameters of the fault geometric model, the fault plane was discretized into several uniform sub-fault planes, and the slip distribution of each sub-fault plane was then inverted to further obtain the fine slip distribution of the entire fault. The slip parameter model coefficient matrix is called Green's function matrix, and we chose the constrained least-squares algorithm to estimate slip distribution [31]. As Green's function matrix is severely rank deficient, the Laplace smoothing matrix was added to the inversion for regularization, and the L-curve method was used to select the best regularization parameters [32].

4. Results

4.1. InSAR Coseismic Deformation Field

In this study, the time-series InSAR processing method was applied to the Yangbi M_W 6.1 earthquake using ESA Sentinel-1 A satellite Single Look Complex (SLC) images as the data source. We acquired 36 SLC images from ascending track AT99 and 34 SLC images from descending track DT135, with an image date interval of 12 days, consistent with the satellite revisit cycle. For the Yangbi M_W 6.1 earthquake occurred on 21 May 2021, the date for the selected master image in the ascending track data is 20 May 2021 and the date of the selected master image in the descending track data is 10 May 2021. Co-registration and differential interferometry were performed using the Swiss interferometry software GAMMA [33], and the corresponding precise orbit ephemerides were used to provide orbit information. ALOS World 3D-30 m DEM data was selected as an external reference DEM to assist SAR image co-registration and topographic phase removal in differential interferometry. We chose the built-in algorithm of StaMPS to perform PS point selection and phase decoupling for all interferograms and then used the time-series processing method adopted in this study to accurately estimate each error. Finally, we extracted the coseismic deformation phases of the Yangbi M_W 6.1 earthquake in the ascending and descending tracks, and obtained the coseismic deformation displacement in the LOS direction after conversion.

It can be seen from the interferograms in Figure 3 that the long axis of the main deformation field of the Yangbi earthquake is roughly in the NW direction, and it can be assumed that the seismic traces divide the deformation area into eastern and western plates, showing an asymmetric distribution. The range of the coseismic deformation field in the ascending track was approximately 22 km in the north–south direction and approximately 24 km in the east–west direction. The maximum LOS displacement was approximately −74 mm in the east plate and 44 mm in the west plate. The range of the descending coseismic deformation field was approximately 22 km in the north–south direction and approximately 20 km in the east–west direction, and the maximum LOS displacement was approximately 43 mm in the east plate and −62 mm in the west plate. A positive value of LOS displacement indicates that the surface deformation is moving towards the satellite, whereas a negative value indicates that the deformation is moving away from the satellite. Differences in the satellite side view and flight attitude in the ascending and descending tracks results in different imaging patterns. Consequently, the coseismic deformations obtained from the ascending and descending tracks differ in their spatial distribution and magnitude. Based on the above results, the observed values of deformation in the east and west plates extracted from the ascending and descending tracks showed exactly opposite trends. Therefore, this earthquake had characteristics consistent with strike-slip type fault deformation. Because the coseismic deformation area extends along the NW–SE direction, it can be tentatively concluded that the Yangbi earthquake caused a rupture along the NW–SE direction. Moreover, there was no obvious decoherence area caused by surface rupture in the deformation area of either track, so the earthquake rupture did not reach the surface.

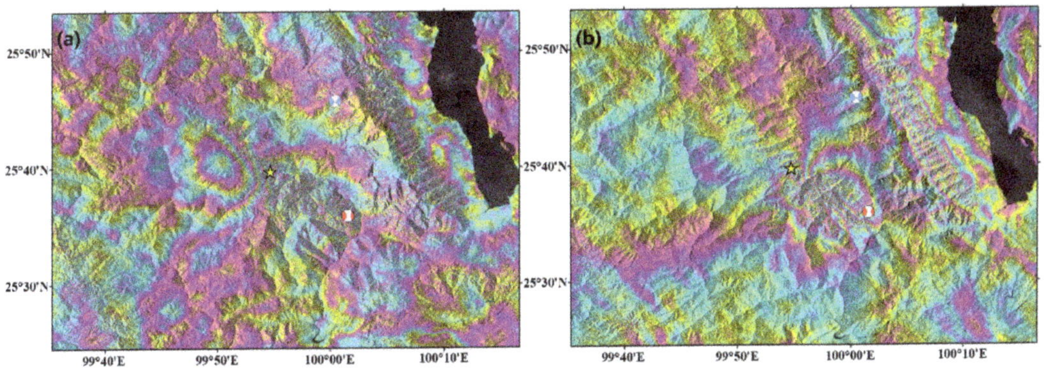

Figure 3. Ascending and descending interferograms of the 2021 Yangbi M_W 6.1 earthquake: (**a**) interferogram in the ascending track; (**b**) interferogram in the descending track.

In this study, we compared the coseismic deformation extracted by time-series InSAR method with that extracted by D-InSAR. The two different coseismic deformations are shown in Figure 4. Figure 4a,d show the LOS coseismic deformation in the ascending and descending tracks after the elimination of each error using time-series InSAR processing, respectively. Figure 4b,e show the LOS coseismic deformations in the ascending and descending tracks extracted from the single differential interferogram without error removal, respectively. Figure 5a,d show the SCLA error values estimated by the time-series analysis, Figure 5b,e show the estimated AOE due to the master image, and Figure 5c,f show the estimated AOE due to the slave images. For the ascending track, the eastern plate of the deformation field without error removal is mainly masked by the master AOE, and the western plate is mainly affected by the SCLA error and the slave AOE. For the descending track, the eastern plate of deformation field without error removal is mainly affected by the master AOE and the SCLA error, and the western plate is affected by the slave AOE. To demonstrate the effectiveness of the above time-series InSAR processing methods in removing errors, we also performed a quantitative analysis of the coseismic deformations extracted by the two different methods. We calculated the differences between the coseismic deformation displacements extracted by the time-series InSAR and the coseismic deformations extracted from the single interferogram without removing the errors. Figure 4c,f show histograms of the frequency distribution of the difference between the two deformations in the ascending and descending tracks, respectively. The difference between the two different coseismic deformations in the ascending track ranges from −20 mm to 70 mm, and the RMS of the difference is 31.05 mm. The difference between the two different coseismic deformations in the descending track ranges from −60 mm to 20 mm, and the RMS of the difference is 16.87 mm. After analysis, we concluded that the time-series InSAR processing method used in this study has a significant effect in estimating errors, which makes the extracted coseismic deformation field more accurate and clarifies the deformation field.

4.2. Fault Geometry Parameters

Before fault inversion, it is necessary to down-sample the coseismic deformation data to reduce the redundancy caused by the large number of data points and reduce the computational cost of inversion. In this study, we chose an adaptive quadtree algorithm to down-sample the ascending and descending LOS deformation data of the Yangbi M_W 6.1 earthquake, and maintained the dense deformation points in the near deformation field [34–36]. The down-sampled data points retained better deformation field characteristics and improved the subsequent calculation efficiency. With the ascending and descending down-sampled deformation points as common constraints, we used the GBIS to invert the fault geometry parameters of the Yangbi Mw 6.1 earthquake. In the framework of the GBIS algorithm, the MCMC method was used to sample the deformation data, and after 10^6 iterations, the burn-in period of the first 2×10^4 iterations was removed to determine the optimal set of model parameters.

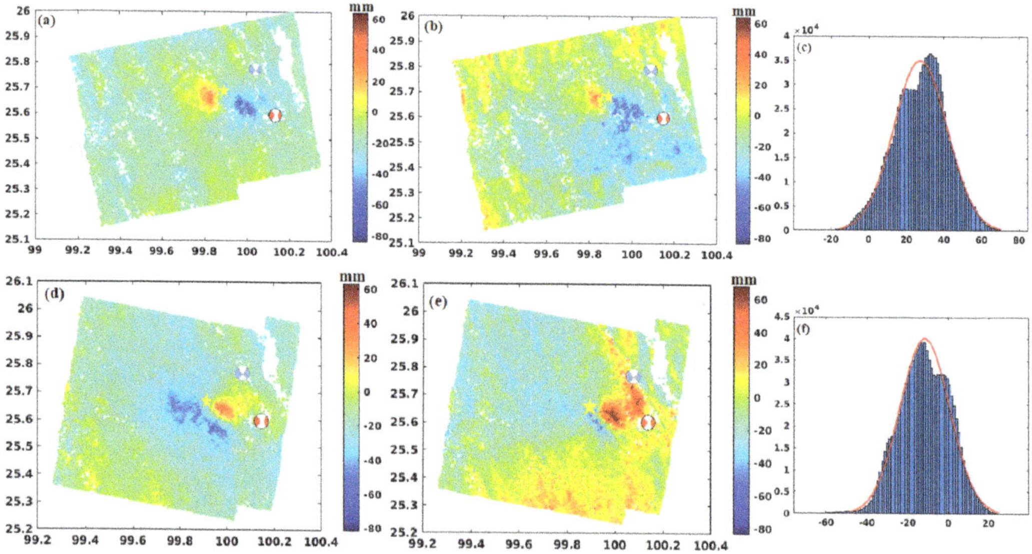

Figure 4. LOS coseismic deformation displacement: (**a**,**d**) represent the LOS coseismic deformation displacements in the ascending and descending tracks extracted by time–series InSAR, respectively; (**b**,**e**) represent the LOS coseismic deformation displacements in the ascending and descending tracks extracted by D–InSAR, respectively; and (**c**,**f**) represent the frequency distribution histograms of the LOS deformation differences in the ascending and descending tracks, respectively.

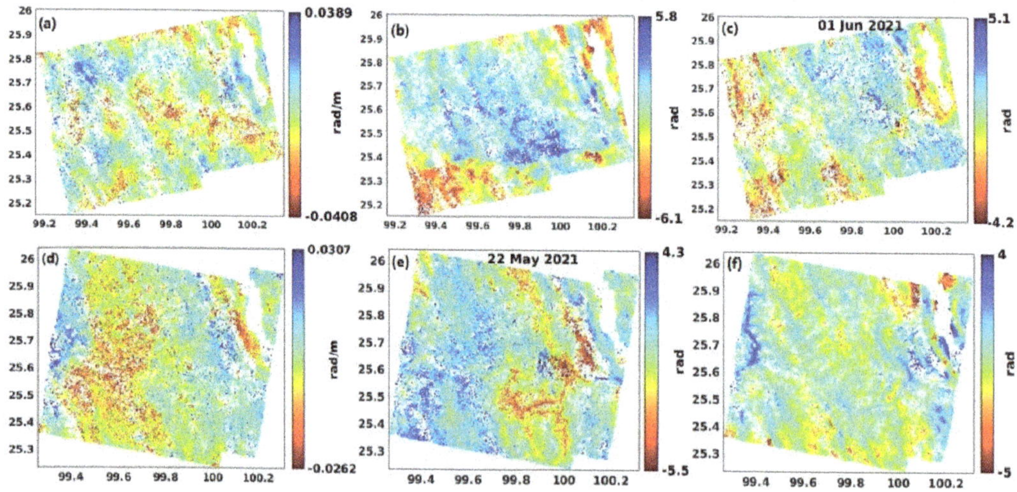

Figure 5. Error diagrams: (**a**–**c**) represent the SCLA error, master AOE, and slave AOE estimated from the ascending data, respectively; and (**d**–**f**) represent the SCLA error, master AOE, and slave AOE estimated from the descending data, respectively.

Based on the Okada elastic half-space dislocation model, we used the GBIS algorithm to obtain the nine source parameters of the Yangbi earthquake, which are listed in Table 2, including the final maximum a posteriori probability solution and a 95% confidence interval. Here, 2.50% and 97.50% represent the upper and lower limits of the confidence intervals, respectively. There was little difference between the different parameters obtained from the inversion, proving the reliability of the inversion results. According to the optimal solution, the fault length of the Yangbi M_W 6.1 earthquake was approximately 17 km and the fault width was approximately 1033 m. The dip and strike of the fault were 76.98° and 313.43°, respectively, which are close to the values provided by the GCMT. The optimal solution of the X and Y coordinates at the midpoint of the lower edge is converted to a geographic coordinate system of 25.6441° N and 99.9261° E, corresponding to a depth of approximately 5.8 km. After inversion, we also obtained the slip components in the dip and strike directions, where the strike-slip component was 2.56 m and the dip-slip component was 0.13 m. Therefore, we can conclude that the slip direction of the fault mechanism is consistent with a large right-lateral strike-slip component and a small thrust component. In conclusion, the optimal fault geometric parameters obtained by GBIS inversion are in agreement with the focal mechanism solutions provided by the USGS and GCMT.

Table 2. Seismic fault geometric parameters of the 2021 Yangbi M_W 6.1 earthquake.

Parameter	Optimal	Mean	Median	2.50%	97.50%
Length (m)	17,233.9	17,295.8	17,275.5	15,465.8	19,263.9
Width (m)	1032.87	1332.03	1023.97	606.288	4089.61
Depth (m)	5498.51	5362.01	5411.48	4384.8	5931.69
Dip (°)	76.9801	78.2016	77.7524	83.8792	75.131
Strike (°)	133.428	133.559	133.544	130.624	136.601
X center (m)	9031.41	8813.28	8828.41	9525.23	8018.04
Y center (m)	10,176.4	10,319.8	10,264.8	10,928.8	10,011.5
Strike slip (m)	2.56309	2.48903	2.55033	0.685637	3.91905
Dip slip (m)	0.133527	0.123478	0.112916	0.0114473	0.287731

4.3. Fault Slip Distribution

In this study, a constrained least-squares algorithm was used to invert the fault slip distribution. First, the fault geometry parameters calculated by GBIS were introduced into the linear inversion model, Green's function coefficient matrix was solved, and a Laplace smoothing matrix was used to constrain the slip roughness. According to the GBIS fault geometry parameters, the strike of the fault model was fixed at 133.43°, and the dip angle was fixed at 76.98°. The fault activity mode was right-lateral strike-slip; therefore, the range of the rake was set to 90°–180°. The fault plane extended to 30 km along the strike, and to 15 km along the dip. This fault plane was divided into 20 × 15 sub-fault planes to analyze the fine slip distribution. The optimal slip factor was determined by plotting the L-curve between the roughness and misfit of the slip distribution, which assessed the reasonableness of the inversion results and obtained the optimal fault slip distribution. Figure 6 shows the roughness and relative misfit corresponding to the optimal smoothing factor determined from the L curve.

Figure 7 shows the fault-slip distribution determined by inversion. The direction of the arrow indicates the movement direction of each sub-fault plane, indicating that the slip direction of the entire fault plane is correct. Figure 7 illustrates that there is a main fracture area in the Yangbi earthquake, concentrated at 6–24 km along the strike direction and 3–12 km down the dip direction. It was also determined that the rupture caused by the earthquake did not reach the surface. The 3D slip distribution indicated that the slip-concentrated area is located at a depth of 5–10 km, and the maximum slip was approximately 0.6738 m, occurring at a depth of 5.5 km. Combined with the geometric parameters of the fault and the results of slip distribution, it can be determined that the fault mechanism of Yangbi earthquake is dominated by a right-lateral component. Assuming that the shear modulus is 30 GPa, the inversion results show that the seismic moment energy released by this earthquake is approximately 1.58×10^{18} N· m, and the corresponding moment magnitude is approximately M_W 6.07, which is slightly lower than the moment magnitude of M_W 6.1 provided by the USGS and GCMT. Finally, it was determined that the epicenter of the Yangbi earthquake is located at 25.6388° N and 99.9175° E, which is close to the focal center provided by the USGS.

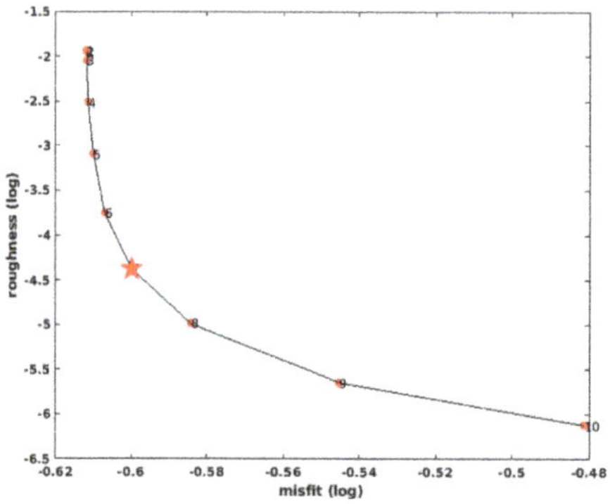

Figure 6. Trade–off curve between the misfit and model roughness. The star indicates optimal smoothing factor.

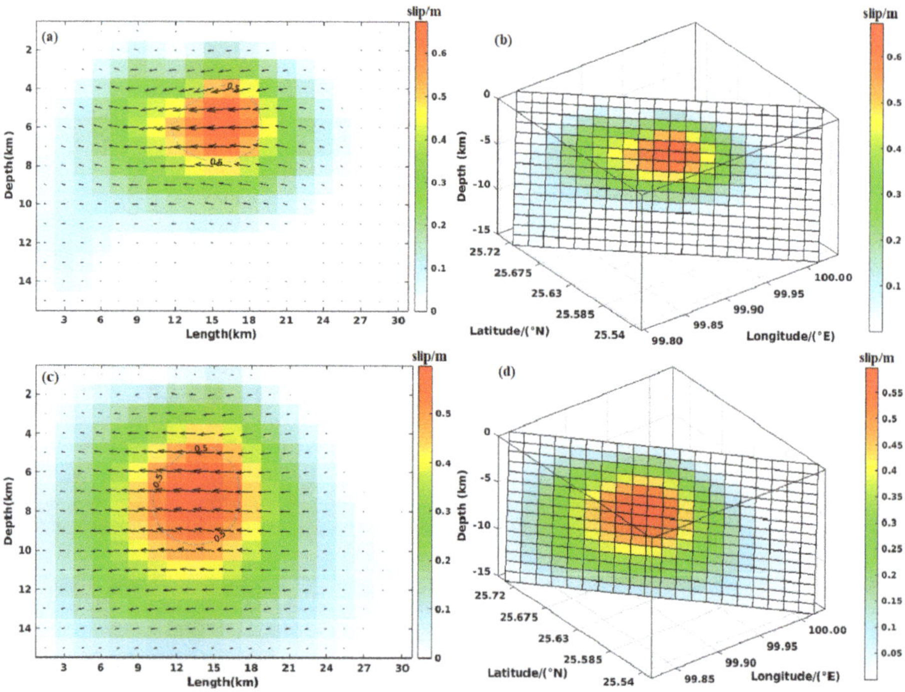

Figure 7. Different fault slip distribution models from two InSAR deformation datasets: (**a**,**b**) are two–dimensional (2D) and three–dimensional (3D) displays of fault slip distribution inverted based on deformation results extracted by time–series InSAR, respectively; (**c**,**d**) are 2D and 3D displays of fault slip distribution based on deformation results extracted from a single interferogram, respectively; the arrow indicates the movement direction of each sub–fault plane.

To evaluate the reliability of the inversion results, we simulated the LOS coseismic deformation fields in the ascending and descending tracks according to the inversion results of the fault slip distribution. After obtaining models of the coseismic deformation fields for the ascending and descending tracks, we calculated the residuals by subtracting them from the deformation observations obtained using the time-series InSAR method and then judged the reliability of the fault inversion based on the magnitude of the residuals. We also performed an inversion of the fault geometry and slip distribution on the coseismic deformation data extracted by a single interferogram to demonstrate that the coseismic deformation extracted by the time-series InSAR method is more accurate than that extracted by D-InSAR. The fault models were then fitted based on the results of the slip distribution inversion, and the fitting residuals between the InSAR observations and the models were calculated.

As shown in Figure 7c, many errors remained in the deformation results extracted using only a single interferogram, and the slip distribution inversion based on this dataset was largely disturbed, and as a result the slip distribution model to explain the data well. Figure 8 shows the observation values of the ascending and descending down-sampled LOS coseismic deformation fields extracted by the two different methods, the model values fitted according to slip distribution inversion, and the residuals between them. Comparing the inversion results of the two different datasets, it can be seen from Figure 8c,f that the LOS deformation field model simulated by the deformation dataset extracted by time-series InSAR is highly consistent with the observation values, and the fitting residual values for ascending and descending tracks are both less than 0.02 m, indicating the high fitting degree between the model and observation values and proving the reliability of the inversion results. As shown in Figure 9a,b, the residuals between the observed LOS deformation in the ascending track extracted by the time-series InSAR method and the model values fitted by the inversion range of −0.03~0.03 m with a residual RMS of 0.79 cm, while the residuals between the observed LOS deformation in the descending track and the model values fitted by inversion range from −0.05~0.03 m with a residual RMS of 0.91 cm. Figure 8i,l show that the fitting residuals between the LOS coseismic deformation fitted model value according to the deformation dataset extracted by D-InSAR and the observation value are large, and the fitting degree is relatively low because the error in the observation data affected the simulation of the model for real data. As shown in Figure 9c,d the residuals between the observed ascending LOS deformation extracted by D-InSAR and the model values fitted by inversion range from −0.04 m to 0.04 m with a residual RMS of 1.34 cm, while the residuals between the observed ascending LOS deformation and the model values fitted by inversion range from −0.05 m to 0.05 m with a residual RMS of 1.32 cm. Therefore, the errors in the deformation data observed by InSAR have a negative impact on seismic inversion, and the comparison proves the effectiveness of the time-series InSAR method used in this study to extract the coseismic deformation.

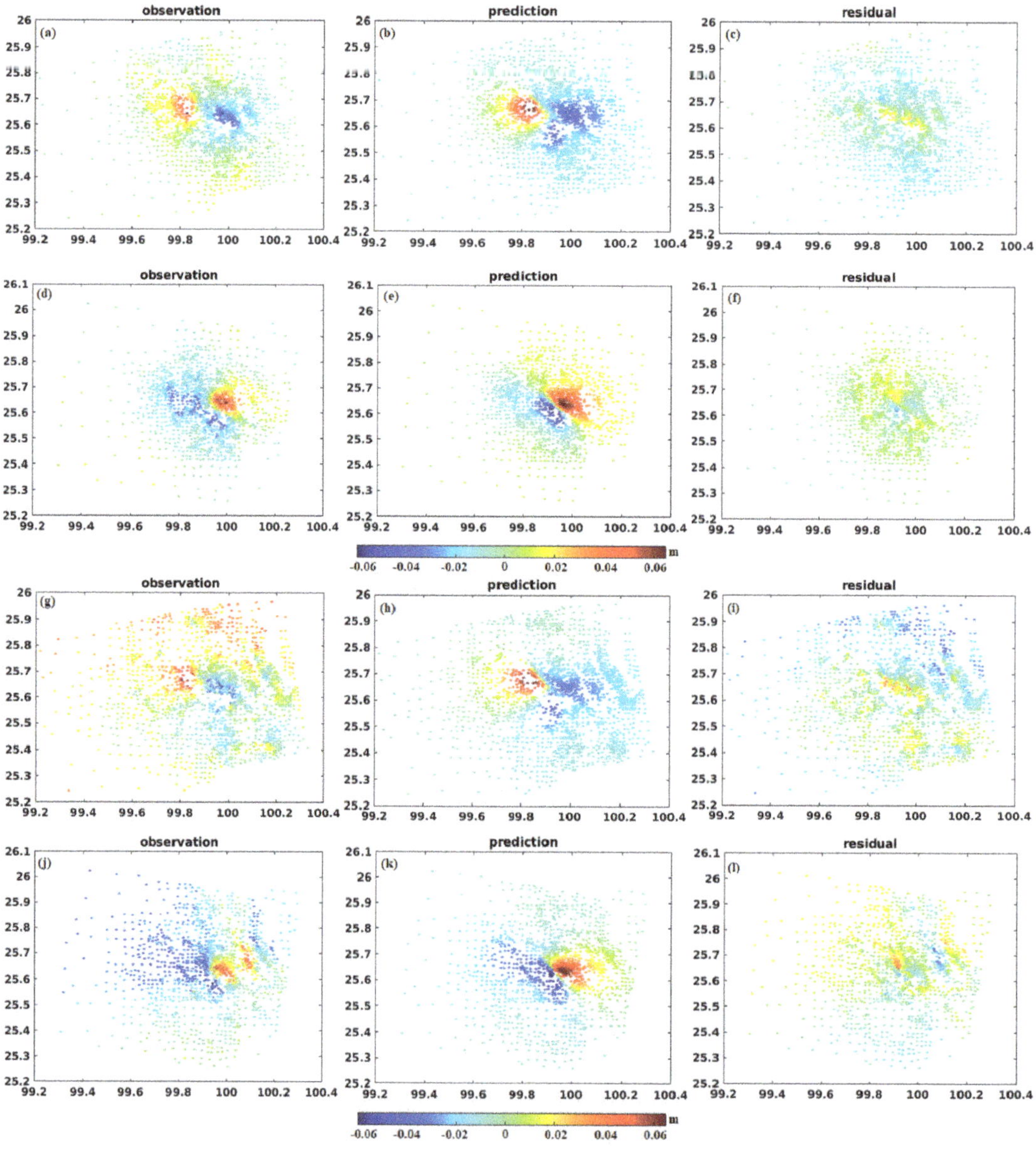

Figure 8. LOS coseismic deformation model fitting results based on the time-series InSAR and D–InSAR deformation datasets: (**a**–**c**) represent ascending LOS deformation observations, predicted model values, and residuals, respectively; all are based on time–series InSAR data; (**d**–**f**) represent descending LOS deformation observations, model values, and residuals, respectively, based on time-series InSAR data; (**g**–**i**) represent ascending LOS deformation observations, model values, and residuals, respectively, based on D–InSAR data; and (**j**–**l**) represent descending LOS deformation observations, model values, and residuals, respectively, based on D–InSAR data.

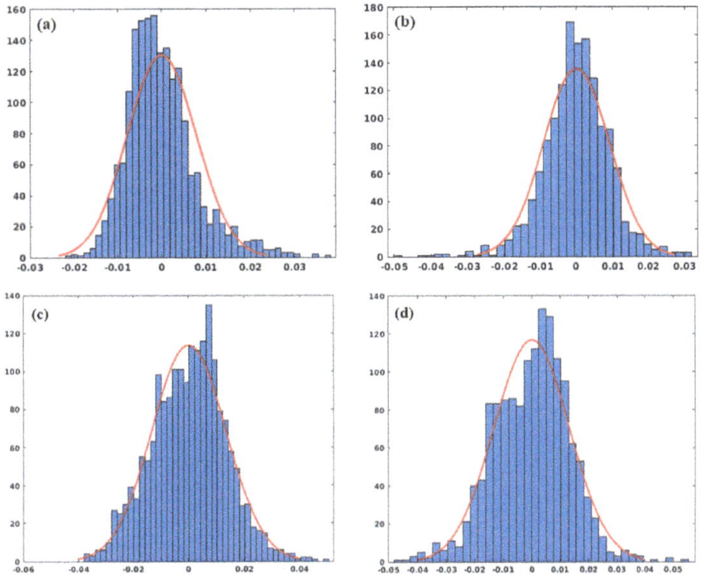

Figure 9. Histograms of residual frequency distributions: (**a**,**b**) indicate the ascending and descending residual distributions between the time–series InSAR–based deformation and its inversion–fitted model values, respectively; and (**c**,**d**) indicate the ascending and descending residual distributions between the D–InSAR–extracted deformation and its inversion-fitted model values, respectively.

5. Discussion

5.1. Validation of Methods Using Synthetic Data

To verify the effectiveness of the time-series InSAR processing method used in this paper to eliminate errors and extract coseismic deformations, synthetic time-series interferograms based on the inversion-fitted LOS coseismic deformations in the descending track. The inverse model of the downsampled LOS coseismic deformation was used as the true coseismic deformation of the synthetic data. Because the model values are the downsampled data, we used an interpolation algorithm to obtain interferograms corresponding to the original data size. First, all deformation phases of the pre-seismic interferograms were set to 0, and all deformation phases of post-seismic interferograms were set to the true deformation phases. Then, we added errors estimated using the time-series analysis as noise to the corresponding interferogram, such that the interferometric phase of the synthesized time-series interferograms were closer to the actual characteristics of the unwrapped interferogram. Finally, we used the time-series InSAR processing method to perform a time-series analysis of the synthesized time-series unwrapped interferograms to estimate errors, and obtained estimates of the LOS coseismic deformation after removing the errors.

Figure 10a shows the interpolated downgraded LOS coseismic deformation model values as the true coseismic deformation, Figure 10b shows the coseismic deformation extracted from the first synthesized post-seismic interferogram, and Figure 10c shows the coseismic deformation extracted from the synthesized time-series interferogram with time-series InSAR processing. It is evident from Figure 10 that the difference between the deformation field including noise and the real deformation simulation is large, and the RMS of the difference between the two is 16.68 mm. However, the deformation extracted by the time-series analysis was similar to the real deformation field pattern. Figure 10e shows the frequency distribution of the difference between the deformation extracted by the time-series processing and the real deformation, and here the difference is no greater than

1 cm and the RMS of the difference is reduced to 8.77 mm, with a very small difference of no more than 1 cm. After validation using the synthetic dataset, it was demonstrated that it is feasible and effective to use the time-series InSAR processing method for error estimation using time-series interferograms and to extract the coseismic deformation. We concluded that the use of time-series analysis can improve estimate the errors in the unwrapped interferograms and as well as the extraction accuracy of the coseismic deformation.

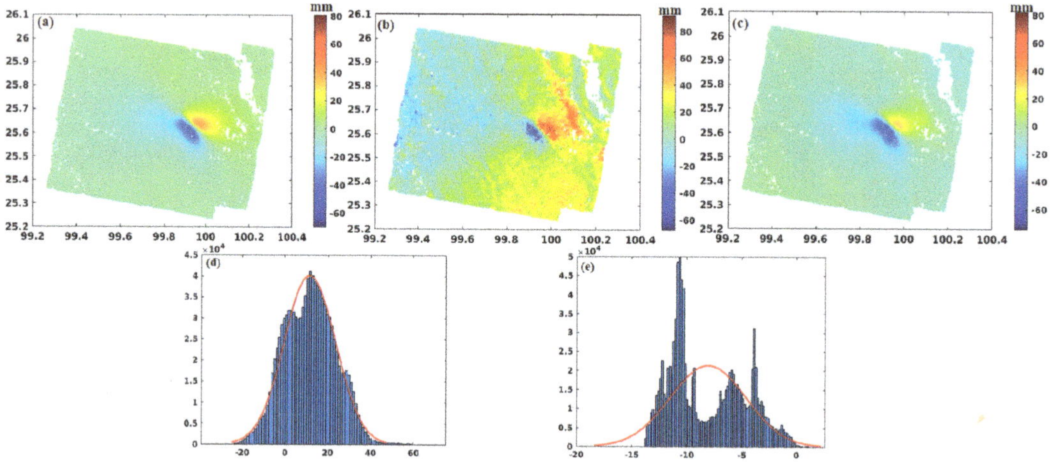

Figure 10. (**a**) represents the true coseismic deformation of the synthetic dataset; (**b**) represents the coseismic deformation extracted from the unwrapped interferogram after adding noise; (**c**) represents the coseismic deformation extracted from the synthetic dataset by the time–series InSAR processing method; (**d**) is the histogram of the difference distribution between the deformation after adding noise and the true deformation; and (**e**) is the histogram of the difference distribution between the deformation extracted by the time–series processing and the true deformation.

5.2. Advantages of Time-Series InSAR Processing Method for Coseismic Deformation Extraction

A major contribution of this study is the use of a specifically designed time-series InSAR technique to extract the coseismic deformation information. A time-series analysis can eliminate disturbance that cannot be resolved using conventional D-InSAR techniques. The errors due to atmospheric delay and the deformation phase both have low-frequency spatial characteristics. It is difficult to distinguish these to eliminate atmospheric effects by using a single interferogram. The D-InSAR technique uses orbital parameters to eliminate orbital errors in the interferogram. However, the residual phase ramp has a significant impact on the final measurement results. The time-series approach used in this study can estimate three disturbance terms: terrain-related residual errors, errors owing to atmospheric delays, and orbital errors. The application of this approach to the Yangbi M_W 6.1 earthquake proves that the time-series InSAR processing technique can accurately resolve certain unavoidable problems associated with the D-InSAR technique. The coseismic deformation phases contained in the post-seismic time-series interferograms have a significant impact on the error estimates, resulting in incorrect results. Only pre-seismic time-series interferograms are used to estimate the SCLA error and the master AOE, which prevents the coseismic deformation of the post-seismic interferograms from influencing the estimation errors. By performing temporal filtering on the pre-seismic and post-seismic interferograms separately, the slave AOE can be accurately separated from the interferograms. Moreover, the coseismic deformations of moderate earthquakes are not as obvious as those of large magnitude earthquakes, and often the various errors in the differential interferograms can often mask the true deformations, resulting in the poor accuracy of the extracted deformations. The time-series InSAR processing method can retrieve weak coseismic signals by eliminating various errors and avoids the leakage of coseismic deformation signals into error components. Compared to the deformation field extracted by D-InSAR, the coseismic deformation extracted by the processing method in this study shows better agreement with the actual pattern of earthquake deformation. By investigating the inversion results, we found that the errors remaining in the D-InSAR results had a significantly negative impact on the inversion. The time-series InSAR proessing method used in this study is based on PS-InSAR technology; therefore, the time-series information of all persistent scatterers can be retained to observe the pre-seismic and post-seismic deformation trends.

6. Conclusions

With the sentinel-1 SAR images, we used a time-series InSAR method to extract the coseismic deformation field. The geometric parameters and slip distribution of the fault were then inverted with the down-sampled deformation datasets as a constraint. The following conclusions were drawn.

(1) The time-series InSAR method can estimate the erroneous terms accurately, including the DEM error and the AOE due to master image and slave images. Analysis of both synthetic and real-world datasets validated the effectiveness of this processing method.

(2) The deformation fields extracted by the time-series InSAR method displayed opposite trends in ascending and descending tracks. In the ascending LOS direction, the maximum displacement of the eastern disk was approximately -74 mm and that of the western disk is approximately 44 mm. In the descending LOS direction, the maximum displacement of the eastern disk was approximately 43 mm, and that of the western disk is approximately -62 mm.

(3) The fault inversion results showed that the strike of the seismic fault was 133.43°, the dip was 76.98°, and the maximum slip was 0.6738 m, located at a depth of 5.5 km. The earthquake was a right-lateral strike-slip event with a moment magnitude (M_W) of 6.07; this finding is close to the source mechanisms proposed by the USGS and GCMT.

Author Contributions: Conceptualization, X.L. and C.W.; methodology, C.W.; software, X.L.; validation, X.L., C.W. and C.Z.; formal analysis, C.W.; investigation, S.W. and L.W.; writing—original draft preparation, X.L.; writing—review and editing, W.L. and W.Z. All authors have read and agreed to the published version of the manuscript.

Funding: This research was funded by Chang'an University (Xi'an, China) through the National Key Re-search and Development Program of China (Grant No. 2020YFC1512001), the National Natural Science Foundation of China (Grant No. 41974006), the Shenzhen Scientific Research and Devel-opment Funding Program (Grant No. KQJSCX20180328093453763 and 20200812164904001), the state Key Laboratory of Earthquake Dynamics (Grant No., LED2021B05), and the Lhasa National Geophysical Observation and Research Station (Grant No. NORSLS20-08).

Institutional Review Board Statement: Not applicable.

Informed Consent Statement: Not applicable.

Data Availability Statement: The Sentinel-1A dataset and precise orbit information are publicly available via Alaska Satellite Facility (ASF, https://search.asf.alaska.edu, accessed on 20 July 2021).

Acknowledgments: The authors would like to thank the European Space Agency (ESA) for providing Sentinel-1 SAR images.

Conflicts of Interest: The authors declare no conflict of interest. The funders had no role in the design of the study; in the collection, analyses, or interpretation of data; in the writing of the manuscript, or in the decision to publish the results.

References

1. Massonnet, D.; Rossi, M.; Carmona, C.; Adragna, F.; Peltzer, G.; Feigl, K.; Rabaute, T. The displacement field of the Landers earthquake mapped by radar interferometry. *Nature* **1993**, *364*, 138–142. [CrossRef]
2. Fialko, Y.; Sandwell, D.; Simons, M.; Rosen, P. Three-dimensional deformation caused by the Bam, Iran, earthquake and the origin of shallow slip deficit. *Nature* **2005**, *435*, 295–299. [CrossRef]

3. Peltzer, G.; Rosen, P. Surface Displacement of the 17 May 1993 Eureka Valley, California, Earthquake Observed by SAR Interferometry. *Science* **1995**, *268*, 1333–1336. [CrossRef]
4. Ferretti, A.; Prati, C.; Rocca, F. Nonlinear subsidence rate estimation using permanent scatterers in differential SAR interferometry. *IEEE Trans. Geosci. Remote Sens.* **2000**, *38*, 2202–2212. [CrossRef]
5. Rogers, A.; Ingalls, R.P. Venus: Mapping the Surface Reflectivity by Radar Interferometry. *Science* **1969**, *165*, 797–799. [CrossRef]
6. Lundgren, P.; Usai, S.; Sansosti, E.; Lanari, R.; Tesauro, M.; Fornaro, G.; Berardino, P. Modeling surface deformation observed with synthetic aperture radar interferometry at Campi Flegrei caldera. *J. Geophys. Res. Solid Earth* **2001**, *106*, 19355–19366. [CrossRef]
7. Ferretti, A.; Prati, C.; Rocca, F. Permanent scatterers in SAR interferometry. *IEEE Trans. Geosci. Remote Sens.* **2001**, *39*, 8–20. [CrossRef]
8. Colesanti, C.; Ferretti, A.; Novali, F.; Prati, C. SAR monitoring of progressive and seasonal ground deformation using the permanent scatterers technique. *IEEE Trans. Geosci. Remote Sens.* **2003**, *41*, 1685–1701. [CrossRef]
9. Ferretti, A.; Prati, C.; Rocca, F. Permanent scatterers in SAR interferometry. In Proceedings of the IEEE 1999 International Geoscience and Remote Sensing Symposium, IGARSS'99 (Cat. No. 99CH36293). Hamburg, Germany, 28 June–2 July 1999; pp. 1528–1530.
10. Ferretti, A.; Savio, G.; Barzaghi, R.; Borghi, A.; Rocca, F. Submillimeter Accuracy of InSAR Time Series: Experimental Validation. *IEEE Trans. Geosci. Remote Sens.* **2007**, *45*, 1142–1153. [CrossRef]
11. Hooper, A.; Zebker, H.; Segall, P.; Kampes, B. A new method for measuring deformation on volcanoes and other natural terrains using InSAR persistent scatterers. *Geophys. Res. Lett.* **2004**, *31*. [CrossRef]
12. Hooper, A.; Segall, P.; Zebker, H. Persistent scatterer interferometric synthetic aperture radar for crustal deformation analysis, with application to Volcán Alcedo, Galápagos. *J. Geophys. Res. Solid Earth* **2007**, *112*. [CrossRef]
13. Zhang, L.; Hu, J.; Ding, X.; Wen, Y.; Liang, H. Estimation of Coseismic Deformation with Multitemporal Radar Interferometry. *IEEE Geosci. Remote Sens. Lett.* **2021**, *19*, 1–5. [CrossRef]
14. Luo, H.; Wang, T.; Wei, S.; Liao, M.; Gong, J. Deriving centimeter-level coseismic deformation and fault geometries of small-to-moderate earthquakes from time-series Sentinel-1 SAR images. *Front. Earth Sci.* **2021**, *9*, 32. [CrossRef]
15. Zhang, K.; Gan, W.; Liang, S.; Xiao, G.; Dai, C.; Wang, Y.; Li, Z.; Zhang, L.; Ma, G. Coseismic displacement and slip distribution of the 2021 May 21. M_S 6.4, Yangbi Earthquake derived from GNSS observations. *Chin. J. Geophys.* **2021**, *64*, 2253–2266.
16. Zhang, B.; Xu, G.; Lu, Z.; He, Y.; Peng, M.; Feng, X. Coseismic Deformation Mechanisms of the 2021 Ms 6.4 Yangbi Earthquake, Yunnan Province, Using InSAR Observations. *Remote Sens.* **2021**, *13*, 3961. [CrossRef]
17. Xu, X.; Ding, Z.; Shi, D.; Li, X. Receiver function analysis of crustal structure beneath the eastern Tibetan plateau. *J. Asian Earth Sci.* **2013**, *73*, 121–127. [CrossRef]
18. Tapponnier, P.; Xu, Z.; Roger, F.; Meyer, B.; Arnaud, N.; Wittlinger, G.; Yang, J. Oblique Stepwise Rise and Growth of the Tibet Plateau. *Science* **2001**, *294*, 1671–1678. [CrossRef]
19. Su, Y.J.; Qin, J.Z. Strong Earthquake Activity and Relation to Regional Neotectonic Movement in Sichuan-Yunnan Region. *Earthq. Res. China* **2001**, *17*, 24–34.
20. Wang, E. The Jinsha River transform fault sedimentary basin. *Sci. Geol. Sin.* **1985**, 33–42.
21. Allen, C.R.; Gillespie, A.R.; Han, Y.; Sieh, K.E.; Zhang, B.; Zhu, C. Red River and associated faults, Yunnan Province, China: Quaternary geology, slip rates, and seismic hazard. *Geol. Soc. Am. Bull.* **1984**, *95*, 686–700. [CrossRef]
22. Hooper, A.J. Persistent Scatter Radar Interferometry for Crustal Deformation Studies and Modeling of Volcanic Deformation. Ph.D. Dissertation, Stanford University, Stanford, CA, USA, 2006.
23. Ding, X.-L.; Li, Z.-W.; Zhu, J.-J.; Feng, G.-C.; Long, J.-P. Atmospheric effects on InSAR measurements and their mitigation. *Sensors* **2008**, *8*, 5426–5448. [CrossRef]
24. Emardson, T.; Simons, M.; Webb, F. Neutral atmospheric delay in interferometric synthetic aperture radar applications: Statistical description and mitigation. *J. Geophys. Res. Solid Earth* **2003**, *108*. [CrossRef]
25. Fialko, Y. Interseismic strain accumulation and the earthquake potential on the southern San Andreas fault system. *Nature* **2006**, *441*, 968–971. [CrossRef]
26. Bagnardi, M.; Hooper, A. Inversion of surface deformation data for rapid estimates of source parameters and uncertainties: A Bayesian approach. *Geochem. Geophys. Geosyst.* **2018**, *19*, 2194–2211. [CrossRef]
27. Mosegaard, K.; Tarantola, A. Monte Carlo sampling of solutions to inverse problems. *J. Geophys. Res.* **1995**, *100*, 12431–12447. [CrossRef]
28. Metropolis, N.; Rosenbluth, A.W.; Rosenbluth, M.N.; Teller, A.H.; Teller, E. Equation of State Calculations by Fast Computing Machines. *J. Chem. Phys.* **1953**, *21*, 1087–1092. [CrossRef]
29. Hastings, W.K. Monte Carlo sampling methods using Markov chains and their applications. *Biometrika* **1970**, *57*, 97–109. [CrossRef]
30. Okada, Y. Surface deformation due to shear and tensile faults in a half-space. *Bull. Seismol. Soc. Am.* **1985**, *75*, 1135–1154. [CrossRef]
31. Wang, C.; Wang, X.; Xiu, W.; Zhang, B.; Liu, P. Characteristics of the Seismogenic Faults in the 2018 Lombok, Indonesia, Earthquake Sequence as Revealed by Inversion of InSAR Measurements. *Seismol. Res. Lett.* **2020**, *91*, 733–744. [CrossRef]
32. O'Leary, D.; Hansen, P.C. The Use of the L-Curve in the Regularization of Discrete Ill-Posed Problems. *SIAM J. Sci. Comput.* **1993**, *14*, 1487–1503.

33. Wegmüller, U.; Werner, C. Gamma SAR processor and interferometry software. In Proceedings of the 3rd ERS Scientific Symposium, Florence, Italy, 17–20 March 1997; pp. 1687–1692.
34. Jónsson, S.; Zebker, H.; Segall, P.; Amelung, F. Fault slip distribution of the 1999 M w 7.1 Hector Mine, California, earthquake, estimated from satellite radar and GPS measurements. *Bull. Seismol. Soc. Am.* **2002**, *92*, 1377–1389. [CrossRef]
35. Simons, M.; Fialko, Y.; Rivera, L. Coseismic deformation from the 1999 M w 7.1 Hector Mine, California, earthquake as inferred from InSAR and GPS observations. *Bull. Seismol. Soc. Am.* **2002**, *92*, 1390–1402. [CrossRef]
36. Decriem, J.; Árnadóttir, T.; Hooper, A.; Geirsson, H.; Sigmundsson, F.; Keiding, M.; Ófeigsson, B.; Hreinsdóttir, S.; Einarsson, P.; Lafemina, P.; et al. The 2008 May 29 earthquake doublet in SW Iceland. *Geophys. J. Int.* **2010**, *181*, 1128–1146. [CrossRef]

remote sensing

MDPI

Article

Surface Deformation of Expansive Soil at Ankang Airport, China, Revealed by InSAR Observations

Shuangcheng Zhang [1,2,3,4], Jinzhao Si [1,2,3,*], Yufen Niu [5], Wu Zhu [1,2,3], Qianyou Fan [1,2,3], Xingqun Hu [6], Changbo Zhang [6], Peng An [1,3], Zhipeng Ren [1,2,3] and Zhenhong Li [1,2,3]

1 School of Geology Engineering and Geomatics, Chang'an University, Xi'an 710054, China; shuangcheng369@chd.edu.cn (S.Z.); zhuwu@chd.edu.cn (W.Z.); 2020026013@chd.edu.cn (Q.F.); anpeng110@chd.edu.cn (P.A.); 2021226068@chd.edu.cn (Z.R.); zhenhong.li@chd.edu.cn (Z.L.)
2 Big Data Center for Geosciences and Satellites (BDCGS), Chang'an University, Xi'an 710054, China
3 Key Laboratory of Western China's Mineral Resource and Geological Engineering, Ministry of Education, Xi'an 710054, China
4 State Key Laboratory of Geo-Information Engineering, Xi'an 710054, China
5 School of Mining and Geomatics Engineering, Hebei University of Engineering, Handan 056038, China; niuyufen@hebeu.edu.cn
6 Shaanxi Jiawei Space Geographic Information Technology Co., Ltd., Ankang 725000, China; hxq@mail.nwpu.edu.cn (X.H.); changbo@mail.nwpu.edu.cn (C.Z.)
* Correspondence: 2020126029@chd.edu.cn

Abstract: Ankang Airport is constructed on an expansive soil-fill platform in Shaanxi Province, Central China. Since its completion in 2020, it has suffered surface deformation caused by the consolidation and settlement of the fill layer and instability of the expansive soil slope. Exploring the special deformation law of expansive soil regions by remote sensing and analyzing the deformation characteristics of airports in mountainous areas have always been key issues in related disaster research. Based on the intensity and phase observation data of 37 Sentinel-1 synthetic aperture radar images, this study obtained the spatio-temporal distribution of the deformation of Ankang Airport from May 2020 to October 2021. First, phase optimization was performed on the original interferograms. Second, the persistent scatterer synthetic aperture radar interferometry (PS-InSAR) method was applied to extract the surface deformation information of Ankang Airport, and the accuracy was evaluated. Finally, the singular spectrum analysis method was introduced to jointly analyze the deformation information obtained by the InSAR technology in combination with geological and climatic data. The results show that the excavation area of Ankang Airport was basically stable, the filling area had obvious surface and uneven deformation, and the expansive soil fill slope exhibits deformation characteristics strongly related to slope, rainfall, and fill depth. The deformation was mainly caused by consolidation and settlement, supplemented by the expansion and shrinkage deformation of the expansive soil.

Keywords: ankang airport; expansive soil; phase optimization; PS-InSAR; singular spectrum analysis (SSA)

Citation: Zhang, S.; Si, J.; Niu, Y.; Zhu, W.; Fan, Q.; Hu, X.; Zhang, C.; An, P.; Ren, Z.; Li, Z. Surface Deformation of Expansive Soil at Ankang Airport, China, Revealed by InSAR Observations. *Remote Sens.* **2022**, *14*, 2217. https://doi.org/10.3390/rs14092217

Academic Editor: Michele Saroli

Received: 22 March 2022
Accepted: 4 May 2022
Published: 5 May 2022

1. Introduction

Ankang Airport is one of the representative projects of Chinese regional aviation network construction, located in Shaanxi Province, China. Expansive soil is mainly distributed in the Ankang Basin where Ankang Airport is located. Therefore, considering the actual cost and transportation situation, the foundation of the airport is built with the expansive soil material. The airport construction lasted 50 months; it is a typical airport located in a mountainous region and, during its construction, excavation, and filling, works with a huge amount of earthwork were carried out. The total filling volume reached 30,000,000 m^3, with excavation and filling works staggered. These factors make the uneven land subsidence process likely to occur in the Ankang Airport area, thereby causing related disasters. In contrast to airports located in mountainous regions, expansive soil is widely distributed in the Ankang Airport area. Expansive soil is a disaster-prone geological constituent with obvious expansion and contraction properties and developed fissures [1]. In summary, the expansion and contraction of expansive soil and the staggered distribution of many excavation and filling works have caused potential hazards to the safe and stable operation of the airport. The early research about the stability of Ankang Airport reflect that it has suffered a certain degree of surface subsidence [2]. Therefore, it is necessary to monitor the deformation of Ankang Airport and analyze its stability.

Traditionally, the surface deformation of large-scale artificial structures, such as airports, can be monitored on-site using measurement methods such as GNSS and leveling. Although these traditional geodetic methods have high accuracy, they tend to have poor spatial resolution, omit a large amount of surface information, and are labor-intensive. In recent years, with the rapid development of remote sensing technology with a higher spatio-temporal resolution, various remote sensing technologies have been widely used in deformation monitoring research of large-scale artificial structures. Examples include laser scanning technologies [3], distributed fiber optic sensing (DFOS) [4], and unmanned aerial vehicle measurements [5]. As a new type of Earth observation method, synthetic aperture radar interferometry (InSAR) technology can overcome the shortcomings of traditional observation methods. It can obtain high-resolution surface deformation information and realize large-scale and precise long-term monitoring [6]. Recent studies have shown that time-series InSAR technology is a reliable Remote Sensing method for airport deformation monitoring. For example, Dai and Gao combined InSAR technology with groundwater, active faults, and other factors to interpret the uneven subsidence of Beijing's Da'xing Airport [7,8]; Marshall combined InSAR technology to study the deformation of the Kuala Lumpur Airport related to the subsidence model and tropical peat intensity [9], and Jiang investigated the long-term reclamation subsidence of Hong Kong's Chek Lap Kok Airport based on the geological data of reclamation [10]. Furthermore, Wu successfully inverted the 20-year subsidence history of the airport using multi-sensor SAR data [11]; Liu and Zhuo used time-series InSAR technology to analyze the subsidence temporal and spatial characteristics of Xiamen Airport in detail [12,13]. Taking Iqaluit Airport as an example, Short proved that InSAR technology has the ability to obtain the periodic deformation related to special soils such as frozen soil in the airport area [14]. Wu used Sentinel-1 images to study the uneven deformation patterns caused by the construction of Yan'an New Airport's cut-and-fill project [15]. These studies made full use of SAR data, climate, hydrology, and geological data from various sensors [16] and, based on InSAR technology, interpreted the land subsidence patterns of major airports. The reliability of the time-series InSAR technique for monitoring surface deformations related to soil properties has been demonstrated [16,17].

The Sentinel-1 series of SAR satellite data, provided worldwide for free by the European Space Agency (ESA), has greatly promoted the development and popularization of InSAR technology. However, using Sentinel-1 SAR data alone to monitor airport surface deformation still poses challenges, for example, the long observation period, surface scattering characteristics, and other factors affect the coherence, and stable persistent scatterer (PS) points can only be acquired in a limited area [18,19]. In this study, PS-InSAR technology considering the phase optimization of homogenous points is used, combined with geological and climatic data, to study the surface deformation of Ankang Airport due to filling engineering and the distribution of expansive soil. This study has two goals. First, we combine the Fast Statistically Homogeneous Pixel Selection-synthetic aperture radar interferometry (FaSHPS-InSAR) method with the conventional StaMPS data processing framework [20,21] to obtain the spatial distribution and deformation information of homogeneous ground scattering points at Ankang Airport, and compare the deformation results with the leveling data measurement results to verify the potential of this data method to study the surface deformation information. Second, multi-source data, such as construction data, geotechnical characteristics, regional rainfall data, and SAR interferometric data, were integrated to explore the acquired temporal and spatial characteristics of deformation and its possible evolution process, the possible causes of deformation were analyzed, and stability analysis was conducted.

The remainder of this paper is organized as follows. First, the geological characteristics of the expansive soil distribution area where Ankang Airport is located and the surface deformation caused by high fill are described. Second, the InSAR data used and key technical details of the data processing strategy are introduced. In the next section, the spatio-temporal distribution of the deformation information from Ankang Airport obtained by InSAR technology is introduced, and the accuracy checked using leveling data. The newly compiled results are interpreted and analyzed in combination with the construction and rainfall data. Finally, the conclusions are presented in the final section and recent work is summarized.

2. Study Area

Ankang Airport is located in Ankang City, Shaanxi Province, China. This area belongs to the intersection of the southern foothills of Qinling Mountains and the northern foothills of Bashan Mountains. There is a regionally active fault here: the Yuehe fault. The Yuehe fault was affected by the Dabashan torsional tectonic system in the late Yanshan period. At present, it is a fault that is mainly tensile, has long-term activity, and is still active today. Therefore, the study area itself is at risk of earthquakes (Figure 1). The airport is 15 km northwest of the city center and designed to have an annual passenger throughput of 300,000 passengers and a cargo throughput of 750 tons. The total planning area of the airport is about 128,000 m^2, the total area of the airport terminal is 5500 m^2, and the airport runway is 2600 m long and 45 m wide. It is a 4C-level civil airport and was built in 2016 and completed in May 2020. It represents China's gradual improvement of the civil airports in recent years [22].

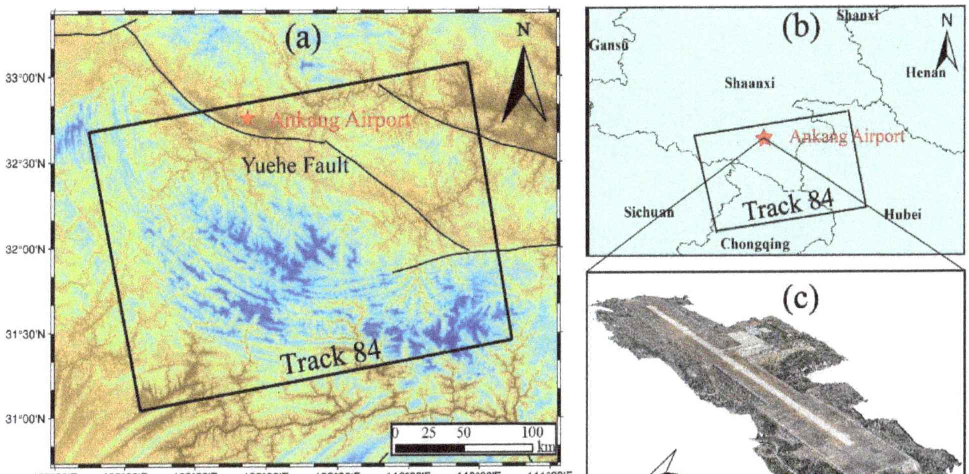

Figure 1. (**a**) Study area location and Synthetic Aperture Radar (SAR) data coverage. In the background is the shaded terrain-generated digital elevation model (SRTM DEM) of the Shuttle Radar Topography Mission, the black rectangles indicate Sentinel-1 SAR data coverage, and the black lines represent the faults location. The fault data is supported by "gmt-china". (**b**) The study area is located in Shaanxi Province, Central China. (**c**) A 3D model of Ankang Airport was acquired using oblique photogrammetry.

The original topography of the airport is dominated by hills, high in the north and low in the south, spanning the Luojia River. Therefore, a large amount of filling work has been carried out in the construction of the airport; the spatial distribution of cut and fill is shown in Figure 2. The volume filled in the study area reached 30,000,000 m^3 and the filling area reached 52,000 m^2, accounting for 41% of the total area of the airport. The maximum filling depth was 48.08 m, and the maximum excavation depth was 45.86 m. The excavation and filling works were staggered, and the maximum vertical height difference in the excavation and filling area reached 93.96 m. The airport fill originated from the undisturbed soil in the adjacent excavation area. The soil was silty clay, dominated by expansive soil. Expansive soil is a naturally formed multi-fissured geological body with significant expansion and contraction characteristics, over-consolidation, multi-fracture, and other undesirable properties [23]. During the construction process of the airport's interior area, expansive soil improvement and soil compaction took place, the area was concreted, and the strength of the structure was significantly improved. However, there was an area of exposed soil on the expansive high-fill slopes on the south, north, and east sides of the airport. The geological body was dominated by medium-strength expansive soil, and obvious surface fissures developed in this area.

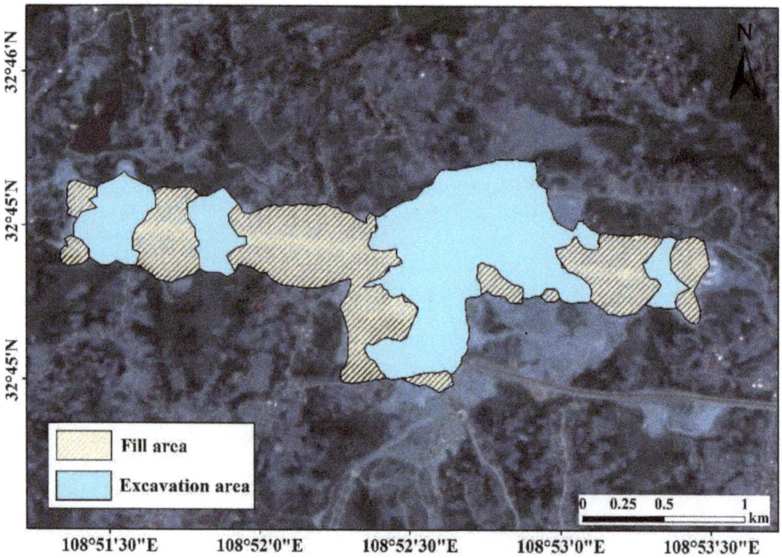

Figure 2. Distribution of excavation and filling works at Ankang Airport.

A comprehensive analysis of the hydrological conditions and geological environment in the study area was conducted which showed that it will increase the possibility of disasters, including the concentration of high-intensity rainfall and a large quantity of expansive soils. During the heavy rainfall period, expansive soil absorbs water and expands; during the drought period, it loses water and shrinks. The alternate wet and dry periods cause repeated expansion and contraction, resulting in surface soil fissures. Second, carrying out high-fill engineering under complex landform conditions leads to potential geological problems such as the settlement and deformation of the high-fill foundation caused by consolidation of the original foundation and compression of the structure itself. Therefore, it is necessary to continuously monitor the ground deformation of the airport in combination with precipitation data, geological information, and InSAR observations.

3. Dataset and Methodology

3.1. Dataset

A set of 40 ascending Sentinel-1 images acquired between May 2020 and December 2021 were used to characterize and monitor the ground surface deformation at Ankang Airport. We chose 20210308 as the master image for co-registration with other images. The thresholds of the temporal and perpendicular baselines were set to 60 days and 160 m, respectively. In total, 36 interferograms were generated (Figure 3).

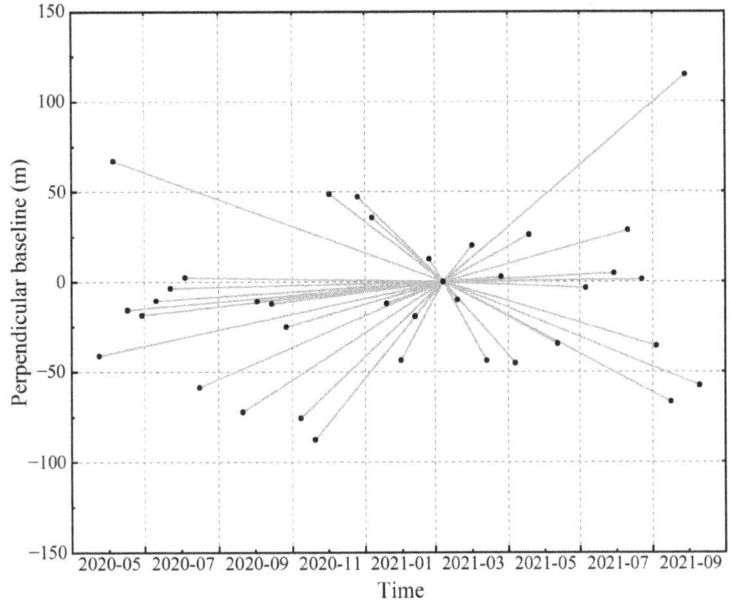

Figure 3. Baseline distribution of high-quality interferograms used in this study.

An unmanned aerial vehicle digital elevation model (UAV DEM) with a grid spacing of 3 m and vertical accuracy of 1 m was provided by Shaanxi Jiawei Spatial Geographic Information Technology Co., Ltd., Ankang, China. This DEM is also available to simulate and remove the topographic phase contribution within the interferograms process. Table 1 lists the primary parameters of the SAR dataset and DEM used in this study.

Table 1. Specific information of Sentinel-1A images and DEM data.

Data	Parameters	Description
Sentinel-1	Type	SLC
	Track number	84
	Orbit number	Ascending
	Azimuth resolution	1
	Range resolution	1
DEM	Source	UAV
	Acquired Date	2020/6/11
	Resolution	3 (m)
	Positioning accuracy	1 (m)
	Elevation accuracy	1 (m)

The available and independently monitored ground deformation ground-truth points we obtained are two leveling points (L1 and L2) located in the central area of the airport (in

Section 4.2). In addition, we used the rainfall data provided by the Ankang Meteorological Station to analyze the features of deformation time series, the observation data of this station are provided by The China Meteorological Data Service Center.

3.2. Methodology

The intensity and phase information of the SAR images were applied to study the spatio-temporal characteristics of the surface deformation at Ankang Airport. The data processing methods used included original image preprocessing, time-series phase optimization, and PS-InSAR deformation calculation.

First, Sentinel-1A image data were preprocessed using GAMMA software, including image co-registration, topographic phase contribution removal, image resampling, and data de-ramping [24]. As Sentinel-1 adopts the progressive scanning (TOPS) observation imaging mode [25], the Doppler centroid rapidly varies along the track, which leads to differences between different bursts. It is prone to obvious phase ramps [26]. Hence, intensity matching was adopted for the burst overlap areas to ensure that all SAR images were accurately co-registered [27].

In the case of varying surface conditions (e.g., planting vegetation or paving runways), the stability of the scattering properties of the target may be compromised, which reduces the number of PSs being detected [18,28]. Therefore, we designed a data-processing strategy to optimize the interference phase by combining intensity information. First, the intensity images were stacked in a time series to select the homogeneous points. The selection of homogeneous points is an algorithm that measures the similarity between the domain and central pixels by means of statistical inference [20]. Assuming that similar objects have the same backscattering characteristics and phase center, the parameter estimation of similar pixels can be used to improve the signal-to-noise ratio of echo signals in complex scenes [29]. In this study, the FaSHPS-InSAR open-source toolbox proposed by Jiang et al. was mainly used for homogeneous point selection [30]. The theoretical basis of FaSHPS was to transform the hypothesis testing problem into a parameter estimation method for confidence interval estimation under the condition that a Gaussian hypothesis was established. Compared with traditional nonparametric hypothesis testing, it has higher computational efficiency and less uncertainty.

The intensity samples of each scatter point p as the central pixel on the N intensity images were $\{A_1, A_2, \ldots, A_N\}$, and the sample points for the mathematical expectation $\mu(p)$ of p were estimated as $\overline{A}(p) = (A_1(p) + A_2(p) + \ldots + A_N(p))/N$. To obtain the expected interval estimate, we determined the distribution of $\overline{A}(p)$ according to the central limit theorem. As the number of samples N increased, $\overline{A}(p)$ gradually approached a Gaussian distribution. Assuming that N was sufficiently large to hold the Gaussian hypothesis, the interval estimate of $\overline{A}(p)$ was obtained as shown in Equation (1).

$$P\{\mu(p) - z_{1-\frac{\alpha}{2}} \cdot \sqrt{\frac{Var(A(p))}{N}} < \overline{A}(p) < \mu(p) + z_{1-\frac{\alpha}{2}} \cdot \sqrt{\frac{Var(A(p))}{N}}\} = 1 - \alpha \quad (1)$$

where $P\{.\}$ indicates the probability of the interval, $z_{1-\frac{\alpha}{2}}$ is the $1 - \frac{\alpha}{2}$ quantile of the standard normal distribution, and $Var\ A(p)$ is the variance of p. It is assumed that when the image is a single look, the intensity image of the homogeneous area obeys the Rayleigh distribution [31]. Its coefficient of variation CV is quantitative and can be expressed as

$$CV = \sqrt{\frac{Var\,A(p)}{E^2}} = \sqrt{\frac{4}{\pi} - 1} \quad (2)$$

Therefore, when the dispersion of the backscattering coefficients in the N-intensity image stacks was not large, the scattered intensity samples were considered stable and homogeneous. At this time, Equation (1) is rewritten as Equation (3), which includes only $\mu(p)$ in the form of an interval.

$$P\left\{\mu(p) - z_{1-\frac{\alpha}{2}} \cdot \mu(p) \cdot \sqrt{\frac{\frac{4}{\pi}-1}{N}} < \overline{A}(p) < \mu(p) + z_{1-\frac{\alpha}{2}} \cdot \mu(p) \cdot \sqrt{\frac{\frac{4}{\pi}-1}{N}}\right\} = 1 - \alpha \quad (3)$$

When the value of $\mu(p)$ was calculated, the interval represented by Equation (3) was also uniquely determined, such that the confidence interval of p can be estimated. When processing the intensity image stack, it was assumed that the reference pixel is p_K, and the number of pixels to be measured is K. First, take $\overline{A}(p_K)$ as the true value of p_K, then estimate the sample mean $\overline{A}(p_K)$ of K-1 pixels and compare them with the confidence interval in Equation (3) one by one. The points that fall within the interval are homogeneous points. To reduce the estimation bias, α = 50%.

After the homogeneous point sample selection was completed, it was used as a parameter, and a time sequence phase optimizer was introduced to optimize the phase of the original interferograms. The core theory of the time series phase optimizer was: in the SAR stack of N SLC images, the time vector of each point p was usually considered to obey a complex multivariate Gaussian model with a covariance matrix [32], as shown in Equation (4).

$$f(g) = \frac{1}{\pi^n \det(\sum)} \exp\left(-g^H \sum\nolimits^{-1} g\right) \quad (4)$$

where superscript H is Hermitian transposition, the temporal vector is $g = [g_1, g_2, \ldots, g_n]^T$, and $\sum$ represents the covariance matrix. One of the main objectives of phase optimization is to estimate $\sum$ using observation matrix $G = [g_1, g_2, \ldots, g_l]$. The maximum likelihood estimator of $\sum$ is:

$$\hat{\sum} = \frac{GG^H}{l} = \hat{\Gamma} \circ \hat{\phi} \quad (5)$$

where $\circ$ is the Hermitian transposition, $\hat{\phi}$ consists of the phase difference between two acquisitions, and $\hat{\Gamma}$ indicates a real symmetric matrix in which each element is the coherence between different acquisitions, also known as spatial coherence, can be expressed as

$$\hat{\Gamma}_{i,k} = \frac{\left|\sum_{p\in\Omega} \exp\left(j\left(\angle g_{i,p} - \angle g_{k,p} - \phi_{fft,p}\right)\right)\right|}{l} \quad (6)$$

where $\hat{\Gamma}_{i,k}$ is the coherence between ith and kth acquisition. Ω denotes a set that includes l homogeneous spatial pixels that satisfy the stationary assumption. $\angle g_{i,p}$ is the observed value of ith SLC images at pixel p. $\phi_{fft,p}$ is a phase compensation used to remove the spatial phase ramp in set Ω [33]. Next, we use Equation (4) to perform a maximum joint likelihood function on the samples in the set Ω to achieve timing phase optimization [34].

Considering the superiority of the PS-InSAR algorithm, we introduce the phase-optimized interferograms into the traditional StaMPS data processing framework. StaMPS method select PS pixels based on amplitude dispersion and estimated phase stability. Subsequently, the three-dimensional phase unwrapping method was applied to unwrapped the wrapped phase. After the atmospheric phase removal and spatio-temporal filtering, we can perform the extraction of deformation phase [35–37]. Finally, we obtain the velocity and deformation time series in LOS direction. The overall data processing flow is shown in Figure 4.

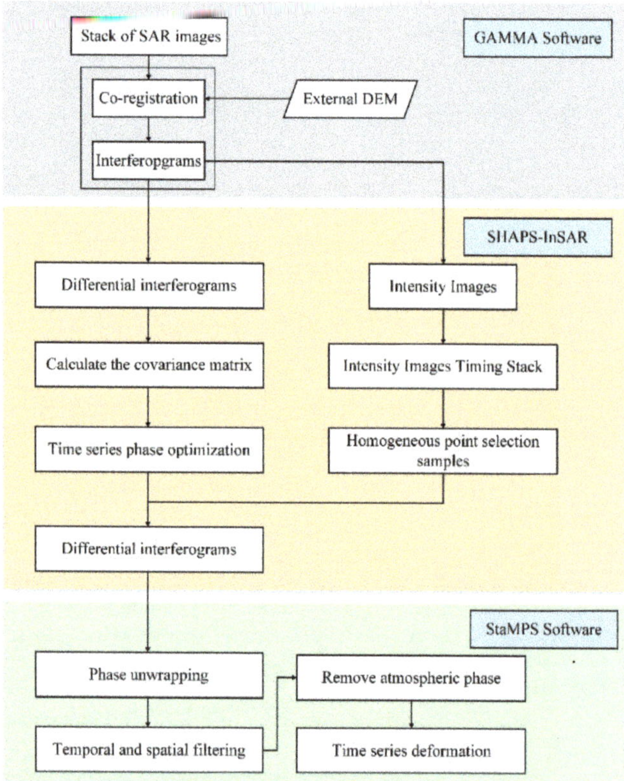

Figure 4. The flowchart for the surface deformation research of Ankang Airport.

To verify the regional stability of the deformation results, the standard deviation of the mean velocity derived from the PS-InSAR technique was calculated using Equation (7) [38,39]:

$$\sigma_{\Delta v}^2 \approx \left(\frac{\lambda}{4\pi}\right)\frac{\sigma_\varphi^2}{M_{\sigma_{Bt}^2}} \tag{7}$$

where λ is the wavelength of the radar wave, σ_φ^2 indicates the phase dispersion, M represents the number of interferograms, and the vertical baseline deviation of the interferograms is σ_{Bt}.

In order to explore the relationship between deformation time series and rainfall, this paper introduces the singular spectrum analysis (SSA) method to analyze the deformation time series of Ankang Airport.

Singular spectrum analysis is an orthogonal analysis method that can decompose a set of relatively complex time series into a few, independent, and easy-to-interpret parts. The main principle is shown in Equation (8) [40].

$$R = XX^T = USV^T \tag{8}$$

where X is the Hankel matrix constructed from the original discrete time series, R is the autocorrelation matrix constructed from X, and U and V are orthogonal matrices belonging to R. $S = [\text{diag}(\lambda_1, \lambda_2, \ldots.), \ldots, 0]$, and 0 is a zero matrix. λ_i is the singular value of the autocorrelation matrix R, $\lambda_1 \geq \lambda_2 \geq \ldots \geq 0$, and the larger the corresponding eigenvalue λ_i of the non-zero singular value, the larger the proportion of the component.

The specific process is as follows: first, remove the noise part of the deformation time series; second, calculate the eigenvalue matrix and decompose the trend deformation and periodic deformation [41]; and finally, analyze the periodic deformation in combination with the regional daily rainfall.

4. Results

4.1. Interferograms Phase Optimization

For the i interferograms obtained by differentially obtaining SLC images, there was a common phenomenon of spatio-temporal decoherence, and it was difficult to obtain sufficient effective PS points. To solve the above problems, this study used a sequential phase optimizer based on the multidimensional complex circular Gaussian distribution described in Section 3.1, to perform phase optimization on the original interferograms stack.

First, spatial samples with similar backscattering properties in space were obtained based on the intensity image set and backscattering coefficient. In this study, the parameter statistics FaSHPS homogenous point selection algorithm was used, which had a high-test effect on the samples. The homogeneous point selection samples are shown in Figure 5.

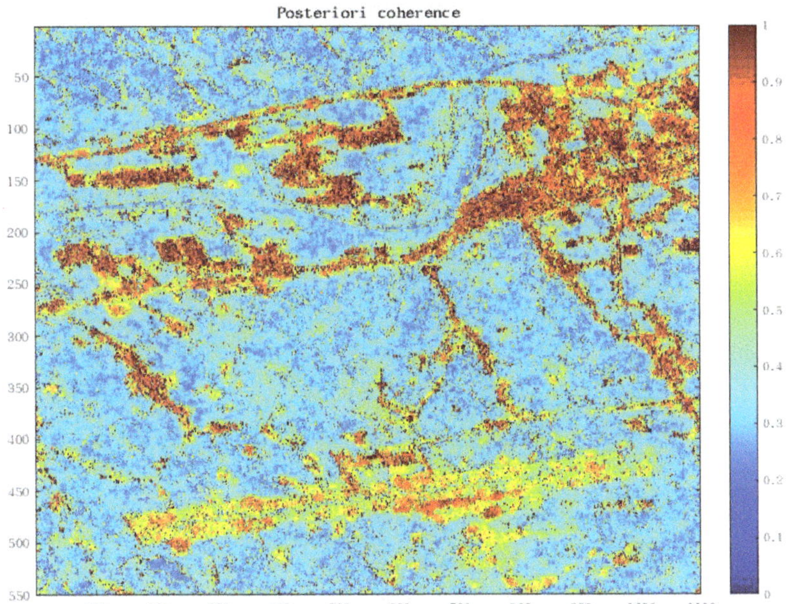

Figure 5. Homogeneous point selection samples with similar scattering properties in space obtained by FaSHPS-InSAR.

Using the timing phase optimizer described in Section 3.2, timing phase optimization was performed on all interferograms. A comparison between the timing-phase optimization results and the original phase is shown in Figure 6. The original phase depicted in Figure 6a has a certain spatial coherence in the high-coherence region, but the signal-to-noise ratio was low in the medium- and low-coherence regions. Notably, the phase optimization method used in this paper not only had good results in the high coherence area but also restored part of the signal in the noise area, and the coherence and integrity of the spatial phase were significantly improved.

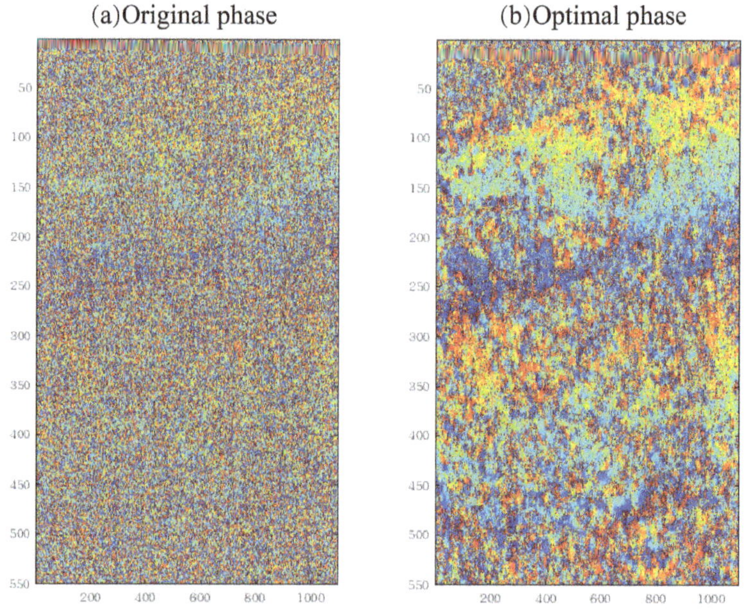

Figure 6. (**a**) Phase distribution of the original interferograms, (**b**) the phase distribution of the interferograms after time series phase optimization.

The coherence distribution of the original interferograms is shown in Figure 7a. The spatial coherence in the airport area was poor, and especially in the high fill slope area, the inner vegetation coverage area, and the runway area, there was a poor coherence distribution. Figure 7b depicts the coherence distribution of the interferograms after the phase optimization. The coherence of the airport area was significantly improved, and the spatial noise had been suppressed to a certain extent; in particular, the spatial continuity of coherence was better in the high fill slope, airport runway area, and terminal area. This illustrates the suitability of the adopted phase optimizer for the phase optimization of the Ankang Airport interferometry.

Using the StaMPS software, we set consistent phase mean and amplitude dispersion threshold parameters and selected stable PS points from the two sets of interferograms. The distribution of PS points identified using the original interferograms set was shown in Figure 7c, and the distribution of PS points identified using the phase-optimized interferograms set is shown in Figure 7d. The identification results of stable PS points showed that, compared with the original interferograms, the phase-optimized interferograms shows higher phase and intensity stability in the study area, and more PS points were identified, especially in low shrubs. Compared to the original interferograms, the high-fill slope area covered by the main vegetation had obvious advantages. Using the data processing strategy in this study, under the same parameter settings, a total of 2129 PS points were identified within the airport range, whereas using the original interferograms, only 1372 PS points were identified. More importantly, in the airport runway and expansive soil high-fill slope area with historical deformation, the number of PS points had increased significantly. This is conducive to better monitoring of the safe operation of the airport functional area and a more comprehensive stability analysis of high-fill slopes.

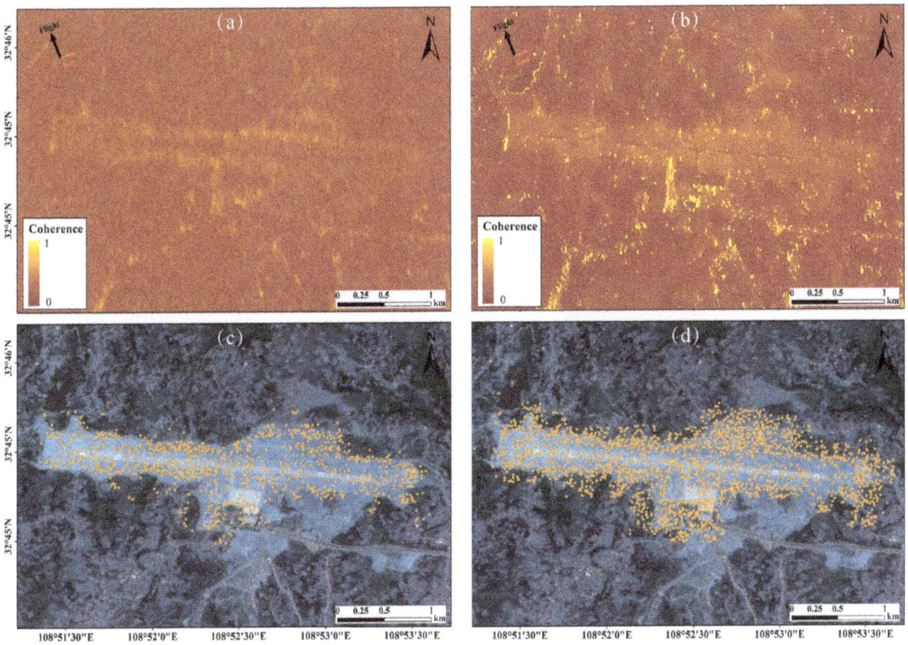

Figure 7. (**a**) Original interferograms average coherence, (**b**) optimized interferograms average coherence, (**c**) original interferograms scatterers identification, and (**d**) optimized interferograms scatterers identification.

4.2. InSAR-Derived Results and Reliability Analysis

In this section, the set of interferograms processed by the phase optimization step described in Section 4.1 was solved by the StaMPS method, and good deformation monitoring results were obtained at Ankang Airport. The annual average deformation rate map of the airport from May 2020 to October 2021 is shown in Figure 8, where the deformation rate was in the direction of LOS, the positive value in blue indicates movement towards the satellite, and the negative value in red indicates movement away from the satellite. The InSAR results showed that most of the airport area was relatively stable, and the deformation rate was between (−15–19 mm/year). However, four distinct deformation areas were identified in the high-fill slope area of the airport. Moderate (−27.5–−35.9 mm/year) deformation areas and severe (−35.9–−44.4 mm/year) deformation areas were concentrated in the slope area of the airport expansive soil filling (I, II, III, and IV). The maximum deformation rate appeared in the I area, and the annual maximum deformation rate was −44.4 mm/year. This area was the slope formed by the area with the largest fill layer in the airport, and expansive soil was used as the main filling soil. Therefore, it was necessary to analyze the time-series deformation in combination with the special hydraulic laws of expansive soil. In addition, combined with the distribution map of the airport cut-and-fill shown in Figure 2, the magnitude of the deformation rate was proportional to the depth of the fill layer. The aforementioned deformation laws are discussed in detail in Section 5.

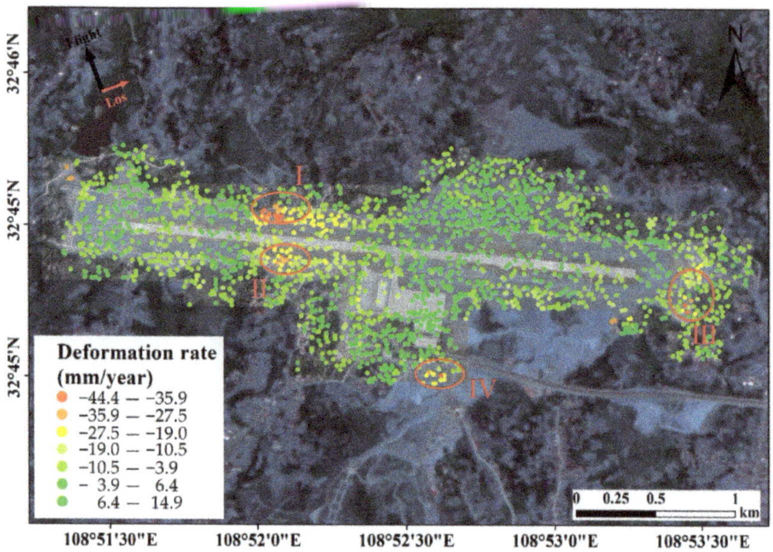

Figure 8. Deformation rate map of Ankang Airport along the LOS direction. Areas I, II, III, and IV are areas with obvious deformation (deformation rate >−30 mm/year).

The standard deviation of the calculated deformation rate is shown in Figure 9a. It can be seen from Figure 9a that the standard deviation of the slope area is lower than 5.3 mm/year, the standard deviation of the runway area is 3.5 mm/year, and the overall standard deviation is lower than 5.5 mm/year, showing the reliability of the data processing strategies.

In addition, the leveling observation data of two ground reference points located on the airport slope and north side of the runway at Ankang Airport (L1 and L2 in Figure 9a) were collected. As the observation value of the leveling point reflects the vertical deformation and the InSAR deformation result mainly reflects the deformation along the LOS direction, according to the incident angle of the pixel corresponding to the leveling point, the leveling deformation result was uniformly projected in the LOS direction for accuracy verification. The comparison results of the deformation time series of points L1 and L2 observed by leveling observation and InSAR technology are shown in Figure 9b,c. The difference in the deformation results of the L1 point in the slope area is less than 3 mm, and the difference in the deformation results of the L2 point in the runway area is less than 5 mm. Although the two observation methods are not strictly registered in time, the deformation trend is consistent, further illustrating the reliability of the InSAR monitoring results.

4.3. Stability Analysis of High Fill Slope

The high-fill expansive soil slope stability influence area was distributed around the airport flight area, and the filling foundation was mainly expansive soil, which may have caused deformation related to the hydraulic properties of the expansive soil. The stability of an expansive soil slope area was also a key issue in airport safety operations.

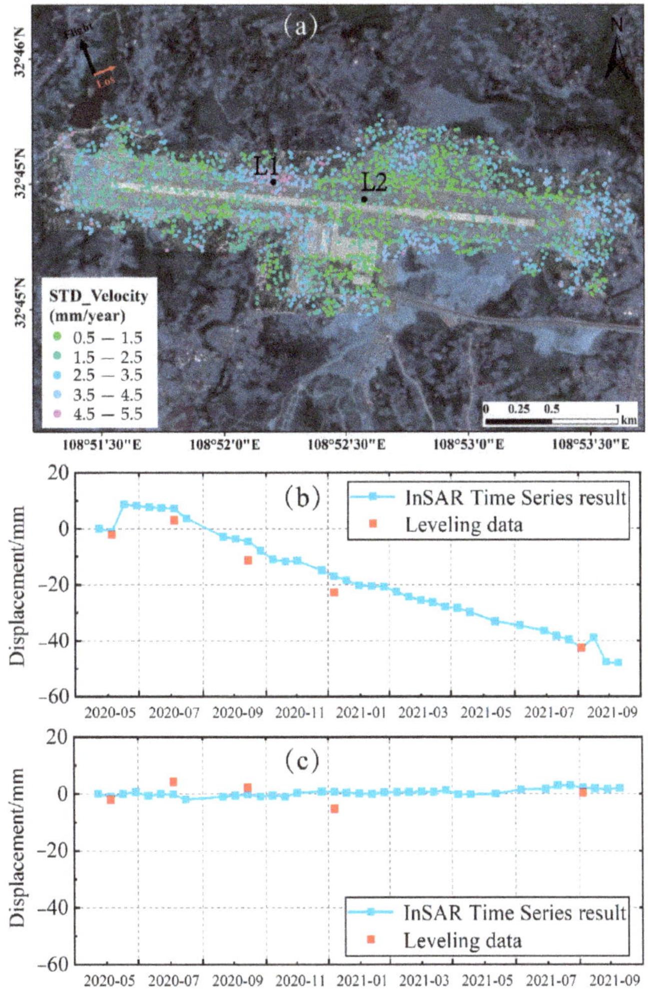

Figure 9. (**a**) Standard deviation of the mean velocity on coherent point targets, (**b**) time series analysis of L1 by leveling and PS-InSAR technique, and (**c**) time series analysis of L2 by leveling and PS-InSAR technique.

According to the excavation and filling construction conditions (Figure 10a), Figure 10b shows the spatial distribution of the deformation rate of the high-fill expansive soil slope at the airport. The results showed that the deformation rate of most slope areas was relatively low, but InSAR technology identified three obvious deformation areas (Figure 10b in I, II, III, and IV). Area I and II are located on the fill slope on the north and south sides of the airport runway. According to the excavation and filling construction conditions (Figure 2), area I was the area with the largest fill layer depth (48.08 m) in the entire airport with a history of slope failure. The maximum deformation rate of area I was −42.19 mm/year, which was the largest deformation in the entire airport area; area III was located on the fill slope on the east side of the airport. The fill depth of this area was 18.9 m, and the maximum deformation rate reached −37.1 mm/year; area IV was located south of the terminal area, and the engineering slope at the airport expressway had a maximum deformation rate of −37.4 mm/year. The filling depth here was only 14.206 m, but according to previous geological data, the filling soil in this area was mainly strong expansive soil. The above deformation areas appeared in the fill slope, indicating that the consolidation and compression of the fill layer may be one of the potential causes of slope instability.

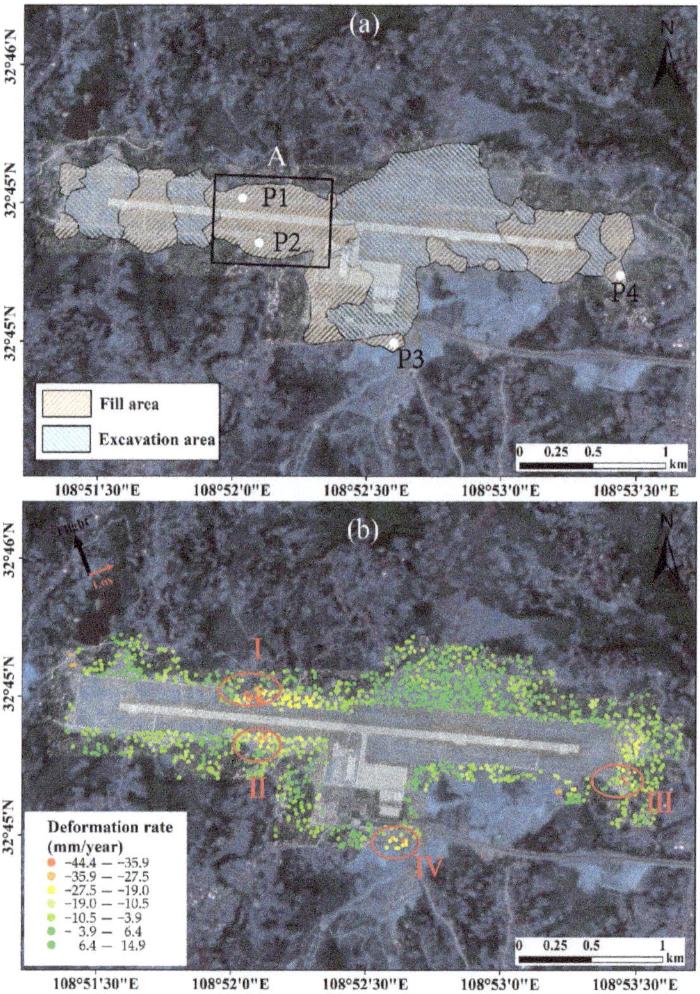

Figure 10. (**a**) The location of characteristic points, and the distribution of fill-excavation, (**b**) deformation rate map of the slopes of Ankang Airport along the LOS direction, regions I, II, III, and IV are the areas of unstable slopes identified by InSAR technology. Area A is located on the fill slope on the north and south sides of the airport runway.

From the site survey data, we found that there was a certain area of exposed expansive soil in the fill slope area of the airport. Considering the expansion, contraction, and crack-prone characteristics of expansive soils that are highly correlated with soil moisture, this study collected regional rainfall data in the study area from May 2020 to July 2021, and, based on the maximum daily rainfall and rainfall time intensity, this was divided into two rainy seasons from May 2020 to July 2020 and April 2021 to August 2021. Four feature points (P1, P2, P3, and P4 in Figure 10a) were extracted from three regions, and the deformation time series and rainy season time distribution obtained by InSAR technology were combined for analysis. A comparison between the deformation time series of the P1–P4 feature points and the distribution during the rainy season is shown in Figure 11.

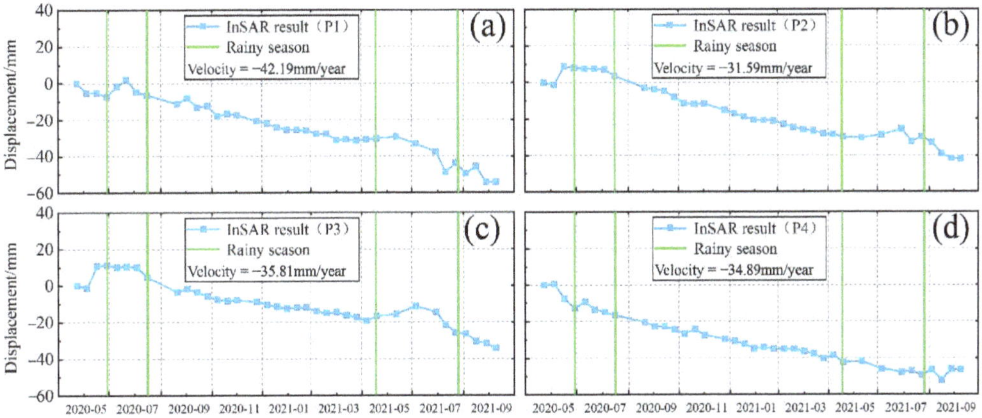

Figure 11. Deformation time series in the LOS direction of regions I, II, III, and IV (the areas of unstable slopes identified by InSAR technology, marked in Figure 10) from May 2020 to October 2021 for P1–P4, which are indicated as white dots in Figure 10a. (**a**) Point P1; (**b**) point P2; (**c**) point P3; and (**d**) point P4.

P1 and P2 were located on the fill slopes on the north and south sides of the runway area in deformation area A, respectively. As shown in Figure 11a,b, the deformation rates reached −42.19 mm/year and −31.59 mm/year, respectively. The overall deformation trend was away from the sensor. However, in the two rainy seasons, the deformation rate slowed down significantly, and an expansion phenomenon appeared, which was consistent with the soil mechanical properties of expansive soil to a certain extent. P3 showed a similar deformation time series to P1 and P2: During the rainy seasons, there was obvious rebound deformation phenomenon along the LOS direction, which was related to the strong expansive soil distributed in this area. The deformation time series of P4 is slightly different, although the deformation phenomenon away from the SAR sensor also occurred in this area, no obvious rebound phenome-non occurred in the rainy season. The maximum cumulative deformation of P4 reached −53.12 mm, and the deformation rate of P4 was −34.89 mm/year.

To describe the spatial distribution pattern of the high-fill slope deformation in area A in detail, a 3D model was constructed using the UAV to obtain the DEM and DSM information of Ankang Airport, as shown in Figure 12a. The annual average deformation rate of the slope in area A obtained using InSAR technology was registered to the 3D model of the airport, as shown in Figure 12b. The fill slope on both sides of the runway was divided into two parts, north and south. In Figure 12b, area N represents the north-side slope, and area S represents the south-side slope. To analyze the correlation between the spatial distribution of the slope and the slope rate, the deformation rate and 3D model registration results of the slope in the north and south of area A obtained by InSAR were enlarged, as shown in Figure 12c,d, respectively.

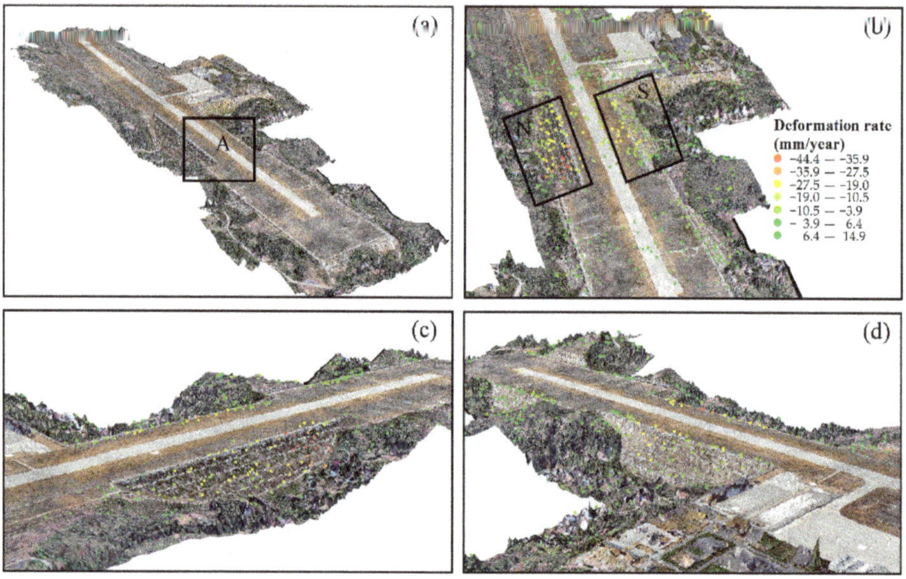

Figure 12. (**a**) Three-dimensional model of area A, (**b**) deformation rate of area A in LOS direction, (**c**) deformation rate of north slope in LOS direction, and (**d**) deformation rate of south slope in LOS direction.

As shown in Figure 12c, the spatial distribution characteristics of the deformation of the northern slope were as follows: the deformation rate gradually increases from east to west along the slope; the deformation rate gradually increased with the increase in elevation; and the larger deformation points gathered in the top area at the west end of the slope. As shown in Figure 12d, the deformation rate of the southern slope of area A increased gradually with increasing elevation but did not show obvious spatial distribution characteristics in the east–west direction, and the larger deformation points were gathered in the top area of the slope. The spatial distribution of slope deformation at N and S presented characteristics related to the slope and elevation.

4.4. Stability Analysis of Runway

Ankang Airport currently has a runway with a total length of 2700 m and a total width of 45 m. The ground of the runway is relatively weak and prone to consolidation and compression, leading to uneven ground settlement. From the InSAR deformation monitoring results of the runway area shown in Figure 13b, the airport runway is in a stable state, and only the maximum fill area of the airport has a moderate deformation area (−19 mm/year to −27.5 mm/year).

According to the distribution of fill and cut described in Section 2, P5–P8 were extracted along the runway of Ankang Airport, and four feature points were used for time-series deformation analysis. Based on a comparative analysis of the deformation time series of the feature points in the filling area, the deformation time series of the four feature points and the distribution of the regionally intensive rainfall period are shown in Figure 14.

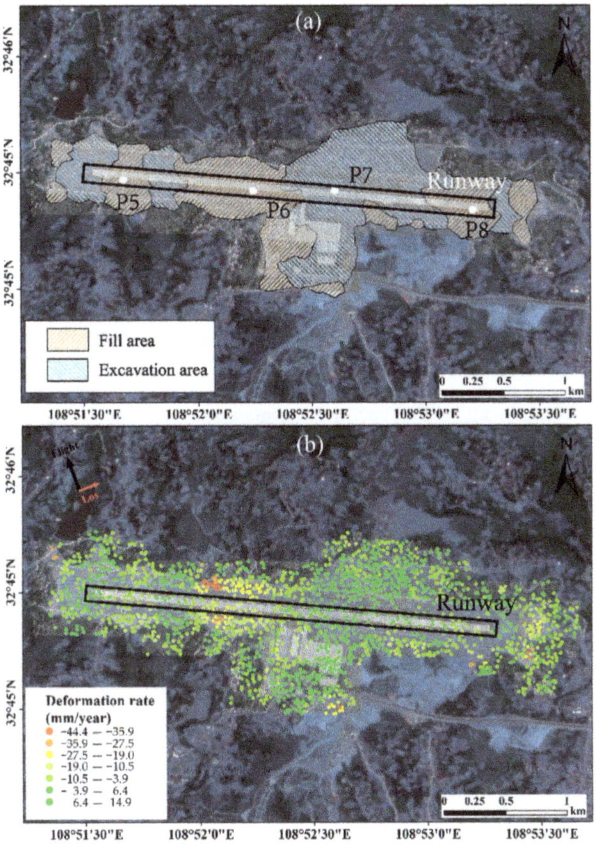

Figure 13. (**a**) Distribution of characteristic points of runway area of Ankang Airport, (**b**) deformation rate map of the runway in Ankang Airport along the LOS direction.

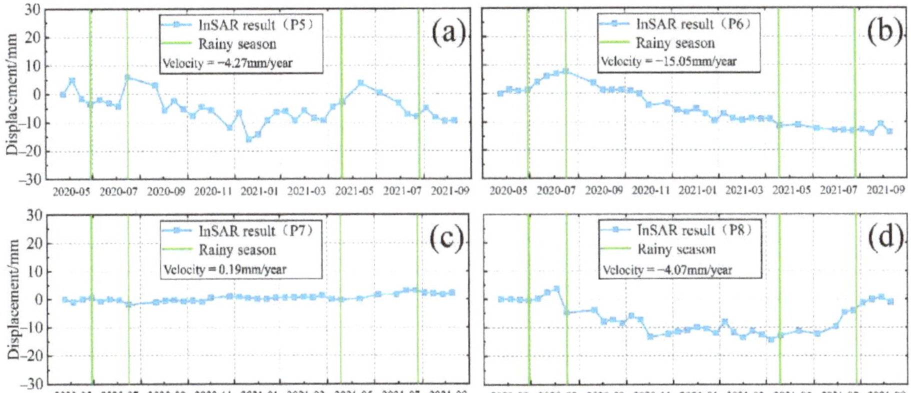

Figure 14. Time series deformation along LOS direction of Runway from May 2020 to October 2021 for P5–P8, which are indicated as white dots in Figure 14a. (**a**) Point P5; (**b**) point P6; (**c**) point P7; and (**d**) point P8.

The deformation rate of P5 is −4.27 mm/year, and the maximum cumulative deformation reaches −18.7 mm. The deformation time series is shown in Figure 14a. In both rainy seasons, P5 had been uplifted to different degrees along the LOS direction, the surface was the largest, the lift amount reached 6.4 mm, and the overall deformation trend tended to be stable. P6 was in the deepest filling area of the airport, the filling depth reached 48.08 m, the deformation rate was −15.05 mm/year, and the maximum cumulative deformation reached −14.13 mm. The deformation time series is shown in Figure 14b. In the rainy season of 2020, there was obvious uplift along the LOS direction, whereas in the rainy season of 2021, the deformation rate slowed down significantly, and the overall deformation trend tended toward subsidence along the LOS direction. Consolidation compression correlation; P7 was in the excavation area of the airport, the deformation rate was 0.19 mm/year, and the maximum cumulative deformation reached 2.3 mm. The deformation time series is shown in Figure 14c. Compared with points P5, 6, and 8, the deformation time series of point P7 is relatively stable and has no obvious correlation with rainfall distribution, and the overall deformation trend fluctuates between ±3 mm. The deformation rate of P8 was −4.07 mm/year, and the maximum cumulative deformation reached 15.4 mm. The deformation time series is shown in Figure 14d. It was similar to P5 and P6, and both have a certain degree of uplift along the LOS direction during the rainy season; however, the fluctuation range of P8 was large, and the cumulative deformation was relatively small (0.79 mm).

We created a section line RR1 (marked in Figure 15a) along the airport runway, obtained the elevation change information of the runway along the section line of RR1 (Figure 15b), and extracted the deformation rate in the 25 m area on both sides of RR1, as shown in Figure 15c. There was obvious uneven deformation along the airport runway. The runway deformation in area A was significantly larger than that in other areas; multiple settlement funnels were observed along the runway, and the maximum deformation rate in the vertical direction was −24 mm/year.

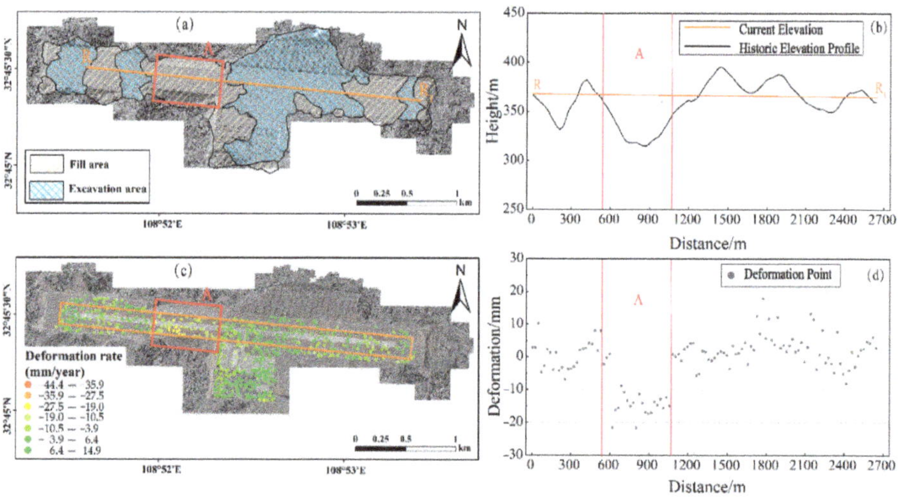

Figure 15. (**a**) The distribution of fill–excavation and location of RR_1 profile, The red rectangle represents the maximum fill area of the airport (marked in Figure 12) (**b**) RR_1 profile elevation information, the current elevation information is provided by the DEM obtained by the UAV, and the historic elevation information is provided by the SRTM, (**c**) RR1 runway area deformation rate distribution, and (**d**) distribution of deformation points along the RR_1 profile.

Based on the above analysis, the following conclusions were drawn: (1) the expansive soil in the excavation area is more stable than the expansive soil in the filling area; and

(2) the overall deformation of the runway area is sufficiently small to meet functional requirements, such as general navigation.

5. Discussion

5.1. Driving Factors of Ankang Airport Deformation

The common deformation types of large-scale high-fill buildings were post-construction settlement, caused by consolidation and compression of filled foundation, and uneven settlement, caused by discontinuous construction [42].

The deformation rate of the PS point in the study area and the depth of the fill volume at its location were extracted to describe the relationship between the distribution of the fill volume in the Ankang expansive soil airport and the surface deformation, as shown in Figure 16.

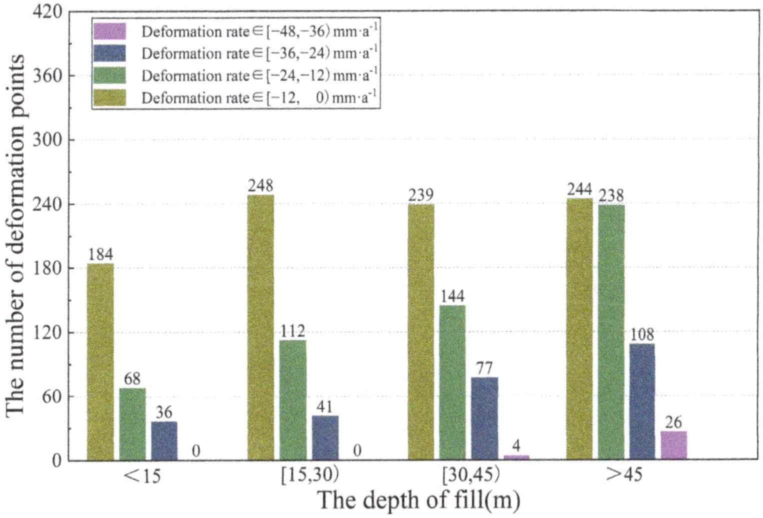

Figure 16. Distribution of the relationship between fill depth and deformation rate at Ankang Airport.

As shown in Figure 16, with an increase in the thickness of the fill body, the number of settlement observation points and the degree of deformation increased significantly, and the distribution and thickness of the fill soil affected the distribution and size of the surface deformation. The areas with the most obvious deformation were all distributed in the area where the depth of the fill layer was >30 m, and the deformation monitoring points in this area accounted for 62% of all deformation and settlement monitoring points. Monitoring points with a deformation rate greater than −24 mm/year appear only in the area where the depth of the fill layer is >15 m. By contrast, in the area where the depth of the fill layer is less than 15 m, there are fewer settlement points and lower deformation rates. As the compressive deformation of the fill soil at different depths and the pressure on the undisturbed soil increases with the depth of the fill layer [43], the consolidation and compression of the fill layer is one of the reasons for the surface deformation of Ankang Airport.

In addition, it is necessary to consider the ground deformation caused by the dry–wet cycle of expansive soil [44,45]. The area has a subtropical monsoon climate, with rainfall concentrated between late June and late August. The analysis in Sections 4.3 and 4.4 shows that the deformation trend of the filling area in the entire airport area is relatively similar and has certain regularity; the deformation rate slows down in the rainy season and then springs back, and there are periodic deformations to different degrees, which is consistent with the hydraulic effect of swelling and shrinking of expansive soil. However, the trend deformation caused by post-consolidation compression is the main component of the deformation time series, and errors are caused by the noise phase in the InSAR settlement process, which hinders the exploration of the relationship between the deformation time series and rainfall. Therefore, this paper introduced the singular spectrum analysis (SSA) method to analyze the deformation time series of Ankang Airport. The P1, P3, P5, and P7 deformation time series for SSA decomposition, and the trend deformation and periodic deformation time series are shown in Figure 17.

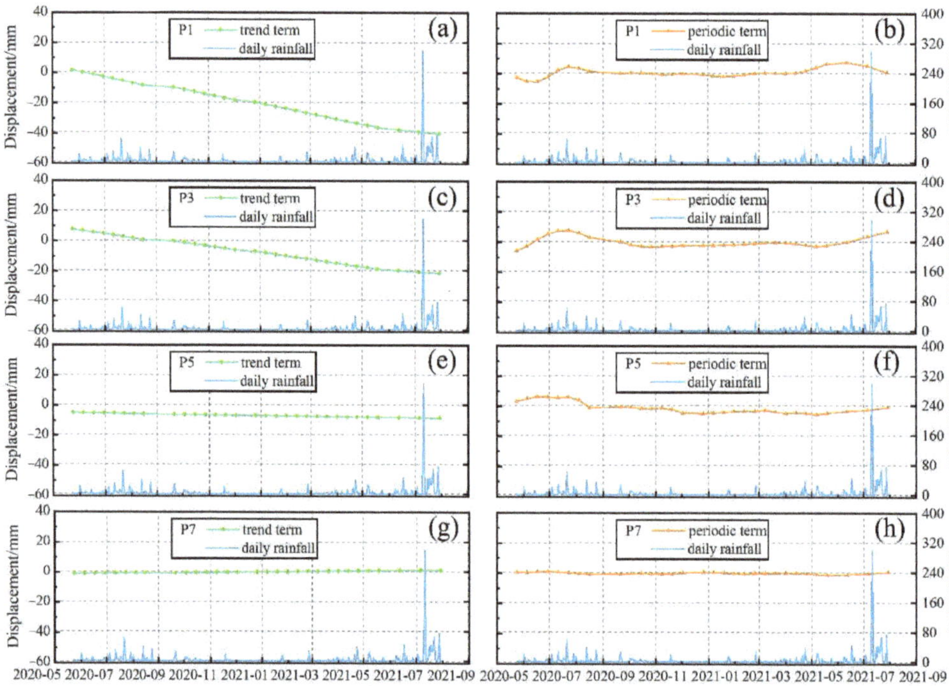

Figure 17. The deformation time series of the four feature points (P1, P3, P5, and P7) is decomposed into two parts, the periodic term and the trend term by the SSA method, (**a**) P1 Trend Item Deformation, (**b**) P1 Period Item Deformation, (**c**) P3 Trend Item Deformation, (**d**) P3 Period Item Deformation, (**e**) P5 Trend Item Deformation, (**f**) P5 Period Item Deformation, (**g**) P7 Trend term deformation, and (**h**) P7 Periodic term deformation.

P1 and P3 are in the slope fill area, P5 is in the runway fill area, and P7 in the runway excavation area, serving as the control group. The first column (a,c,e,g) of Figure 17 describes the deformation trend of the four points. The deformation trends of P1 and P3 located on the filled slope were similar, showing an obvious settlement trend. The trend deformation of P5 in the fill area of the runway is small, and the trend deformation of P7 does not appear obvious and tends to be stable. The periodic deformation was obtained after removing the trend deformation and high-frequency noise (shown in Figure 17b,d,f,h). When the rainfall increased significantly, P1, P3, and P5 showed an obvious uplift phenomenon; P1 and P3 located on the expansive soil fill slope show obvious periodic deformation, which is similar to the monitoring results of similar cohesive soils using InSAR technology [46]. However, the P7 point located in the excavation area remained stable during the observation period, with no periodic deformation clearly related to rainfall. This is because the expansive soil in the airport runway, terminal building, and other areas has been mixed with lime, cement, or other chemical additives to inhibit the expansion and contraction characteristics of the expansive soil. In addition, the foundation of the runway area has been compacted several times and covered with concrete as a water barrier. However, the expansive soil slope around the airport still has some areas of bare leakage area, which are also the focus of this research [47].

5.2. Potional of InSAR in Reaveling Surface Deformation in Expansive Soil Airport

The above analysis confirms that the deformation information obtained by the data processing method adopted in this paper is consistent with the characteristics of the geological environment.

Monitoring the surface deformation of the airport in expansive soil region by InSAR Technology, the low-coherence effect in runway and vegetation area is the first problem to be solved in data processing [19,48]. This paper adopted the timing phase optimization method described in Section 3.2 to improve the coherence of interferograms. As the result shown in Figure 7, Section 4.1, our method has much denser PS points in the whole airport area, especially in the low-coherence area such as slope and runway. The results are similar to the experimental results of using the likelihood function to estimate the phase at Shenzhen Airport [49]. This illustrates the great potential of adopting methods for airport deformation monitoring. The second problem is that airport deformation is often a complex manifestation of multiple deformations [50]. In order to reasonably interpret the monitoring results of InSAR, the SSA method is introduced in this paper to separate the periodic and trend deformations of the deformation time series. As shown in Figure 17, the relationship between the periodic term deformation and rainfall is more pronounced. This method can help us understand the development law of expansive soil in engineering construction.

However, the proposed method still has its limitations. First, in the selection of homogeneous point samples, the window size needs to be set carefully. A window size that is too small will lead to biased estimates due to few samples, while a window size that is too large will cause the interference signal to be non-stationary and underestimate coherence. Second, the determination of eigenvalues in the SSA method still requires subjective judgment, and we will establish an adaptive SSA method for analysis in future research. In addition, this paper does not consider the topographic error and the effect of temperature on the deformation of concrete structures [11,51]. Therefore, the monitoring of surface deformation of expansive soil at Ankang Airport still needs further research.

6. Conclusions

In this study, interferogram phase optimization and PS-InSAR technologies were used to monitor the ground deformation of Ankang Airport from its completion to the initial stage of operation. Furthermore, SSA technology was introduced to decompose the trend deformation of the fill area and the periodic deformation of the expansive soil from the InSAR time series and analyzed with rainfall information. A total of 39 interferograms were used, and the monitoring period was from May 2020 to October 2021. It fully reveals the temporal and spatial distribution characteristics of the surface deformation after the overall completion of Ankang Airport and provides important deformation information for it. The research results have important guiding significance for airport safety navigation, stability analysis, and future construction. The main conclusions drawn from this study are as follows:

In this study, a method for selecting homogenous points under the assumption of a Gaussian complex circle is introduced, and phase optimization of the interferograms is carried out according to the samples of homogenous points. Compared to the original interferograms, this method identifies more persistent scatterers in the airport. At Ankang Airport, the method used in this paper identified 2129 persistent scatterer pixels, while the original Stamps only obtained 1372 persistent scatterer pixels. The results show that this method can significantly improve the coherence of interferograms and verify the reliability of the monitoring results according to the level monitoring data and standard deviation distribution.

From May 2020 to October 2021, three relatively obvious deformation areas were identified at Ankang Airport, all of which were located on the high-fill slopes of the airport. The deformation trend of these filled slopes is positively correlated with the slope and depth of the fill layer. The runway area of the airport is generally stable, but small-scale land subsidence occurs in the fill area of the runway, which is mainly caused by consolidation and compression of the fill layer in this area. In addition, the analysis results of the deformation time series in different areas show that the expansive soil located in the fill area is more sensitive to rainfall, particularly in the high-fill slope area without integral concrete pouring. There is bare soil in this area, which is conducive to expansive soil disasters. The deformation results showed an obvious periodic deformation.

In general, the monitoring results of the time-series InSAR technology, based on phase optimization, in the Ankang Airport area show that the airport area is dominated by the trend deformation caused by the consolidation and compression of the fill. In the exposed slope and high-fill area, the deformation time series is consistent with the rainfall distribution, and there are characteristics of expansive soil disaster development. At present, the Ankang Airport is still in a stage of rapid deformation. Therefore, it is necessary to monitor its long-term deformation.

Author Contributions: Conceptualization, S.Z. and J.S.; methodology, J.S. and Y.N.; software, Y.N. and W.Z.; validation, J.S., Q.F. and Z.R.; formal analysis, W.Z.; investigation, P.A. and X.H.; resources, C.Z.; data curation, X.H.; writing—original draft preparation, S.Z. and J.S.; writing—review and editing, S.Z. and Z.L. All authors have read and agreed to the published version of the manuscript.

Funding: This research was funded by the National Key Research and Development Program of China (Grant No. 2019YFC1509802,2020YFC1512000); The National Natural Science Foundation of China Projects (Grant No.42074041); State Key Laboratory of Geo-Information Engineering (Grant No. SKLGIE2019-Z-2-1), the Shaanxi Natural Science Research Program (Grant No.2020JM-227), This research was also supported in part by the Fundamental Research Funds for the Central Universities, Chang'an University, (Grant No. 300102260301), in part by the Shaanxi Province Science and Technology Innovation Team (Grant No. 2021TD-51), and in part by the European Space Agency through the ESA-MOST DRAGON-5 Project (Grant No. 59339).

Acknowledgments: Sentinel-1A data used in this study were provided by European Space Agency (ESA) through the Sentinel-1 Scientific Data Hub. The digital elevation model (DEM) data and Geological data of Ankang Airport is provided by Shaanxi Jiawei Space Geographic Information Technology Co., Ltd. The measured rainfall data is provided by China Meteorological Administration and China Meteorological Data Network. We are very grateful for the above support. In addition, the authors express their deep gratitude to Mi Jiang for providing the FaSHPS-InSAR toolbox.

Conflicts of Interest: The authors declare no conflict of interest.

References

1. Xu, Y.F.; Liu, S.Y. Fractal characteristic of grain-size distribution of expansive soil. *Fractals* **1999**, *74*, 359–366. [CrossRef]
2. Zhang, S.C.; Si, J.Z.; Xu, Y.F.; Niu, Y.F.; Fan, Q.Y.; Zhu, W.; An, P.; Guo, Y.H. Time-Series InSAR for Stability Monitoring Re-search of Ankang Airport with Expansive Soil. *Geomat. Inf. Sci. Wuhan Univ.* **2021**, *46*, 1519–1528. (In Chinese)
3. Soilan, M.; Sanchez-Rodriguez, A.; Rio-Narral, P.; Perez-Collazo, C.; Arias, P.; Riveiro, B. Review of laser scanning technol-ogies and their applications for road and railway infrastructure monitoring. *Infrastructures* **2019**, *44*, 58. [CrossRef]
4. Soga, K.; Luo, L. Distributed fiber optics sensors for civil engineering infrastructure sensing. *J. Struct. Integr. Maint.* **2018**, *3*, 1–21. [CrossRef]
5. Escobar-Wolf, R.; Oommen, T.; Brooks, C.N.; Dobson, R.J.; Ahlborn, T.M. Unmanned aerial vehicle (UAV)-based assessment of concrete bridge deck delamination using thermal and visible camera sensors: A preliminary analysis. *Res. Nondestruct. Eval.* **2017**, *29*, 183–198. [CrossRef]
6. Hooper, A.; Zebker, H.; Segall, P.; Kampes, B. A new method for measuring deformation on volcanoes and other natural terrains using InSAR persistent scatterers. *Geophys. Res. Lett.* **2004**, *31*. [CrossRef]
7. Dai, K.; Shi, X.; Gou, J.; Hu, L.; Chen, M.; Zhao, L.; Dong, X.; Li, Z. Diagnosing Subsidence Geohazard at Beijing Capital International Airport, from High-Resolution SAR Interferometry. *Sustainability* **2020**, *12*, 2269. [CrossRef]
8. Gao, M.; Gong, H.; Li, X.; Chen, B.; Zhou, C.; Shi, M.; Guo, L.; Chen, Z.; Ni, Z.; Duan, G. Land Subsidence and Ground Fissures in Beijing Capital International Airport (BCIA): Evidence from Quasi-PS InSAR Analysis. *Remote Sens.* **2019**, *11*, 1466. [CrossRef]

9. Marshall, C.; Large, D.J.; Athab, A.; Evers, S.L.; Sowter, A.; Marsh, S.; Sjögersten, S. Monitoring tropical peat related settlement using ISBAS InSAR, Kuala Lumpur International Airport (KLIA). *Eng. Geol.* **2018**, *244*, 57–65. [CrossRef]
10. Jiang, L.; Lin, H. Integrated analysis of SAR interferometric and geological data for investigating long-term reclamation settlement of Chek Lap Kok Airport, Hong Kong. *Eng. Geol.* **2010**, *110*, 77–92. [CrossRef]
11. Wu, S.; Yang, Z.; Ding, X.; Zhang, B.; Zhang, L.; Lu, Z. Two decades of settlement of Hong Kong International Airport measured with multi-temporal InSAR. *Remote Sens. Environ.* **2020**, *248*, 111976. [CrossRef]
12. Liu, X.J.; Zhao, C.Y.; Zhang, Q.; Yang, C.S.; Zhang, J. Characterizing and Monitoring Ground Settlement of Marine Recla-mation Land of Xiamen New Airport, China with Sentinel-1 SAR Datasets. *Remote Sens.* **2019**, *11*, 585. [CrossRef]
13. Zhuo, G.C.; Dai, K.R.; Huang, H.N.; Li, S.P.; Shi, X.L.; Feng, Y.; Li, T.; Dong, X.J.; Deng, J. Evaluating Potential Ground Sub-sidence Geo-Hazard of Xiamen Xiang'an New Airport on Reclaimed Land by SAR Interferometry. *Sustainability* **2020**, *12*, 6991. [CrossRef]
14. Short, N.; LeBlanc, A.M.; Sladen, W.; Oldenborger, G.; Dufour, V.M.; Brisco, V. RADARSAT-2 D-InSAR for ground displacement in permafrost terrain, validation from Iqaluit Airport, Baffin Island, Canada. *Remote Sens. Environ.* **2014**, *141*, 40–51. [CrossRef]
15. Wu, Q.; Jia, C.; Chen, S.B.; Li, H.Q. SBAS-InSAR Based Deformation Detection of Urban Land, created from Mega-Scale Mountain Excavating and Valley Filling in the Loess Plateau: The Case Study of Yan'an City. *Remote Sens.* **2019**, *11*, 1673. [CrossRef]
16. Zhang, J.M.; Zhu, W.; Cheng, Y.; Li, Z.H. Landslide Detection in the Linzhi–Ya'an Section along the Sichuan–Tibet Railway Based on InSAR and Hot Spot Analysis Methods. *Remote Sens.* **2021**, *13*, 3566. [CrossRef]
17. Li, Z.-W.; Zhao, R.; Hu, J.; Wen, L.; Feng, G.; Zhang, Z.; Wang, Q. InSAR analysis of surface deformation over permafrost to estimate active layer thickness based on one-dimensional heat transfer model of soils. *Sci. Rep.* **2015**, *5*, 15542. [CrossRef]
18. Zebker, H.; Villasenor, J. Decorrelation in interferometric radar echoes. *IEEE Trans. Geosci. Remote Sens.* **1992**, *30*, 950–959. [CrossRef]
19. Jiang, Y.N.; Liao, M.S.; Wang, H.M.; Zhang, L.; Balz, T. Deformation Monitoring and Analysis of the Geological Environment of Pudong International Airport with Persistent Scatterer SAR Interferometry. *Remote Sens.* **2016**, *8*, 1021. [CrossRef]
20. Jiang, M.; Ding, X.; Hanssen, R.F.; Malhotra, R.; Chang, L. Fast Statistically Homogeneous Pixel Selection for Covariance Matrix Estimation for Multitemporal InSAR. *IEEE Trans. Geosci. Remote Sens.* **2014**, *53*, 1213–1224. [CrossRef]
21. Hooper, A.; Segall, P.; Zebker, H. Persistent scatterer interferometric synthetic aperture radar for crustal deformation analysis, with application to Volcán Alcedo, Galápagos. *J. Geophys. Res.* **2007**, *112*, B7. [CrossRef]
22. Ankang Fuqiang Airport. Available online: https://baike.baidu.com/item/%E5%AE%89%E5%BA%B7%E5%AF%8C%E5%BC%BA%E6%9C%BA%E5%9C%BA/12030673 (accessed on 21 February 2022).
23. Yang, G.L.; Teng, K.; Qin, C. In-situ Experimental Study on Lateral Swelling Force of Expansive Soil. *J. Cent. South Univ. (Nat. Sci. Ed.)* **2014**, *45*, 2326–2332. (In Chinese)
24. Werner, C.; Wegmüller, U.; Strozzi, T.; Wiesmann, A. GAMMA SAR and interferometric processing software. In Proceedings of the ERS-Envisat Symposium, Gothenburg, Sweden, 16–20 October 2000.
25. Sowter, A.; Moh, B.C.A.; Cigna, F.; Marsh, S.; Athab, A.; Alshammari, L. Mexico City land subsidence in 2014–2015 with Sen-tinel-1 IW TOPS: Results using the Intermittent SBAS (ISBAS) technique. *Int. J. Appl. Earth Obs. Geoinf.* **2016**, *52*, 230–242.
26. González, P.J.; Bagnardi, M.; Hooper, A.J.; Larsen, Y.; Marinkovic, P.; Samsonov, S.V.; Wright, T.J. The 2014–2015 eruption of Fogo volcano: Geodetic modeling of Sentinel-1 TOPS interferometry. *Geophys. Res. Lett.* **2015**, *42*, 9239–9246. [CrossRef]
27. Yague-Martinez, N.; Prats-Iraola, P.; Gonzalez, F.R.; Brcic, R.; Shau, R.; Geudtner, D.; Eineder, M.; Bamler, R. Interferometric Processing of Sentinel-1 TOPS Data. *IEEE Trans. Geosci. Remote Sens.* **2016**, *54*, 2220–2234. [CrossRef]
28. Ciampoli, L.B.; Gagliardi, V.; Ferrante, C. Displacement Monitoring in Airport Runways by Persistent Scatterers SAR In-terferometry. *Remote Sens.* **2020**, *12*, 3564. [CrossRef]
29. Ferretti, A.; Fumagalli, A.; Novali, F.; Prati, C.; Rocca, F.; Rucci, A. A new algorithm for processing interferometric datastacks: SqueeSAR. *IEEE Trans. Geosci. Remote Sens.* **2011**, *49*, 3460–3470. [CrossRef]
30. Jiang, M.; Ding, X.L.; Li, Z.W. Homogeneous pixel selection algorithm for multitemporal InSAR. *Chin. J. Geo-Physics* **2018**, *61*, 4767–4776.
31. Lee, J.S.; Pottier, E. *Polarimetric Radar Imaging: From Basics to Applications*; Thompson, B.J., Ed.; University of Rochester: New York, NY, USA, 2009; pp. 1–372.
32. Jiang, M.; Andrea, M. Distributed Scatterer Interferometry with The Refinement of Spatiotemporal Coherence. *IEEE Trans. Geosci. Remote Sens.* **2020**, *58*, 3977–3987. [CrossRef]
33. Jiang, M.; Ding, X.; Li, Z. Hybrid approach for unbiased coherence estimation for multitemporal InSAR. *IEEE Trans. Geosci. Remote Sens.* **2013**, *52*, 2459–2473. [CrossRef]
34. Guarnieri, A.M.; Tebaldini, S. On the exploitation of target statistics for SAR interferometry applications. *IEEE Trans. Geosci. Remote Sens.* **2008**, *46*, 3436–3443. [CrossRef]
35. Ferretti, A.; Prati, C.; Rocca, F.L. Permanent scatterers in SAR interferometry. *SPIE Proc. 1999* **2001**, *39*, 8–20. [CrossRef]
36. Liao, M.S.; Wang, T. *Time Series InSAR Technique and Application*; Science Press: Beijing, China, 2014; pp. 69–71. (In Chinese)
37. Yu, C.; Li, Z.; Penna, N. Interferometric synthetic aperture radar atmospheric correction using a GPS-based iterative tropospheric decomposition model. *Remote Sens. Environ.* **2018**, *204*, 109–121. [CrossRef]
38. Wang, T.; Perissin, D.; Rocca, F.; Liao, M.-S. Three Gorges Dam stability monitoring with time-series InSAR image analysis. *Sci. China Earth Sci.* **2011**, *54*, 720–732. [CrossRef]

39. Colesanti, C.; Ferretti, A.; Novali, F.; Prati, C.; Rocca, F. SAR monitoring of progressive and seasonal ground deformation using the Permanent Scatterers Technique. *IEEE Trans. Geosci. Remote Sens.* **2003**, *41*, 1685–1701. [CrossRef]
40. Chen, Q.; Dam, T.V.; Sneeuw, N.; Collilieux, X.; Weigelt, M.L.; Rebischung, P. Singular spectrum analysis for modeling sea-sonal signals from gps time series. *J. Geodyn.* **2013**, *72*, 25–35. [CrossRef]
41. Vautard, R.; Ghil, M. Singular spectrum analysis in nonlinear dynamics, with applications to paleoclimatic time series. *Phys. D Nonlinear Phenom.* **1989**, *35*, 395–424. [CrossRef]
42. Terzaghi, K.; Peck, R.B.; Mesri, G. *Soil Mechanics in Engineering Practice;* John Wiley and Sons: Hoboken, NJ, USA, 1996; pp. 1–592.
43. Zhao, Q.; Lin, H.; Gao, W.; Zebker, H.A.; Chen, A.; Yeung, K. InSAR detection of residual settlement of an ocean reclamation engineering project: A case study of Hong Kong International Airport. *J. Oceanogr.* **2011**, *67*, 415–426. [CrossRef]
44. Han, Z.; Zhao, G.T.; Lin, J.; Fan, K.W.; Zou, W.L. Influences of temperature and moisture histories on the hydro structural characteristics of a clay during desiccation. *Eng. Geol.* **2022**, *297*, 106533. [CrossRef]
45. Shi, B.; Jiang, H.; Liu, Z.; Fang, H. Engineering geological characteristics of expansive soils in China. *Eng. Geol.* **2002**, *67*, 63–71. [CrossRef]
46. Gabriel, A.K.; Goldstein, R.M.; Zebker, H.A. Mapping Small Elevation Changes over Large Areas: Differential Radar Interferometry. *J. Geophys. Res. Solid Earth* **1989**, *94*, 9183–9191. [CrossRef]
47. Wang, W.L. Study on Deformation control and Slope Stability of High Filled Foundtion with Expansive Soil. Ph.D. Thesis, Chang'an University, Xi'an, China, 2018.
48. Gagliardi, V.; Ciampoli, L.B.; Trevisani, S.; D'Amico, F.; Alani, A.M.; Benedetto, A.; Tosti, F. Testing Sentinel-1 SAR Interferometry Data for Airport Runway Monitoring: A Geostatistical Analysis. *Sensors* **2021**, *21*, 5769. [CrossRef] [PubMed]
49. Wang, C.S.; Wang, X.S.; Xu, Y.P.; Zhang, B.C.; Jiang, M.; Xiong, S.T.; Zhang, Q.; Li, W.; Li, Q. A new likelihood function for consistent phase series estimation in distributed scatterer interferometry. *IEEE Trans. Geosci. Remote Sens.* **2022**. [CrossRef]
50. Haghighi, H.M.; Motagh, M. Ground surface response to continuous compaction of aquifer system in Tehran, Iran: Results from a long-term multi-sensor InSAR analysis. *Remote Sens. Environ.* **2019**, *221*, 534–550. [CrossRef]
51. Zhang, W.T.; Zhu, W.; Tian, X.D.; Zhang, Q.; Zhao, C.Y.; Niu, Y.F.; Wang, C.S. Improved DEM Reconstruction Method based on Multi Baseline InSAR. *IEEE Geosci. Remote Sens. Lett.* **2022**, *19*, 4011505.

remote sensing

MDPI

Article

Thaw Settlement Monitoring and Active Layer Thickness Retrieval Using Time Series COSMO-SkyMed Imagery in Iqaluit Airport

Deying Ma [1,2], Mahdi Motagh [3,4], Guoxiang Liu [1,*], Rui Zhang [1], Xiaowen Wang [1], Bo Zhang [1], Wei Xiang [1] and Bing Yu [2,5]

1 Faculty of Geosciences and Environmental Engineering, Southwest Jiaotong University, Chengdu 611756, China; 200331010056@swpu.edu.cn (D.M.); zhangrui@swjtu.edu.cn (R.Z.); insarwxw@swjtu.edu.cn (X.W.); rsbozh@gmail.com (B.Z.); xiangwei@my.swjtu.edu.cn (W.X.)

2 School of Civil Engineering and Geomatics, Southwest Petroleum University, Chengdu 610500, China; 201599010140@swpu.edu.cn

3 Remote Sensing and Geoinformatics, GFZ German Research Centre for Geosciences, 14473 Potsdam, Germany; motagh@gfz-potsdam.de

4 Faculty of Civil Engineering and Geodetic Science, Institute of Photogrammetry and Geoinformation, Leibniz University Hannover, 30167 Hannover, Germany

5 Institute of Petroleum and Natural Gas Spatial Information Engineering, Southwest Petroleum University, Chengdu 610500, China

* Correspondence: rsgxliu@swjtu.edu.cn

Citation: Ma, D.; Motagh, M.; Liu, G.; Zhang, R.; Wang, X.; Zhang, B.; Xiang, W.; Yu, B. Thaw Settlement Monitoring and Active Layer Thickness Retrieval Using Time Series COSMO-SkyMed Imagery in Iqaluit Airport. *Remote Sens.* **2022**, *14*, 2156. https://doi.org/10.3390/rs14092156

Academic Editor: Nereida Rodriguez-Alvarez

Received: 26 January 2022
Accepted: 29 April 2022
Published: 30 April 2022

Abstract: Thaw consolidation of degrading permafrost is a serious hazard to the safety and operation of infrastructure. Monitoring thermal changes in the active layer (AL), the proportion of the soil above permafrost that thaws and freezes periodically, is critical to understanding the conditions of the top layer above the permafrost and regulating the construction, operation, and maintenance of facilities. However, this is a very challenging task using ground-based methods such as ground-penetrating radar (GPR) or temperature sensors. This study explores the integration of interferometric measurements from high-resolution X-band Synthetic Aperture Radar (SAR) images and volumetric water content (VWC) data from SoilGrids to quantify detailed spatial variations in active layer thickness (ALT) in Iqaluit, the territorial capital of Nunavut in Canada. A total of 21 SAR images from COSMO Sky-Med (CSK) were first analyzed using the freely connected network interferometric synthetic aperture radar (FCNInSAR) method to map spatial and temporal variations in ground surface subsidence in the study area. Subsequently, we built an ALT retrieval model by introducing the thaw settlement coefficient, which takes soil properties and saturation state into account. The subsidence measurements from InSAR were then integrated with VWC extracted from the SoilGrids database to estimate changes in ALT. For validation, we conducted a comparison between estimated ALTs and in situ measurements in the airport sector. The InSAR survey identifies several sites of ground deformation at Iqaluit, subsiding at rates exceeding 80 mm/year. The subsidence rate changes along the runway coincide with frost cracks and ice-wedge furrows. The obtained ALTs, ranging from 0 to 5 m, vary significantly in different sediments. Maximum ALTs are found for rock areas, while shallow ALTs are distributed in the till blanket (Tb), the intertidal (Mi) sediments, and the alluvial flood plain (Afp) sediment units. The intersection of taxiway and runway has an AL thicker than other parts in the glaciomarine deltaic (GMd) sediments. Our study suggests that combining high-resolution SAR imagery with VWC data can provide more comprehensive ALT knowledge for hazard prevention and infrastructure operation in the permafrost zone.

Keywords: permafrost; ALT; FCNInSAR; CSK images; thaw settlement coefficient; VWC; SoilGrids

1. Introduction

Permafrost regions are defined as grounds that remain continuously frozen for two or more consecutive years. They underlie about 20% of the land surface of the Earth and are located mainly in the Arctic, Antarctic, and high mountainous regions [1]. The seasonal thawing and freezing of the active layer (AL), the surface layer atop permafrost, results in ground motion, making infrastructure in such regions vulnerable to varying degrees of damage. Therefore, accurate monitoring of active layer thickness (ALT) and related ground surface change is vital to characterize permafrost degradation and regulate the construction, operation, and maintenance of facilities [2–7].

Commonly used in situ investigation methods for ALT include point measurements obtained by recording soil temperatures of probes on the ground at different sparse locations [8] and profile exploration of the underground geophysical status using ground-penetrating radar (GPR) [9–11]. These methods are reliably accurate for estimating ALT at a single point or local area, but are extremely limited in space and are costly and inefficient for large-scale mapping in severe cold environments in the permafrost areas. Another method of acquiring ALT information is using analytical models, which evaluate permafrost thermal dynamics using air temperature, vegetation, snow, and soil properties [12–14]. However, the results of such models are inadequate, especially in data-poor areas. Remote-sensing-based assessment of ALT fulfills the urgent need to extend point observations to a broader spatial domain in permafrost areas. Similarly to GPR, the dielectric contrast between the thawed and frozen soils results in a strong scattering of electromagnetic waves, providing a signature in backscattering measurements for freeze/thaw boundary detection in radar remote sensing imagery. However, due to the lack of proper satellite sensors, few studies have explored the potential of such imagery for monitoring the processes operating in permafrost [15].

The retrieval of ALT from surface displacement measurements detected by the InSAR has been promising due to the technique's capability for accurately mapping ground displacements on the scale of millimeters to centimeters [16–20]. There are two main types of ALT retrieval methods from InSAR-derived subsidence maps. The first works based on soil's one-dimensional heat transfer process [16]. The basic idea is that ALT could be determined if the time intervals in which the maximal temperature diffuses from the ground surface downward to the bottom of the active layer are known. Considering the difficulty in measuring the time intervals directly, the model replaces them with the lag time between the periodic feature of InSAR-observed surface deformation over permafrost and the meteorologically recorded temperatures. The second method is based on the idea that the volume change from ice to the water of the AL leads to surface subsidence [17–22]. It can be represented by the Stefan formula or the Berggren model. In these models, soil moisture, as the most sensitive factor, is quantified as volumetric water content (VWC), defined as the volume of water divided by the total soil volume. It determines the soil heat capacity and thermal conductivity, influencing the flow of energy into and out of the soil, thus, ALT and ground temperature [23]. The more accurate the soil moisture data (i.e., VWC) as an input, the more accurate the predictions of ALT are to be expected.

Several studies used the second retrieval approach discussed above to monitor ALT with the subsidence derived from InSAR. According to the stratification of soil profiles, Liu et al. proposed a vertical distribution model containing organic, mineral, pure water, or mixed layers to estimate ALT over Alaska using uniform soil moisture values [17]. However, soil moisture is highly variable in both space and time [23]. Wang et al. built a multilayered groundwater model for different land cover types using field data, then integrated it into the Stefan formula to retrieve ALT over the Qinghai–Tibet Plateau [20]. Anne et al. estimated ALT using the modified Berggren solution over Barrow Peninsula at a resolution of 30 × 30 m [24]. Both studies required detailed in situ measurements, e.g., surface meteorological variables and soil thermal parameters, which are not available everywhere. Zhang et al. used soil moisture data from ERA-Interim reanalysis and SMAP L4 for ALT estimation. The spatial resolutions of ERA-Interim reanalysis and SMAP are

0.1° (9 km) and 0.125° (13 km), respectively [24]. The data are too coarse to be used for integration with submeter level SAR images.

However, the active layer freezes, sometimes with no uplift of the surface and at other times with an uplift of even 100 percent of the depth of freezing [25–28]. The soil is an open system in that the frost heaving can be much greater than the volume change of pore water due to the available water supply [25]. In a summer season, the magnitude of the subsidence resulting from the soil thawing downward from the surface is mainly related to the soil thermal properties and ice content formed in the last winter. As Yanagiya commented in [29], the retrieval models based on the volume change of pore water are inadequate for obtaining more realistic results in most cases.

The objective of this study was to use a realistic model to quantify the meter-level resolution ALT in Iqaluit based on the accurate thaw subsidence and explicit knowledge of soil moisture from a global product. Here, we demonstrate the ALT retrieval solution combining both the six-layer VWC dataset from the SoilGrids database at 250 m spatial resolution and the subsidence field derived from FCNInSAR [30,31]. First, we explored 21 SAR images from COSMOSky-Med (SLC) using the FCNInSAR method to quantify and map spatial and temporal variations in ground surface subsidence in the study area. Secondly, the thaw-settlement coefficient was introduced to establish the relationship between the thaw settlement and ALT. Soil properties and saturation state are considered in the model. Then, ALT over the study area was obtained by integrating the high-resolution subsidence rates measurements from InSAR with VWC. The results are then interpreted against maps of surface geology to analyze the spatial distribution of ALT and find possible reasons for the differences in ALTs.

2. Materials and Methods

2.1. Study Area

Iqaluit is the territorial capital of Nunavut, Canada, as shown in Figure 1. In the far north latitude of 63, the city at the northern shore of Frobisher Bay is located in the permafrost zone, extending in the subarctic from a few meters to 1500 m into the ground, and is therefore extremely sensitive to climate change [32]. This area is under the subarctic-tundra climate and has a series of typical climate characteristics, e.g., short growing seasons, extremely low temperatures, strong winds, and variable sunlight periods. The summers here are very short and humid. The period that air temperatures drop below 0 centigrade is at least eight months per year. The active layer thaws during hot summers (usually in June to September) but heaves in winter when the air temperature slides dramatically to negative 30–40 centigrade. According to the temperature recordings in July and August, the highest temperature reached 21 centigrade in 2016, and it has been generally warmer than before in recent years.

The subsurface sediments under the city are mainly composed of lacustrine (Lv) sediments, fluvial (coarse-grained) and glaciomarine deltaic (GMd) deposits, bedrock (R), and rock with till cover (Tv) [33]. Except for the bedrocks that are relatively stable, all subsurface sediments are considered to have an ice-rich content. As a result, in response to seasonal alternation, the ground surface in frozen soil areas undergoes dramatic freezing uplift and thawing settlement, as much as 10 cm [34], so the ALTs increase in thickness.

The elevation difference over the study area is about 70 m. Iqaluit airport was built on a flat basin surrounded by hills and rock plateaus. It serves as the transport hub for the eastern Canadian Arctic due to the lack of highway or railway there. The available alternative forms of transportation are mainly ships in summer to connect the city to other northern communities or the rest of Canada. The airport has been expanded with the increasing mining and tourism in addition to serving nearly 8000 residents of the city. Influenced by climate change in recent years, the airport and the houses built in this area are becoming more vulnerable to the freeze–thaw cycle. The taxiway and runway of the airport sustain continuous cracking due to the ground deformation. It is vital to monitor uneven subsidence at the airport to guarantee safe operation.

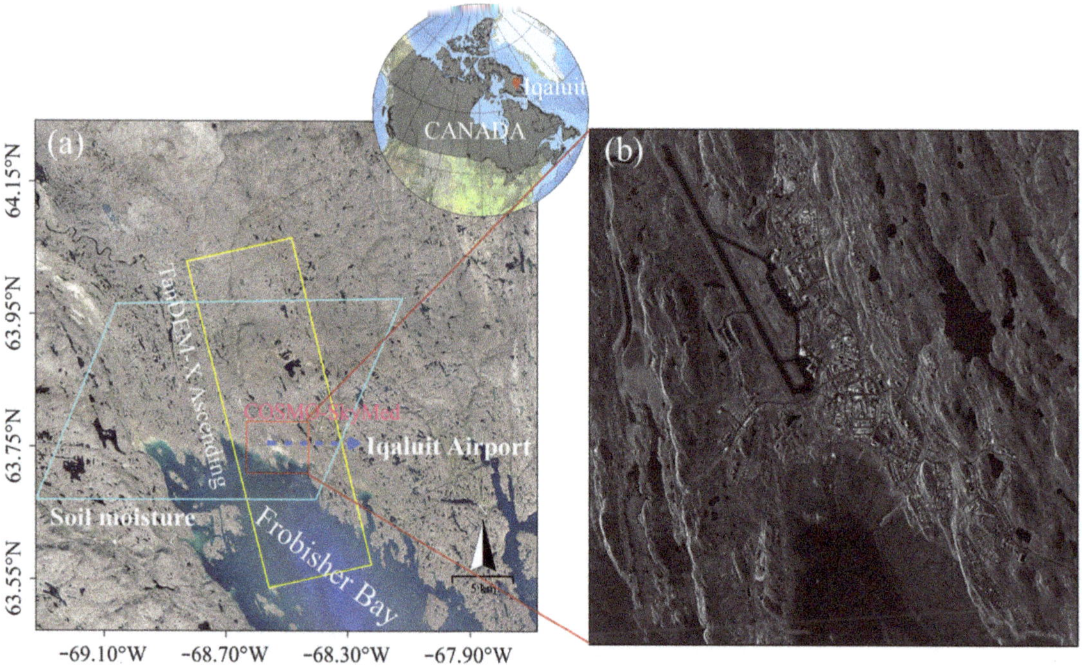

Figure 1. (**a**) Location of the study area and coverage of SAR images and soil moisture; (**b**) a CSK amplitude image.

2.2. Dataset

2.2.1. SAR Data

For the InSAR analysis, 21 COSMO-SkyMed images in Spotlight mode acquired from June to September 2014 with an incidence angle of 25.4° and "HH" polarization were used. The pixel spacing of the single-look complex image in the slant range and azimuth direction is 0.312 and 0.702 m, respectively. The coverage and the amplitude of the SAR images are shown in Figure 1. The acquisition dates of all images can be found in Table 1.

Table 1. Iqaluit COSMO-SkyMed acquisition dates.

Date	Date	Date
10 June 2014	24 July 2014	29 August 2014
18 June 2014	28 July 2014	6 September 2014
26 June 2014	5 August 2014	14 September 2014
4 July 2014	13 August 2014	22 September 2014
5 July 2014	21 August 2014	23 September 2014
20 July 2014	22 August 2014	26 September 2014
21 July 2014	25 August 2014	30 September 2014

Two TanDEM-X images in bistatic mode acquired in July 2012 were used to produce a DEM with 5 m spatial resolution and 2 m relative height accuracy for acquiring more precise differential interferograms.

2.2.2. Soil Moisture Data

Publicly available soil profile data from SoilGrids™ [35] provide global predictions for standard numeric soil properties including pH, soil organic carbon content, bulk density, coarse fragment content, sand content, silt content, clay content, cation exchange capacity (CEC), total nitrogen soil organic carbon density, and soil organic carbon stock, as well as volumetric water content (VWC) at six standard depths (0–5, 5–15, 15–30, 30–60, 60–100, and 100–200 cm) with the highest scale of 250 m resolution. In this study, we used the VWC in 1500 kPa predictions with the prediction uncertainty at 95th percentiles. Figure 2a–f shows the spatial distribution of soil moisture over Iqaluit and the vicinity. Figure 3 gives examples of the vertical profile of random points located in different sediments.

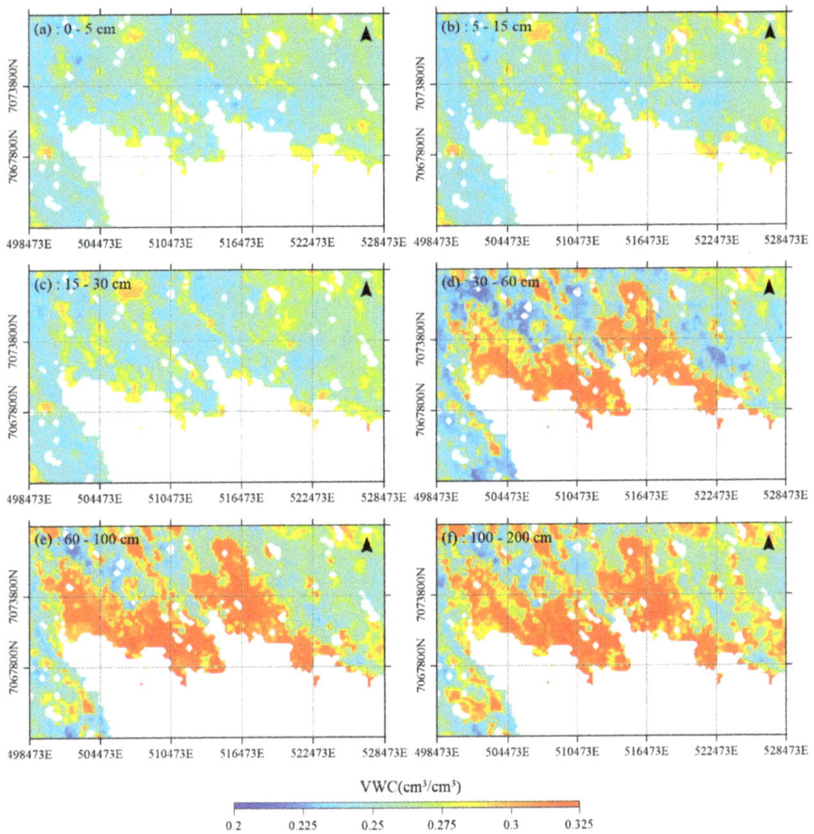

Figure 2. Spatial distribution of soil moisture over Iqaluit at six standard layers under UTM projection.

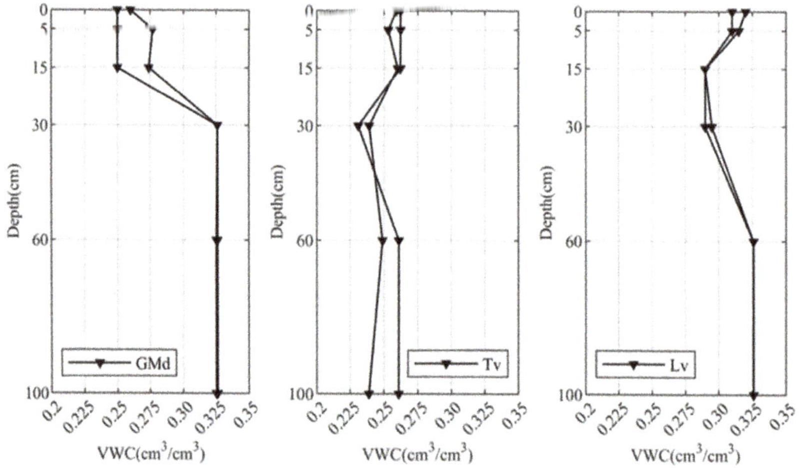

Figure 3. Profiles of soil moisture at points in GMd, Tv, and Lv sediments.

2.3. Methodology

The aim of this study was to acquire high spatial resolution subsidence maps of Iqaluit using InSAR and estimate ALTs on the basis of seasonal thaw subsidence. The main two steps were (1) subsidence observation and (2) ALT retrieval.

2.3.1. Subsidence Observation from CSK Spotlight SAR Images

We applied the FCNInSAR algorithm to CSK images to map the subsidence in Iqaluit. This is a robust method taking advantage of both PSInSAR [36] and SBAS [37] methods. The main processing steps in FCNInSAR are similar to PSInSAR. The main differences are as follows: (1) Forming interferograms following the SBAS method to increase the measurements by connecting all possible images instead of using only one single master image. This is useful especially when the number of available images is limited. (2) More observations of arcs improve the stability of deformation acquisition. (3) The best estimation of linear deformation rates and DEM errors can be found by searching in a predefined value space rather than phase unwrapping. (4) SVD must be adopted to reconstruct the time series of nonlinear deformation due to the SBAS method of forming interferograms.

We used a linear deformation model to estimate velocities since all selected images were acquired in summer. However, the solution of FCNInSAR fully considers the extraction of the nonlinear deformation component. In order to select a sufficient number of persistent scatters (PSs) and obtain reliable subsidence, this study only processed SAR images spanning the thawing season without snow cover. In Iqaluit, as mentioned above, the average temperature from June to September is greater than 0 °C, so the thawing settlement of frozen soil mainly occurs throughout these four months. Therefore, 21 images in June and September (see Table 1) were selected to map the subsidence. The image acquired on 21 August was chosen to be the reference image for co-registration. All images were resampled into the reference image space. Then, the mean and the standard deviation (SD) of amplitudes were calculated based on the co-registered amplitude images. A total of 1,000,561 PSs were selected in the study area, of which the amplitude dispersion index is less than 0.25.

We also set the perpendicular baseline and temporal thresholds of 500 m and 90 days to generate 167 interferograms. We visually inspected the quality of all interferograms and removed those with temporal baselines longer than 30 days due to their low coherence. A total of 72 interferograms remained for deformation analysis (see Figure 4). The phase at each pixel consists of many components, such as topography, flat-earth trend, deformation, atmosphere, and decorrelation noise. We removed the topographic effects and the flat-earth trend from interferograms using the newly made DEM and the orbital data and then acquired 72 differential interferograms. The differential phase at each pixel (x, y) of differential interferograms can be modeled as:

$$\Delta\phi_{i(x,y)} = \Delta\phi_i^{ldef}{}_{(x,y)} + \Delta\phi_i^{dem}{}_{(x,y)} + \Delta\phi_i^{res}{}_{(x,y)} \tag{1}$$

$$\Delta\phi_i^{res}{}_{(x,y)} = \Delta\phi_i^{nldef}{}_{(x,y)} + \Delta\phi_i^{atm}{}_{(x,y)} + \Delta\phi_i^{\text{noise}}{}_{(x,y)} \tag{2}$$

where i denotes the ith interferogram, and $\Delta\phi_i^{ldef}{}_{(x,y)}$, $\Delta\phi_i^{dem}{}_{(x,y)}$, and $\Delta\phi_i^{res}{}_{(x,y)}$ are the differential phase of linear deformation, elevation error, and residual phase along the radar line of sight (LOS), respectively. The residue phase includes nonlinear deformation phase, atmospheric phase, and noise.

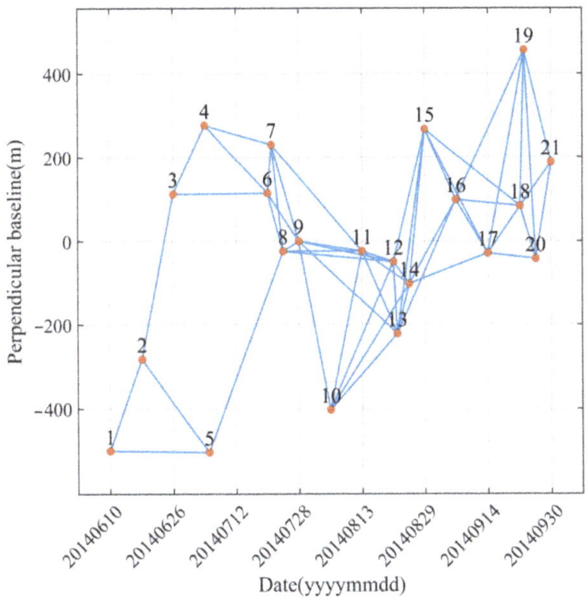

Figure 4. Spatiotemporal baselines of interferograms.

We freely linked the neighborhood PSs (e.g., within 1 km distance) to form a strong network. By differencing the phase between two adjacent PSs, the residue phase can be largely subtracted due to spatial autocorrelation [38]. Therefore, the differential phase between two adjacent PSs (A and B) for the ith interferogram can be illustrated as:

$$\begin{aligned}
&\Delta\phi_i(x_A, y_A; x_B, y_B) = \tfrac{4\pi}{\lambda R \sin\theta} B_{\perp i} \Delta\varepsilon(x_A, y_A; x_B, y_B) + \tfrac{4\pi}{\lambda} T_i \Delta v(x_A, y_A; x_B, y_B) + \delta_i^{arc_{AB}}, \\
&\Delta\phi_i(x_A, y_A; x_B, y_B) = \Delta\phi_i(x_A, y_A) - \Delta\phi_i(x_B, y_B) \\
&\Delta\varepsilon(x_A, y_A; x_B, y_B) = \varepsilon(x_A, y_A) - \varepsilon(x_B, y_B) \\
&\Delta v(x_A, y_A; x_B, y_B) = v(x_A, y_A) - v(x_B, y_B)
\end{aligned} \tag{3}$$

$$\delta_i^{arc_{AB}} = \Delta\phi_i^{nldef}(x_A, y_A; x_B, y_B) + \Delta\phi_i^{atm}(x_A, y_A; x_B, y_B) + \Delta\phi_i^{noi}(x_A, y_A; x_B, y_B) \tag{4}$$

where $\delta_i^{arc_{AB}}$ denotes the residue phase on the arc AB, and $\Delta\phi_i^{nldef}(x_A, y_A; x_B, y_B)$, $\Delta\phi_i^{atm}$ $(x_A, y_A; x_B, y_B)$, and $\Delta\phi_i^{noi}(x_A, y_A; x_B, y_B)$ are the nonlinear deformation and the atmospheric and noisy differential phase between AB, respectively. ε and v denote the DEM errors and linear displacement velocities at pixels, and $\Delta\varepsilon$ and Δv are the increment of elevation errors and the increment of linear displacement velocities between two connected

PSs (e.g., A and B), respectively. $B_{\perp i}$ and T_i are the perpendicular (spatial) and temporal baseline of the ith interferogram, respectively; λ, R, and θ are the radar wavelength, sensor-to-target distance, and radar incident angle, respectively.

Due to the phase ambiguity in the differential interferograms, the two unknowns $\Delta\varepsilon$ and Δv of each pair of PS points can be determined by maximum likelihood estimation, i.e., by searching $\Delta\varepsilon$ and Δv in a pre-defined solution space (e.g., (−5, 5) m, (−200, 50) mm/year in this study) to maximize the following equation:

$$\Upsilon = \left| \frac{1}{N} \sum_{i=1}^{N} \left(\cos w_i^{arc_{AB}} + j \bullet \sin w_i^{arc_{AB}} \right| , j = \sqrt{-1} \tag{5}$$

$$w_i^{arc_{AB}} = \Delta\phi_i(x_A, y_A; x_B, y_B) - \frac{4\pi}{\lambda R \sin\theta} B_{\perp i} \Delta\varepsilon(x_A, y_A; x_B, y_B) - \frac{4\pi}{\lambda} T_i \Delta v(x_A, y_A; x_B, y_B) \tag{6}$$

where Υ is the model coherence, and $w_i^{arc_{AB}}$ denotes the difference between measured and estimated phase values. After the increments of velocities and DEM errors are calculated, we interpolated linear deformation rates and DEM errors at the regular grid points based on the irregular PS samples. The residual phases of every pixel were obtained by subtracting the gridded phase from each of the differential interferograms. As the interferograms were freely generated, the SVD was applied to reconstruct the time series of residual phases corresponding to the SAR acquisitions. We further isolated the nonlinear and atmospheric phase by using the empirical mode decomposition (EMD) method [38] on the unwrapped residue phase as the two terms have different frequencies in space and time. The deformation of each PS is the sum of the linear and the nonlinear deformation. A bedrock area close to the airport was chosen as the reference point for the velocity calculation. The velocities of all PSs can be calculated by spatial integration in the freely connected network with respect to the given reference point. InSAR measurements were along the direction of the line of sight (LOS). We converted all results from LOS to vertical displacement using the following equation:

$$V_{vertical} = V_{LOS} / \cos\theta \tag{7}$$

2.3.2. ALT Estimation

The maximal thawing settlement throughout the year should be used as input for inversion of ALT from subsidence data. Considering Iqaluit, the monthly average temperatures from January to May or from October to December are all below 0 degrees Celsius. The daily temperature was almost negative in those months by 2015, and therefore, the frozen soil rarely thaws. In addition, the temperature decreases so rapidly in October that while the soil close to the permafrost table remains in a continuous thawing state, the surface starts freezing downward when the ground surface temperature becomes negative. Therefore, in this study, we assumed that the thawing settlement from June to the end of September equals the yearly thawing settlement, and the water totally changes to ice in winter.

The total settlement is mainly affected by three terms: one from thawing settlement and two volume-compressibility terms due to a surcharge load and the self-weight [39]. The volume compressibility can be ignored since consolidation comprises only a fraction of the surface settlement that occurs as a result of drainage from the lower boundary [25]. Thus, the total thawing settlement can be simplified as:

$$H = A_0 Z \tag{8}$$

where H is the thaw-induced settlement of the frozen soil, A_0 is the thaw settlement coefficient, and Z is the depth to thaw front from the original surface. The maximum thawing depth corresponding to the largest settlement is the ALT. Thus, the relationship between ALT and the maximal thaw subsidence is:

$$Z = \frac{H}{A_0} \tag{9}$$

The main factors that affect the thawing settlement coefficient are soil types, ice content, and the dry density of frozen soil [40,41]. With the same moisture content, sandy soil has a smaller thaw coefficient than clay. The thawing settlement coefficient increases with the decrease in the dry density when the soil is saturated. If the ice content is large, the thawing settlement coefficient and settlement of frozen soil are both significant because the settlement results not only from the volume shrink from the ice-water phase change but also the drainage of saturated or oversaturated soil thawing. Owing to extremely low temperatures in winter at Iqaluit (around −40 °C), we simply assumed that the ice content is equated to the water content in this study. Based on the aforementioned ideas, the thaw settlement coefficient can be calculated separately in the following conditions:

- Soil in an unsaturated state

It is generally assumed that when unsaturated soil freezes and thaws in situ, the volume expansion cannot fill the pores fully. As a result, there is no heave uplift or settlement. Experiments indicated that there is still observable uplifting upon freezing which results from the volume expansion of bound water [40,41]. The combining rate C_r is introduced to denote the ratio of water attending the freezing expansion:

$$\begin{aligned} C_r &= \tfrac{W}{W_b} \\ W_b &= \tfrac{\rho_w}{r_d}\left(1 - \tfrac{r_d}{r_s}\right) \end{aligned} \tag{10}$$

where W is the water content of soil by mass and W_b is the water content of saturated soil by mass. ρ_w refers to the water density, while r_d is the dry density of the soil, and r_s is the specific gravity of soil particle. The thaw settlement coefficient of unsaturated soil is:

$$A_0 = 0.09 C_r \frac{r_d W}{\rho_w} \tag{11}$$

- Soil in a saturated state

For a saturated soil, $C_r = 1$. The thaw settlement coefficient is:

$$A_0 = 0.09 \frac{r_d W}{\rho_w} \tag{12}$$

- Soil in an oversaturated state

When the oversaturated soil freezes in winter, some water is pulled through the soil to build up layers of segregated ice. When the AL thaws in summer, part of the water becomes free water and discharges under good drainage conditions [25–29,39]. This part of the water contributes its whole volume to the subsidence. Therefore, the difference in volume between the frozen and thawed state is equated to the change in volume associated with melting the ice plus the volume of water expelled from the soil. The void ratio of oversaturated soil n can be solved by:

$$n = 1 - \frac{\rho_w}{(r_s W + \rho_w) r_s} \tag{13}$$

The relationship of the height of the free water in unit area and the height of the thawing soil is:

$$H_W = \frac{n - n_1}{1 - n_1} H \tag{14}$$

where n_1 is the void ratio under the liquid limit, H is the total height of the thawing soil, and H_W is the height of the free water in unit area. The height of pore water that approaches the liquid limit in unit area is:

$$\Delta h_1 = \frac{1-n}{1-n_1} n_1 H \tag{15}$$

The thaw settlement coefficient can be expressed as:

$$A_0 = 1.09\frac{H_w}{H} + 0.09\frac{\Delta h_1}{H} \tag{16}$$

It also can be described as:

$$A_0 = \frac{1.09n - n_1 - 0.09nn_1}{1-n_1} \tag{17}$$

Thawing settlement coefficients for all soil types can be determined given values of soil properties and VWC. In this study, the values of parameters [42] we used in the ALT estimation are listed in Table 2. The VWC at 1500 kPa is transferred to the absolute VWC in the soil by multiplying the ratio of VWC from remote sensing and the mean of VWC at 1500 kPa over the study site.

Table 2. Values for thaw settlement coefficient calculation.

Sediment Type	n_1	r_d (g/cm^{-3})	r_s (g/cm^{-3})
R (bedrock)	0.42	2.4	2.75
Other sediments	0.42	1.9	2.65

3. Results

3.1. Deformation Results

Figure 5a illustrates mean displacement velocities in summer 2014. The results, superimposed on Google Earth over the entire region, show that the city undergoes thaw subsidence in summer. Ground subsidence varies significantly in size and shape over the study area. It is larger at the airport and surrounding infrastructure than the natural ground. Five subsidence funnels can be seen in Figure 5a. The most conspicuous settlement (labeled as blue circle A) is centered the intersection of taxiway alpha and runway 17/35 of the airport, with a diameter of approximately 450 m. The vertical displacement rate is up to 163 mm/year which is so detrimental to the airport's performance that the taxiway stopped working for some time. Another elliptical subsidence area (the ellipse where GPS station IQAC [43] is located) with a major axis of about 420 m elongated along the NW–ES direction was observed at the immediate south side of the runway. The mean velocity around this area is 80–88 mm/year. In the vicinity of oil depots and the city landfill (the ellipse where thaw tube T2 [44] is located), there is a significant settlement region of approximately 1 km^2 with a 110 mm/year subsidence rate. Another settlement region appears in the south of the city (labeled as B), with a similar rate to the oil depot area. In the northern end of the runway (labeled as E) and the residential area, the subsidence rates are between 44 and 88 mm/year.

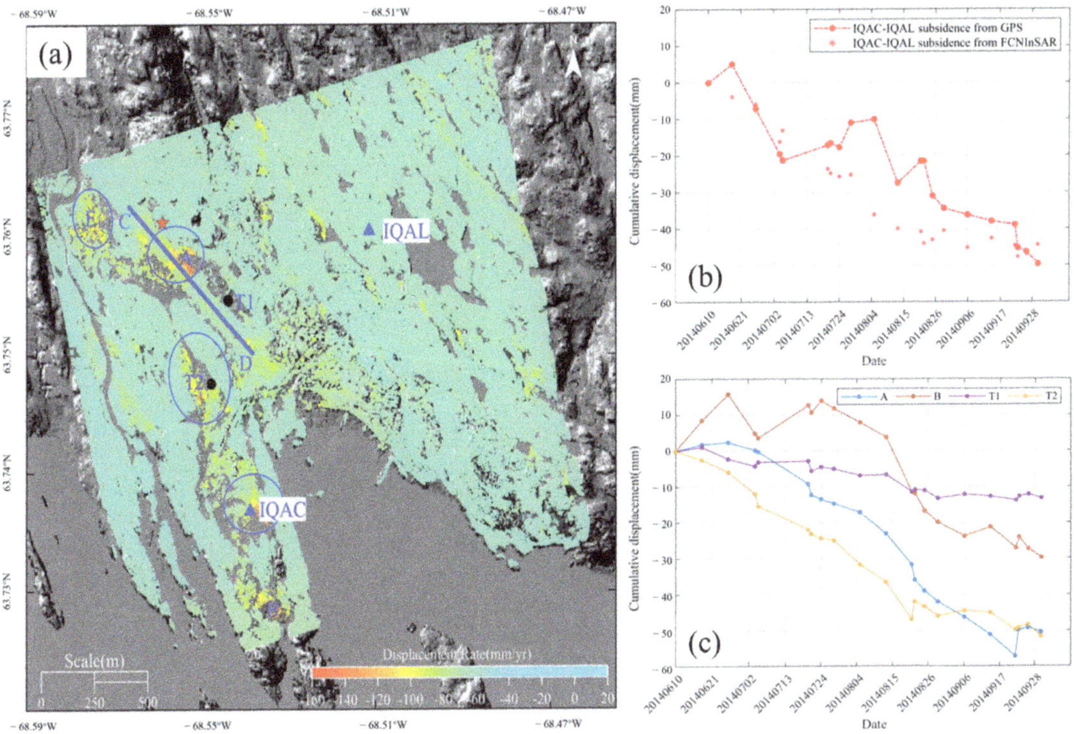

Figure 5. (**a**) Displacement velocities over the study area. The red star is the reference point for velocity calculation; blue triangles denote GPS stations; black dots denote two thaw tubes; black polygons represent infrastructure and facilities; the blue line labeled as CD is the profile site along runway 17/35 for further discussion; (**b**) cumulative subsidence between IQAL and IQAC from GPS measurement and FCNInSAR time series; (**c**) time series subsidence of four points labeled in (**a**).

For further analysis of the dynamic evolution of the active layer, time series displacements at four points are plotted in Figure 5c. The points are selected from the different areas as presented in Figure 5a. A similar subsidence trend, albeit at different rates, is observed for all points. The ground surface subsides relatively slowly in June and in September, even with random uplifts, but much faster in July and August. Across four months, the settlement of point A on the runway of the airport amounted to 57 mm. The total subsidence of the thaw tube T2 installed in a marshy area [44] with seasonally ponded water is 51.5 mm. It has higher rates compared to B and T1. The thaw tube T1 has little sign of displacement of approximately 13 mm.

3.2. Estimated ALT Results

Figure 6 shows the ALT distribution over the entire study area. The magnitudes of estimated ALTs are between 0 and 5 m and exhibit considerable regional heterogeneity. ALTs in bedrock areas are generally larger than other sediments. The maximal ALTs are located at the southern part of the study area. The 1–2 m ALTs are widely spread in the Lv, GMd, and Mn sediments, while ALTs are smaller than 1 m in the Afp and the Tb sediments. Stable areas without subsidence also have shallow ALTs in rock area. A deeper AL of 1.2 to 2.5 m dominates the intersection of the runway and taxiway alpha.

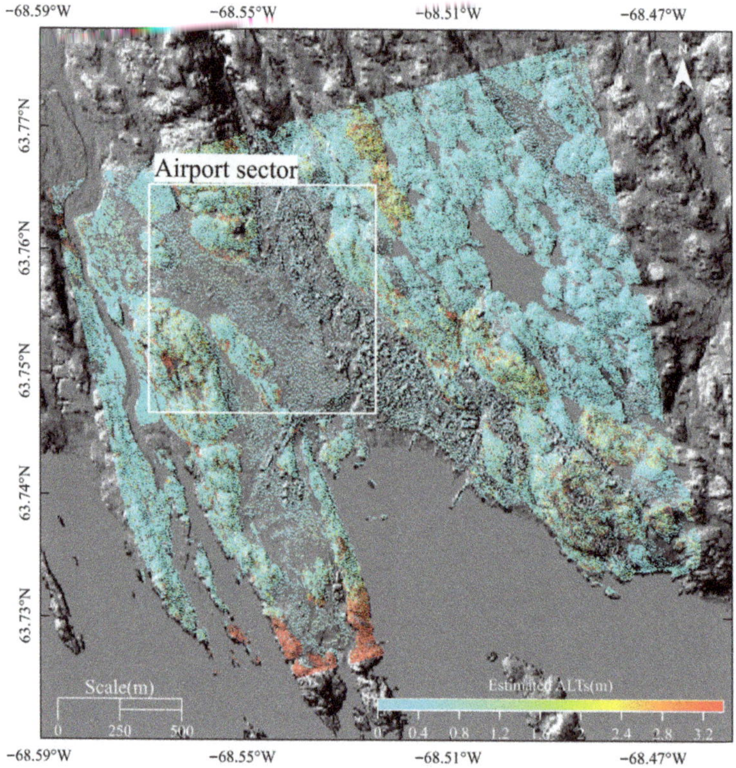

Figure 6. Estimated ALTs.

Closer inspection of the airport sector (Figure 7) shows that the AL ranges between 0 and 2.5 m with a paved surface. The natural ground in the airport section shows a slightly smaller ALT trend with most part of it having an AL of less than 1 m. Only the bare surface between the taxiway and the runway has an AL similar to the paved ground which is even deeper than 2.5 m.

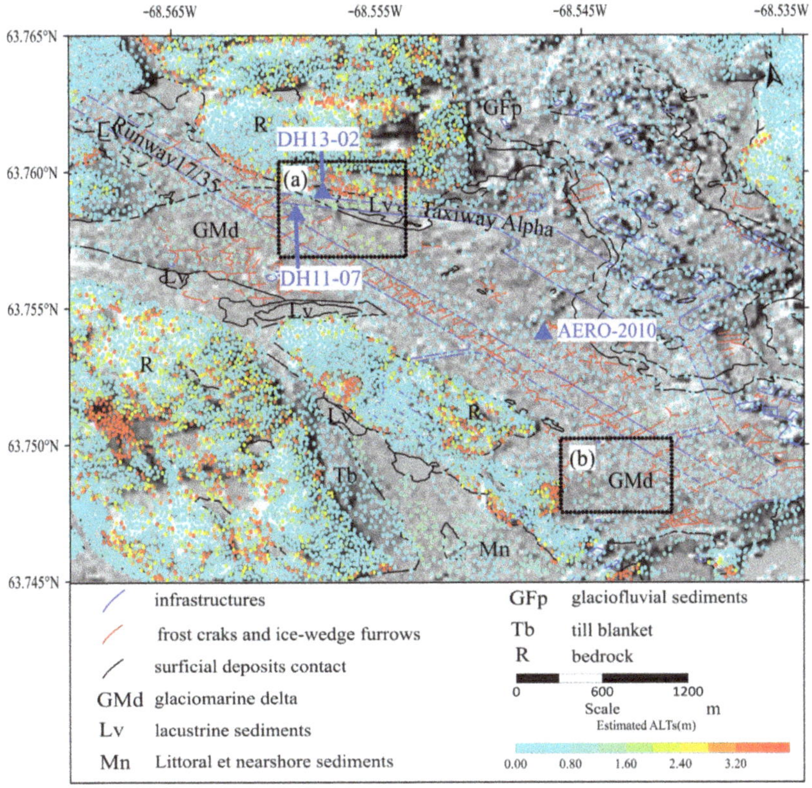

Figure 7. The ALT distribution in the airport sector. Blue triangles denote boreholes.

4. Analysis and Discussion

4.1. Comparison of Subsidence with GPS Data and Thaw Tubes

In order to assess the accuracy of subsidence detection, GPS data and thaw-tube recordings were used for validation. We downloaded observations from the Nevada Geodetic Lab for two GPS stations that were contemporaneous with the SAR data used in this study [43]. The locations of GPS stations are listed in Table 3.

Table 3. GPS stations used for comparison.

GPS Station	Coordinates
IQAC	63.737 N, −68.540 W
IQAL	63.756 N, −68.510 W

Figure 5b presents the temporal evolution of displacements relevant to the points labeled as IQAC and IQAL. GPS station IQAL was set close to Geraldine Lake, and it is underlain by till veneer sediments. The other GPS station IQAC was installed in Sylvia Grinnell Park. Cumulative thawing settlements between IQAC and IQAL from FCNInSAR and GPS measurements are 4.5 and 5 cm, respectively. The discrepancies' mean value and standard deviation are 7.06 and 8.76 mm. The FCNInSAR subsidence trend generally agrees well with GPS displacement, but there is a significant difference in July and August, up to 2.7 cm.

We also compared our subsidence to the results measured on site. Two thaw tubes located in well-drained glaciomarine deposits (T1) and poorly drained marine sediments

(T2) were available [44,45]. As plotted in Figure 5c, the subsidence of T1 in the whole summer from FCNInSAR is 1.32 cm, while that recorded from the thaw tube is 0.8 cm. Due to the construction work, T1 was removed in July 2014. Therefore, only part of the subsidence was observed. Our InSAR result is somewhat consistent with that observed at T2, in which we see a difference of 0.15 cm in 2014 (see Table 4). Compared to the report published by Short in 2017 which stacked the DInSAR results for deformation analysis, as shown in Figure 8, the FCNInSAR deformation trend better matches the deformation from the measurement of the thaw tube [46].

Table 4. Subsidence at the location of thaw tubes.

Thaw Tubes	In Situ Subsidence (cm)	Subsidence from FCNInSAR (cm)	Year
T1 (removed in July)	0.8	1.32	2014
T2	5.3	5.15	2014

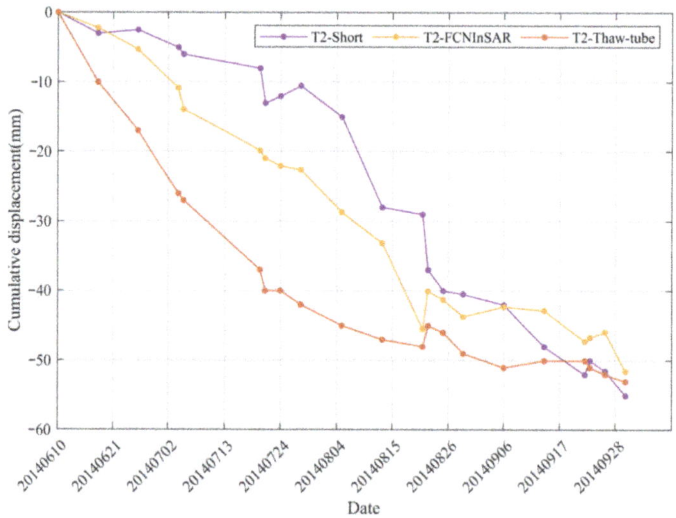

Figure 8. Ground displacement at T2 from Short et al. (2017), in situ measurement and FCNInSAR.

4.2. ALT Validation with Thermistor Cable Measurements

Mathon et al. studied the ALTs over Iqaluit by installing thermistor cables in boreholes and recording schematic stratigraphy from boreholes [47]. The results show that the thickness of the active layer varies between 90 cm for sectors covered with vegetation and more than 2 m below paved surfaces [47]. In our study, the AL in the natural ground of the airport sector within GMd sediments ranges between 0 and 2 m, and it is up to 2.504 m under the paved taxiway, which shows good consistency with previous studies.

We averaged the ALTs of all PSs within the distance of 5 m to each thermistor cable for point-to-point validation. The ALT of 1.5 m was observed in the field in 2012 at thermistor cable AERO-2010 (with the same location as T1) [48], while it was 1.01 m in our results. The ALT of DH13-02 under the paved taxiway is 2.7 m, and the result of this study is 2.5 m. The ALT of thermistor DH11-07 installed on the taxiway shoulder increased from 1.5 m in 1992 to 2.2 m in 2012 [45,47,48]. The retrieved ALT in 2014 of this study is 1.8 m. All three datasets are generally consistent with each other. The slight difference could be partially attributable to the uncertainties in soil moisture data. Especially during the summertime, the soil is observed to be saturated or flooded in shallow layers [45], but the soil moisture data used to retrieve ALTs generally had values under 0.35. Subsidence detection errors and the different acquisition times of results for comparison are also possible reasons for the discrepancy.

4.3. Uneven Trend of Subsidence and ALT along the Runway

Both subsidence velocity and ALT profile along CD were plotted to assess the heterogeneous changes of the ground along runway 17/35. As shown in Figure 9, the subsidence velocity and ALT fluctuate considerably. The section from 230 to 800 m is recognized as the part close to the runway entrance that exhibits more subsidence and a thicker active layer than other parts of the airport. There were dense cracks on the asphaltic surface which can be easily seen on optical images. Allard [33] mapped them in 2012, as shown by red lines in Figure 7. Locations of cracks designated by dashed lines as shown in Figure 9 are consistent with sudden changes of displacement velocity and ALT along the runway. With the temperature rise, different soil characteristics of changing sediments introduce variable subsidence, which results in ground surface cracking. Due to the high resolution of SAR images, we obtained detailed subsidence fields and ALTs, showing many turning points on the profile CD. The geophysical results proved that ice wedges appeared along the runway [44]. Ice wedges are another possible reason for cracks. The v-shaped ice wedges that vertically taper down into the permafrost grow as ground cracks occur in response to thermal contraction of the ground during winter and that water entering the open cracks freezes [49]. As the temperature rises, vein ice and ice wedges melt. Surrounding rocks and upper soil collapse afterwards.

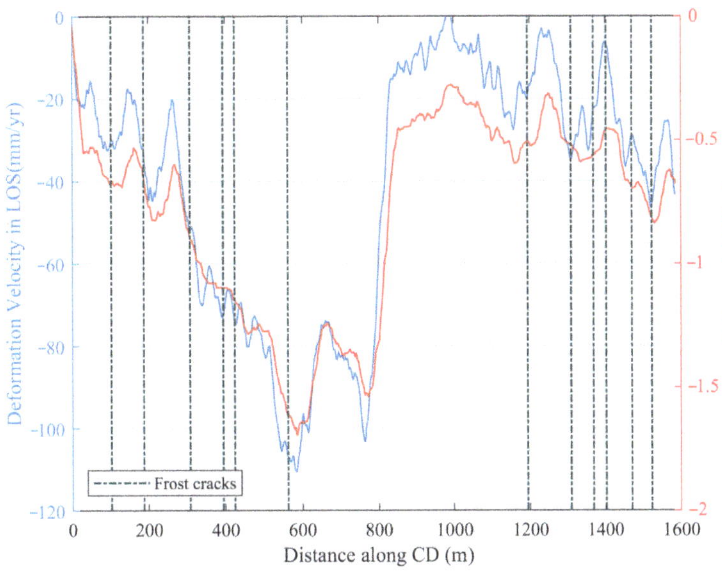

Figure 9. Deformation and ALT profile along CD.

4.4. ALTs under Infrastructure and Natural Ground

The histograms of ALT (Figure 10) display the difference between the AL under infrastructure and the natural ground within the airport sector. This shows that the AL under the paved surface (the runway, aprons, and the taxiway) ranges from 0 to 2.504 m. As shown in Figure 10a, a part of taxiway alpha experienced significant subsidence, which led to several resurfacings and diversions. In total, 96% of the natural ground has an AL of less than 1.5 m. Only the tiny part mainly located at the shoulder of the runway has an AL of thicker than 2 m. The surface paving and the construction work probably increase the ground temperature, leading to the thicker ALs. According to previous field tests, the intersection part of the runway and taxiway alpha is wetter than neighboring parts [44,45]. Although the previous lake (where the Lv sediments on taxiway alpha are located) and drainage channels were modified and covered by the airport's construction, a water rich region extended from taxiway alpha across the runway remaining there. This indicates that there are errors in VWC data. We found that the subsidence is the dominant factor of the ALT in the paved area due to the subtle difference in VWC.

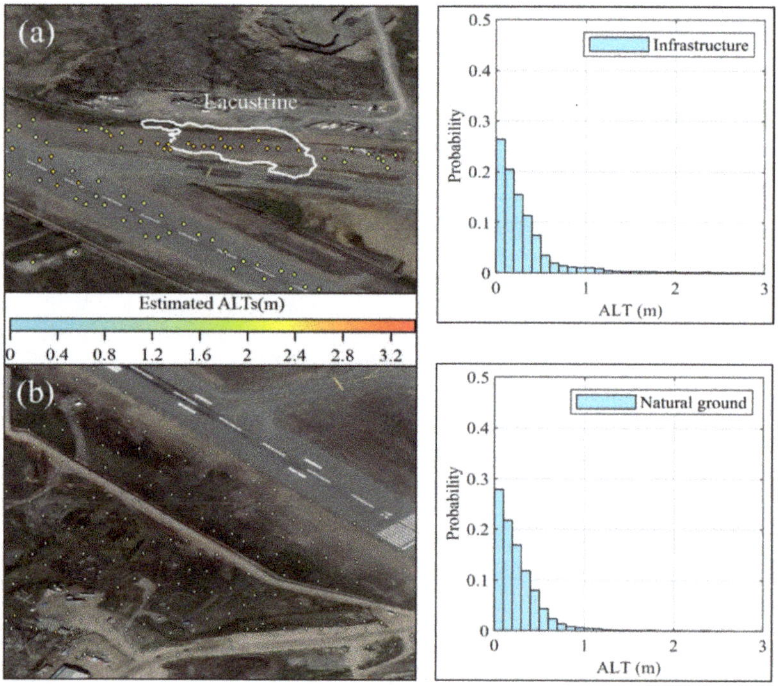

Figure 10. (**a**) Enlarged view and the statistic histogram of ALTs under infrastructure; (**b**) enlarged view and the statistic histogram of ALTs under natural ground.

4.5. Relation between ALT and Geology

An ALT–geology combined analysis over the entire study area was conducted for a further ALT interpretation. The surface geology map [29] over Iqaluit (Figure 11) shows that the area mainly consists of glaciomarine deltaic (GMd) sediments, alluvial flood plain (Afp) sediments, alluvial terraced (At) sediments, lacustrine (Lv), Littoral and nearshore (marked as Mn) sediments, intertidal (marked as Mi) sediments, till blanket (Tb), till veneer (Tv), and bedrock (R).

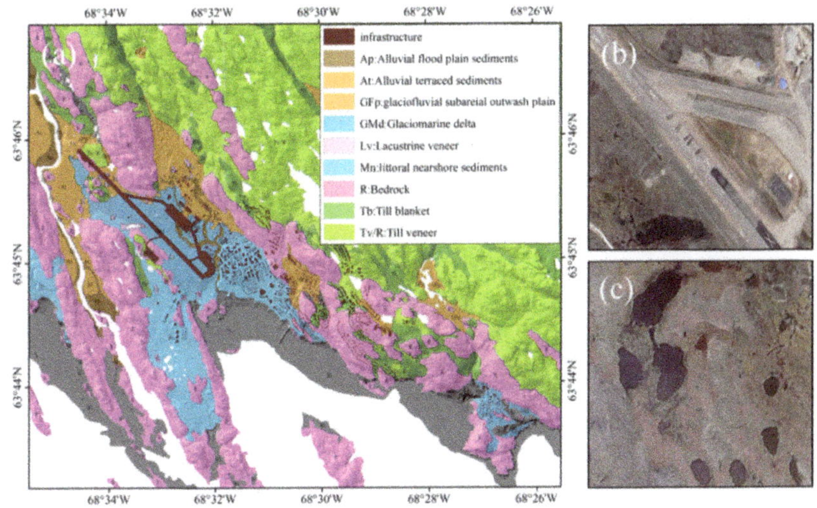

Figure 11. (**a**) Surficial geology map over Iqaluit; (**b**) optical image of the intersection of the runway and taxiway alpha; (**c**) optical image of landscape in Mn sediments.

The solid and tightly bound bedrock is considered stable, as reflected by the slightly upward ground and patches without subsidence shown in Figure 5a. ALTs of almost 0 or small values exist in these patches resulting from their positively proportional relationship with the subsidence. Thick ALs up to 5 m shown at the southern tip of the study area agree well with the large settlement and the small thaw settlement coefficient of bedrock. Bedrocks are usually dry and prone to thaw deeper and quicker due to their high thermal conductivity. The considerable settlement was observed in the Tb unit which is composed of ice-rich coarse sediment. The ALTs in this area are generally less than 0.5 m (Figure 12). The underlying deposits on which large downward movements occur at the airport and nearby infrastructure (Figure 11b,c) are mainly related to the GMd and Mn shown in dark and light blue in Figure 11a, which agrees with previous studies [44,45]. The GMd was deposited in shallow water and formed adjacent to the grounded margin of a tidewater glacier since it formed when marine submergence occurred [49]. Therefore, it is dominantly filled with silt composition and salinity. Thus, the AL likely subsides more in summer. Subsequently, the ALT estimated from the thawing subsidence is larger (Figure 12), and thick estimated ALTs around 2 m are aligned well with the maximal subsidence in GMd sediments. They may be overestimated, however, in well-drained parts with low water content since ice-wedge melting is considered as soil thawing. Field campaign results [48,50,51] show that the sediments are composed of 10 cm organic and 90 cm ice-boned medium brown sand with grey silty sand layers. The higher the ground ice content, the larger the seasonal subsidence or shallower corresponding ALT.

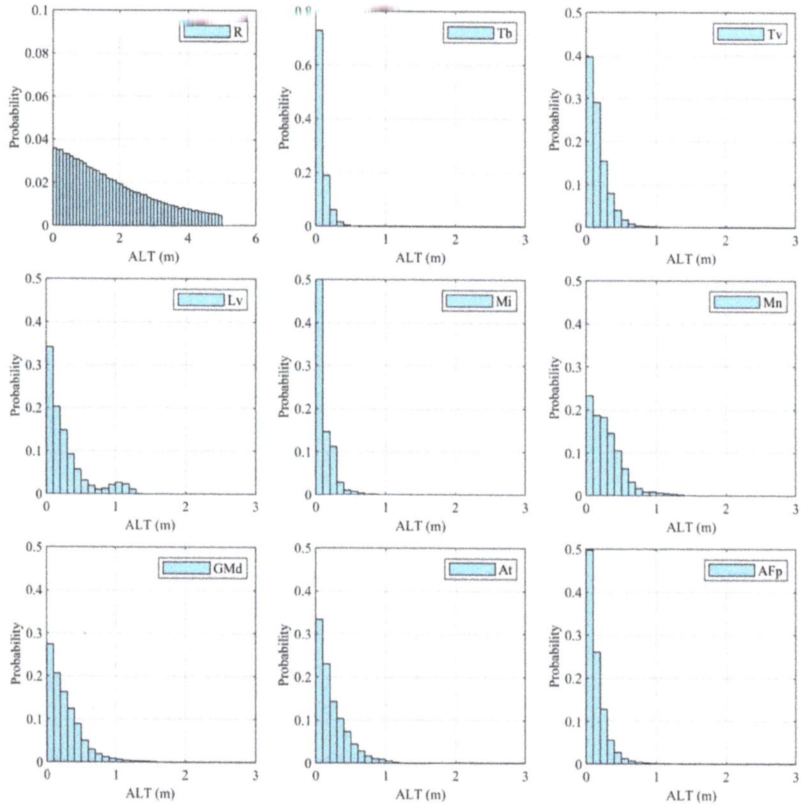

Figure 12. Statistic histograms of ALT in different sediments.

ALT values reaching up to 2.504 m appear in the Lv, with a mean thickness of 0.5 m, and 20% of these kinds of sediments have an AL of more than 1 m (Figure 12). We can see several patches of Lv distributed at the northwest end of the runway and the immediate south side of the runway. In Figure 10, a notable patch of Lv could be the reason for the malfunction of taxiway alpha. The taxiway and marshy regions at the west side of Sylvia Grinnell Road have significant subsidence since the Lv sediment is generally fine-grained material with high moisture content [50,51].

Afp and At units exhibit an ALT less than 1 m (Figure 12). This is reasonable for sediments with grain size ranging from boulders to silt particles [52,53]. The saturated water content of this deposit is higher than other sediments. Therefore, to thaw or freeze fully, more heat is required in the same height layer of soil. This is why subsidence and ALTs are both subtle in this region.

4.6. Limitations

Even though the comparison of estimated ALTs and in situ measured results from other studies shows differences of less than 0.5 m, some limitations should be noted. The ice content is critical for ALT retrieval other than soil properties. The soil moisture data used in this study reveal changes in water content across space but do not include ice-wedge information. The data may introduce substantial errors when the water content and the ice content are considered the same thing. Furthermore, the VWC from SoilGrids at 1500 kPa is not identical to the VWC in the soil throughout the summer. The ratio between them should be appropriately examined before estimating the ALT. In SoilGrids, the explained soil property variance ranges between 30% and 70% [35]. As a result, the uncertainties of ALT estimation are possibly up to meter-level. In addition, the thaw settlement coefficient can properly indicate the thawing conditions of frozen soil. However, more in situ data are necessary for the precise determination of soil moisture and other parameters needed in this retrieval model.

5. Conclusions

In this study, we retrieved high-resolution ALTs at Iqaluit airport by integrating the subsidence derived from X-band interferometric measurements and volumetric water content (VWC) datasets with a resolution of 250 m. The InSAR survey using 21 CSK images in Spotlight mode identified several sites of ground deformation at Iqaluit, subsiding at rates exceeding 80 mm/year. The largest settlement occurs at the intersection of runway 17/35 and taxiway alpha, which is up to 163 mm/year. We determined the saturation state and the corresponding thaw settlement coefficient for all measurement points in different sediments using the VWC data from SoilGrids and soil dry density. Our results suggest ALTs that range from 0 to 5 m. Combined analysis of the ALT and surface geology indicates that the retrieved ALT distribution is highly related to the subsurface geological conditions. Maximal ALTs are distributed in bedrock areas. In the airport sector, thicker ALTs are found in large settlement sites due to the thaw settlement coefficient being positively proportional to the soil moisture where the soil is in an unsaturated state according to the VWC data.

We should emphasize that the detailed ALT map is valuable not only for supporting infrastructure design and maintenance over the permafrost zone but also for comprehensive understanding of the AL conditions with global warming. The proposed ALT solution can provide denser observations over large areas than traditional point-based methods.

Author Contributions: Conceptualization, D.M., G.L., X.W., R.Z., and M.M.; methodology, D.M., X.W., G.L., and M.M.; validation, D.M., R.Z., and B.Z.; data curation, B.Z. and W.X.; writing—original draft preparation, D.M.; writing—review and editing, R.Z., M.M., G.L., B.Z., and X.W.; funding acquisition, G.L., R.Z., B.Y., and X.W. All authors have read and agreed to the published version of the manuscript.

Funding: This research was jointly funded by the National Natural Science Foundation of China (Grant No. 42171355, 51774250, 41804009, 41801399, 42071410); the Sichuan Science and Technology Program (Grant No. 2019ZDZX0042, 2020JDTD0003, 2020YJ0322); and the Open fund of State Key Laboratory of Geodesy and Geodynamics(Grant No. SKLGED2020-5-1-E).

Institutional Review Board Statement: Not applicable.

Informed Consent Statement: Not applicable.

Data Availability Statement: The data presented in this study are available on request from the corresponding author. The data are not publicly available due to compliance with commercial data confidentiality requirements.

Acknowledgments: We thank ASI (Italian Space Agency) and DLR (German Aerospace Center) for providing SAR images.

Conflicts of Interest: The authors declare no conflict of interest.

References

1. Obu, J. How Much of the Earth's Surface Is Underlain by Permafrost? *J. Geophys. Res. Earth Surf.* **2021**, *126*, e2021JF006123. [CrossRef]
2. Streletskiy, D.A.; Suter, L.J.; Shiklomanov, N.I.; Porfiriev, B.N.; Eliseev, D.O. Assessment of Climate Change Impacts on Buildings, Structures and Infrastructure in the Russian Regions on Permafrost. *Environ. Res. Lett.* **2019**, *14*, 025003. [CrossRef]
3. Streletskiy, D.; Anisimov, O.; Vasiliev, A. Permafrost Degradation. In *Snow and Ice-Related Hazards, Risks and Disasters*; Elsevier: Amsterdam, The Netherlands, 2015; pp. 303–344.
4. Schneider von Deimling, T.; Lee, H.; Ingeman-Nielsen, T.; Westermann, S.; Romanovsky, V.; Lamoureux, S.; Walker, D.A.; Chadburn, S.; Trochim, E.; Cai, L.; et al. Consequences of Permafrost Degradation for Arctic Infrastructure–Bridging the Model Gap between Regional and Engineering Scales. *Cryosphere* **2021**, *15*, 2451–2471. [CrossRef]

5. Karjalainen, O.; Aalto, J.; Luoto, M.; Westermann, S.; Romanovsky, V.E.; Nelson, F.E.; Etzelmüller, B.; Hjort, J. Circumpolar Permafrost Maps and Geohazard Indices for Near-Future Infrastructure Risk Assessments. *Sci. Data* **2019**, *6*, 1–16. [CrossRef]
6. Niu, F.; Yin, G.; Luo, J.; Lin, Z.; Liu, M. Permafrost Distribution along the Qinghai-Tibet Engineering Corridor, China Using High-Resolution Statistical Mapping and Modeling Integrated with Remote Sensing and GIS. *Remote Sens.* **2018**, *10*, 215. [CrossRef]
7. Chen, F.; Lin, H.; Li, Z.; Chen, Q.; Zhou, J. Interaction between Permafrost and Infrastructure along the Qinghai–Tibet Railway Detected via Jointly Analysis of C- and L-Band Small Baseline SAR Interferometry. *Remote Sens. Environ.* **2012**, *123*, 532–540. [CrossRef]
8. Rickard, W.; Brown, J. The Performance of a Frost-Tube for the Determination of Soil Freezing and Thawing Depths. *Soil Sci.* **1972**, *113*, 149–154. [CrossRef]
9. Grasmueck, M.; Weger, R.; Horstmeyer, H. Full-Resolution 3D GPR Imaging. *Geophysics* **2005**, *70*, K12–K19. [CrossRef]
10. Park, B.; Kim, J.; Lee, J.; Kang, M.-S.; An, Y.-K. Underground Object Classification for Urban Roads Using Instantaneous Phase Analysis of Ground-Penetrating Radar (GPR) Data. *Remote Sens.* **2018**, *10*, 1417. [CrossRef]
11. Jonard, F.; Weihermüller, L.; Schwank, M.; Jadoon, K.Z.; Vereecken, H.; Lambot, S. Estimation of Hydraulic Properties of a Sandy Soil Using Ground-Based Active and Passive Microwave Remote Sensing. *IEEE Trans. Geosci. Remote Sens.* **2015**, *53*, 3095–3109. [CrossRef]
12. Goodrich, L.E. The Influence of Snow Cover on the Ground Thermal Regime. *Can. Geotech. J.* **1982**, *19*, 421–432. [CrossRef]
13. Nelson, P.; Oades, J. Organic Matter, Sodicity and Soil Structure. In *Sodic Soils: Distribution, Properties, Management and Environmental Consequences*; Topics in Sustainable Agronomy; Oxford University Press: Oxford, UK, 1998; pp. 51–75.
14. Vuik, C. *Some Historical Notes about the Stefan Problem*; Delft University of Technology: Delft, The Netherlands, 1993.
15. Chen, R.H.; Tabatabaeenejad, A.; Moghaddam, M. Retrieval of Permafrost Active Layer Properties Using Time-Series P-Band Radar Observations. *IEEE Trans. Geosci. Remote Sens.* **2019**, *57*, 6037–6054. [CrossRef]
16. Li, Z.; Zhao, R.; Hu, J.; Wen, L.; Feng, G.; Zhang, Z.; Wang, Q. InSAR Analysis of Surface Deformation over Permafrost to Estimate Active Layer Thickness Based on One-Dimensional Heat Transfer Model of Soils. *Sci. Rep.* **2015**, *5*, 1–9. [CrossRef] [PubMed]
17. Liu, L.; Schaefer, K.; Zhang, T.; Wahr, J. Estimating 1992–2000 Average Active Layer Thickness on the Alaskan North Slope from Remotely Sensed Surface Subsidence. *J. Geophys. Res. Earth Surf.* **2012**, *117*. [CrossRef]
18. Chang, T.; Han, J.; Li, Z.; Wen, Y.; Hao, T.; Lu, P.; Li, R. Active Layer Thickness Retrieval Over the Qinghai-Tibet Plateau from 2000 TO 2020 Based on Insar-Measured Subsidence and Multi-Layer Soil Moisture. *ISPRS-Int. Arch. Photogramm. Remote Sens. Spat. Inf. Sci.* **2021**, *43*, 437–442. [CrossRef]
19. Rouyet, L.; Liu, L.; Strand, S.M.; Christiansen, H.H.; Lauknes, T.R.; Larsen, Y. Seasonal InSAR Displacements Documenting the Active Layer Freeze and Thaw Progression in Central-Western Spitsbergen, Svalbard. *Remote Sens.* **2021**, *13*, 2977. [CrossRef]
20. Wang, C.; Zhang, Z.; Zhang, H.; Zhang, B.; Tang, Y.; Wu, Q. Active Layer Thickness Retrieval of Qinghai–Tibet Permafrost Using the TerraSAR-X InSAR Technique. *IEEE J. Sel. Top. Appl. Earth Obs. Remote Sens.* **2018**, *11*, 4403–4413. [CrossRef]
21. Clayton, L.K.; Schaefer, K.; Battaglia, M.J.; Bourgeau-Chavez, L.; Chen, J.; Chen, R.H.; Chen, A.; Bakian-Dogaheh, K.; Grelik, S.; Jafarov, E. Active Layer Thickness as a Function of Soil Water Content. *Environ. Res. Lett.* **2021**, *16*, 055028. [CrossRef]
22. Zhang, X.; Zhang, H.; Wang, C.; Tang, Y.; Zhang, B.; Wu, F.; Wang, J.; Zhang, Z. Active Layer Thickness Retrieval Over the Qinghai-Tibet Plateau Using Sentinel-1 Multitemporal InSAR Monitored Permafrost Subsidence and Temporal-Spatial Multilayer Soil Moisture Data. *IEEE Access* **2020**, *8*, 84336–84351. [CrossRef]
23. Luo, D.; Jin, H.; Wu, Q.; Bense, V.F.; He, R.; Ma, Q.; Gao, S.; Jin, X.; Lü, L. Thermal Regime of Warm-Dry Permafrost in Relation to Ground Surface Temperature in the Source Areas of the Yangtze and Yellow Rivers on the Qinghai-Tibet Plateau, SW China. *Sci. Total Environ.* **2018**, *618*, 1033–1045. [CrossRef]
24. Klene, A.E.; Nelson, F.E. Urban Geocryology: Mapping Urban–Rural Contrasts in Active-Layer Thickness, Barrow Peninsula, Northern Alaska. *Ann. Am. Assoc. Geogr.* **2019**, *109*, 1394–1414. [CrossRef]
25. Taber, S. The Mechanics of Frost Heaving. *J. Geol.* **1930**, *38*, 303–317. [CrossRef]
26. Beskow, G. *Soil Freezing and Frost Heaving with Special Application to Roads and Railroads*; Osterberg, J.O., Translator; The Swedish Geological Society, C, no. 375, Year Book no. 3; Technological Institute, Northwestern University: Evanston, IL, USA, 1935.
27. Black, P.B.; Hardenberg, M.J. *Historical Perspectives in Frost Heave Research: The Early Works of S. Taber and G. Beskow*; Special Report (Cold Regions Research and Engineering Laboratory (U.S.)); Cold Regions Research and Engineering Laboratory, Engineer Research and Development Center: Vicksburg, MS, USA, 1991; pp. 23–91.
28. Dash, J.G.; Rempel, A.W.; Wettlaufer, J.S. The Physics of Premelted Ice and Its Geophysical Consequences. *Rev. Modern Phys.* **2006**, *78*, 695–741. [CrossRef]
29. Yanagiya, K.; Furuya, M. Post-wildfire surface deformation near Batagay, Eastern Siberia, detected by l-band and c-band InSAR. *J. Geophys. Res. Earth Surface.* **2020**, *125*, e2019JF005473. [CrossRef]
30. Liu, G.X.; Buckley, S.M.; Ding, X.L.; Chen, Q.; Luo, X.J. Estimating Spatiotemporal Ground Deformation With Improved Persistent-Scatterer Radar Interferometry. *IEEE Trans. Geosci. Remote Sens.* **2009**, *47*, 3209–3219. [CrossRef]
31. Liu, G.; Jia, H.; Zhang, R.; Zhang, H.; Jia, H.; Yu, B.; Sang, M. Exploration of Subsidence Estimation by Persistent Scatterer InSAR on Time Series of High Resolution TerraSAR-X Images. *IEEE J. Sel. Top. Appl. Earth Obs. Remote Sens.* **2010**, *4*, 159–170. [CrossRef]
32. Murton, J.B. Permafrost and Climate Change. In *Climate Change*; Elsevier: Amsterdam, The Netherlands, 2021; pp. 281–326.

33. Allard, M.; Doyon, J.; Oldenborger, G.A.; Sladen, W.; LeBlanc, A.-M.; Mathon-Dufour, V.; L'Hérault, E.; Mate, D. *Surficial Geology, Iqaluit, Nunavut*; Geological Survey of Canada: Ottawa, ON, Canada, 2012.
34. International Permafrost Association. Frozen Ground-The News Bulletin of the International Permafrost Association, No. 25. *Frozen Ground* **2001**, *25*.
35. Hengl, T.; Mendes de Jesus, J.; Heuvelink, G.B.; Ruiperez Gonzalez, M.; Kilibarda, M.; Blagotić, A.; Shangguan, W.; Wright, M.N.; Geng, X.; Bauer-Marschallinger, B.; et al. SoilGrids250m: Global Gridded Soil Information Based on Machine Learning. *PLoS One* **2017**, *12*, e0169748. [CrossRef]
36. Ferretti, A.; Prati, C.; Rocca, F. Permanent scatterers in SAR interferometry. *IEEE Trans. Geosci. Remote Sens.* **2001**, *39*, 8–20. [CrossRef]
37. Lanari, R.; Mora, O.; Manunta, M.; Mallorquí, J.J.; Berardino, P.; Sansosti, E. A small-baseline approach for investigating deformations on full-resolution differential SAR interferograms. *IEEE Trans. Geosci. Remote Sens* **2004**, *42*, 1377–1386.
38. Huang, N.E.; Shen, Z.; Long, S.R.; Wu, M.C.; Shih, H.H.; Zheng, Q.; Yen, N.C.; Tung, C.C.; Liu, H.H. The Empirical Mode Decomposition and the Hilbert Spectrum for Nonlinear and Non-Stationary Time Series Analysis. *Proc. Math. Phys. Eng. Sci.* **1998**, *454*, 903–995. [CrossRef]
39. Watson, G.; Slusarchuk, W.; Rowley, R. Determination of Some Frozen and Thawed Properties of Permafrost Soils. *Can. Geotech. J.* **1973**, *10*, 592–606. [CrossRef]
40. Ping, H.; Guodong, C.; Chengsong, Y. The Evaluation of Thawing-Settlement Coefficient of Frozen Soils. *J. Glaciol. Geocryol.* **2003**, *25*, 608–613.
41. Xifa, Z.; Ji, C.; Dongqing, Z. Application of Thawing Settlement Coefficient to the Research on the Roadbed Frost Damage of Freeway in Seasonal Frost Region. *J. Glaciol. Geocryol.* **2002**, *24*, 634–638.
42. Brooks, H.; Doré, G.; Lemieux, C. *Quantitative Risk Assessment of the Iqaluit International Airport, Iqaluit, Nu-navut.*; Prepared for The Northern Transportation Adaptation Initiative from Transport Canada; Université Laval: Québec, QC, Canada, 2017; p. 53.
43. Blewitt, G.; Hammond, W.C.; Kreemer, C. Harnessing the GPS data explosion for interdisciplinary science. *Eos* **2018**, *99*, 485. [CrossRef]
44. LeBlanc, A.M.; Short, N.; Oldenborger, G.A.; Mathon-Dufour, V.; Allard, M. Geophysical Investigation and InSAR Mapping of Permafrost and Ground Movement at the Iqaluit Airport. In *Cold Regions Engineering 2012: Sustainable Infrastructure Development in a Changing Cold Environment*; American Society of Civil Engineers: Reston, VA, USA, 2012; pp. 644–654.
45. Short, N.; LeBlanc, A.M.; Sladen, W.; Oldenborger, G.; Mathon-Dufour, V.; Brisco, B. RADARSAT-2 D-InSAR for ground displacement in permafrost terrain, validation from Iqaluit Airport, Baffin Island, Canada. *Remote Sens. Environ.* **2014**, *141*, 40–51. [CrossRef]
46. Short, N. *RADARSAT Constellation Mission DInSAR Potential in Permafrost Terrain*; Natural Resources Canada: Ottawa, ON, Canada, 2017. [CrossRef]
47. Mathon-Dufour, V.; Allard, M.; Leblanc, A.; L'Hérault, E.; Oldenborger, G.; Sladen, W. Mapping Surficial Geology and Assessment of Permafrost Conditions under the Iqaluit Airport, Nunavut, Canada. In Proceedings of the AGU Fall Meeting Abstracts, San Francisco, CA, USA, 3 December 2012.
48. LeBlanc, A.M.; Mathon-Dufour, V.; Allard, M.; Oldenborger, G.A.; Short, N.; L'Hérault, E.; Sladen, W.E. Permafrost characterization at the Iqaluit International Airport, Nunavut, in support of decision-making and planning. *Summ. Act.* **2012**, 131–142.
49. Rémillard, A.M.; Hétu, B.; Bernatchez, P.; Buylaert, J.-P.; Murray, A.S.; St-Onge, G.; Geach, M. Chronology and Palaeoenvironmental Implications of the Ice-Wedge Pseudomorphs and Composite-Wedge Casts on the Magdalen Islands (Eastern Canada). *Boreas* **2015**, *44*, 658–675. [CrossRef]
50. Mathon-Dufour, V. *Caractérisation Du Pergélisol En Vue de La Réfection et de l'adaptation Aux Changements Climatiques de L'aéroport D'Iqaluit*; Université Laval: Québec, QC, Canada, 2014.
51. LeBlanc, A.M.; Short, N.; Mathon-Dufour, V.; Allard, M.; Tremblay, T.; Oldenborger, G.A.; Chartrand, J. DInSAR seasonal surface displacement in built and natural permafrost environments, Iqaluit, Nunavut, Canada. In Proceedings of the GEO, Québec, QC, Canada, 25–28 February 2015; pp. 20–23.
52. Martini, I.P. Deglaciation Stages of the Laurentide Ice Sheet in Canada and Related Glaciomarine and Glaciolacustrine Deposits. *Rev. Sel. Features* **2007**. Available online: https://minerva.usc.es/xmlui/handle/10347/3757 (accessed on 26 January 2022).
53. Hillier, S. Erosion, Sedimentation and Sedimentary Origin of Clays. In Origin and mineralogy of clays. In *Origin and Mineralogy of Clays*; Springer: Berlin/Heidelberg, Germany, 1995; pp. 162–219.

remote sensing

MDPI

Article

Monitoring of Land Subsidence and Ground Fissure Activity within the Su-Xi-Chang Area Based on Time-Series InSAR

Chengsheng Yang [1], Sen Lv [1,*], Zuhang Hou [1], Qin Zhang [1], Tao Li [2] and Chaoying Zhao [1]

[1] College of Geology Engineering and Geomatics, Chang'an University, Xi'an 710054, China; yangchengsheng@chd.edu.cn (C.Y.); 2020226043@chd.edu.cn (Z.H.); dczhangq@chd.edu.cn (Q.Z.); cyzhao@chd.edu.cn (C.Z.)

[2] Land Satellite Remote Sensing Application Center, Beijing 100048, China; lit@lasac.cn

* Correspondence: 2020226012@chd.edu.cn; Tel.: +86-29-8233-9261

Abstract: Serious land subsidence and ground fissure (GF) disasters have brought huge economic losses to the Su-Xi-Chang area (China) and threatened the safety of its residents. To better understand the development of these disasters, it is urgent to carry out long-term and large-scale deformation monitoring in this region. In this study, based on time-series interferometric synthetic aperture radar (InSAR) technology, ground deformation characteristics were obtained at different periods. Meanwhile, Fast Lagrangian Analysis of Continua in Three Dimensions (FLAC3D) version 5.00 was used to study the stress, seepage field, and displacement changes in the soil layers caused by pumping activities at the bedrock bulge. The results showed that three subsidence centers were located in Suzhou, Wuxi, and Changzhou from 2007 to 2010. The ground fissures in Guangming village had obvious differential settlements and intense activities. The land subsidence in the Su-Xi-Chang area was under control from 2018 to 2021, while there was a relative rebound in most areas. Combined with numerical simulation and geological data, we demonstrated that pumping activities would accelerate and intensify the land subsidence process, and differential subsidence was prone to occur at the buried hill, which in turn led to the formation of ground fissures. By comparing the characteristics of ground deformation in different periods, it was proven that banning groundwater exploitation is an effective measure for preventing and controlling such disasters.

Keywords: Su-Xi-Chang area; InSAR time-series; ground fissure; land subsidence; FLAC3D

Citation: Yang, C.; Lv, S.; Hou, Z.; Zhang, Q.; Li, T.; Zhao, C. Monitoring of Land Subsidence and Ground Fissure Activity within the Su-Xi-Chang Area Based on Time-Series InSAR. *Remote Sens.* **2022**, *14*, 903. https://doi.org/10.3390/rs14040903

Academic Editor: Nicola Casagli

Received: 29 December 2021
Accepted: 9 February 2022
Published: 14 February 2022

1. Introduction

Overexploitation of groundwater resources may result in compression of aquifer systems (aquifers and permeable aquifers), which causes rapid land subsidence [1]. Many major cities around the world are facing such geological disasters due to the excessive extraction of groundwater, such as central Mexico, Las Vegas in the USA, and the United Arab Emirates [2–4]. Moreover, ground subsidence disasters are often accompanied by the formation and development of ground fissures (GFs). A GF is a type of geological disaster that occurs following the fracture of rock and soil [5,6]. Ground subsidence disasters pose an enormous threat to underground engineering, surface construction, and the safety and properties of human life [7–9]. Land subsidence in coastal areas or around lakes is also prone to causing disasters such as waterlogging and flooding due to the loss of elevation [10,11].

Previous investigations have shown that at least 3000 GFs are distributed throughout the Northern China Plain, and approximately 80% of the GFs are non-structural [6,12,13]. In some regions, although land subsidence disasters have been alleviated to a certain extent owing to natural and artificial groundwater recharges, the trend has not yet reversed because of excessive groundwater extraction [14].

The mechanisms of land subsidence can be divided into two categories: the tectonic control type and non-tectonic control type [6]. Tectonic factors such as faults, earthquakes, basements, basins and sediments affect the formation and development of surface deformation to a great extent, while non-tectonic factors [15,16] such as the over-exploitation of groundwater, rainfall conditions and climate also have a great impact on the breeding process of land subsidence. The inducing factors of ground fissures can be divided into three categories [5,9,16,17]—they are: (1) internal dynamic action, such as fault creep, earthquakes and volcanic eruptions; (2) extraction of underground fluids, such as water, oil and gas; (3) semiarid climates. Based on the ground fissures found in China, the main causes of such disasters are mainly tectonic activities and groundwater exploitation.

To better understand the internal relationship between land subsidence and GFs in the study area, a wide range of ground subsidence characteristics must be investigated. Many methods can be used for land subsidence monitoring, such as: (1) taking measurements with standard land surveying methods (precise leveling, total stations or global navigation satellite systems (GNSS)) over the established stable benchmarks at different times at the same site, while obtaining the elevation differences; (2) making comparisons of measurements from borehole extensometers; (3) making comparisons of elevations from light detection and ranging (LIDAR) data acquired at different times or by using other elevation datasets [11,17–20]. In addition to mapping technology for directly obtaining surface deformation, some commonly used technologies in engineering exploration are also used to study land subsidence and ground fissure disasters, such as pore water pressure gauges [17], convergence measurements, layered standards and distributed optical fibers [21], which can measure changes in the strata below the surface. However, these technologies generally have the characteristics of a high cost, a large workload in the field, and poor continuity of deformation.

The Su-Xi-Chang area mainly includes the Suzhou, Wuxi, and Changzhou cities of China (Figure 1), which—especially Jiangyin City (subordinate to Wuxi)—are typical areas with land subsidence and ground fissure phenomena. It lies in the southern part of Jiangsu Province, around the lower reaches of the Yangtze River, and east of Shanghai City (Figure 1). The Su-Xi-Chang region has various types of groundwater, complex burial conditions, and an uneven distribution of groundwater, with obvious regional characteristics. Pore water is the main groundwater type in unconsolidated materials in plain areas [22,23]. The aquifers are partitioned into phreatic aquifers and confined aquifers—marked as I, II, and III, respectively, from top to bottom based on their genesis, ages, burial distributions, hydraulic connections, and hydro-chemical characteristics [24,25]. Confined aquifer II is the main groundwater exploitation layer, consisting of Middle Pleistocene loam, medium coarse sand, fine sand, sandy loam, coarse sand with pebbles and gravel [14,22,24].

Due to the rapid development of industry and the large water demand for residential daily life, as well as irrigated agriculture, groundwater is consumed quickly and in large quantities, and land subsidence disasters began to emerge in the Su-Xi-Chang region in the early 1960s [26,27]. Over the next few decades, the groundwater level has continued to decrease, with a wider range of uneven land subsidence. The rapid development of land subsidence has brought on the discovery of several GFs [22]. The government has made considerable efforts to mitigate regional land subsidence since 2005, including prohibiting groundwater withdrawal. Currently, the land subsidence rate in the study area has declined to less than 10 mm/year in most areas [25,26].

GFs such as those in Yangshuli (YSL), Guangming village (GM), and Wuxi City are the most typical, appearing in early 1998 and further developing after 2000. In the next few years, the inclination of the paddy fields increased, forcing villagers to level the paddy fields every year and divide the paddy fields into small pieces for cultivation [28]. With the development of the JY subsidence field, the occurrence of GF disasters in the region peaked after 2007. Although the groundwater ban has been implemented for a long time, the ground rebound has a hysteresis effect. Thus, the ground rebound rate is extremely slow. It was not until surface monitoring in recent years that a tiny rebound phenomenon was observed [25].

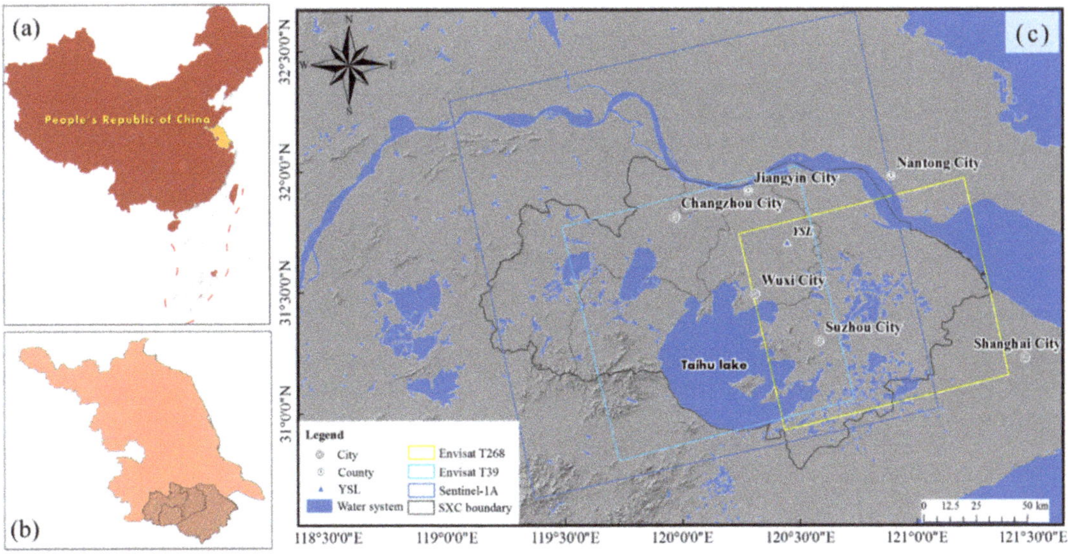

Figure 1. Geographical location of Su-Xi-Chang area. (**a**) Geographical position of Jiangsu Province in China. (**b**) Geographical location of Su-Xi-Chang area in Jiangsu Province. (**c**) SAR image coverage. YSL stands for Yangshuli where the GF disaster occurred.

As a space-to-earth observation technology, interferometric synthetic aperture radar (InSAR) has the advantage of obtaining surface deformations over a large area with high precision. Combined with geological data, the corresponding relationship between the surface deformation information obtained by InSAR and the geological background can be obtained, such as the temporal and spatial variation relationship between the deformation funnel and the fluctuation of bedrock.

In this study, land subsidence and GF activities in the Su-Xi-Chang area were monitored using time-series InSAR technology based on SAR images from different periods. To study surface deformation from 2007 to 2010, we used two sets of Envisat advanced synthetic aperture radar (ASAR) images from different orbits: Envisat T39 and T268. The surface deformation of this area was observed based on Sentiel-1A images from 2018 to 2021. The Stanford method for persistent scatterers (StaMPS) was used to process SAR data [29]. We then analyzed and discussed the deformation information of the Su-Xi-Chang region. Fast Lagrangian Analysis of Continua in Three Dimensions (FLAC3D) software was used to simulate the surface deformation caused by pumping activities to verify the causes of land subsidence and GF disasters. Our research can support the government's decision making in response to disasters and the effective allocation of resource development and provide a reference for studying surface subsidence and GF disasters.

2. Research Status

Since land subsidence and GF disasters have appeared in this region, scholars have done a lot of work to analyze the mechanisms of land subsidence and ground fissure, including model experiments, numerical simulations and data measurements. Bu et al. analyzed the developmental process of land subsidence with pumping in the Su-Xi-Chang area [24]. Wu et al. observed the difference in land subsidence on both sides of the ground fissure through leveling, and found that the uneven distribution of loose layers determined the uneven land subsidence through geological drilling and a controllable source geodetic audio sounding method [22]. Using distributed fiber optical sensing technologies and microstructure analysis, Gu et al. found that the larger the proportion of soil voids, the easier it is to produce land subsidence [26]. Gu conducted a pumping experiment by analyzing a ground fissure in a buried hill area via a macro model experiment, and he found that the more the groundwater level drops, the more serious are the land subsidence and ground fissures that are likely to occur [28]. Through leveling, groundwater head measurements and a borehole extensometer, Wang et al. found that there was a time lag between land subsidence or uplift and changes in groundwater level in the Su-Xi-Chang area [25], and she analyzed the Changjing ground fissure located in Jiangyin City [30]. Many studies have surmised that the development of GFs is affected by various factors, such as bedrock undulation, heterogeneity of the Quaternary strata, groundwater over-extraction and differential land subsidence [27,31–33]. Chen et al. found that excessive extraction of groundwater is the principal factor in the occurrence of land subsidence disasters. As the depth of the strata increases, the compression of the soil layer caused by the weight of the soil body increases, and the settlement caused by the upper loads increases with increases in the upper load intensity; this phenomenon was verified using a column element settlement model [34]. Liu obtained the same conclusion using a numerical simulation method based on the numerical simulation software MatDEM [35].

At present, the rebound phenomenon of the ground is mainly explained by measured data and model experiments. Hu concluded from the measured data that the effective stress in the stratum disappeared owing to the sharp rise in the groundwater level, the elastic deformation in the stratum was partially restored owing to the unloading effect, and the ground rebounded [14]. Wang et al. also confirmed the existence of ground rebound through measured data [25]. Xu et al. confirmed that land subsidence and GF disasters caused by groundwater extraction are reversible processes by building physical models and related experimental systems in the laboratory [13].

Predecessors have done a lot of research on the disaster mechanisms in this area, but there is a lack of corresponding research results on the developmental process of disasters and the changes in surface deformation after groundwater mining prohibition. For example, the measured data of these studies are only based on individual points and there is a lack of deformation monitoring of the entire area in previous studies. Therefore, it is difficult to conduct a more refined analysis and description of surface deformation. To better understand the causes and development of land subsidence and the formation of GFs, we used different bands of synthetic aperture radar (SAR) images to study the surface deformation of this area—namely, Envisat ASAR and Sentinel 1A—using time-series InSAR technology.

3. Methods and Data Processing

3.1. Methods

The differential InSAR technology used for land surface deformation monitoring is subject to various factors such as the residual phase of the look angle error, satellite orbit error, decoherence, phase unwrapping error, digital surface model (DSM) error and atmospheric delay error [36,37]. Among the factors mentioned above, the accuracy and precision of InSAR technology for monitoring surface deformation are mainly limited by coherence and atmospheric delay errors. By utilizing permanent scatterer (PS) InSAR technology, the aforementioned errors can be effectively reduced.

The PS-InSAR method is based on PS points, which exhibit stable scattering characteristics. The PS points were obtained by analyzing the phase and amplitude variations of the SAR data in the same region during a certain period of time. By utilizing the StaMPS method firstly provided by Hooper in 2007, slowly decorrelating filtered phase (SDFP) pixels with smaller amplitudes and relatively stable phases were selected. PS/SDFP pixels were refined using multiple iterative calculations. Based on the selected PS/SDFP pixels, deformation information can be analyzed and extracted. The number of PS points in non-urban areas can be enhanced using the StaMPS method [29]. Figure 2 shows the technical process of StaMPS technology.

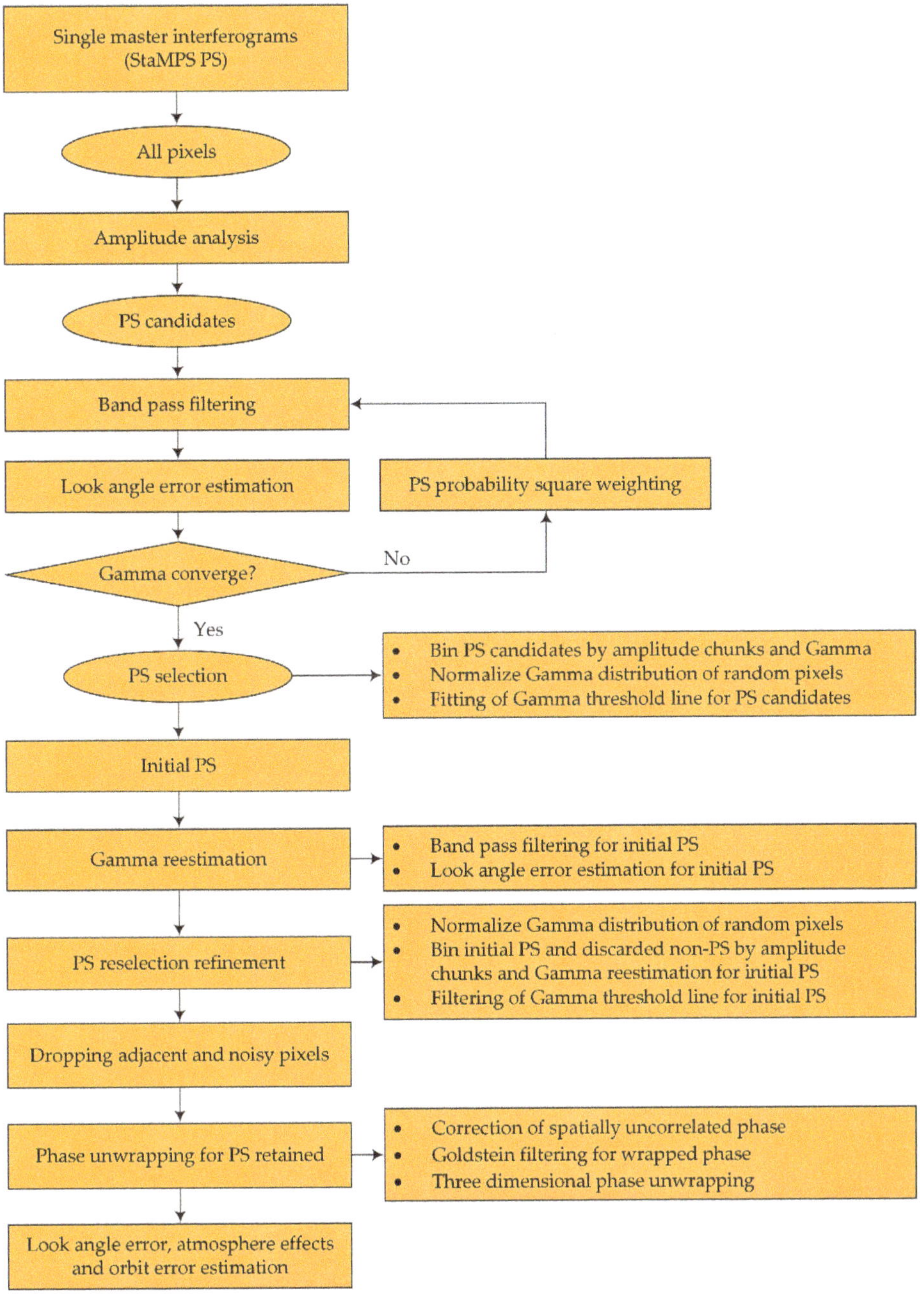

Figure 2. StaMPS technology flow chart.

However, this measurement technology based on interference points has high requirements for the phase stability of point targets in the study area. For rural areas dominated by farmland, it is difficult to obtain sufficient monitoring data. In addition, in the data processing, due to the differences in the selection of unwrapping reference points, the deformation results may have an overall deviation. As we analyzed the deformation characteristics on a large scale in our study, we ignored the problem of insufficient data in farmland areas. In order to solve the deviation caused by different unwrapping reference points, we set the same reference area for different orbital SAR images.

3.2. InSAR Data Processing

In this study, we monitored the surface deformation of two different time periods based on the data obtained from different sensors. Envisat ASAR images were used to study ground deformation from 2007 to 2010. In addition, data from Sentinel-1 were collected to observe the ground settlement from 2018 to 2021. Specific information, the number of interferograms, and the acquisition time of the main image involved in the calculation of the different sets of SAR images are listed in Table 1. The repeat orbit interferometry PACkage (ROI_PAC) was used to convert the Envisat ASAR raw images into single-look complex images [38]. Doris software V4.06 was used to produce interferograms for the Envisat ASAR images [39]. Interferograms were generated using GAMMA Remote Sensing software version 201807 for Sentinel-1 images. Time-series analysis can be utilized with a single master image, assuming that the number of SAR images is sufficient. Therefore, the ground deformation was obtained using the StaMPS PS-InSAR technique. To reduce the perpendicular and temporal baselines and Doppler difference, and considering the degree of vegetation cover, we chose the SAR images taken in the middle of the time span and taken in winter as the master image SAR images (Table 1). The distribution information of the spatiotemporal baselines of the interferograms obtained from different sensors is shown in Figure 3.

Table 1. Detailed information for the SAR images used in this study.

Sensor	Envisat T268	Envisat T39	Sentinel 1A (T69)
Band	C	C	C
Orbit direction	ascending	ascending	ascending
Heading (°)	−13.2	−13.2	−12.7
Incidence angle (°)	23.2	23.2	39.3
Polarization	VV	VV	VV
Number of interferograms	21	23	42
Master image	14 February 2009	29 January 2009	22 February 2020
Time span	March 2008–February 2009	November 2007–April 2010	January 2019–June 2021

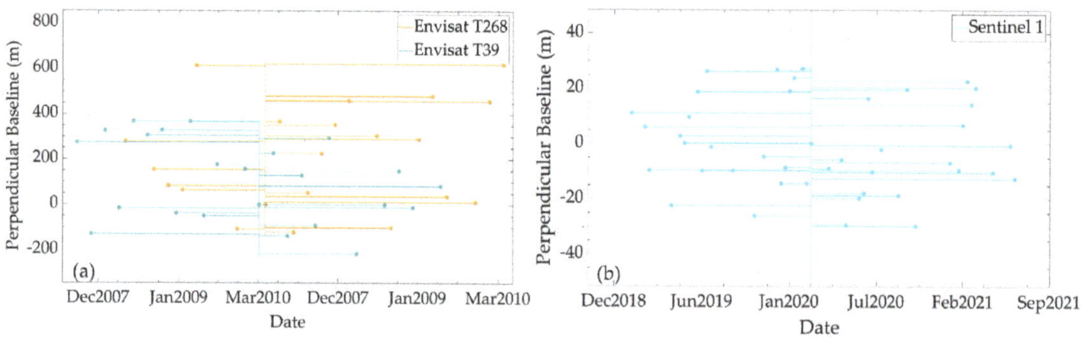

Figure 3. Space-time baseline distributions of the interferograms from the three sets of SAR images. (**a**) Envisat ASAR (2007–2010), (**b**) Sentinel 1A (2018–2021).

To subtract the terrain phase from the interference phase more accurately, a shuttle radar topography mission with 30 m resolution (SRTM1) was utilized. In this study, 21, 23 and 42 interferograms from Envisat T268, T39 and Sentinel-1 images, respectively, were used. Subsequently, phase iterative filtering was performed on the selected interferograms to optimize the DEM error. Points with phase standard deviations less than 0.8 rad were selected for the refinement of coherent points [40]. Generally, the difference in the atmospheric delay between passes has a more significant effect on the unwrapped phase quality [41]. Based on the spatial and temporal correlations of differential interferograms, this type of error caused by the atmosphere was reduced from each differential interferogram by filtering. The deformation velocity and time-series deformation information in the Su-Xi-Chang area were obtained through the decomposition of the unwrapping phase and the removal of the error phase.

3.3. Constitutive Relationship Model

When using FLAC3D software for numerical simulations to study the influence of pumping activities on surface deformation, it is very important to select an appropriate constitutive model to describe the mechanical response between soil particles. This model is an empirical expression of the mechanical characteristics of materials. Given the discrepancy of different material properties, to show the mechanical response of geotechnical materials with different external loads in different situations, the FLAC3D software developed 12 different models. The Mohr-Coulomb model is frequently used in geotechnical mechanics studies [42,43]; it is more appropriate for the analysis of loose and cemented granular materials [44]. We used this model in our study to obtain the variation in the geotechnical materials around the GFs under pumping activities. The physical parameters of the soil play a key role in the improvement of the simulation accuracy of stratum activity, including porosity (n), permeability coefficient (k), Young's modulus (E), Poisson's ratio (v), internal friction angle (u), cohesion (c), expansion angle (w), tensile strength (T), normal stiffness (Kn) and tangential stiffness [18]. The model analysis assists in understanding the relationship between land subsidence, ground fissure activity and pumping in this area, and provides guidance for regional groundwater exploitation.

4. Results and Analysis

In order to interpret the deformation information monitored by time-series InSAR, spatial analysis technology was applied, and time-series points of typical deformation area were extracted in our study. Through these analyses, the temporal and spatial characteristics of surface deformation in the study area can be clearly demonstrated. The specific research method is shown in Figure 4.

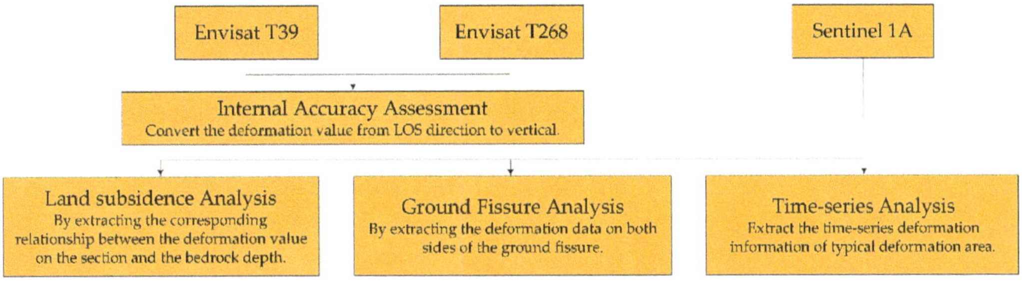

Figure 4. Flow chart of deformation field analysis.

4.1. Results from SAR Images

The annual deformation rates calculated from the two bands of the SAR images are shown in Figure 5a,b, which reveal the ground displacement toward or away from the sensor in the line-of-sight (LOS) by the positive or negative value of the deformation velocity, respectively. However, differences in imaging geometry parameters may influence the results, such as the resolution of SAR images, azimuth angle, and incident angle, which lead to differences in the results. Through comparison and comprehensive analysis of the results, we found that the southern part of JY was the main subsidence area. Therefore, we mainly demonstrate JY where severe land subsidence and GFs had occurred and continued to develop. From 2007 to 2010, the largest deformation rate was more than 20 mm/year, which is consistent with the results from previous research [14,22].

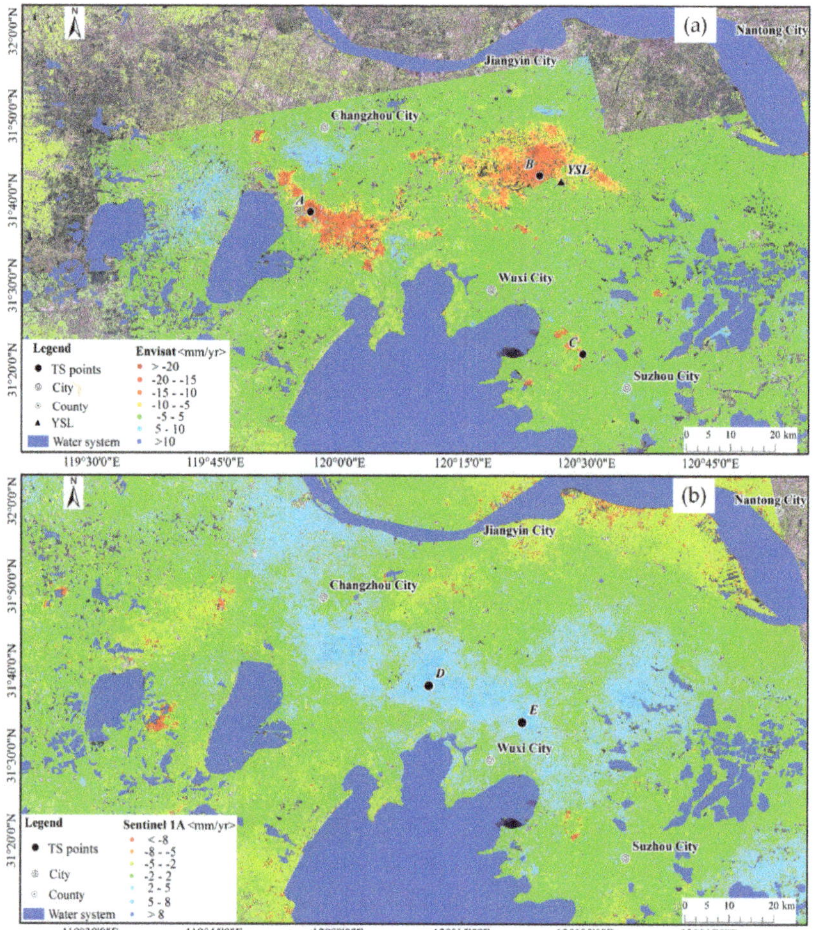

Figure 5. Su-Xi-Chang area deformation rates derived from SAR data. InSAR results inferred from (**a**) E nvisat T39 and Envisat T268, and (**b**) Sentinel 1A.

Three time-series (TS) points were selected from the subsidence regions (Wuxi, Suzhou, and Changzhou) to obtain the time-series deformation, as shown in Figures 5a and 6a. From November 2007 to May 2010, point A, which is located in the settlement center of Changzhou, had a cumulative settlement of 68 mm; point B was located in the settlement center of Wuxi, south of JY, with a cumulative settlement of 67 mm; and point C was in Suzhou City, with a cumulative settlement of 45 mm.

After examining the deformation monitoring results from 2018 to 2020 (Figure 5b), we found that land subsidence in the entire Su-Xi-Chang area was effectively controlled, and a large area of surface rebound occurred in Changzhou and Wuxi. The maximum rebound rate was approximately 10 mm/year. To analyze the deformation process of the ground during this period, two TS points were selected (Figure 5b); specifically, D and E, between

Changzhou and Wuxi. The TS points showed that the surface deformation during this period slowed down, and there was a slight rebound after July each year. This may be related to an increase in rainfall and short-term groundwater supplementation (Figure 6b).

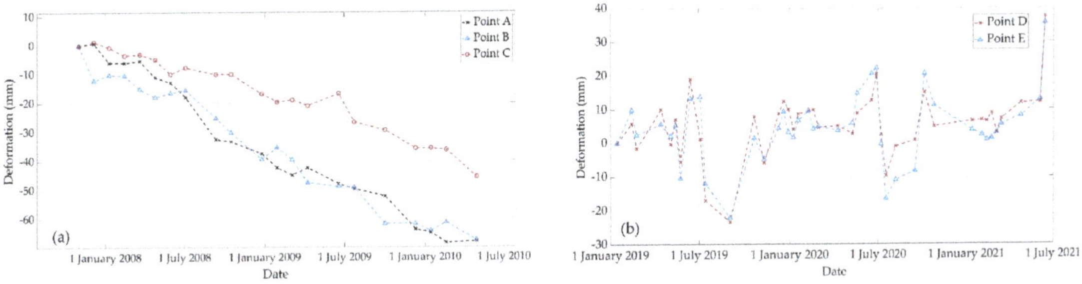

Figure 6. Time-series deformation information at the typical deformation area. (**a**) Obtained from Envisat T39 (Points A, B, and C are displayed in Figure 5a). (**b**) Obtained from Sentinel 1A (Point D and E are displayed in Figure 5b).

Since the external observation materials applied to verify the InSAR observation results are lacking, such as GNSS observation or leveling data, an internal compliance accuracy assessment is utilized to evaluate the quality of the InSAR monitoring results [44,45]. As the two sets of Envisat ASAR data obtained from different orbits are consistent over time, the monitoring results of their common areas corresponding to the same pixel are extracted and compared. Since the results of the InSAR monitoring are along the line of sight (LOS) of the sensor, the InSAR results obtained at different incidence angles differ from each other in theory [46–48]. Before comparing the two sets of data, the monitoring results must be transferred in the vertical direction according to the incident angle of the sensor using the following equation:

$$d_u = d_{los} / \cos\theta \quad (1)$$

where the θ represents the incidence angle and d_u and d_{los} are the deformations in the vertical and LOS directions, respectively. The vertical deformation rate calculated from the two sets of Envisat data was linearly fitted (Figure 7a) and the correlation coefficient was 0.88. The difference in the vertical deformation rate values of the common pixels between the two sets of Envisat ASAR data is shown in Figure 7b. The absolute value of the difference between the two sets of data, which was less than 5 mm, was 95%. These two results have a high degree of consistency, which also demonstrates the reliability of our results.

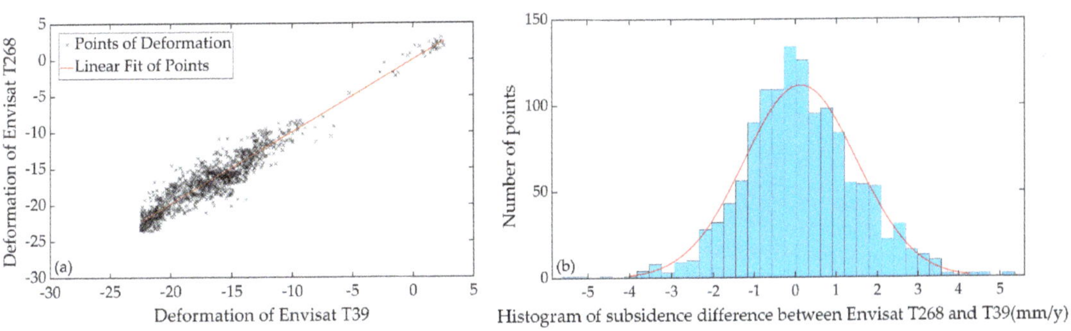

Figure 7. Internal accuracy verification. (**a**) Scatter plot and linear fit of deformation rate for the two sets of results, and (**b**) difference between the two sets of results in the vertical direction.

4.2. Land Subsidence Analysis

Through the deformation zone, Profile S1 was created to observe the relationship between the settlement funnel and the undulation of the bedrock, as shown in Figure 8. The surface deformation rate and bedrock depth along S1 were extracted as shown in Figure 9a. We can intuitively see that compared with shallow areas, the land subsidence is more serious where the bedrock is deeply buried. In other words, the trend of the deformation field was in good agreement with the undulation of the bedrock. The distribution of subsidence areas was spatially consistent with the distribution of ancient river channels (Figure 8). This verifies that fluctuations in bedrock control the formation and development of ground subsidence disasters.

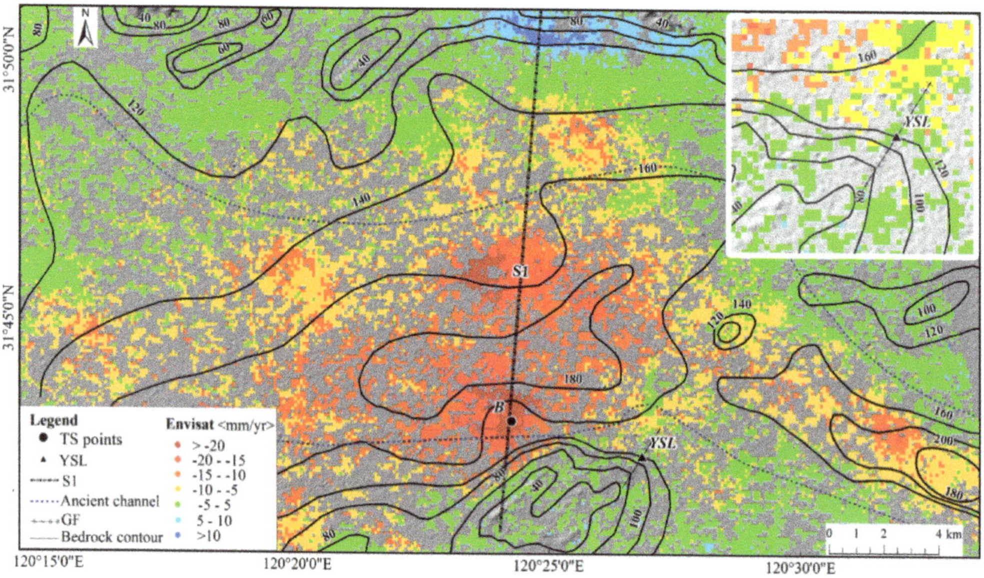

Figure 8. The land subsidence field in southern Jiangyin city (the small figure in the upper right corner is an enlarged view of the ground fissure area).

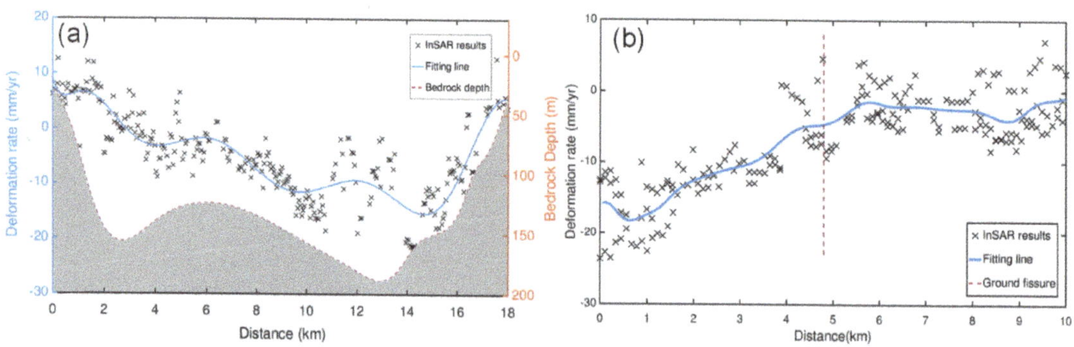

Figure 9. Profile analysis. (**a**) Comparison of InSAR results with the variation of bedrock undulations along profile S1 whose position is indicated in Figure 8. (**b**) Comparison of deformation information on both sides of the ground fissure.

4.3. Ground Fissure Analysis

The first GF in the Su-Xi-Chang plain area was formed in 1989, and GF disasters increased every year before 2004 [49]. The specific spatial distribution of GFs in this area is disarranged and closely related to paleogeography. Among the many paleo-geographic factors, the bedrock buried hill is the dominant one [28]. According to existing investigations, GFs in the YSL, GM Village, corresponding to the bedrock ridge in the area, have the following similarities with other GFs in Changzhou and Wuxi: the geological background is similar, the ground subsidence is severe in places where the GF disasters occurred and the time of their occurrence was similar [22,28]. Therefore, only representative and typical YSL GFs were analyzed, and numerical simulations and analyses were carried out.

The GFs in the YSL are shown as clusters of GFs, with the main cracks trending at 48°, approximately 100 m in length, and 20 m in width [28]. To analyze the causes and development of GFs in this area, we extracted InSAR monitoring data within a range of 5 km on each side of the GF. As the GF is too small relative to the study area, we used the black dashed line to indicate the location and direction of the GF (Figure 9b).

In Figure 9b, the black 'x' symbol represents the InSAR results, the blue solid line is obtained by fitting the InSAR results, and the red dashed line represents the location of the GF. We can intuitively see that the gradient of the settlement rate on both sides of the GF has changed significantly, the footwall of the GF is relatively stable (i.e., the part on the right side of the dashed line), and the settlement rate is relatively slow, approximately 5 mm/year. The settlement rate of the hanging wall is relatively large (i.e., the part on the left side of the dotted line)—above 15 mm/year—and locally exceeds 20 mm/year. The monitoring results proved that differential settlement led to the development of the GFs.

5. Deformation Modeling

Long-term groundwater exploitation is an essential factor in inducing GFs. Fluctuation of the bedrock surface is the internal cause of the birth of GFs. Excessive exploitation of groundwater is a direct condition [13,25]. To understand the impact of pumping activities on GFs above bedrock undulations, we conducted a numerical simulation of the process using the Moore–Coulomb model with the FLAC3D software.

5.1. Model Establishment

We simulated the influence of groundwater extraction activities on the formation and development of GFs based on the bedrock bulge. First, a simple model was built using the built-in extrusion module of FLAC3D software. The model was 400 m in length, 150 m in depth and 250 m in height. The undulations of the buried hill in the model were designed based on an existing geological map (Figure 10a). The pumping well was set above the highest point of the bedrock. The depth of the pumping well was set to 90 m, and the pumping strength was 15 m^3/s. As shown in Figure 10b, the model was divided into four layers: the first layer was the loess layer and the second was the clay layer; the third one was composed of sand and the fourth was bedrock. Due to the lack of detailed geological exploration data in this area, we designed the specific parameters of each soil layer in our experimental model (Table 2), referring to the published research results [50–52] and investigation reports [13,28,35] in this area. Figure 10c shows the displacement boundary conditions. Blue, green and red represent fixed X-direction, Y-direction and Z-direction displacements, respectively.

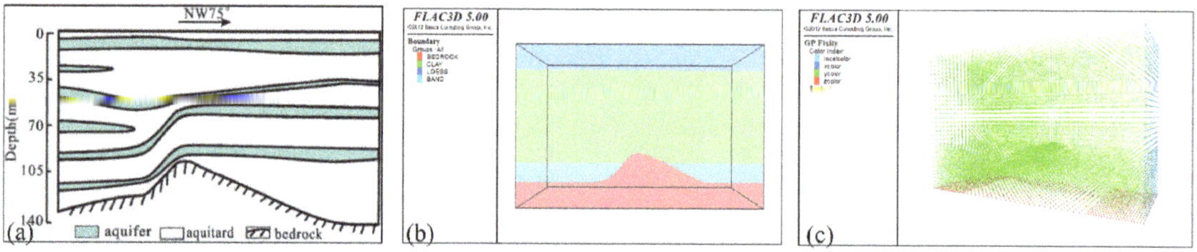

Figure 10. (**a**) Stratigraphic cross section. (**b**) Model based on actual bedrock fluctuations. (**c**) Boundary condition setting.

Table 2. Physical parameters of each soil layer.

Parameter	Value			
	Loess	**Clay**	**Sand**	**Bedrock**
Cohesion, c (kPa)	35	800	39	4620
Dilation angle, Ψ (°)	14	9.5	12.5	22.5
Internal friction angle, φ (°)	28	19	25	45
Tensile strength, T (kPa)	0	0	400	4500
Permeability coefficient, k ($m^2/Pa/s$)	5.1×10^{-10}	0.3×10^{-7}	2.04×10^{-9}	1.02×10^{-15}
Elastic modulus, E (MPa)	40	150	120	5400
Porosity, n	0.48	0.38	0.4	0.1
Poisson's ratio, υ	0.35	0.3	0.23	0.29

The bulk modulus K and shear modulus G must be calculated from the elastic modulus E and Poisson's ratio ν:

$$\text{bulk modulus}: \ K = \frac{E}{3(1-2\nu)} \tag{2}$$

$$\text{shear modulus}: \ G = \frac{E}{2(1+\nu)} \tag{3}$$

5.2. State of Equilibrium before Pumping

Figure 11a shows the initial stress field, and the color scale on the left shows the magnitude of the stress. The negative values indicate compressive stress and the positive values indicate tensile stress. It is the effect of pressure accumulation caused by the action of gravity on the rock and soil that causes the initial stresses to all be compressive stresses. Furthermore, we found that the vertical downward pressure on the left side of the ridge was greater than that on the right. This is because the overlying soil layers on the left side of the bedrock are more affected by gravity than those on the right side and have a tendency to produce differential settlement. This also shows that undulation of the bedrock is the internal cause of the GFs.

The same conclusion from the initial displacement field can also be obtained. (Figure 11b,c). The X-direction displacement on the left side of the bedrock protrusion was negative, indicating that the soil on the left had a displacement away from the bedrock protrusion. Similarly, the soil on the right was displaced away from the bedrock protrusion. The Z-direction displacement of the soil on both sides of the bedrock protrusion was negative, which indicates a downward displacement. Combining the displacement in the X and Z directions, the ground above the bedrock protrusion will be an area highly prone to GFs, which also verifies that the undulation of the bedrock surface is the internal cause of GFs.

Figure 11d shows the initial seepage field; the color scale on the left indicates the amount of seepage flow and the direction of the arrow indicates the direction of seepage. It shows that the seepage in the upper soil was mainly concentrated in the lower soil, indicating that the movement of pore water in the middle and upper layers is less affected by external forces. Moreover, the seepage flow on the upper surface of the bedrock was almost zero, which indicates that the bedrock has a blocking effect on the plane distribution of the aquifer. However, this isolation was not absolute, and a small amount of pore water was still observed to be moving along the model boundary.

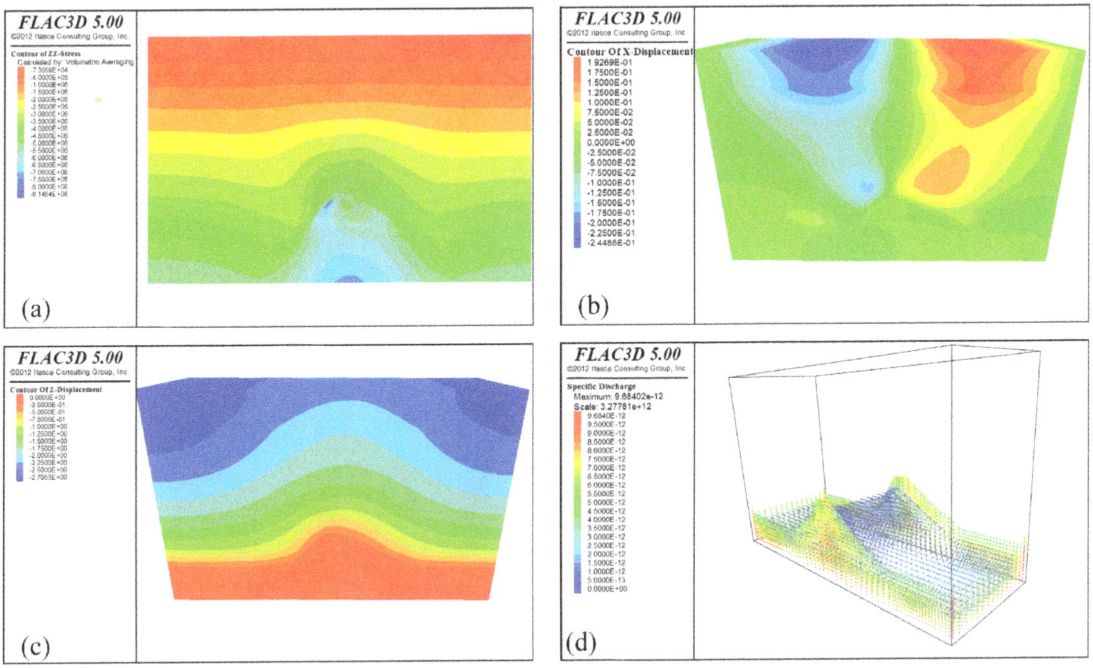

Figure 11. The initial state of equilibrium. (**a**) Initial stress field. (**b**) Initial displacement field (X direction). (**c**) Initial displacement field (Z direction). (**d**) Initial seepage field.

5.3. Pumping Simulation

Figure 12a,b show the stress distribution during the pumping activity. The magnitude of the x- and z-directions stress is increased compared with the initial equilibrium state, indicating that the pumping activity causes the stress on the soil particles around the well to change significantly. Figure 12c shows the total displacement, with the arrow indicating the direction of the displacement, indicating that pumping activities have aggravated the surface deformation near the pumping well. Figure 12d shows the displacement component in the X-direction; we found that the soil was displaced away from the bedrock bridge and that the deformation area on the left side of the buried hill is larger than that on the right side, indicating that the top of the bedrock ridge has become an area prone to ground fissures. Figure 12e shows that the vertical compression around the pumping well is more serious, and the settlement on the left side is greater at the same level. As is shown in Figure 12f, in the seepage field in an equilibrium state after the pumping activity, it can be seen from the color scale on the left that the maximum seepage flow rate was lower than that in the initial seepage field, indicating that decreases in the aquifer water and the compression of the soil layer reduced the seepage velocity after the pumping.

According to the principle of effective stress, when the pore water pressure changes, the porosity between the soil skeleton increases after the pore water is discharged. Due to the large horizontal seepage in the horizontal direction of the pumped aquifer, the seepage force will drive the solid particles to migrate and compact, which is macroscopically manifested as the horizontal displacement of the formation. In the vertical direction, the aquifer loses water. When the total stress remains unchanged, the stress borne by the pore water decreases, and the stress borne by the soil skeleton will increase, resulting in the increase of effective stress, which will compress the soil particles downward to decrease the porosity, form vertical displacement in the stratum, and macroscopically show vertical settlement; the seepage activity of the pore water in the underlying soil was restrained. Therefore, the seepage rate was lower than that before pumping.

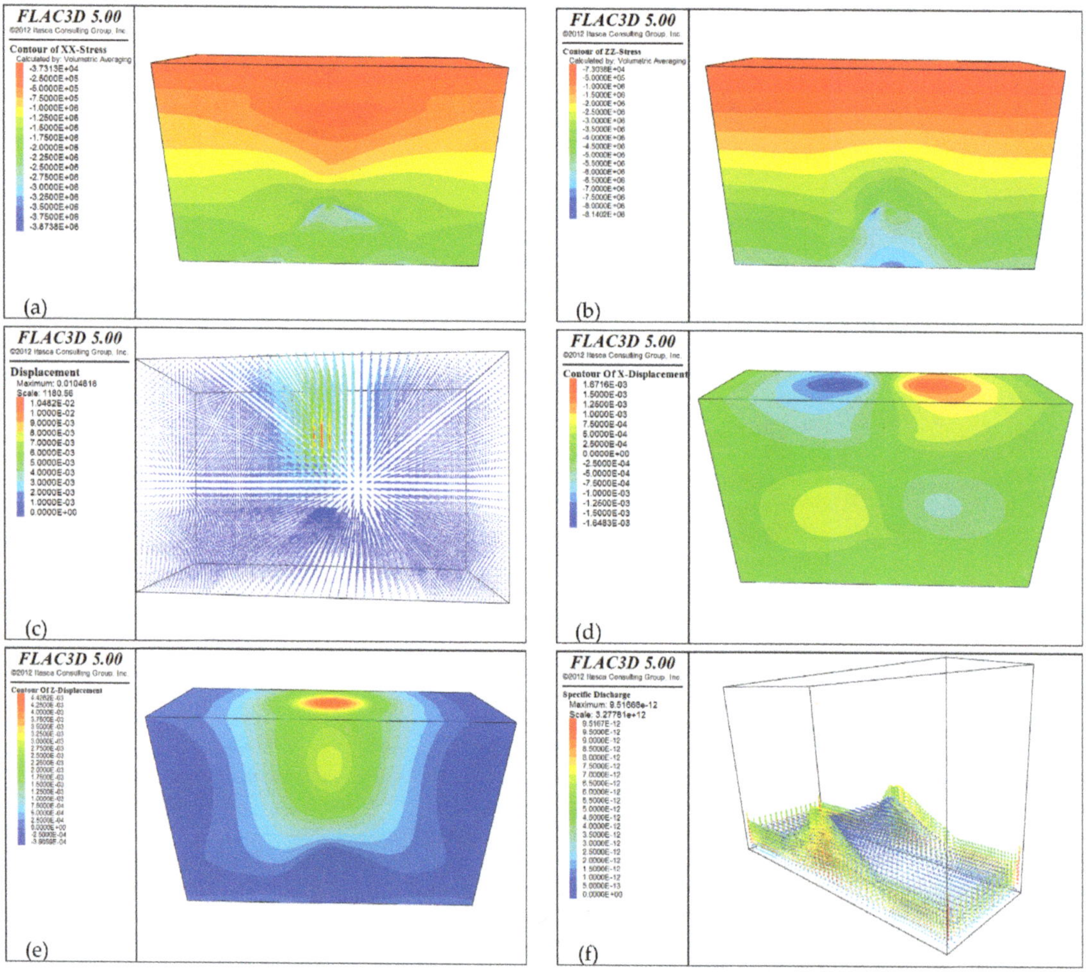

Figure 12. Distribution of stress field, deformation field, and seepage field after pumping. (**a**) Stress field distribution (x direction). (**b**) Stress field distribution (z direction). (**c**) Total displacement. (**d**) Displacement field distribution (x direction). (**e**) Displacement field distribution (Z direction). (**f**) Pumping seepage field.

6. Discussion

6.1. Deformation and Geological Factors

The Su-Xi-Chang area has suffered severe land subsidence due to the long-term over-exploitation of groundwater. Differences in bedrock undulations, quaternary sedimentary structures, and hydrogeological conditions are the internal conditions that trigger GF disasters. The long-term excessive exploitation of groundwater is an external factor that induces

GF disasters. The excessive extraction of groundwater by human activities causes the compression of soil particles, forming an underground sedimentation funnel and resulting in ground subsidence [30].

By comparing the relationship between bedrock undulations and the InSAR results, we can intuitively see that bedrock undulations control land subsidence. In addition, the settlement field in southern JY had a large degree of correlation with the distribution of ancient river channels (Figure 8). The monitoring results from 2007 to 2010 showed that although the extraction of groundwater was banned in 2005, subsidence was still serious and rapidly developing. However, with the implementation of water prohibition measures, land subsidence in most areas of the Su-Xi-Chang area improved. By observing the InSAR results from 2018 to 2021, we found that land subsidence in the Su-Xi-Chang area has stabilized as a whole, and there has been an obvious rebound in some areas.

6.2. Ground Fissure Activity and Pumping

Field investigations have shown that the YSL GF is located just at the uplift of the bedrock, and because of the uneven distribution of the aquifer, under the action of pumping water in this area, there is too much water loss in the thick aquifer and less water loss in the thin aquifer, resulting in uneven differential settlement [22,28]. Subsequently, GF disasters occurred in this area.

Through the monitoring results of the InSAR data, we intuitively found that the area had obvious differential settlements on both sides of the GF. The footwall of the GF is relatively stable, but it still has a settlement rate of 5 mm/year. The settlement of the hanging wall was more serious, with a deformation rate difference of nearly 15 mm/year with the footwall; this verifies the conclusions of the aforementioned field investigation. Through the analysis of deformation rate from 2018 to 2021, the difference of overall deformation rate in southern Jiangyin is less than 5 mm/year (Figure 5b), and the ground deformation has been basically restrained. Due to the small deformation magnitude, the current situation of ground fissures is not analyzed.

Land subsidence and the formation of GFs in this area are mainly controlled by bedrock and affected by pumping activities, which accelerate the process of land subsidence. Through the process of FLAC3D numerical simulation and analysis, we found that because of the shape of the bedrock undulation in the area and the uneven thickness of the soil on both sides of the bedrock, the soil tends to compress and become uneven under the action of its own weight, and there is a potential danger of uneven settlement. Pumping activities accelerated and aggravated this process, leading to GF disasters.

7. Conclusions

To analyze the land subsidence and GF activities in the Su-Xi-Chang area, we obtained the ground deformation characteristics at different time periods using the time-series InSAR technique. The maximum land subsidence rate exceeded 25 mm/year from 2007 to 2010. By extracting the characteristic points from the three subsidence centers in the Su-Xi-Chang area, we found that land subsidence in Suzhou was relatively well controlled. The land subsidence in Changzhou and Wuxi was serious, and the cumulative subsidence at the characteristic points exceeded 75 mm. It is verified that even after the prohibition of groundwater extraction in this area, the land subsidence disaster was still serious in the following years owing to the lag effect of aquifer recovery.

Through the comparison between the deformation rate and the corresponding bedrock depth, it is concluded that areas with a relatively thick soil layer have larger compressible space, and so the corresponding deformation rate may also be larger. We also found that formation and development are related to the spread of the ancient channel. By comparing the deformation rate of the hanging and foot walls of the GF, we proved that the direct cause of the ground fissure is differential settlement (Figure 9). Land subsidence in most areas of the Su-Xi-Chang area has decelerated, and a large area of rebound has occurred from 2018 to 2021; the rebound rate has been approximately 10 mm/year. This verifies the effectiveness of the prohibition of groundwater extraction.

Through the simulation of pumping activities using FLAC3D, we proved that at the bedrock buried hill, the soil is prone to ground fissure disaster under the influence of its own gravity. By comparing the equilibrium state before and after pumping activities, we found that pumping activities accelerate and aggravate the process of land subsidence. Pumping activity leads to the release of water from soil particles, and the soil layer compresses and consolidates with increases in the effective stress of soil particles. Due to the uneven distribution of the soil layer caused by the existence of bedrock buried hill, differential settlement easily occurs in the process of pumping, which makes the upper part of bedrock buried hill an area prone to ground fissures.

Unfortunately, we have not been able to obtain more detailed geological data on this area from public data, such as the thickness of aquifers and a larger range of bedrock undulations. Therefore, we were not able to perform further analyses in some areas. Furthermore, when FLAC3D is used for numerical simulations, only simple models can be built. Hence, there was a difference between the simulated and real results. These limitations necessitate that in-depth research be conducted in the future.

Author Contributions: C.Y. conceived and designed the experiments; C.Y. and S.L. performed the experiments and drafted the manuscript; Z.H., Q.Z., T.L. and C.Z. contributed to the discussion of the InSAR results. All authors have read and agreed to the published version of the manuscript.

Funding: This research was funded by the National Natural Science Foundation of China (NSFC), (Nos. 42174032, 41731066).

Institutional Review Board Statement: Not applicable.

Informed Consent Statement: Not applicable.

Data Availability Statement: The sentinel 1 data in the experiment were obtained by ESA Copernicus data hub. SRTM1 data can be downloaded at https://earthexplorer.usgs.gov/ (accessed on 28 December 2021).

Acknowledgments: We also thank JPL/Caltech for the ROI PAC, TU-Delft for DORIS, and Andy Hooper for StaMPS. We are grateful for the helpful corrections and suggestions of the anonymous reviewers.

Conflicts of Interest: The authors declare no conflict of interest.

References

1. Chaussard, E.; Wdowinski, S.; Cabral-Cano, E.; Amelung, F. Land subsidence in central Mexico detected by ALOS InSAR time-series. *Remote Sens. Environ.* **2014**, *140*, 94–106. [CrossRef]
2. Hernandez-Marin, M.; Pacheco-Martinez, J.; Ramirez-Cortes, A.; Burbey, T.J.; Ortiz-Lozano, J.A.; Zermeno-de-Leon, M.E.; Guinzberg-Velmont, J.; Pinto-Aceves, G. Evaluation and analysis of surface deformation in west Chapala basin, central Mexico. *Environ. Earth Sci.* **2014**, *72*, 1491–1501. [CrossRef]
3. Ojha, C.; Werth, S.; Shirzaei, M. Recovery of aquifer-systems in southwest US following 2012–2015 drought: Evidence from InSAR, GRACE and groundwater level data. *J. Hydrol.* **2020**, *587*, 16. [CrossRef]
4. El Kamali, M.; Papoutsis, I.; Loupasakis, C.; Abuelgasim, A.; Omari, K.; Kontoes, C. Monitoring of land surface subsidence using persistent scatterer interferometry techniques and ground truth data in arid and semi-arid regions, the case of Remah, UAE. *Sci. Total Environ.* **2021**, *776*, 11. [CrossRef] [PubMed]
5. Peng, J.B.; Xu, J.S.; Ma, R.Y.; Wang, F.Y. Characteristics and mechanism of the Longyao ground fissure on North China Plain, China. *Eng. Geol.* **2016**, *214*, 136–146. [CrossRef]
6. Xu, J.; Meng, L.; An, H.; Wang, L. The bending mechanism of Anping ground fissure in the Hebei Plain, North China. *Environ. Earth Sci.* **2015**, *74*, 6859–6870. [CrossRef]
7. Chen, B.B.; Gong, H.L.; Chen, Y.; Lei, K.C.; Zhou, C.F.; Si, Y.; Li, X.J.; Pan, Y.; Gao, M.L. Investigating land subsidence and its causes along Beijing high-speed railway using multi-platform InSAR and a maximum entropy model. *Int. J. Appl. Earth Obs. Geoinf.* **2021**, *96*, 15. [CrossRef]
8. Li, M.-G.; Chen, J.-J.; Xu, Y.-S.; Tong, D.-G.; Cao, W.-W.; Shi, Y.-J. Effects of groundwater exploitation and recharge on land subsidence and infrastructure settlement patterns in Shanghai. *Eng. Geol.* **2021**, *282*, 105995. [CrossRef]
9. Xiong, Z.M.; Wang, Y.W.; Chen, X.; Xiong, W.Y. Seismic behavior of underground station and surface building interaction system in earth fissure environment. *Tunn. Undergr. Space Technol.* **2021**, *110*, 14. [CrossRef]

10. Cigna, F.; Tapete, D. Sentinel-1 big data processing with P-SBAS InSAR in the geohazards exploitation platform: An experiment on coastal land subsidence and landslides in Italy. *Remote Sens.* **2021**, *13*, 885. [CrossRef]
11. Fabris, M. Monitoring the coastal changes of the Po River delta (Northern Italy) since 1911 using archival cartography, multi-temporal aerial photogrammetry and LiDAR data: Implications for coastline changes in 2100 AD. *Remote Sens.* **2021**, *13*, 529. [CrossRef]
12. Wang, J. *Theory and Application of Ground Fissures and Their Disasters*; Shaanxi Science and Technology Press: Xi'an, China, 2000.
13. Xu, J.; Peng, J.; An, H. Experimental study on Su-Xi-Chang earth fissures induced by repeated groundwater pumping and impounding. *Geomat. Nat. Hazards Risk* **2019**, *10*, 2051–2068. [CrossRef]
14. Hu, J.P. A Study on the Land Subsidence Effect After Prohibiting Extraction of Ground Water in Suzhou-Wuxi-Changzhou Area. Ph.D. Thesis, Nanjing University, Nanjing, China, 2011.
15. Budhu, M. Earth Fissure formation from the mechanics of groundwater pumping. *Int. J. Geomech.* **2011**, *11*, 1–11. [CrossRef]
16. Jia, Z.J.; Qiao, J.W.; Peng, J.B.; Lu, Q.Z.; Xia, Y.Y.; Zang, M.D.; Wang, F.Y.; Zhao, J.Y. Formation of ground fissures with synsedimentary characteristics: A case study in the Linfen Basin, northern China. *J. Asian Earth Sci.* **2021**, *214*, 104790. [CrossRef]
17. Nespoli, M.; Cenni, N.; Belardinelli, M.E.; Marcaccio, M. The interaction between displacements and water level changes due to natural and anthropogenic effects in the Po Plain (Italy): The different point of view of GNSS and piezometers. *J. Hydrol.* **2021**, *596*, 126112. [CrossRef]
18. Tiwari, A.; Narayan, A.B.; Dwivedi, R.; Swadeshi, A.; Pasari, S.; Dikshit, O. Geodetic investigation of landslides and land subsidence: Case study of the Bhurkunda coal mines and the Sirobagarh landslide. *Surv. Rev.* **2020**, *52*, 134–149. [CrossRef]
19. Zhang, Y.Q.; Gong, H.L.; Gu, Z.Q.; Wang, R.; Li, X.J.; Zhao, W.J. Characterization of land subsidence induced by groundwater withdrawals in the plain of Beijing city, China. *Hydrogeol. J.* **2014**, *22*, 397–409. [CrossRef]
20. Damoah-Afari, P. Detecting Ground Settlement of Megacities Using InSAR Techniques. Ph.D. Thesis, Hong Kong Polytechnic University, Hong Kong, China, 2009.
21. Wan, J.; Bin, L.I.; Tan, C.; Zhang, P.; Feng, C. Characteristics and mechanism of earth fissures in China: A case study of Fenhe River-Weihe River basin, Hebei plain and Suzhou-Wuxi-Changzhou plain. *J. Geol. Rev.* **2019**, *65*, 1383–1396. [CrossRef]
22. Jian-Qiang, W.U.; Zhou, G.J.; Shu-Liang, W.U.; Wei, L.I.; Ming, W. Research on earth fissures in Guangming Village of Wuxi City. *J. Geol.* **2013**, *37*, 203–207.
23. Shi, X.-Q.; Xue, Y.-Q.; Ye, S.-J.; Wu, J.-C.; Zhang, Y.; Yu, J. Characterization of land subsidence induced by groundwater withdrawals in Su-Xi-Chang area, China. *Environ. Geol.* **2007**, *52*, 27–40. [CrossRef]
24. Bu, J.W.; Sun, Z.Y.; Ma, R.; Liu, Y.D.; Gong, X.L.; Pan, Z.; Wei, W.H. Shallow groundwater quality and its controlling factors in the Su-Xi-Chang region, Eastern China. *Int. J. Environ. Res. Public Health* **2008**, *17*, 1267. [CrossRef] [PubMed]
25. Wang, G.-Y.; Zhu, J.-Q.; Zhang, D.; Wu, J.-Q.; Yu, J.; Gong, X.-L.; Gou, F.-G. Land subsidence and uplift related to groundwater extraction in Wuxi, China. *Q. J. Eng. Geol. Hydrogeol.* **2020**, *53*, 609–619. [CrossRef]
26. Gu, K.; Shi, B.; Liu, C.; Jiang, H.; Li, T.; Jinghong, W. Investigation of land subsidence with the combination of distributed fiber optic sensing techniques and microstructure analysis of soils. *Eng. Geol.* **2018**, *240*, 34–47. [CrossRef]
27. Zhao, W.-T.; Li, L. The mechanism of land subsidence and its prevention measures in Suzhou-Wuxi-Changzhou area. Chinese. *J. Geol. Hazard Control* **2009**, *20*, 88–93. [CrossRef]
28. Gu, C. Experimental Study on the Development and Evolution of Ground Fissures Caused by Extraction of Groundwater in Suzhou-Wuxi-Changzhou Region. Master's Thesis, China University of Mining and Technology, Xuzhou, China, 2017.
29. Hooper, A.; Bekaert, D.; Spaans, K.; Arikan, M. Recent advances in SAR interferometry time series analysis for measuring crustal deformation. *Tectonophysics* **2012**, *514–517*, 1–13. [CrossRef]
30. Wang, G.-Y.; You, G.; Zhu, J.-Q.; Yu, J.; Gong, X.-L.; Wu, J.-Q. Investigations of Changjing earth fissures, Jiangyin, Jiangsu, China. *Environ. Earth Sci.* **2016**, *75*, 502. [CrossRef]
31. Wu, Z.; Yu, Q.; Zhang, Y. Forming process of earth fissure hazard in the Suzhou-Wuxi-Changzhou area. *Hydrogeol. Eng. Geol.* **2003**, *30*, 67–72. [CrossRef]
32. Yang, Y. The Influence of the Land Subsidence Around Su-Xi-Chang Area to the Bejing-Shanghai High-Speed Railway. Master's Thesis, Southwest Jiaotong University, Nanjing, China, 2012.
33. Yuan, Y.; Zhang, T. Prediction of ground fissures risk in Suxichang area based on principal component analysis and support vector machine. *J. Catastrophol.* **2019**, *34*, 57–63. [CrossRef]
34. Chen, Y.; Zhao, W.; Huang, Y.; Jia, P. Investigation of land subsidence based on the column element settlement model in a soft-soil area. *Geofluids* **2019**, *2019*, 9530135. [CrossRef]
35. Liu, Y.; Zhang, D.; Wang, G.Y.; Liu, C.; Zhang, Y. Discrete element method-based prediction of areas prone to buried hill-controlled earth fissures. *J. Zhejiang Univ. Sci. A* **2019**, *20*, 794–803. [CrossRef]
36. Gabriel, A.K.; Goldstein, R.M.; Zebker, H.A. Mapping small elevation changes over large areas: Differential radar interferometry. *J. Geophys. Res.* **1989**, *94*, 9183–9191. [CrossRef]
37. Zebker, H.A.; Rosen, P.A.; Goldstein, R.M.; Gabriel, A.; Werner, C.L. On the derivation of coseismic displacement fields using differential radar interferometry: The Landers earthquake. *J. Geophys. Res.* **1994**, *99*, 19617–19634. [CrossRef]
38. Rosen, P.A.; Henley, S.; Peltzer, G.; Simons, M. Updated repeat orbit interferometry package released. *Eos Transact. Am. Geophys. Union* **2013**, *85*, 47. [CrossRef]
39. Kampes, B.M. *Radar Interferometry: Persistent Scatterer Technique*; Springer: Dordrecht, The Netherlands, 2006.

40. Hooper, A.; Bekaert, D.; Spaans, K. *StaMPS/MTI Manual Version 3.3 b1*; University of Leeds: Leeds, UK, 2013.
41. Likoka, A.S. Atmospheric effects estimation on SAR images using a pure statistical method. In *Computational Imaging II, Proc. SPIE 10222*; SPIE: Bellingham, WA, USA, 2017. [CrossRef]
42. Budhu, M. Mechanics of earth fissures using the Mohr–Coulomb failure criterion. *Environ. Eng. Geosci.* **2008**, *14*, 281–295. [CrossRef]
43. Martinez-Noguez, I.; Hinkelmann, R. Land subsidence caused by a single water extraction well and rapid water infiltration. In *Prevention and Mitigation of Natural and Anthropogenic Hazards due to Land Subsidence*; Daito, K., Galloway, D., Eds.; Copernicus Gesellschaft Mbh: Gottingen, Germany, 2015; pp. 33–38. [CrossRef]
44. Yang, C.; Zhang, F.; Liu, R.; Hou, J.; Zhao, C. Ground deformation and fissure activity of the Yuncheng Basin (China) revealed by multiband time series InSAR. *Adv. Space Res.* **2020**, *66*, 490–504. [CrossRef]
45. Qu, F.; Zhang, Q.; Lu, Z.; Zhao, C.; Yang, C.; Zhang, J. Land subsidence and ground fissures in Xi'an, China 2005–2012 revealed by multi-band InSAR time-series analysis. *Remote Sens. Environ.* **2014**, *155*, 366–376. [CrossRef]
46. Bechor, N.; Zebker, H.A. Measuring two-dimensional movements using a single InSAR pair. *Geophys. Res. Lett.* **2006**, *33*, 275–303. [CrossRef]
47. Samsonov, S.; Tiampo, K. Analytical optimization of a DInSAR and GPS dataset for derivation of three-dimensional surface motion. *IEEE Geosci. Remote Sens. Lett.* **2006**, *3*, 107–111. [CrossRef]
48. Wright, T.J.; Parsons, B.E.; Lu, Z. Toward mapping surface deformation in three dimensions using InSAR. *Geophys. Res. Lett.* **2004**, *31*, L01607. [CrossRef]
49. Wang, Y.; Shujun, Y.E.; Jun, Y.U.; Gong, X.; Hydrosciences, D.O. Features and mechanisms of earth fissures induced groundwater withdrawal in China. *Geol. J. China Univ.* **2016**, *22*, 741. [CrossRef]
50. Li, C.-Q.; Liang, Q.G.; Xu-Yang, W.U.; Wang, L.L.; Shan-Chang, X.U. Study on the test of tensile strength of remolded loess. *China Earthq. Eng. J.* **2014**, *233–238*, 248. [CrossRef]
51. Yao, Y.; Zhang, B.; Zhu, J. Behaviors, constitutive models and numerical simulation of soils. *Tumu Gongcheng Xuebao/China Civil Eng. J.* **2012**, *45*, 127–150. [CrossRef]
52. Zhang, Y. Study on The Engineering Mechanic Properties of Loess-Paleosol. Master's Thesis, Xi'an University of Architecture and Technology, Xi'an, China, 2011.

remote sensing

MDPI

Article

InSAR Modeling and Deformation Prediction for Salt Solution Mining Using a Novel CT-PIM Function

Xuemin Xing [1,2,3], Tengfei Zhang [1,3], Lifu Chen [1,4,*], Zefa Yang [5], Xiangbin Liu [1,3], Wei Peng [1,3,5] and Zhihui Yuan [1,4]

1 Laboratory of Radar Remote Sensing Applications, Changsha University of Science & Technology, Changsha 410114, China; xuemin.xing@csust.edu.cn (X.X.); tengfei@stu.csust.edu.cn (T.Z.); liuxb0219@foxmaul.com (X.L.); pengwei@csust.edu.cn (W.P.); yuanzhihui@csust.edu.cn (Z.Y.)
2 Hunan Key Laboratory of Remote Sensing of Ecological Environment in Dongting Lake Area, Changsha 410114, China
3 School of Traffic and Transportation Engineering, Changsha University of Science & Technology, Changsha 410114, China
4 School of Electrical and Information Engineering, Changsha University of Science & Technology, Changsha 410114, China
5 School of Geosciences and Info-Physics, Central South University, Changsha 410083, China; yangzf@csu.edu.cn
* Correspondence: lifu_chen@csust.edu.cn

Abstract: Deformation prediction for a salt solution mining area is essential to mining environmental protection. The combination of Synthetic Aperture Radar Interferometry (InSAR) technique with Probability Integral Method (PIM) has proven to be powerful in predicting mining-induced subsidence. However, traditional mathematical empirical models (such as linear model or linear model combined with periodical function) are mostly used in InSAR approaches, ignoring the underground mining mechanisms, which may limit the accuracy of the retrieved deformations. Inaccurate InSAR deformations will transmit an unavoidable error to the estimated PIM parameters and the forward predicted subsidence, which may induce more significant errors. Besides, theoretical contradictory and non-consistency between InSAR deformation model and future prediction model is another limitation. This paper introduces the Coordinate-Time (CT) function into InSAR deformation modeling. A novel time-series InSAR model (namely, CT-PIM) is proposed as a substitute for traditional InSAR mathematical empirical models and directly applied for future dynamic prediction. The unknown CT-PIM parameters can be estimated directly via InSAR phase observations, which can avoid the error propagation from the InSAR-generated deformations. The new approach has been tested by both simulated and real data experiments over a salt mine in China. The root mean square error (RMSE) is determined as ±10.9 mm, with an improvement of 37.2% compared to traditional static PIM prediction method. The new approach provides a more robust tool for the forecasting of mining-induced hazards in salt solution mining areas, as well as a reference for ensuring the environment protection and safety management.

Keywords: InSAR; mine; land subsidence; time series deformation; model

Citation: Xing, X.; Zhang, T.; Chen, L.; Yang, Z.; Liu, X.; Peng, W.; Yuan, Z. InSAR Modeling and Deformation Prediction for Salt Solution Mining Using a Novel CT-PIM Function. *Remote Sens.* **2022**, *14*, 842. https://doi.org/10.3390/rs14040842

Academic Editors: Chisheng Wang, Daqing Ge, Guohong Zhang, Wu Zhu and Siting Xiong

Received: 7 January 2022
Accepted: 8 February 2022
Published: 10 February 2022

1. Introduction

By the end of 2018, the reserves of mirabilite deposits in China had reached up to 117.297 billion tons [1]. For mirabilite production in the salt mining deposits, water-soluble exploitation, with comprehensively multi-propulsion through groups of drilling wells, is the primary mining mode in salt solution mining area. Attributing to the multi-direction of water-soluble mining activities, the upper layer of the rock salt cavern is prone to overburden, or even serious collapse [2]. As long as the roof of the cavern uplifts to the ground surface, a large pit, which may cause potential damage to the nearby infrastructures, will be formed [3]. In addition, accumulated water-soluble mining activities may cause significant lithological changes of the underground rock, even inducing a serious pumping of brine and water salinization. Therefore, long-term spatio-temporal deformation prediction of salt solution mines are of considerable importance to the environmental protection and safety management. The multi-temporal interferometric synthetic aperture radar (MT-InSAR) is an advanced space-to-earth observation technology developed in recent years. Compared with traditional InSAR technology, MT-InSAR technology has been reported to be more effective and applicable for deriving ground deformations (i.e., mining area or infrastructures) [4–13]. However, the single MT-InSAR has a limitation in that it can only obtain the deformation sequences during SAR acquisition dates; the subsequent future displacement beyond the span of the SAR observations cannot be acquired. To compensate for the limitation, researchers have incorporated InSAR deformation observations into the Probability Integral Method (PIM), which has successfully solved the problems, such as surface deformation space calculation and overlying strata movement prediction [14–22]. The basic idea for those studies is firstly using the traditional MT-InSAR technique to generate the line-of-sight (LOS) deformation results as input dataset to estimate the PIM parameters; then, the corresponding subsidence sequences for the future period can be predicted based on the PIM prediction model. Those researchers have successfully retrieved the spatio-temporal deformation characteristics and achieved the forward prediction of mining-induced deformations, which proves that the combination with PIM theory is an efficient and feasible complementation to single MT-InSAR technique.

However, there still exist limitations: Firstly, according to previous studies, pure empirical mathematical models (i.e., linear model or linear model combined with periodical function) are mostly utilized to generate the InSAR observations, and the underground mining subsidence mechanism is not considered in the InSAR modeling procedure. In MT-InSAR data processing, the deformation modeling procedure determines the temporal functional relationship between the interferometric phase observations and the parameters. The evolution of mining-induced surface displacement is a complex physical non-linear process, varying with time. Apparently, modeling a physical process with a single empirical mathematical function cannot accurately describe the underground deformation mechanisms and, thus, may have a great impact on the accuracy of the generated InSAR deformation observations. Secondly, inaccurate InSAR deformation results will transmit an unavoidable error to the estimated PIM parameters, then, secondly, propagating to the future subsidence prediction, leading even larger errors in the predicted displacements. Finally, the traditional mathematical models (i.e., linear model or linear model combined with periodical function) used in the deformation monitoring is not consistent with the PIM used in the future deformation prediction, which is based on random medium theory. Using InSAR deformation observations generated via a simple linear model to estimate parameters for a complex non-linear model may not be reasonable. Consequently, contradiction and inconsistency between the InSAR deformation model and the future prediction model is the third limitation.

Based on the above analysis, in order to improve the accuracy of deformation prediction, we propose a novel InSAR deformation model, which can better describe the dynamic evolution disciplines of the underground mining subsidence in the MT-InSAR deformation modeling procedure, as well as be directly used to the forward deformation prediction. The novel model is based on the Coordinate-Time (CT) function (namely CT-PIM) and used as a substitute for traditional InSAR models. CT-PIM improves the static PIM via integrating the temporal parameter to achieve the description of the temporal dynamical characteristics of the mining-induced subsidence. Considering the physical subsurface mechanism, CT-PIM can describe the temporal dynamical characteristics of the mining-induced subsidence more realistically. The PIM parameters are introduced into InSAR time-series phase functions and can be estimated via the phase observations, which can avoid the secondary error propagation from the inaccurate InSAR deformations to the subsequent deformation prediction. Subsidence can be predicted directly based on CT-PIM, which implements in opposition to the deficiency of inconsistency and lack of reasonability between the InSAR model and the forward prediction model. Consequently, the new method is expected not only to enrich the time series InSAR modeling theory but also provide a more robust tool for forecasting mining-induced hazards in mining areas.

2. Materials and Methods

The CT-PIM, including the Coordinate-Time function, and the PIM parameters are introduced in Section 2.1. The InSAR modeling based on CT-PIM is derived in Section 2.2, together with the final functional relationship between the time series phase observations and the PIM parameters.

2.1. *Dynamic Coordinate-Time Probability Integration Model*

The Coordinate-Time (CT) function was proposed in Reference [23], which improved the traditional static PIM with introducing the temporal parameter t, thus being referred as CT-PIM hereinafter. It has proven to be effective and reliable in the application for the dynamic ground deformation prediction caused by mining activities. The surface subsidence described in CT-PIM function is as follows [23]:

$$W\,(x,\,y,\,t) = \frac{1}{w_0} W\,(x,\,t) W\,(y,\,t), \tag{1}$$

where $(x,\,y)$ denotes the coordinate of the surface point, and $W\,(x,\,y,\,t)$ is the surface vertical subsidence related to mining activities at time t; w_0 is the maximum settlement, which can be written as $w_0 = mq{\cdot}cos\alpha$, where m is the thickness of the mine, q is the subsidence factor, and α is the dip angle of the layer. $W\,(x,\,t)$ and $W\,(y,\,t)$ can be expressed as follows:

(1) When the dip direction of the working face is under critical extraction, whereas the strike is under subcritical extraction. This happens in the practical condition that the advancing distance is longer than the depth along the dip direction. The ground subsidence can be written as (the schematic diagram is shown as Figure 1b) [23]:

$$W(x,t) = \begin{cases} \frac{w(x)}{2}\left[erf(\sqrt{\pi}) + erf\left(\sqrt{\pi}\frac{v_m t - nH - r}{r}\right)\right] \\ x \le nH \\ \frac{w(x)}{2}\left[erf(\sqrt{\pi}) + erf\left(\sqrt{\pi}\left[\frac{v_m t - r - x + \frac{H}{a+bx} - \frac{H(a+bx)}{3}}{r}\right]\right)\right] \\ nH < x < 1.3r \\ \frac{w_0}{2}\left[erf\left(\frac{\sqrt{\pi}}{r}\right) + erf\left(\sqrt{\pi}\frac{v_m t + H/tan\omega - 2x}{r}\right)\right] \\ 1.3r \le x \end{cases}, \tag{2}$$

where $w(x)$ can be calculated based on the principles of static PIM following the equation: $w(x) = \frac{w_0}{2}\left[erf(\sqrt{\pi}\frac{x}{r}) + 1\right]$; erf is a probability integral function: $erf(\frac{\sqrt{\pi}}{r}x) = \frac{2}{\sqrt{\pi}}\int_0^{\frac{\sqrt{\pi}}{r}x} e^{-u^2}\mathrm{du}$ (where u is the integral parameter, r is the main influence radius, $r = \frac{H}{tan\beta}$; H is the mining depth and $tan\beta$ is the tangent of the main influence angle); nH is the starting distance, describing the advancing distance of the working face when the ground point starts to move; n is the starting factor; ω is the static leading influence angle (in degrees); v_m is the mining advancing speed; and the coefficients a and b can be expressed as $a = \frac{117r - \omega nH}{1.3r - nH}$ and $b = \frac{\omega - 90}{1.3r - nH}$, respectively.

(2) When the strike direction of working face is under critical extraction, whereas the dip is under subcritical extraction. This happens in the most practical condition of the coal mining activities, since the strike direction is better to contain a longer advancing distance (even up to several kilometers, much deeper than the depth along the strike),

which can easily satisfy the critical extraction condition. The surface subsidence can be written as (the schematic diagram is shown as Figure 1c) [16]:

$$W(y, t) = \frac{w_0}{2}\left[erf\left(\sqrt{\pi}\frac{y}{r_1}\right) - erf\left(\sqrt{\pi}\frac{y-L}{r_2}\right)\right], \tag{3}$$

where y denotes the vertical coordinate of the ground point in the mining area; r_1 and r_2 represent the main influence radius along the down-dip, and up-dip directions, respectively, which can be expressed as $r_1 = \frac{H_1}{tan\beta_1}$ and $r_2 = \frac{H_2}{tan\beta_2}$. H_1 and H_2 are the mining depths along the down-dip, and up-dip directions, respectively. Here, $H_1 = H - \frac{D_1}{2}sin\alpha$ and $H_2 = H + \frac{D_1}{2}sin\alpha$; L is the estimated length of strike working face, which can be expressed as $L = (D_1 - s_1 - s_2)\frac{sin(\theta_0-\alpha)}{sin\theta_0}$; D_1 is the inclined length of working face; s_1 and s_2 are the offsets of inflection point along down-dip and up-dip directions of the working panel; θ_0 is the mining influence angle, which can be calculated following $\theta_0 = 90° - k\alpha$.

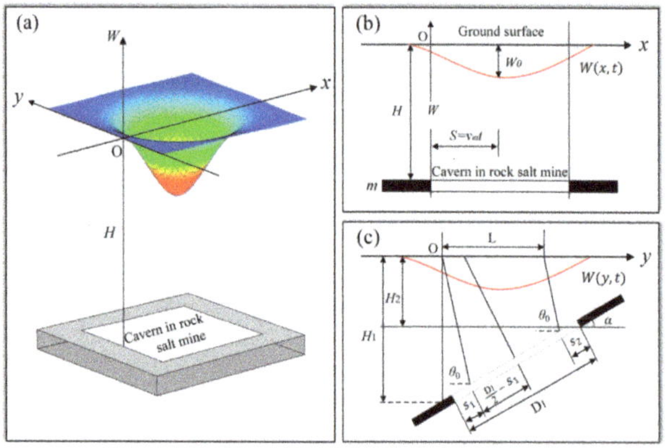

Figure 1. Schematic diagrams for the coordinate system: (**a**) 3D view of the coordinate system; (**b**) profile along strike direction; (**c**) profile along dip direction.

It is worth noting that, when the strike of the working face is under subcritical extraction, the final deformation should be estimated via multiplying a coefficient: $C_{xm} = \frac{W_{xm}}{W_0} < 1$, where $W_0 = mq{\cdot}cos\theta$. W_{xm} is the maximum value of $W(x, t)$, which can be calculated according to the first equation in Equation (2), under the assumption that the tendency has reached an extreme exploitation [16]. Similarly, when the dip of the working face is under subcritical extraction, then a dip mining degree coefficient, $C_{ym} = \frac{W_{ym}}{W_0} < 1$, should be multiplied with it, where W_{ym} denotes the maximum value of $W(y, t)$ in Equation (3).

2.2. InSAR Modeling Based on CT-PIM (InSAR-CTPIM)

For each high coherence pixel in the i-th interferogram [24–26], its corresponding interferometric phase can be expressed as:

$$\begin{aligned}\delta\varphi^i &= \delta\varphi^i_{def} + \delta\varphi^i_{topo} + \delta\varphi^i_{orbit} + \delta\varphi^i_{atm} + \delta\varphi^i_{noise} + \Delta\varphi^i_{non} \\ &\approx \frac{4\pi}{\lambda}\Delta d^i + \frac{4\pi B_i}{\lambda R sin\theta}\Delta H^i + \Delta\varphi^i_{res},\end{aligned} \tag{4}$$

where λ is the radar wavelength; Δd is the LOS accumulated deformation spanning the time period of the i-th interferogram, referred as the low-pass (LP) component hereinafter; $\Delta\varphi^i_{topo}$ represents the residual topographic phase component; $\Delta\varphi^i_{topo} = \frac{4\pi B_i}{\lambda R sin\theta}\Delta H^i$, where B_i defines the perpendicular baseline, θ the incident angle, R the sensor-target distance, and ΔH the residual elevation, which is an unknown parameter; $\Delta\varphi^i_{res}$ denotes the final residual phase, which includes the phase noise, atmospheric delay, and the high-pass (HP) deformation component [27].

During the extraction of the brine from the wellhead, the deformation of the surface caused by the change of the top pressure in the cavern is dominantly along the vertical direction. Besides, the upper dissolution rate of the drilling solution mining process is approximately twice of the side dissolution rate [28]. Therefore, compared with the vertical subsidence, the horizontal displacement caused by water-soluble mining is relatively minor, and omitted hereinafter. Then, the functional relationship between the LOS deformation and the ground subsidence can be expressed as:

$$\Delta d^i = d_{LOS}\left(t_B^i\right) - d_{LOS}\left(t_A^i\right) = [W\ (x,\ y,\ t_B) - W\ (x,\ y,\ t_A)]\cdot cos\theta, \tag{5}$$

where $W\ (x,\ y,\ t_B)$ and $W\ (x,\ y,\ t_A)$ are the corresponding dynamic subsidence at t_B and t_A, respectively. Substituting Equation (5) into (4), the functional relationship between the InSAR phases and PIM parameters can be written as:

$$\delta\varphi^i = \frac{4\pi}{\lambda}[W\ (x,\ y,\ t_B) - W\ (x,\ y,\ t_A)]\cdot cos\theta + \frac{4\pi B_i}{\lambda Rsin\theta}\Delta H^i + \Delta\varphi^i_{res}, \tag{6}$$

where the geological parameter $GP = [m,\ \alpha,\ H,\ D_1,\ \omega]$ of the rock salt mine can be determined according to the in-situ investigation of the study area or the materials provided by the mining company, which can be regarded as known parameters. The unknown parameters to be solved here are $UP = [q, tan\beta,\ s_1,\ s_2,\ k,\ n]$. Among them, q is within the range of [0.01, 1], $tan\beta$, $tan\beta_1$, and $tan\beta_2$ are all distributed within the range of [1, 3.8], and both s_1 and s_2 are within the range of [0.05H, 0.3H]. The propagation influence angle of mining is $\theta_0 = 90^\circ - k\alpha$. Here, k ranges from 0.5 to 0.8, and the starting distance coefficient n ranges from 1/7 to 1/2 [14,16,23]. Equation (6) describes the temporal functional relationship between the InSAR time series phases and the PIM parameters, which is referred as InSAR-CTPIM, and will be applied for the time series deformation generation.

2.3. CT-PIM Parameters Estimation Based on InSAR Phases

If $N+1$ scenes covering the whole study area are acquired at dates $(t_0,\ t_1,\ \ldots,\ t_N)$, then M differential interferometric pairs are generated accordingly. For each pixel $(x,\ y)$ on the differential interferograms, M InSAR functions (in Equation (6)) based on the phase observations can be established. If M is higher than six, the unknown parameters $UP = [q, tan\beta,\ s_1,\ s_2,\ k,\ n]$ can be estimated. The ill-conditioned nature of the phase functions (introduced as Equation (6)) has also been tested to ensure the robustness. For each pixel $(x,\ y)$ on the differential interferograms, M InSAR functions based on the phase observations can be established. The condition number $Cond_\infty(A) = ||A||_\infty||A^{-1}||_\infty$ is estimated as 18 here, where A defined the coefficient matrix of the phase functions. The value of $Cond_\infty(A)$ implies that no ill-conditioned phenomenon exists in the functions.

The solving of Equation (6) is obviously a non-linear parameter estimation problem. The Genetic Algorithm (GA), combined with the Simplex Method (SM), is introduced here to estimate UP. GA processes the advantages of global optimization searching ability and unrestricting to different model forms [29], and it has been proven to be feasible in solving the problems of non-linear parameter estimation [30]. It generates the optimized estimations in form of population individuals following the principle of fitness function minimization [31]. However, the disadvantages of slow convergence speed and low accuracy are also prominent for GA. SM has proven to be an efficient tool in solving the non-linear estimation problems when combined with GA, which can refine the GA generated results effectively [32]. Consequently, the combination of GA and SM, namely GASM, is adopted here to effectively estimate the unknown parameters.

As for Equation (6), each individual gene is $UP = [q, tan\beta,\ s_1,\ s_2,\ k,\ n]$, and the fitness function here is:

$$f = \left|\left|\Delta\varphi^i_{res}\right|\right| = min, \tag{7}$$

where $\Delta\varphi_{res}^{i}$ denotes the residual phase in Equation (6). After the iteration of population selection, crossing over, and evolution for each individual gene until the minimum fitness function value condition is satisfied, the final obtained individual genes, which are treated as the initial GA-solutions for PIM parameters, will be taken as the input initial values of SM. Then, SM is utilized to improve the accuracy of GA estimations after the global searching, and the corresponding SM-output results are determined as the final solutions for PIM parameters [33].

2.4. Flow Chart and Processing Steps

Figure 2 displays the processing flow of deformation prediction using the CT-PIM function. The steps are as follows:

1. Differential Interferometry according to the $N+1$ SAR images covering the study area.
2. High-coherent candidates extraction considering both the average coherence and the amplitude dispersion index.
3. InSAR modeling following the CT-PIM function, which establishes the functional relationship between InSAR phases and unknown PIM parameters.
4. PIM parameter estimation based on Genetic Algorithm and Simplex Method (GASM) [32,33], which includes GA global searching to generate the initial values of PIM parameters and SM to optimize the GA solutions.
5. Forward dynamic subsidence prediction beyond the spans of SAR acquisitions based on CT-PIM function introduced as Equation (1).

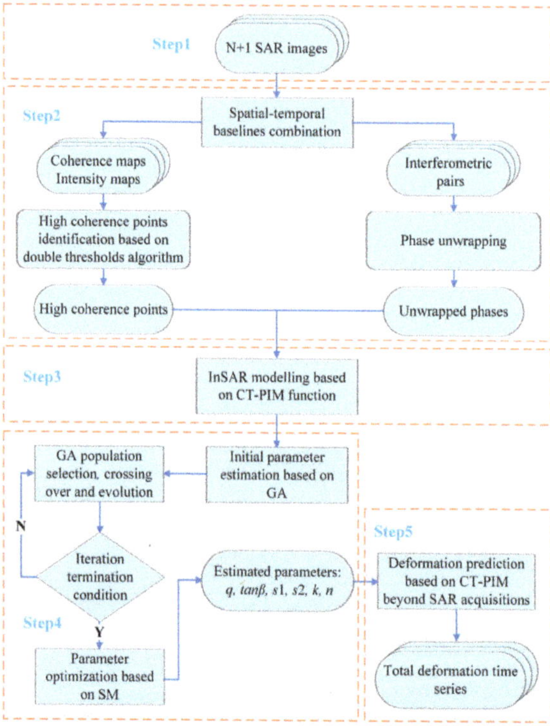

Figure 2. Flow chart of deformation prediction based on CT-PIM function.

3. Results

3.1. Simulated Experiment

A simulated experiment was designed and executed to evaluate the feasibility and accuracy of the proposed method. Satellite parameters used in the simulated experiment were set according to Sentinel-1 A spaceborne parameters, which was to keep consistent with the real data experiment. The geological parameters m, α, H, D_1, ω of the strata overlying the saline layer were set as 0.66 m, 4.5°, 600 m, 300 m, 80° according to real data collected by the company in charge of the mining activity in the area. The advancing speed v_m at the simulated experiment was set as 0.24 m/d, which was also according to the real data (will be discussed in Section 3.2.1). The simulated true values of UP were set primarily as 0.504, 1.560, 30.310 m, 16.080 m, 0.524, 0.167.

After setting those parameters, the time series settlement field could be generated according to Equation (1), which would be used as the real settlement field compared with the subsidence generated based on the InSAR-CTPIM processing. The selected images of the simulated real settlement field are shown in Figure 3. InSAR phase functions can be simulated according to Equation (6), and the residual elevation ΔH was simulated via Gaussian random simulator with its magnitude within [−20, 20] m. The random noise with a variance from 0 rad to 0.65 rad was added in the simulation. The random noise we added here can be expressed as $Noi = sqrt\ (0.65) * randn\ (60, 60)$, where Noi represents the noise; $sqrt$ (0.65) represents the variance of 0.65 rad; $randn$ is the random noise function in MATLAB; and (60, 60) is total simulated size of the phase function. Totally, 600 indexes of pixels were randomly extracted for the subsequent quantitative comparison with the CT-PIM generated deformation. In order to evaluate the impacts of different numbers of interferometric pairs on the corresponding generated deformation results, the simulation experiments with a multi-master connection were designed [34]. Six subgroups of simulation were carried out, with the number of interferometric pairs as 10, 15, 20, 25, 30, and 35, respectively, which also represented the number of InSAR phase functions. The corresponding GASM procedure was performed to obtain the CT-PIM parameters based on the simulated InSAR phase observations, and the final estimated subsidence could be obtained via Equation (1). The parameters, such as the spatio-temporal baselines, used in the simulation were set according to the real data experiment (in Section 3.2.2).

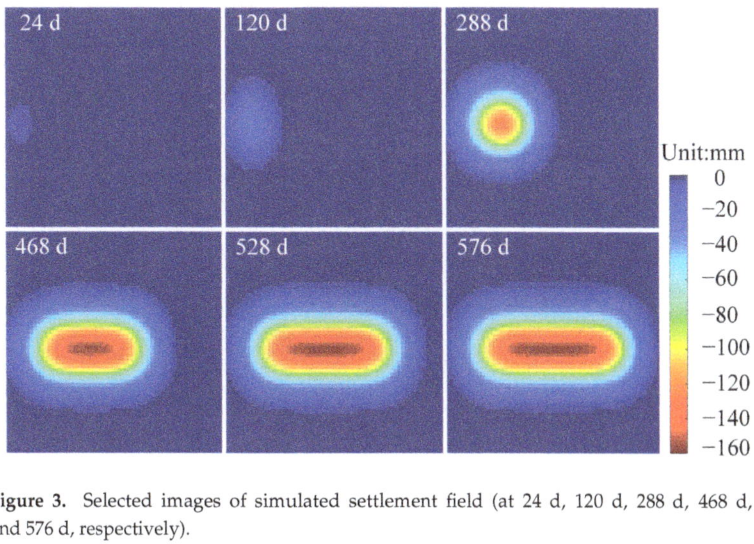

Figure 3. Selected images of simulated settlement field (at 24 d, 120 d, 288 d, 468 d, 528 d, and 576 d, respectively).

To estimate the unknown parameters of CT-PIM, the maximum iteration number, the population size, the crossover probability, and the mutation probability in GASM algorithm were set as 400, 60, 0.5, and 0.2, empirically. After the SM secondary optimization, the final optimal estimations of the CT-PIM parameters could be generated for each subgroup of simulation. Substituting the obtained parameters into Equation (1), the time series settlement field could be generated.

Figure 4 illustrates the relative errors for each parameter between the CT-PIM generated values and the simulated real ones under a different number of interferometric pairs. The lower the relative error is, the higher is accuracy suggested for the CT-PIM estimated parameter. As can be seen from the figure, generally, the relative errors decrease with the increase of interferometric numbers. Under a multi-master connection, a stack of 20 interferograms is acceptable in our cases. It can also be noted from Figure 4 that, among the six parameters, q and $tan\beta$ were more sensitive with higher curvature slopes than the remaining four parameters.

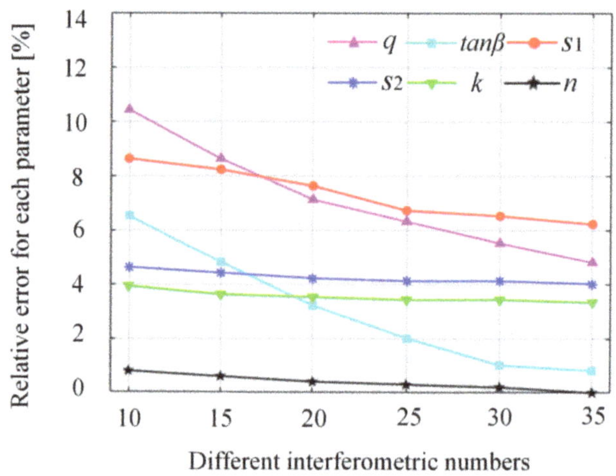

Figure 4. Relative errors with different interferometric numbers under multi-master connection.

We extracted the interferometric pairs of 10 and 35, respectively, to show a quantitively comparison with the simulated real values (shown as Tables 1 and 2). The results show that the use of 35 interferometric pairs under the multi-master connection to estimate the unknown parameters is evidently more accurate (for instance, the relative error for q is only 4.8% under 35 pairs compared with that of 10.4%). Therefore, we finally determined to use 35 interferometric pairs (noise level with a variance of 0.65 rad) under the multi-master connection in the real data experiment. It can also be noted that, among the six parameters, the relative error of s_1 is relatively large, both in Tables 1 and 2, with 8.6 % and 6.2%, respectively. This is consistent with the results of the sensitivity analysis in Section 3.3.3.

Table 1. Estimated CT-PIM parameters (under the 10 interferometric pairs).

Parameters	Simulated Real Value	CT-PIM Generated Value	Error	Relative Error
q	0.504	0.559	0.055	10.4
$tan\beta$	1.560	1.662	0.102	6.4
$s_1(m)$	30.310	27.801	2.509	8.6
$s_2(m)$	16.080	15.340	0.740	4.5
k	0.524	0.504	0.020	3.9
n	0.167	0.166	0.001	0.6

Table 2. Estimated CT-PIM parameters (under the 35 interferometric pairs).

Parameters	Simulated Real Value	CT-PIM Generated Value	Error	Relative Error
q	0.504	0.528	0.024	4.8
$tan\beta$	1.560	1.572	0.012	0.8
$s_1(m)$	30.310	28.431	1.879	6.2
$s_2(m)$	16.080	15.437	0.643	4.0
k	0.524	0.507	0.017	3.3
n	0.167	0.167	0	0

Figure 5 shows the comparison results between the selected time series settlement generated by CT-PIM and the simulated settlement with 600 pixels (35 interferometric pairs, multi-master connection, with a noise variance level of 0.65 rad), and the differences are illustrated in Figure 5f in red polylines. It can be noted that the two groups of results demonstrate good agreement even under a relatively high noise level of 0.65 rad, and the max deviation increases with the time span. According to the quantitative statistics for Figure 5a–f, the number of points with the deviations within [−5, 5] mm account for 100%, 100%, 97%, 90%, 87.6%, and 85.8% of the total number of points for spans of 24 days, 120 days, 288 days, 468 days, 528 days, and 576 days, respectively, and the corresponding maximum deviation is 0.1 mm, 1.3 mm, 7.9 mm, 9.3 mm, 9.5 mm, and 9.6 mm, respectively. Compared with the maximum subsidence for each period (i.e., 165 mm at 576 days), the deviation was relatively minor (a 9.6 mm deviation occupy only 5.8% approximately). To quantitatively evaluate the distribution of the deviations, statistical analysis for the probability distribution was carried out. Figure 6 demonstrates the probability distribution of the average deformation deviations with 600 pixels. As determined, the magnitudes of errors are concentrated within a relatively low ranges, with 83.8% of points distributed within [−2, 2] mm and all the pixels distributed within [−6, 6] mm, which implies that the proposed method is of promising accuracy. The RMSE between the CT-PIM generated value and the simulated real settlement is estimated as ±4.5 mm.

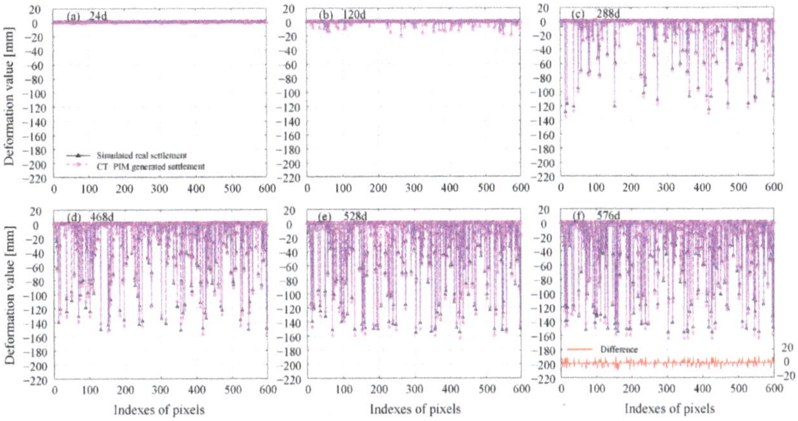

Figure 5. Comparison between the CT-PIM generated deformation sequences (six periods are selected) and the simulated real settlement.

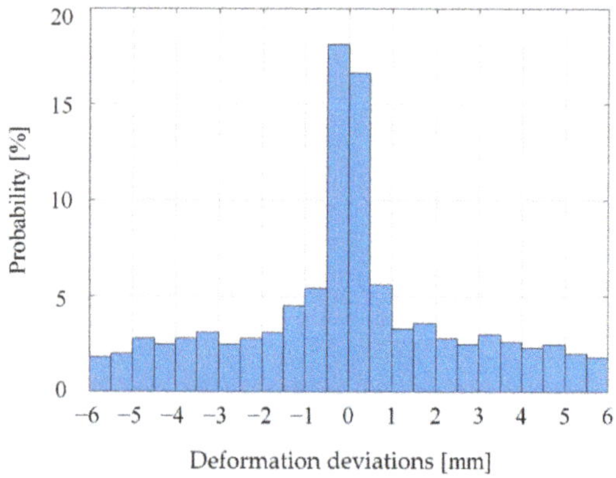

Figure 6. Probability distribution for deformation deviations.

3.2. *Real Data Experiment*

3.2.1. Study Area and SAR dataset

A salt mine, located in China, was selected as the study area. Figure 7 displays the location of our test area in Hunan Province and the corresponding coverage of SAR satellite images. The red rectangle in Figure 7b represents the spatial coverage of Sentinel-1 A image obtained by a satellite ascending geometry, and the purple rectangle represents the area of interest. The mine is located in the central part of Liyang Plain, with the characteristics of abundant natural water systems, numerous large and densely distributed ponds, crisscrossing artificial channels, and extensive surface paddy fields. Long-term drilling soluble mining activities have caused serious land salinization, which imposed significant influences to the surrounding environment (see Figure 8a). The mining-induced accumulated subsidence also led to cracks on the ground surface of roads, which imposed potential damage to infrastructures built on the nearby residential area (see Figure 8b,c). Gradually accumulated ground subsidence may also generate an obvious stagnant water pit (shown as Figure 8d, with a settlement approximately of 1.5 m).

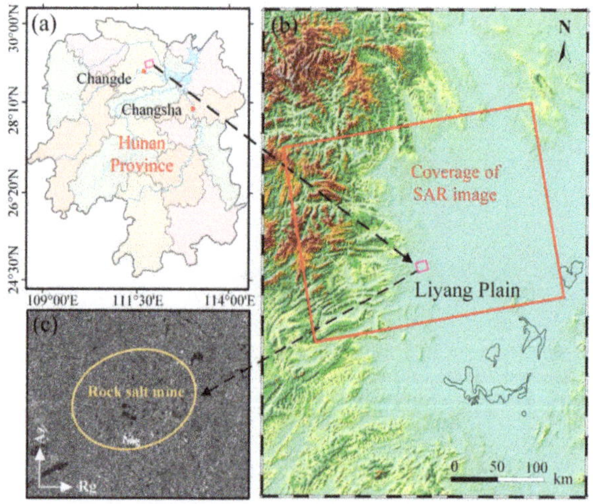

Figure 7. Study area and SAR spatial coverage: (**a**) map of Hunan Province; (**b**) coverage of SAR images; (**c**) mean amplitude image of the study area.

Figure 8. Field pictures of the residential area near the study area: (**a**) area with land salinization; (**b**) uneven displacement on the road; (**c**) cracks on the wall of a house; (**d**) a water pit in the salinized area.

Figure 9 shows the local geological asset. The underground stratums mainly include Lower Tertiary and Quaternary. The lithological components over this area mainly consist of mudstone, dolomite, siltstone, gypsum, mirabilite, and glauberite. Anhydrous Glauber salt (Na_2SO_4, 62.76–78.8%) produced in this area exist in the salt-bearing section (E_2x^3) of Neogene Formation. Argillaceous dolomitic mirabilite is the dominate lithological component, which distributes over the up layer, bottom layer, and interlayer and belongs to weak layered rock mass. The poor stability of the rocks in this area makes the ground surface vulnerable to soften and collapse.

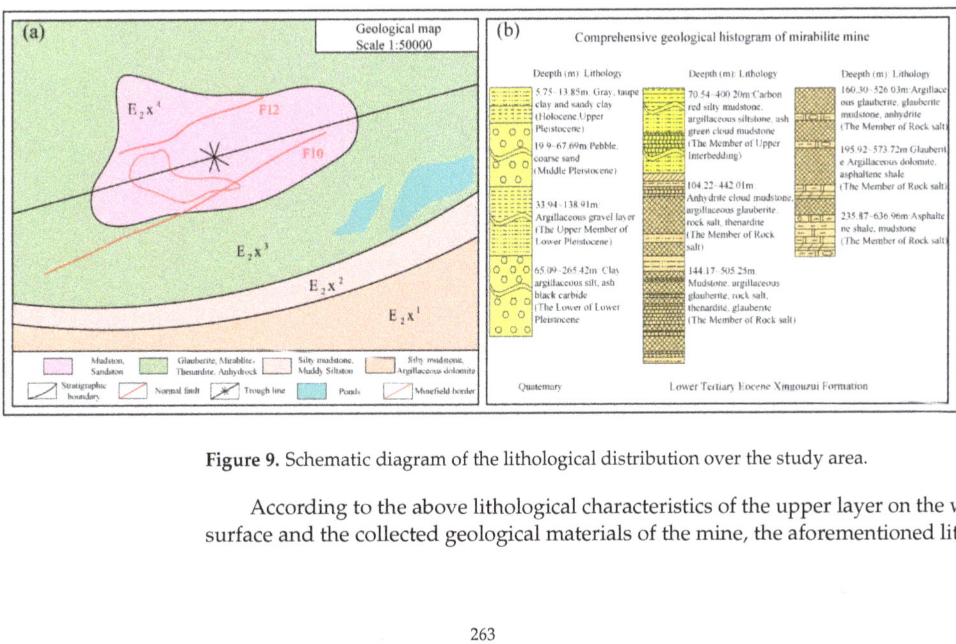

Figure 9. Schematic diagram of the lithological distribution over the study area.

According to the above lithological characteristics of the upper layer on the working surface and the collected geological materials of the mine, the aforementioned lithologic geological parameters GP (see Table 3) is set as $GP = [3\ \text{m}, 5.7°, 240\ \text{m}, 258\ \text{m}, 80°]$ according to the geological materials provided by the mining company. The advancing speed v_m is practically calculated according to properties of the sedimentation funnel in the mining area, which follows the equation $v_m = \frac{S}{t} = 0.056\ \text{m/day}$, where S denotes the mean radius for the sedimentation funnels, and t denotes the time span of the mining activities.

Table 3. Geological parameters.

Geological Parameters	m (m)	α (°)	H (m)	D_1 (m)	ω (°)	v_m (m/Day)
value	3	5.7	240	258	80	0.056

3.2.2. SAR Data Acquisition and Preprocessing

A total of 32 Sentinel-1A scenes with ascending geometry spanning the period from 15 June 2015 to 15 August 2017 were collected for the differential interferometric dataset processing. The unwrapped interferometric pairs with small spatial-temporal baselines were generated via SARScape 5.2 and ENVI 5.3. The SAR sensor parameters and spatial -temporal baselines are shown in Table 4 and Figure 10, and the subsequent high coherence points extraction, InSAR deformation modeling, CT-PIM parameter estimation, and deformation generation were all executed via MATLAB software. In the preprocessing of differential interferometry, the multi-look ratio of SAR image was set as 5:1, with the sampling resolutions as 9.022 m along azimuth, and 7.368 m along range direction, respectively. Spatial-temporal interferometric baselines were set lower than 150 m and 420 days, respectively, with all the images registered and resampled to a super master image (acquired at 3 July 2016). External 30 m resolution Shuttle Radar Topography Mission Digital Elevation Model (SRTM DEM) data was used to filter the terrain phase [35]. The procedure of phase unwrapping was executed with the Minimum Cost Flow (MCF) method. After artificially eliminating the unsatisfied interferometric pairs with poor unwrapping effect, totally, 58 unwrapped differential interferograms were generated [36,37]. Thresholds for moderate to high coherent points were set as 0.45 for coherence, and 0.35 for mean amplitude dispersion index. Finally, a total of 6027 candidates in the mining area were extracted.

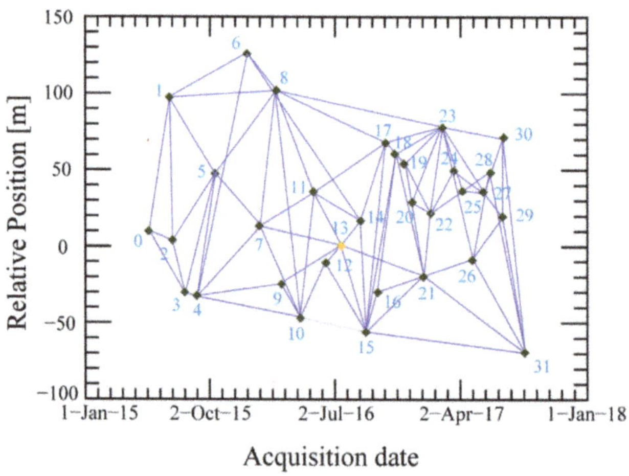

Figure 10. Spatio-temporal baselines.

Table 4. Interferometric pairs and sensor parameters (Ascending, Orbit No. 11).

Image No.	Acquisition Date (yyyy/mm/dd)	Perpendicular Baseline (m)	Temporal Baseline (Days)
0	2015/06/15	10.32	−384
1	2015/07/09	94.46	−360
2	2015/08/02	8.65	−336
3	2015/08/26	−25.59	−312
4	2015/09/19	−27.87	−288
5	2015/10/13	42.76	−264
6	2015/12/24	125.03	−192
7	2016/01/17	19.96	−168
8	2016/02/10	99.87	−144
9	2016/03/05	−21.75	−120
10	2016/03/29	−48.25	−96
11	2016/04/22	37.12	−72
12	2016/05/16	−9.33	−48
13	2016/07/03	0	0
14	2016/08/20	15.15	48
15	2016/09/25	−54.32	84
16	2016/10/07	−27.33	96
17	2016/10/19	63.93	108
18	2016/10/31	57.18	120
19	2016/11/12	52.15	132
20	2016/11/24	26.62	144
21	2016/12/18	25.12	168
22	2016/12/30	20.94	180
23	2017/01/11	76.93	192
24	2017/02/16	48.78	228
25	2017/03/12	34.61	252
26	2017/04/05	−16.41	276
27	2017/04/29	34.59	300
28	2017/05/23	47.07	324
29	2017/06/28	16.92	360
30	2017/07/22	63.26	384
31	2017/08/15	−69.95	408

3.2.3. PIM Parameter Estimation Based on InSAR-CTPIM

Before the novel InSAR-CTPIM processing being applied for the study area, the traditional SBAS-InSAR with multi-linear velocity model was performed to capture the spatial-temporal deformation characteristics over the rock salt mine. As shown in Figure 11, it can also be determined that apparent multi subsidence bowls developed during the observations from a spatial view. This phenomenon is related to drilling water-soluble mining activities. As in situ investigated, the mining mode for the test rock salt mine is multi-well groups connectivity, where more than two wells are connected for each well group during the drilling soluble mining; thus, multi settlement funnels will be formed around each well group. Since the CT-PIM model is based on a single subsidence funnel, it is difficult to accurately cover the whole test area containing complex multi settlement funnels with a single CT-PIM. Consequently, we divided the area into seven sub-settlement funnels, which are shown as Figure 11b, with the subsidence image at 15 August 2017 as the background.

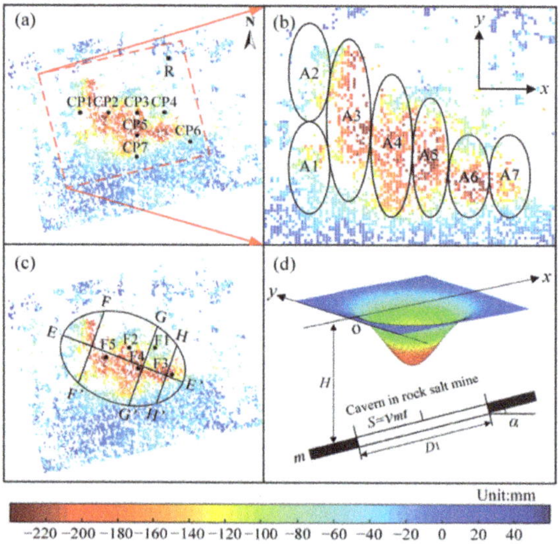

Figure 11. Schematic map of sub-settlement funnels distribution (with the settlement at 15 August 2017 generated via SBAS as the background): (**a**) the leveling points; (**b**) distribution of the seven sub-settlement funnels; (**c**) the distribution of the traverse-longitudinal profiles of the funnels; (**d**) 3D version of the mining coordinate system.

In InSAR-CTPIM processing, 24 images from 15 June 2015 to 11 January 2017 were extracted to estimate the CT-PIM parameters with GASM algorithm, which would be used to generate the corresponding deformation sequences following the step 4 of the procedures introduced in Section 2.4. Based on the obtained CT-PIM parameters, the remaining eight images (from 16 February 2017 to 15 August 2017) were reserved to validate the forward predicted deformation generated by CT-PIM following the step 5. During the GASM procedure, seven sub-funnels were processed separately, with seven groups of parameters generated, which is shown in Table 5.

Table 5. Estimated CT-PIM parameters.

Parameters	A1	A2	A3	A4	A5	A6	A7
q	0.832	0.882	0.919	0.741	0.957	0.750	0.974
$tan\beta$	3.060	3.128	3.473	3.362	3.732	3.608	2.995
$s_1\ (m)$	39.170	40.694	44.933	49.888	42.297	47.468	31.116
$s_2\ (m)$	36.481	38.631	32.001	39.813	48.184	47.374	26.701
k	0.518	0.623	0.743	0.693	0.658	0.702	0.718
n	0.150	0.162	0.182	0.196	0.145	0.193	0.166

3.2.4. Deformation Prediction Based on CT-PIM

Based on the seven groups of obtained CT-PIM parameters in Table 5, the deformation sequences covered by SAR data acquisition can be derived by InSAR-CTPIM. Figure 12 shows the time series deformation generated spanning from 15 June 2015 to 11 January 2017, which were treated as the deformation monitoring results. The deformation beyond SAR data acquisition can be predicted according to CT-PIM (expressed as Equation (1)). Figure 13 shows the eight 3D scenes (from 16 February 2017 to 15 August 2017) of the predicted subsidence based on CT-PIM, which were treated as the deformation prediction results. From 9 July 2015 to 10 February 2016, the overall deformation of the whole area was relatively stable, with the maximum estimated subsidence as only 52 mm. Obvious subsidence began to occur at 5 March 2016 after an 8-month temporal lag, and the subsiding velocities have increased rapidly since then. Since 20 August 2016, the boundary of the seven sub-settlement bowls has become apparently more distinct, with the maximum subsidence increasing from 110 mm to 205 mm on 11 January 2017. As of 15 August 2017, the maximum subsidence had accumulated up to 294 mm.

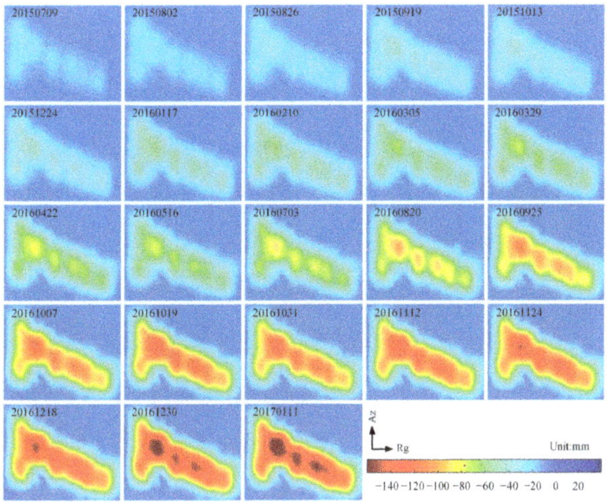

Figure 12. Derived deformation sequences based on InSAR-CTPIM (reference date: 15 June 2015).

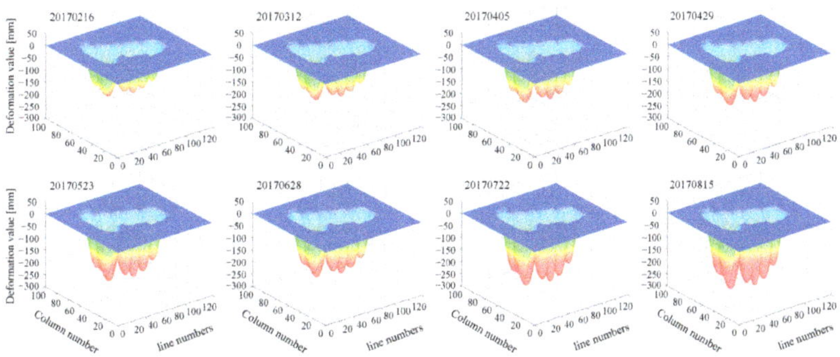

Figure 13. Three-dimensional scenes of the predicted subsidence based on CT-PIM (reference date: 15 June 2015).

3.3. *Accuracy Analysis*

3.3.1. Accuracy Evaluation for InSAR-CTPIM Monitored Subsidence

The historical leveling measurements from 2 August 2015 to 18 December 2016 for the test mine were collected, which can be used to test the accuracy of the deformation results during the SAR acquisitions obtained by InSAR-CTPIM. The locations of the leveling points (CP1, CP2, ... , CP7) are marked in Figure 11a, with the reference point marked as *R* in the north-east corner. The leveling measurements which covered consistent periods with the SAR acquisition dates were extracted to conduct a more precise comparison with the InSAR-CTPIM monitored subsidence.

Figure 14 shows the comparison between the results of InSAR-CTPIM, SBAS-InSAR, and leveling measurements. The average RMSE of InSAR-CTPIM results was estimated as ±6.9 mm, whereas that of traditional SBAS-InSAR was ±8.5 mm, with an increase of approximately 20.8%. It can be determined that the dynamic settlement obtained by InSAR-CTPIM shows better consistency with the leveling measurements. According to our experiment, almost all the leveling points in the mining area were under a considerable subsiding period. The most serious settlement occurred at CP3, with a value of 156 mm. The RMSE accounts for 4.4% of the maximum settlement, which implies the promising accuracy of InSAR-CTPIM in deriving time series subsidence of the rock salt mine.

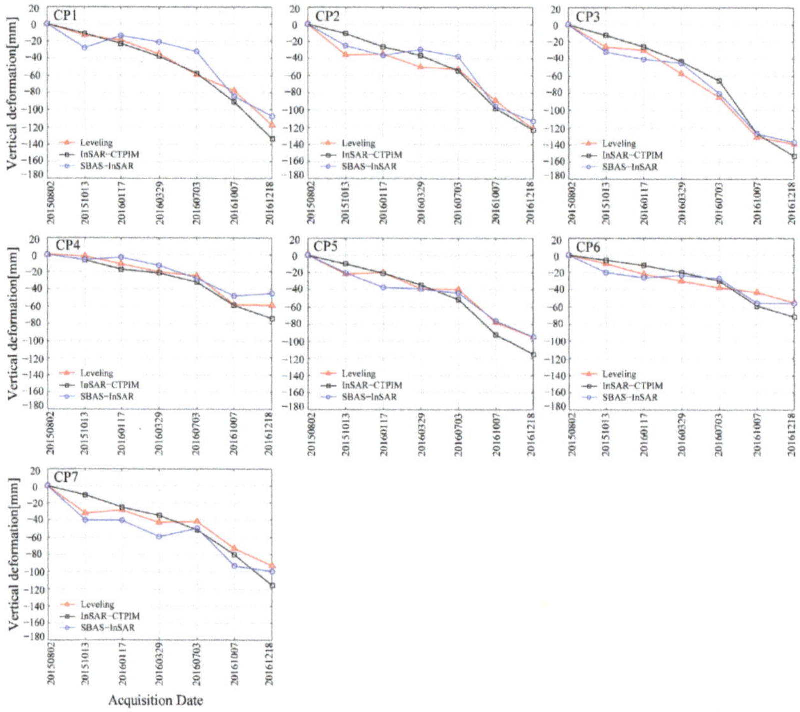

Figure 14. Time series deformation at leveling points (reference date: 2 August 2015).

3.3.2. Accuracy Evaluation for CT-PIM Predicted Subsidence

As discussed above, the remaining eight images from 16 February 2017 to 15 August 2017 were reserved to evaluate the forward subsidence prediction results. To compensate for the unavailability of simultaneous periods of leveling measurements, SBAS-derived results were used to be compared with the predicted subsidence. The traditional static PIM prediction method introduced in References [20,38], which inversed the PIM parameters based on a single interferometric pair, was carried out preliminarily to be compared with the CT-PIM predicted results. Four periods of predicted subsidence (spanning from 15 June 2015 to 26 February 2017, 5 April 2017, 28 June 2017, and 15 August 2017, respectively) were extracted, which is shown as Figure 15. The maximum subsidence predicted via static PIM were relatively lower than those of SBAS-InSAR. On 15 August 2017, the maximum subsidence predicted by static PIM was estimated as 325 mm, with a 39-mm difference with the SBAS-derived result, while the difference of CT-PIM is only 8 mm.

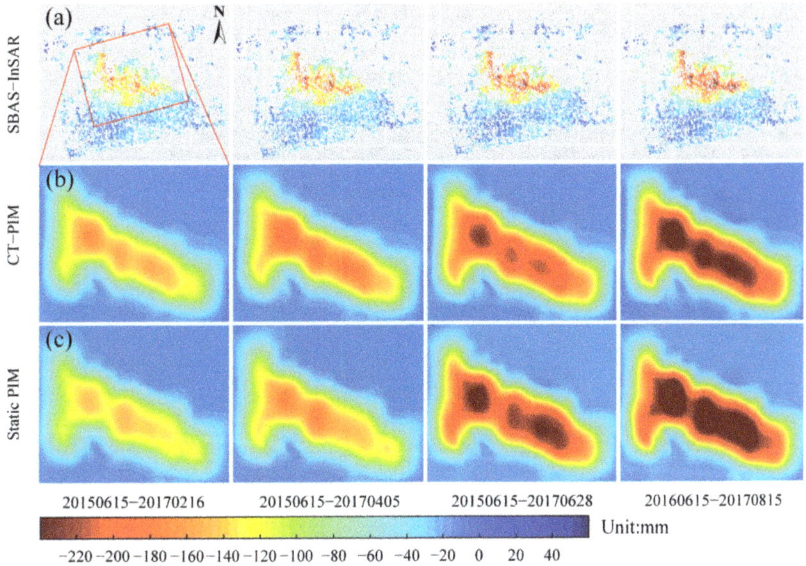

Figure 15. Subsidence comparison: (**a**) SBAS-InSAR results; (**b**) CT-PIM predicted; (**c**) static PIM predicted (with reference to 15 June 2015).

For further quantitative analysis, the points located in the subsidence funnels were extracted from Figure 15. The probability distribution was statistically analyzed, and the distribution histograms are shown as Figures 16 and 17. As Figure 16 shows, the residual errors for all the sampled points between CT-PIM predicted subsidence and those of the SBAS results during the four temporal periods are mainly distributed within the range of [−25, 25] mm. The RMSE is estimated as ±12.8 mm, ±12.7 mm, ±12.9 mm, ±13.2 mm, respectively. In contrast, those residual errors for static PIM (shown as Figure 17) are mainly distributed within a more disperse range of [−40, 40] mm, with its RMSE estimated as ±20.2 mm, ±18.5 mm, ±20.7 mm, ±15.8 mm, respectively, which are much higher. Apparently, the distribution of CT-PIM errors is better concentrated than those of static PIM, which indicates a better consistency with the SBAS technique and a higher predictive accuracy. The total STD of those errors for CT-PIM is estimated as ±12.9 mm. It can be inferred that the CT-PIM predicted results show better consistency with the traditional SBAS-derived subsidence than that of static PIM predicted ones, with an increase of 37.2% compared to that of static PIM.

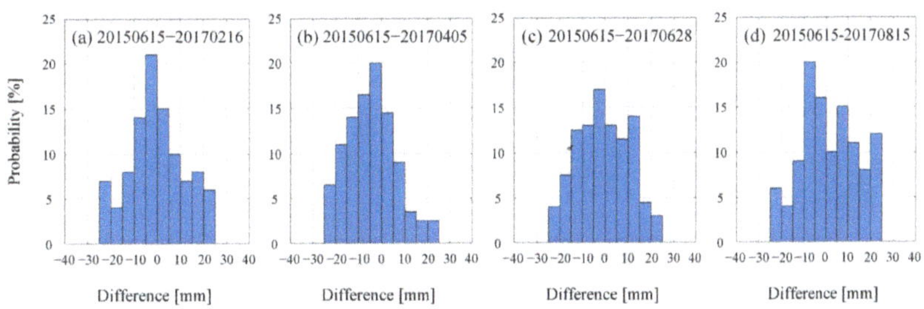

Figure 16. Differences of CT-PIM predicted subsidence with comparison of SBAS-InSAR results (with reference to 15 June 2015).

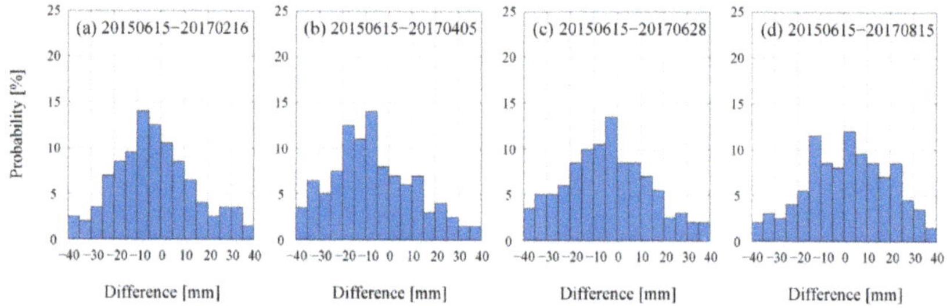

Figure 17. Differences of static PIM predicted subsidence with comparison of SBAS-InSAR results (with reference to 15 June 2015).

3.3.3. Sensitivity Analysis on CT-PIM Parameters

Since CT-PIM is directly used to predict the forward dynamic deformation subsequent to the SAR acquisitions, the accuracy of the predicted mining subsidence directly depends on the accuracy of CT-PIM parameters. Consequently, it is necessary to analyze the sensitivity of CT-PIM parameters. Sensitivity analysis can assist us in understanding how the parameters influence the predicted deformation, i.e., which parameter impacts the most and to what extent it impacts the results. The sensitivity analysis method based on Sobol indices is introduced here to execute a simulated experiment [39].

Sobol indices describe the correlation between each parameter and its global sensitivity on the objective function. The higher the sensitivity index, the higher the result error of the parameter on the deformation. Firstly, the objective function is decomposed based on variance analysis; then, the first-order indices (S_α, where α represents the certain parameter) and total-effect indices (S_α^{Tot}) of each parameter is calculated, and the sensitivity of the interactive input parameters can also be generated. Here, S_α represents the partial variance, which describes the main contribution of a sample of the certain parameter for the output variances, while S_α^{Tot} represents the total variance, which describes the percentage of a group of samples for a certain parameter on the output variances. Finally, the disturbance analysis is used to quantitatively analyze the influences of different parameters on the final predicted deformation. The estimated two kinds of indices for each CT-PIM parameter are shown in Figure 18, where X axis represents the input parameters ($UP = [q, tan\beta, s_1, s_2, k, n]$), and Y axis represents the Sobol indices for each parameter. The closer the magnitude of each corresponding Sobol index to 1, the higher sensitivity the parameter possesses. Table 6 lists the ranges of different importance extent. From it, we can determine that all the parameters are beyond the range of "Not correlated", which implies their impacts on the accuracy of predicted deformation cannot be ignored. Among the six parameters, q and $tan\beta$ are more sensitive to the model, with the Sobol indices S_α^{Tot} of 0.95 and 0.90, respectively, which can be treated as "Very important" parameters. Consequently, we should pay more attention to the error control during the parameter estimation procedure for those two parameters. In contrast, s_1 and s_2 show relatively lower sensitivity than the remaining four parameters, with the Sobol indices S_α^{Tot} of 0.60 and 0.65, respectively. The two parameters can be treated within the united range of "Unimportant" and "Important".

Table 6. Correlation extent for different ranges of sensitivity indices [39].

Correlation Extent	Ranges of Sensitivity Indices
Very important	$0.8 < S_\alpha < S_\alpha^{Tot} \leq 1$
Important	$0.5 < S_\alpha < S_\alpha^{Tot} \leq 0.8$
Unimportant	$0.3 < S_\alpha < S_\alpha^{Tot} \leq 0.5$
Not correlated	$0 < S_\alpha < S_\alpha^{Tot} \leq 0.3$

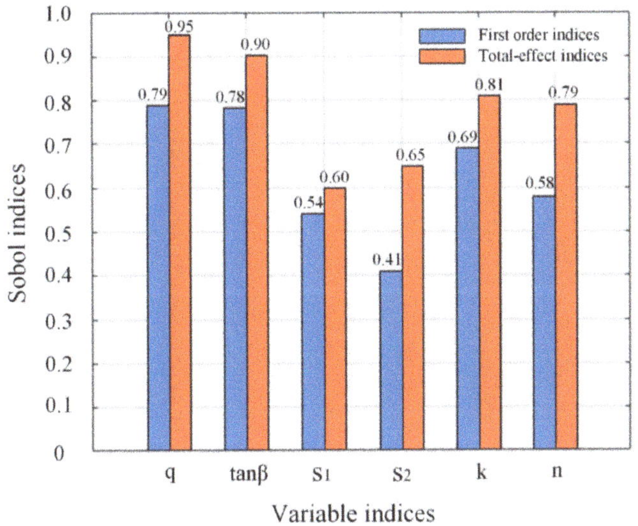

Figure 18. Estimated Sobol indices for CT-PIM parameter.

4. Discussion

Subsidence along the traverse and longitudinal lines marked in Figure 11c (EE′, FF′, GG′, HH′) were extracted for a profile analysis. The profile analysis results based on the SBAS-InSAR derived time series subsidence are shown in Figure 19 (as shown the dotted line), while the CT-PIM results are shown as Figure 19 (as shown the solid line). From the figures, we can see similar spatial-temporal characteristics and approximate same locations of pixels with separate subsidence bowls along EE′, FF′, GG′, and HH′, for both the two groups of generated results. The discrepancy between the SBAS results and CT-PIM predicted ones, in the case of Figure 19c (particularly in the black rectangle), was suggested to be related to the following reasons: (1) The systematic uplifts in the black rectangle in Figure 19c for the SBAS results were related to the residual atmospheric delay and orbital errors. During CT-PIM processing, the mining area was divided into several subsidence bowls artificially. (2) We treated the GG′ profile that only traversed a single subsidence funnel, and only one group of CT-PIM parameters was used to predict the subsidence. Some unrevealed deformation characteristics maybe hidden in the predicted results, which can be revealed by the SBAS-InSAR monitored results. As we determined for the InSAR-CTPIM derived results, the apparent three peak values were detected at 172 m, 322 m, and 414 m along EE′, with the subsidence of 232 mm, 249 mm, and 205 mm, respectively, and peak values at 149 m along the FF′ section of 219 mm, whereas, at 172 m, along GG′ of 218 mm and, at 80 m, along the HH′ of 247 mm. The multi peaks of the subsidence along traverse and longitudinal directions were due to the water-soluble mining activities with groups of drilling wells simultaneously.

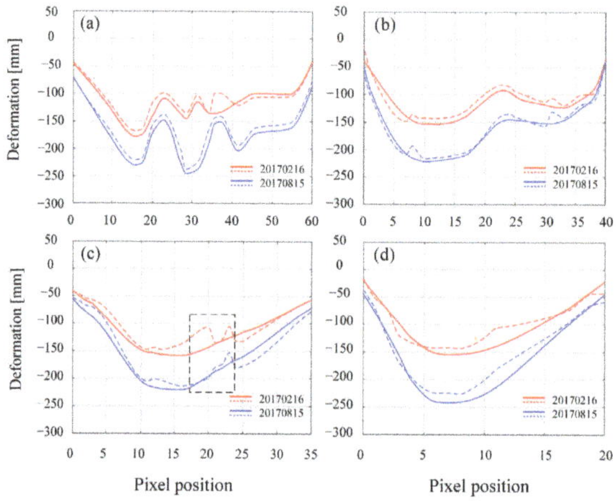

Figure 19. Profile analysis based on SBAS (dotted line) and CT-PIM (solid line) predicted subsidence: (**a**) Profile EE′; (**b**) FF′; (**c**) GG′; (**d**) HH′.

Five feature points (F1–F5, marked in Figure 11c) are extracted for quantitative analysis, and the results are shown in Figure 20. It can be seen from Figure 20 that, during the whole period, the five feature points displayed similar temporal evolution trends: slow settlement period from 15 June 2015 to 29 March 2016, and then a rapid settlement period from 29 March 2016 to 15 August 2017. Maximum subsidence was determined at F5 with its accumulated settlement of 289 mm until 15 August 2017. In contrast, F1, which was far away from the wellhead, was relatively stable, with its maximum deformation accumulated to 98 mm until 15 August 2017. As introduced in Section 3.2.4, an 8-month temporal lag of the subsidence can be found in Figure 20, which is suggested to be mainly related to the dissolution processing of the mirabilite by the solvent, with a certain longer time from injection to the surface displacement emerged. In addition, the depth of rock salt mine is generally deeper than that of common coal mines, which also leads to the time delay from the underground displacement cavity being accumulated on the ground surface.

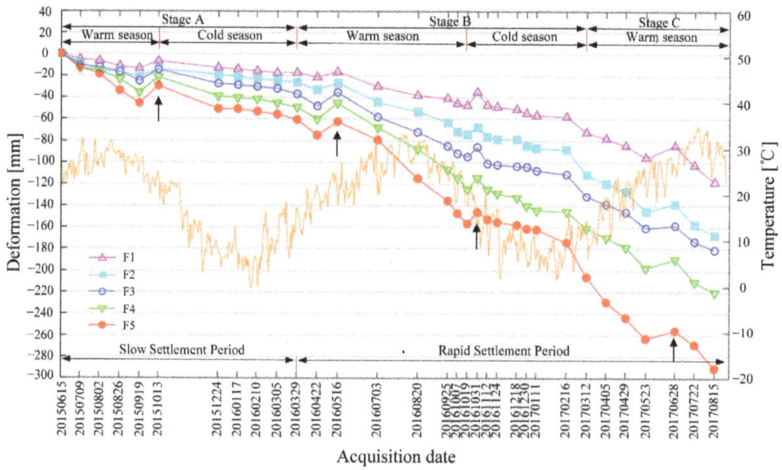

Figure 20. Time series settlement on feature points based on InSAR-CTPIM.

Another phenomenon of seasonal related variations was determined from our experiments. For a warm season (29 March 2016 to 19 October 2016 in Stage B and 12 March 2017 to 15 August 2017 in stage C), a rapid subsiding trend occurred for all the five feature pixels, with the cumulative maximum subsidence reaching up to 98 mm and 83mm, while, for a cold season (13 October 2015 to 29 March 2016 of Stage A and 19 October 2016 to 12 March 2017 of Stage B), a relatively slow developing trend dominated those points, with a maximum subsidence of 31 mm and 45 mm, respectively, for the two periods. The potential reasons are suggested to be related to the air temperature, which influences the dissolution rate of mirabilite. The higher temperature of warm periods can accelerate the dissolution, and, accordingly, increased the subsiding velocities. Comparatively, lower temperature in cold periods slowed down the dissolution processing, which induced a stable trend of deformation. It should be noted that the dissolution rate of mirabilite in water is directly affected by solvent temperature (as shown in Table 7); however, it is indirectly affected by the air temperature. Hot water is used as solvent in drilling soluble production, which is delivered from the processing plant to the injection well after being measured and distributed at the control station. In the process of transportation, the solvent temperature is easily affected by the outside air temperature, which supports the above temperature-related interpretations.

Table 7. Seven groups of obtained InSAR-CTPIM parameters [28,40].

Mineral	Temperature (°C)										
	0	10	20	30	40	50	60	70	80	90	100
Thenardite	–	–	–	50.4	48.8	46.7	45.3	44.1	43.7	42.9	42.5
Glauber's Salt	5.0	9.0	19.4	40.8	–	–	–	–	–	–	–
Glauberite	0.18	0.19	0.20	0.21	0.21	–	0.21	0.20	0.20	–	0.16

It can also be determined, in Figure 20, that four minor jumps occurred for all the five feature points (marked by black arrows). According to the precipitation and temperature information of the Hunan Lixian Meteorological Bureau, the precipitation increased significantly on 13 October 2015, 16 May 2016, 31 October 2016, and 28 Jun 2017, as shown in Figure 21 [41]. Besides the decrease of solvent solubility caused by the aforementioned low temperature in winter, an increase of rainfall also contributed to the recoveries of subsidence.

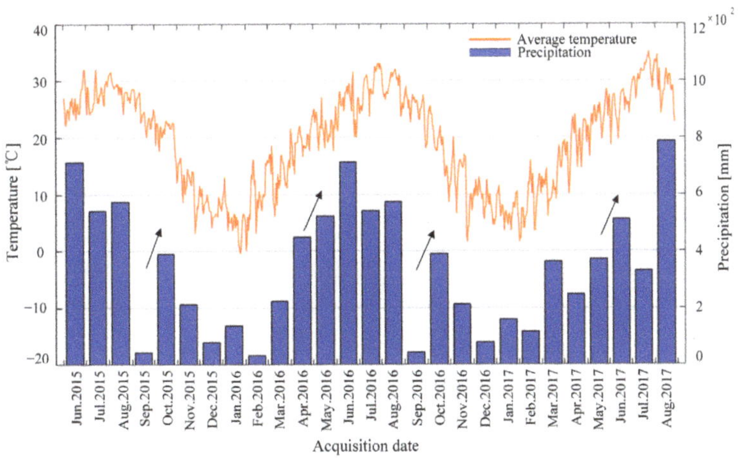

Figure 21. Air temperature and precipitation of the mining area (6 June 2016 to 31 January 2017) [42].

5. Conclusions

A novel InSAR deformation model, namely CT-PIM function, was proposed and applied for predicting the dynamic deformation over a drilling solution rock salt mine. The CT-PIM function was used as a substitute for traditional InSAR pure empirical models, which can also provide as potential combined use with other high-productive inspection methodologies, which has the following advantages: (1) with consideration of the physical underground mechanism, CT-PIM can describe the temporal dynamical characteristics of the mining-induced subsidence more realistically; (2) CT-PIM can be directly applied for the forward deformation prediction, which implements to the deficiency of non-consistency and unreasonableness between the traditional InSAR model (i.e., linear model or linear model combined with periodical function) and the forward prediction model (i.e., PIM); (3) it provides an alternative PIM parameters estimation method directly based on the InSAR phase observations, which can avoid the secondary error propagation from the InSAR inaccurate deformation to the subsequent deformation prediction; and, (4) as for the computational burden, since the GASM are based on the input InSAR unwrapped phases (not the final generated deformation sequences), we saved the preprocessing time for InSAR Geocoding compared to the static PIM.

The feasibility and improvement of our method was verified by both simulated and real-data experiments. The simulated results showed that the RMSE between the deformation results estimated by the new model and the simulated true value was ±4.5 mm, even under a high noise level of 0.65 rad. In the real data experiment, 32 Sentinel-1A SAR images were used to carry out the experiment over a water-soluble rock salt mine. As we determined, the maximum settlement was estimated as 294 mm. The experimental results were compared with external leveling results, and the RMSE was determined as ±10.9 mm, which accounts for only 6.9% of the maximum settlement. During the deformation prediction procedures, CT-PIM showed a considerable improvement of 32.7% than the traditional static PIM prediction approach, with an STD of ±12.9 mm compared to SBAS-generated results.

6. Patents

There are patents resulting from the work reported in this manuscript. Our research content has applied for a Chinese patent (ZL202010122691.8) entitled "A deformation monitoring method in mining area".

Author Contributions: X.X. designed the experiments and produced the results; T.Z. carried out the experiment. X.X. and T.Z. analyzed the experiment results; L.C., X.L. and W.P. analyzed the precipitation data; X.X., W.P. and Z.Y. (Zefa Yang) helped to collect and analyze the leveling measurement in the real data experiment; X.L. and Z.Y. (Zhihui Yuan) contributed to the discussion of the results; X.X. and T.Z. drafted the manuscript. All authors have read and agreed to the published version of the manuscript.

Funding: This work was funded by the National Natural Science Foundation under Grant of China (Grant No. 42074033, 41904003, 41701536, 61701047); the Open Project Program of the Hunan Key Laboratory of remote sensing of ecological environment in Dongting Lake Area (Grant No. 2021-011); the Natural Science Foundation of Hunan Province, China (Grant No. 2019JJ50639, 2020JJ5571); and Key Project of Education Department of Hunan Province, China (Grant No. 18A148, 19C0042).

Data Availability Statement: The Sentinel-1 data presented in this study are openly available in ESA/Copernicus at https://scihub.copernicus.eu (accessed on 16 December 2021). The SRTM data presented in this study are openly available in the National Aeronautics and Space Administration (NASA, United States).

Acknowledgments: The Sentinel-1 A images used in this work are provided by the European Space Agency. Thanks for He Y.G. of Transportation Engineering College, Changsha University of Science and Technology, China, for providing the leveling deformation data of the research area, which provides available validation data for this work.

Conflicts of Interest: The authors declare no conflict of interest.

References

1. Ministry of Natural Resources. *China Mineral Resources Report*; Geological Publishing House: Beijing, China, 2019; pp. 2–6.
2. Zhang, M.G.; Wang, Z.S.; Wang, L.J.; Chen, Y.L.; Wu, Y.; Ma, D.; Zhang, K. Mechanism of collapse sinkholes induced by solution mining of salt formations and measures for prediction and prevention. *Bull. Eng. Geol. Environ.* **2019**, *78*, 1401–1415. [CrossRef]
3. Buchignani, V.; Avanzi, G.D.; Giannecchini, R.; Puccinelli, A. Evaporite karst and sinkholes: A synthesis on the case of Camaiore (Italy). *Environ. Geol.* **2008**, *53*, 1037–1044. [CrossRef]
4. Solari, L.; Montalti, R.; Barra, A.; Monserrat, O.; Crosetto, M. Multi-temporal satellite interferometry for fast-motion detection: An application to salt solution mining. *Remote Sens.* **2020**, *12*, 3919. [CrossRef]
5. Wang, S.; Jiang, G.; Weingarten, M.; Niu, Y. InSAR evidence indicates a link between fluid injection for salt mining and the 2019 Changning (China) earthquake sequence. *Geophys. Res. Lett.* **2020**, *47*, e2020GL087603. [CrossRef]
6. Furst, S.L.; Doucet, S.; Vernant, P.; Champollion, C.; Carme, J.L. Monitoring surface deformation of deep salt mining in vauvert (france), combining InSAR and leveling data for multi-source inversion. *Solid Earth* **2021**, *12*, 15–34. [CrossRef]
7. Jung, J.; Kim, D.J.; Vadivel, P.S.K.; Yun, S.H. Long-term deflection monitoring for bridges using X and C-Band time-series SAR interferometry. *Remote Sens.* **2019**, *11*, 1258. [CrossRef]
8. Tosti, F.; Gagliardi, V.; D'Amico, F.; Alani, A.M. Transport infrastructure monitoring by data fusion of GPR and SAR imagery information. *Transp. Res. Procedia* **2020**, *45*, 771–778. [CrossRef]
9. Gagliardi, V.; Bianchini Ciampoli, L.; Trevisani, S.; D'Amico, F.; Alani, A.M.; Benedetto, A.; Tosti, F. Testing Sentinel-1 SAR interferometry data for Airport Runway monitoring: A Geostatistical Analysis. *Sensors* **2021**, *21*, 5769. [CrossRef]
10. Zheng, M.; Deng, K.; Fan, H.; Du, S. Monitoring and analysis of surface deformation in mining area based on InSAR and GRACE. *Remote Sens.* **2018**, *10*, 1392. [CrossRef]
11. Xiao, L.; He, Y.G.; Xing, X.M.; Wen, D.B.; Tong, C.G.; Chen, L.F.; Yu, X.Y. Time series subsidence analysis of drilling solution mining rock salt mines based on Sentinel-1 data and SBAS-InSAR technique. *Int. J. Remote Sens.* **2019**, *23*, 501–513.
12. Ma, T.; Zhao, Y.J.; Zhang, W. Application of SBAS-InSAR technology in settlement monitoring of mining area. *Geom. Spat. Inf. Tech.* **2020**, *11*, 210–212.
13. Chen, Y.; Tong, Y.X.; Tan, K. Coal mining deformation monitoring using SBAS-InSAR and offset tracking: A case study of Yu County, China. *IEEE J. Sel. Top. Appl. Earth Obs. Remote Sens.* **2020**, *13*, 6077–6087. [CrossRef]
14. Yang, Z.F.; Li, Z.W.; Zhu, J.J.; Hu, J.; Wang, Y.J.; Chen, G.L. InSAR-Based model parameter estimation of probability integral method and its application for predicting mining-induced horizontal and vertical displacements. *IEEE Trans. Geosci. Remote Sens.* **2016**, *54*, 4818–4832. [CrossRef]
15. Fan, H.; Lu, L.; Yao, Y. Method combining probability integration model and a small baseline subset for time series monitoring of mining subsidence. *Remote Sens.* **2018**, *10*, 1444. [CrossRef]
16. Zhu, J.J. A Mining Monitoring Method Based on InSAR Technology. Chinese Patent ZL201310011306.2, 8 May 2013.
17. Li, Z.W.; Yang, Z.F.; Zhu, J.J.; Hu, J.; Wang, Y.J.; Li, P.X.; Chen, G.L. Retrieving three-dimensional displacement fields of mining areas from a single InSAR pair. *J. Geod.* **2015**, *89*, 17–32. [CrossRef]
18. Yang, Z.F.; Li, Z.W.; Zhu, J.J.; Preusse, A.; Yi, H.W.; Wang, Y.J.; Papst, M. An extension of the InSAR-Based probability integral method and its Application for Predicting 3-D Mining-Induced Displacements Under Different Extraction Conditions. *IEEE Trans. Geosci. Remote Sens.* **2017**, *55*, 3835–3845. [CrossRef]
19. Yang, Z.F.; Li, Z.W.; Zhu, J.J.; Feng, G.C.; Wang, Q.J.; Hu, J.; Wang, C.C. Deriving time-series three-dimensional displacements of mining areas from a single-geometry InSAR dataset. *J. Geod.* **2018**, *92*, 529–544. [CrossRef]
20. Xing, X.M.; Zhu, Y.K.; Yuan, Z.H.; Xiao, L.; Liu, X.B.; Chen, L.F.; Xia, Q.; Liu, B. Predicting mining-induced dynamic deformations for drilling solution rock salt mine based on probability integral method and weibull temporal function. *Int. J. Remote Sens.* **2020**, *42*, 639–671. [CrossRef]
21. Wang, L.Y.; Deng, K.Z.; Zheng, M.N. Research on ground deformation monitoring method in mining areas using the probability integral model fusion D-InSAR, sub-band InSAR and offset-tracking. *Int. J. Appl. Earth Obs.* **2020**, *85*, 101981. [CrossRef]
22. Xing, X.M.; Xiao, L.; Bao, L.; Chen, L.F.; Yuan, Z.H. An InSAR Prediction Method for Mining Subsidence in Drilling Water-Soluble Salt Mines. Chinese Patent ZL201910168502.8, 16 March 2021.
23. Zhu, G.Y.; Shen, H.X.; Wang, L.G. Study of dynamic prediction function of surface movement and deformation. *Chin. J. Rock Mech. Eng.* **2011**, *30*, 1889–1895.
24. Berardino, P.; Fornaro, G.; Lanari, R.; Sansosti, E. A new algorithm for surface deformation monitoring based on small baseline differential SAR interferograms. *IEEE Trans. Geosci. Remote Sens.* **2002**, *40*, 2375–2383. [CrossRef]
25. Lauknes, T.R.; Zebker, H.A.; Larsen, Y. InSAR Deformation Time Series Using an L1-Norm Small-Baseline Approach. *IEEE Trans. Geosci. Remote Sens.* **2010**, *49*, 536–546. [CrossRef]
26. Li, D.; Deng, K.Z.; Gao, X.X.; Niu, H.P. Monitoring and analysis of surface subsidence in mining areas based on SBAS-InSAR. *Geom. Inf. Sci. Wuhan Univ.* **2018**, *43*, 1531–1537.
27. Zhang, Y.; Meng, X.M.; Jordan, C.; Novellino, A.; Dijkstra, T.; Chen, G. Investigating slow-moving landslides in the Zhouqu region of China using InSAR time series. *Landslides* **2018**, *5*, 1299–1315. [CrossRef]
28. Wang, Q.M. *Water-Soluble Salt Mine Mining and Design*; Chemical Industry Press: Beijing, China, 2016; pp. 46–71.
29. Yao, L.; Sethares, W.A. Nonlinear parameter estimation via the genetic algorithm. *IEEE Trans. Signal Process.* **1994**, *42*, 927–935.

30. Xing, X.M.; Chang, H.C.; Chen, L.F.; Zhang, J.H.; Yuan, Z.H.; Shi, Z.N. Radar Interferometry Time Series to Investigate Deformation of Soft Clay Subgrade Settlement—A Case Study of Lungui Highway, China. *Remote Sens.* **2019**, *11*, 429. [CrossRef]
31. Tian, Y.G. *Study on Genetic Algorithm for Nonlinear Least Squares Estimation*; Wuhan University Press: Wuhan, China, 2003.
32. Xiao, H.F.; Tan, G.Z. Study on Fusing Simplex Search into Genetic Algorithm. *Comput. Eng. Appl.* **2008**, *44*, 30–33.
33. Yang, Z.F.; Li, Z.W.; Zhu, J.J.; Yi, H.W.; Feng, G.C. Deriving Dynamic Subsidence of Coal Mining Areas Using InSAR and Logistic Model. *Remote Sens.* **2017**, *9*, 125. [CrossRef]
34. Duan, M.; Xu, B.; Li, Z.W.; Wu, W.H.; Liu, J.H. Adaptively selecting interferograms for SBAS-InSAR based on graph theory and turbulence atmosphere. *IEEE Access* **2020**, *99*, 112898–112909. [CrossRef]
35. Goldstein, R.M.; Werner, C.L. Radar interferogram filtering for geophysical applications. *Geophys. Res. Lett.* **1998**, *25*, 4035–4038. [CrossRef]
36. Costantini, M.; Rosen, P.A. A generalized phase unwrapping approach for sparse data. In Proceedings of the IEEE International Geoscience and Remote Sensing Symposium, Hamburg, Germany, 28 June–2 July 1999; pp. 267–269.
37. Li, Z.W.; Ding, X.L.; Zheng, D.W.; Huang, C. Least squares-based filter for remote sensing image noise reduction. *IEEE Trans. Geosci. Remote Sens.* **2008**, *46*, 2044–2049. [CrossRef]
38. Fan, H.D.; Cheng, D.; Deng, K.Z.; Chen, B.Q.; Zhu, C.G. Subsidence monitoring using D-InSAR and probability integral prediction modelling in deep mining areas. *Surv. Rev.* **2015**, *47*, 438–445. [CrossRef]
39. Cannavo, F. Sensitivity analysis for volcanic source modeling quality assessment and model selection. *Comput. Geosci.* **2012**, *44*, 52–59. [CrossRef]
40. Ma, Z.Y.; Nie, C.X.; Wang, R.T. *Geological Basis and Mining Technology of Well and Mineral Salt*; National Mine Salt Industry Science and Technology Information Station: Chengdu, China, 1992; pp. 82–115.
41. Fomelis, M.; Papageorgiou, E.; Stamatopoulos, C. Episodic ground deformation signals in Thessaly Plain (Greece) revealed by data mining of SAR interferometry time series. *Int. J. Remote Sens.* **2016**, *37*, 3696–3711. [CrossRef]
42. Precipitation and Temperature Data of Lixian City, China. Available online: https://tianqi.911cha.com/lixian/2016.html (accessed on 15 December 2021).

remote sensing

Article

Monitoring and Stability Analysis of the Deformation in the Woda Landslide Area in Tibet, China by the DS-InSAR Method

Youfeng Liu [1], Honglei Yang [1,*], Shizheng Wang [2], Linlin Xu [3] and Junhuan Peng [1]

[1] School of Land Science and Technology, China University of Geosciences, Beijing 100083, China; jasonliuyf@163.com (Y.L.); pengjunhuan@163.com (J.P.)
[2] Zhejiang Huadong Mapping and Engineering Safety Technology Co., Ltd., Hangzhou 310005, China; wang_sz2@hdec.com
[3] The Department of Systems Design Engineering, University of Waterloo, Waterloo, ON N2L 3G1, Canada; linlinxu618@gmail.com
* Correspondence: hongleiyang@cugb.edu.cn

Abstract: The Woda area in the upper Jinsha River has steep terrain and broken structures, causing landslide disasters frequently. Here, we used the distributed scatterer interferometric SAR (DS-InSAR) method to monitor and analyze the Woda landslide area. With the DS-InSAR method, we derived the deformation of the Woda landslide area from 106 Sentinel-1A ascending images acquired between 5 November 2014 and 4 September 2019 and 102 Sentinel-1A descending images acquired between 31 October 2014 and 11 September 2019. The obvious advantage of the DS-InSAR method compared to the persistent scatterer (PS) InSAR (PS-InSAR) method is that the densities of the monitoring points were increased by 25.1% and 22.9% in the ascending and descending images, respectively. The two-dimensional deformation of the landslide area shows that the maximum surface deformation rate in the normal direction was −80 mm/yr, and in the east–west direction, 118 mm/yr. According to the rescaled range (R/S) analysis, the Hurst index values of the deformation trends were all greater than 0.5, which means the deformation trend will continue for some time. In addition, we analyzed the influencing factors and the deformation mechanism of the Woda landslide area and found that the surface deformation is closely related to the geological structure and precipitation, among which precipitation is the main factor triggering the deformation. Our monitoring results will help the local government to conduct regular inspections and strengthen landslide disaster prevention in low-coherence mountainous areas.

Keywords: landslide; DS-InSAR; deformation monitoring; stability analysis

Citation: Liu, Y.; Yang, H.; Wang, S.; Xu, L.; Peng, J. Monitoring and Stability Analysis of the Deformation in the Woda Landslide Area in Tibet, China by the DS-InSAR Method. *Remote Sens.* **2022**, *14*, 532. https://doi.org/10.3390/rs14030532

Academic Editors: Chisheng Wang, Daqing Ge, Guohong Zhang, Wu Zhu and Siting Xiong

Received: 29 December 2021
Accepted: 21 January 2022
Published: 23 January 2022

1. Introduction

Vigorous tectonic movements and natural erosions shaped the complex landform of Southwest China, where mountains, plateaus, and vertical and horizontal surface gullies are widely distributed. The Qinghai–Tibet Plateau, known as the 'roof of the world', has obvious crustal uplift and strong and rapid rivers, resulting in frequent geological disasters on both sides of the rivers [1,2]. Landslide disasters are easily triggered in areas with lush natural or planted forests or in steep areas, especially after extreme events. In eastern Tibet, continuous landslide disasters have seriously affected social and economic development and people's lives [3]. The region along the Jinsha River and its tributaries has steep terrains and broken structures, so geological disasters such as ground fissures and landslides have occurred frequently. Since the 1980s, 61 landslides have been recorded [4]. The Woda landslide area in Yanbi Township, Jomda County, eastern Tibet, is located in the upper reaches of the Jinsha River. It mainly includes two landslide bodies, which started slowly deforming in 1985. Remote sensing interpretation and ground survey both show risks of further deformation and sliding that may cause river closure [5]. On 11 October 2018, a major landslide occurred in Baige village in the lower reaches of the Jinsha River, about

80 km from the Woda village [6]. The slope blocked the Jinsha River and damaged many roads, bridges, and buildings. Such major geological disasters may impact the areas hundreds of kilometers upstream and downstream. Therefore, monitoring the long-term surface deformation of the Woda landslide area is of great significance for preventing and controlling the occurrence of geological disasters.

In the past few decades, leveling [7], three-dimensional laser scanning [8], global navigation satellite system (GNSS) [9], and unmanned aerial vehicle (UAV) photogrammetry [10] have been widely used in deformation monitoring. However, they cannot provide satisfactory results in the identification and monitoring of landslides in large areas due to their long measurement cycles, small spatial coverage, large demand of human and material resources, and high-risk factors for close-range work. Synthetic aperture radar (SAR) remote sensing has the advantages of wide coverage as well as all-day and all-weather observation, so it has been widely used in geological disaster monitoring [11,12]. With the deepening of interferometric SAR (InSAR) research, differential InSAR (D-InSAR) uses SAR images before and after the deformation to monitor the surface deformation [13,14], but it has low monitoring accuracy due to time and geometric decoherence and atmospheric delay, meaning it cannot meet the long-term monitoring requirements. In 2000, Ferretti et al. proposed the theory and method of persistent scatterer InSAR (PS-InSAR) [15], suitable for monitoring targets with stable radar scattering characteristics, such as buildings and bridges in urban areas, but it is difficult to effectively apply it in areas such as farmland and woodland. For this reason, subsequently, Berardino et al. contributed many theoretical innovations and algorithm improvements to the PS-InSAR method and proposed some new time series InSAR methods, which greatly enriched the time series InSAR technical theory system [16–19]. In recent years, the time series InSAR method has been widely used in landslide investigation and identification [20], landslide sliding evaluation [21], the establishment of early warning systems for landslides [22], and landslide surface deformation monitoring [23,24], which is powerful technical support for landslide disaster research.

In 2011, Ferretti et al. proposed the second-generation permanent scatterer technology (SqueeSAR) and related scholars around the world began to shift their research focus to distributed targets with relatively weak backscatter [25]. In recent years, the distributed scatterer InSAR (DS-InSAR) method based on SqueeSAR technology was introduced to the research in low-coherent areas with weak backscattering. Goel et al. proposed the Anderson–Darling (A–D) test method to select statistical homogeneous pixels (SHPs), which effectively solved the high error rate of the K–S test in small sample data [26]. Gianfranco et al. decomposed the covariance matrix of DS points and selected the eigenvector with the largest value as the optimal phase of the DS point [27]. Jiang et al. proposed fast SHPs, modified fast SHPs, and hypothesis test of confidence interval (HTCI) algorithms, which effectively improved the SHP selection efficiency and phase optimization quality and greatly promoted the progress and development of the DS-InSAR method [28,29]. This method combines PS points and DS points to significantly increase the distribution density of monitoring points. Compared to other time series InSAR methods, it has obvious advantages and is more suitable for the surface deformation monitoring of landslides in mountainous areas [30,31].

In addition, time series InSAR technology can only extract the deformation rate and the deformation time series along the line of sight (LOS) direction [32]. In fact, real space deformation is three-dimensional, namely in the vertical, east–west, and north–south directions. The vertical direction is usually longer than the other two directions. Since the flight direction of the Sentinel-1A satellite is close to the north–south direction, the satellite heading angle is relatively small, and it is insensitive when calculating the north–south deformation [33]. However, it is not enough to only know the deformation information in the LOS direction, and it is difficult to meet the needs of surface deformation research. Generally, for research on surface deformation, it is best to perform a comprehensive analysis with ascending and descending SAR data and calculate the vertical and east–west two-dimensional deformation, further enriching and expanding the research work of the surface deformation [34,35]. In landslide areas where data sources are scarce and it is difficult to obtain leveling, GNSS, and other data, the two-dimensional monitoring results combining ascending and descending SAR data can help distinguish different landslide deformation directions, reflect different deformation rate modes, and reduce the influence of geometric distortion with high reliability and practicability [36,37].

In view of the fact that the time series and two-dimensional studies of the Woda landslide area are relatively few, in this study, the DS-InSAR method was applied to monitor and analyze the deformation in the Woda landslide area using the ascending and descending Sentinel-1A images. The HTCI algorithm with the highest computational efficiency was used to select SHPs, and the adaptive spatial nonlocal filtering method was constructed based on the results of the SHP selection to perform phase optimization. We analyzed the deformation characteristics and inducing factors of the landslide area, which can provide reference and guidance for the early investigation, monitoring, and timely warning of densely vegetated landslides.

2. Study Area and Datasets

2.1. Study Area

The study area is the Woda village and its surrounding areas in the upper reaches of the Jinsha River. The Woda village is located in Yanbi Town, Changdu City, Tibet, China, with a longitude of 98°49′38.63″ E and latitude of 31°25′56.31″ N (Figure 1a). There are two closely distributed landslides in the area, named landslide bodies 1 and 2, along the Jinsha River (Figure 1b). The front edges of the landslides are steep and reach the Jinsha River valley at the toe, but the middle and rear parts are relatively flat (Figure 1c). The elevation of the Jinsha River surface is about 2950 m, and that of the trailing edge apex of the landslide area is about 4000 m. Such a great elevation difference may contribute to geological disasters, such as large-scale landslides, for example, the Baige landslide, which blocked the main stream of the Jinsha River to form a dammed lake, causing huge property losses.

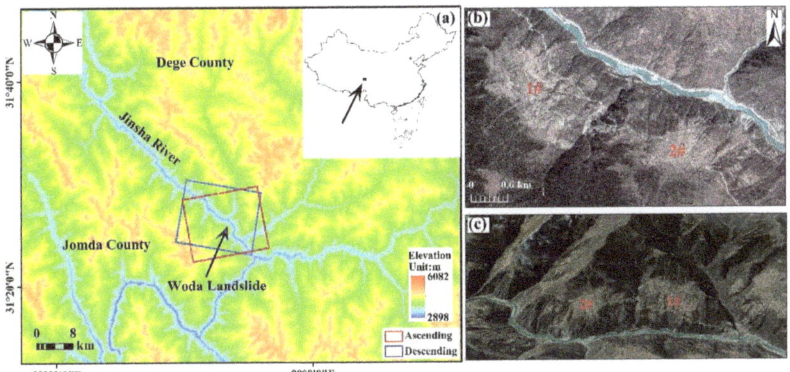

Figure 1. (**a**) Location of the study area in Tibet Autonomous Region, China, and SRTM-derived topographic map of the Woda landslide area (Jomda County). The red and blue boxes indicate the coverage of Sentinel-1A ascending and descending data, respectively. (**b**) Two-dimensional Google map imagery of the study area. (**c**) Three-dimensional Google map imagery of the study area.

2.2. Datasets

We used the Sentinel-1A data in the ascending and descending orbits. Sentinel-1A is a C-band synthetic aperture radar (SAR) satellite with a new TOPSAR imaging mode. It has a 12-day return cycle and a large-scale spatial coverage of 250 km × 250 km at medium resolution. At present, the latest global real-time observation data are free of charge. A total of 208 scenes of Sentinel-1A interferometric wide (IW)-swath mode data were adopted in this study. Detailed information is shown in Table 1.

Table 1. Detailed information of the dataset.

Sensor	Temporal Coverage	Image Number	Orbit	Track	Polarization
Sentinel-1A	5 November 2014–4 September 2019	106	Ascending	99	VV
Sentinel-1A	31 October 2014–11 September 2019	102	Descending	33	VV

In addition, the SRTM1 30 m DEM with a 1 arc-second spatial resolution from National Aeronautics and Space Administration (NASA) was used to remove topographic phase and geocoding. A Google optical image was also used to show the final results and the analysis of the landslide area.

3. Methods

In this study, we used the DS-InSAR method to monitor the Woda landslide area, and the specific data processing flowchart is shown in Figure 2. Firstly, SAR images were registered and clipped by using precise orbit and DEM data, and then differential interferogram pair sequences were generated. When setting the temporal and space baseline thresholds to 72 days and 200 m, the 106-scene ascending SAR images generated 209 interference pairs; when setting the temporal and space baseline thresholds to 96 days and 200 m, the 102-scene descending SAR images generated 201 interference pairs. The generated interference pair information is shown in Figure 3. Then, we selected the PS and DS points and extracted their phase and elevation information to calculate the deformation. Finally, we combined the precipitation data and geological map to analyze the deformation of the landslide area in detail. The precipitation data in this paper can be downloaded from the official Soil and Water Assessment Tool (SWAT) website at https://swat.tamu.edu/ (11 October 2021), and the geological map is from the hydrogeological map of China (1:200,000).

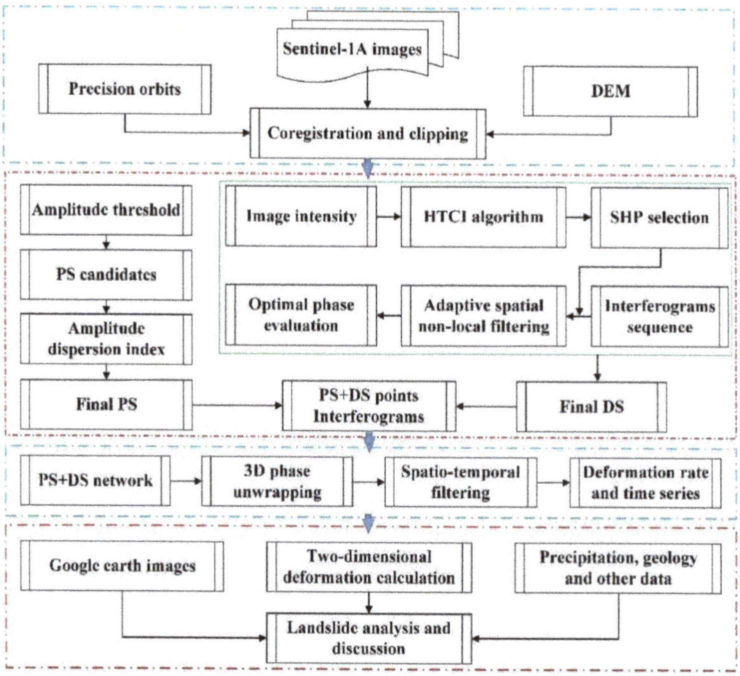

Figure 2. The flowchart of landslide monitoring by the DS-InSAR method.

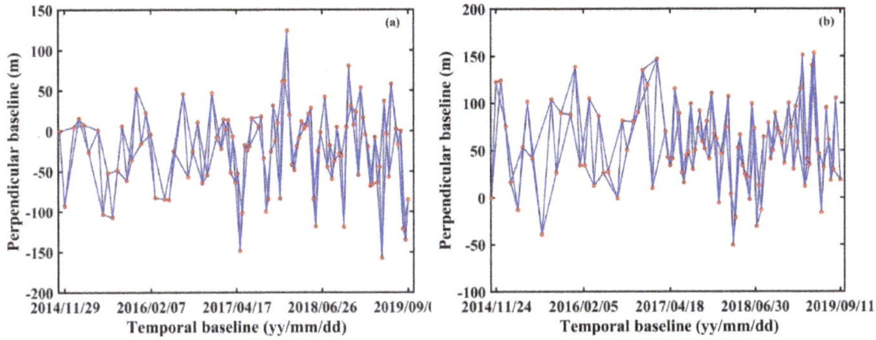

Figure 3. Spatial and temporal baselines of the interferograms obtained by (**a**) ascending data; (**b**) descending data.

3.1. PS Selection

The amplitude threshold and the amplitude dispersion index threshold were jointly used to select the PS point. The amplitude threshold was used to select pixels with high amplitude values as candidate points, from which those with smaller amplitude deviation values were determined as the final PS points. The pixels with high amplitude values indicate better coherence, and a smaller amplitude deviation index indicates better stability. Therefore, the selected PS points have stable radar scattering characteristics, which can improve the accuracy of the results.

3.2. DS Selection

DS point selection includes SHP selection and phase optimization. SHPs were selected according to the SAR image intensity information, which was used to optimize the interference phase of the SAR image. The phase optimization was evaluated by calculating the temporal coherence and setting the temporal coherence threshold to 0.4 to select high-quality DS points [12,38].

3.2.1. SHP Selection

There are dozens of SHP selection algorithms with different accuracies. Assuming that the SAR intensity dataset obeys the exponential distribution and the amplitude dataset obeys the Rayleigh distribution, the algorithm can be divided into two categories—nonparametric hypothesis testing methods, such as the KS test [39] and the BWS test [40], and the parameter test algorithm, such as HTCI. The HTCI algorithm has higher selection efficiency and accuracy than the KS and BWS test methods. In this study, we used the HTCI algorithm to select SHPs.

Assuming that N SAR images obey complex Gaussian distribution in the time dimension and the variance of random variables is $\sigma^2/2$, the SAR image intensity sequence obeys Rayleigh distribution.

$$f(I) = \frac{1}{\sigma^2} e^{-\frac{I}{\sigma^2}}, I \geq 0 \tag{1}$$

According to the central limit theorem, the sample mean obeys a normal distribution of $N(\sigma^2, \sigma^4/(NL))$. The interval estimation analytical equation is as follows:

$$P\left\{\sigma^2 - z_{\alpha/2} \cdot \sigma^2/\sqrt{N \cdot L} < \overline{\sigma} < \sigma^2 + z_{\alpha/2} \cdot \sigma^2/\sqrt{N \cdot L}\right\} = 1 - \alpha \tag{2}$$

where $Z_{\alpha/2}$ is the percentage of the standard normal probability density function $\alpha/2$. Defining the center pixel in the window as the reference pixel and the remaining pixels as the neighboring pixels, if the sample mean of the neighboring pixels is within the given confidence interval, the neighboring pixels are considered the SHPs of the reference pixel. However, in actual data processing, the available SAR images are very few, and the Gaussian hypothesis is often difficult to establish. The Gamma distribution of the accumulated N scene intensity images is $N\sigma \sim G(Na, b)$, where $a = L, b = \sigma^2/L$ (where L denotes the multi-look number). Set $Y = N\sigma/b$, so Y obeys the standard Gamma distribution with the shape parameter Na, which is $Y \sim G(Na, 1)$ [41]. Let $g_{\alpha;Na}$ be the α percentage of the standard Gamma distribution of the known parameter Na, and we get the following:

$$P\{g_{\alpha/2;Na} < Y < g_{1-\alpha/2;Na}\} = 1 - \alpha \tag{3}$$

From which, we obtain the fixed interval analytic equation as

$$P\left\{g_{\alpha/2;NL} \cdot \sigma^2/(N \cdot L) < \overline{\sigma} < g_{1-\alpha/2;NL} \cdot \sigma^2/(N \cdot L)\right\} = 1 - \alpha \tag{4}$$

If there is less available time series SAR image data, Equation (4) is better than Equation (2), as the interval value is asymmetric and the tail feature is emphasized. In summary, the algorithm defines a fixed window for each pixel of the image. Then, the mean temporal intensity of the reference pixel and that of the neighborhood pixel are taken as the values of σ^2 and $\overline{\sigma}$ in Equation (4), respectively. Finally, the neighborhood pixels located in the confidence interval are selected as the SHPs of the reference pixels. The algorithm selects SHPs according to the intensity mean of the local spatial pixels and the local minimum mean square error, which effectively improves its robustness and selection accuracy.

3.2.2. Optimal Phase Estimation

In the selection window, the interference phases of SHPs are affected similarly by noise, are highly correlated, and have distributed characteristics. In this study, we used the adaptive spatial nonlocal filtering method based on the results of SHP selection to perform phase optimization. In a certain window, the reference pixels were filtered by adaptive spatial nonlocal filtering using the selected SHPs to reduce the phase noise of the interferogram.

At any pixel p in the interferogram, the estimated value of the adaptive spatial nonlocal filtering is as follows [42]:

$$NL(p) = \sum_{m \in \Omega} w(p, m) u(m) \tag{5}$$

$NL(p)$ is the phase value of the filtered pixel p, and is the phase value of the homogeneous pixel m in the search range Ω. The weight value $w(p, m)$ depends on the similarity between pixels p and m, and satisfies the following conditions: $0 \leq w(p, m) \leq 1, \sum_m w(p, m) = 1$.

$$d(N_p, N_m) = \frac{\|u(N_p) - u(N_m)\|_{2,a}^2}{1 + \|u(N_p) - u(N_m)\|_{2,a}^2} \tag{6}$$

$$Z(p) = \sum_m \exp(-\frac{d(N_p, N_m)}{h^2}) \tag{7}$$

$$w(p, m) = \exp(-\frac{d(N_p, N_m)}{h^2})/Z(p) \tag{8}$$

where $d(N_p, N_m)$ is the Gaussian weighted Euclidean distance, which is defined as the similarity between the distance weighted windows; $Z(p)$ is the normalization parameter; h is the attenuation factor of the control exponential function; a is the standard deviation of the Gaussian kernel function. The larger the $d(N_p, N_m)$ value is, the greater the weight. The value of h is usually determined according to the noise level. A value that is too large may filter out some useful features and textures, but a value too small makes it difficult to filter out noises. Therefore, selecting the value of h becomes a key factor that affects the nonlocal filtering. In this paper, it is defined as the following:

$$h = \frac{1}{1 + D_A} \tag{9}$$

where D_A is the amplitude deviation index of the estimation window. In areas with large terrain undulations, a small h will be selected to retain as much texture information as possible; in areas with relatively flat terrain, a large h will be used to remove noise.

After phase optimization, the quality of the optimized phase was evaluated by Equation (10). The difference between the optimized interference phase and the original interference phase was fitted to obtain a fitting metric value of γ_{DS} [25].

$$\gamma_{DS} = \frac{2}{N^2 - N} \mathrm{Re} \sum_{n=1}^{N} \sum_{k=n+1}^{N} e^{i\phi_{nk}} e^{-i(\vartheta_n - \vartheta_k)} \tag{10}$$

ϑ_n and ϑ_k are the optimized interference phases of a single master image, and ϕ_{nk} is the original interference phase. γ_{DS} can also be regarded as the temporal coherence of the interference pair sequence and was used to describe the quality of the optimized phase. By setting the temporal coherence threshold, the points with high temporal coherence were selected as the DS points in the subsequent analysis.

3.3. Combined Data Processing

Combining the selected PS and DS points can effectively improve the insufficient point selection density of using the PS-InSAR method alone. The phases of the combined coherent points are differentiated and unwrapped to construct triangulation networks. Through iterative calculation and regression analysis, we separated the linear deformation, the elevation residual, and the residual phase. According to the different characteristics of the residual phase in the time and space domains, multiple filtering in the frequency domain can separate nonlinear deformation, atmospheric delay phase, and noise phase. The singular value decomposition was used to obtain the deformation rate and accumulative deformation during the study period.

Then, using the deformation obtained from the ascending and descending orbit data, we calculated the vertical and east–west deformation characteristics in the landslide area. Finally, the stability analysis of the study area was carried out according to the two-dimensional deformation results. The details are as follows.

3.3.1. Two-Dimensional Deformation Calculation

According to the geometric relationship of radar imaging, the surface deformation monitored by the ascending and descending SAR images were both in the LOS direction (D_{LOS}). D_{LOS} can be decomposed into the deformation in the vertical (D_v), east–west (D_e), and north–south (D_n) directions (Figure 4).

D_{LOS} can be expressed as [43,44] the following:

$$D_{LOS} = \begin{bmatrix} \cos\theta & -\sin\theta\cos\alpha & \sin\theta\sin\alpha \end{bmatrix} \begin{bmatrix} D_v \\ D_e \\ D_n \end{bmatrix} \tag{11}$$

where θ represents the radar incident angle, and α represents the satellite heading angle. Since the flight direction of the Sentinel-1A satellite is close to the north–south direction, the deformation monitoring results in this direction are not sensitive. Therefore, we calculated the deformation in the vertical and east–west directions.

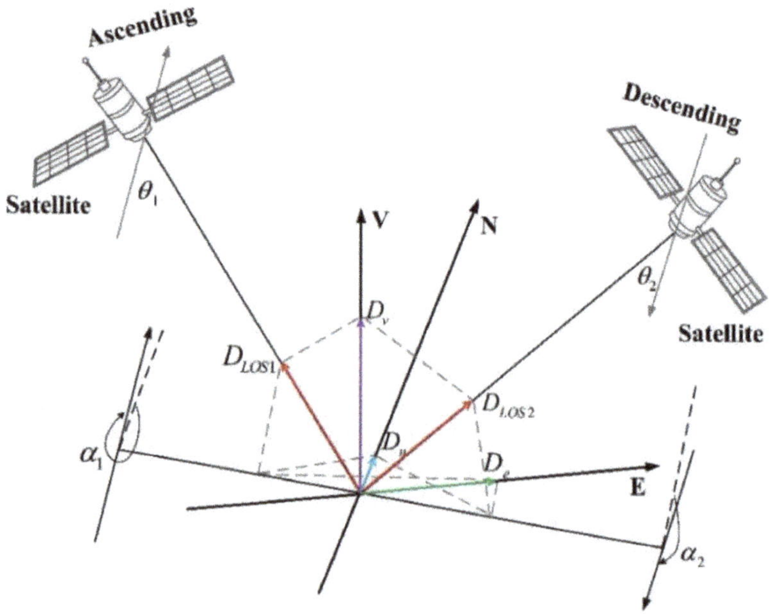

Figure 4. Three-dimensional spatial decomposition model of the LOS deformation. D_{LOS1} and D_{LOS2} are the LOS deformations of the ascending and descending orbits, respectively; θ_1 and θ_2 are the incidence angles of the ascending and descending SAR images, respectively; α_1 and α_2 are the heading angles of the ascending and descending SAR satellites.

$$\begin{bmatrix} D_{LOS1} \\ D_{LOS2} \end{bmatrix} = \begin{bmatrix} \cos\theta_1 & -\sin\theta_1\cos\alpha_1 \\ \cos\theta_2 & -\sin\theta_2\cos\alpha_2 \end{bmatrix} \begin{bmatrix} D_v \\ D_e \end{bmatrix} \quad (12)$$

$$A_{v,e} = \begin{bmatrix} \cos\theta_1 & -\sin\theta_1\cos\alpha_1 \\ \cos\theta_2 & -\sin\theta_2\cos\alpha_2 \end{bmatrix} \quad (13)$$

where $A_{v,e}$ is the coefficient matrix. The radar incidence angle information of each pixel of the ascending and descending SAR images is shown in Figure 5. In areas with large terrain undulations, the incidence angles of neighboring pixels are greatly different, and even the incidence angles of the same pixel in the ascending and descending orbits are significantly different. In this case, the deformation results are invalid in the overlapped and shadow areas, and the reliability in the perspective shrinkage area is reduced [45]. Considering the actual situation, we used the incidence angle of each pixel to calculate the two-dimensional deformation by the least square method as follows.

$$D_{v,e} = \left(A_{v,e}^T P A_{v,e}\right)^{-1} A_{v,e}^T P D_{LOS1,2} \quad (14)$$

where P is the weight matrix, the reciprocal of the residual standard deviation of the monitoring result is used as the weight, and $D_{v,e} = [D_v\ D_e]^T$; $D_{LOS1,2} = [D_{LOS1}\ D_{LOS2}]^T$. Finally, we obtained the deformation in the vertical direction and east–west direction of the monitoring point of its tangent plane. However, it should be noted that since the radar incident angle information of each pixel was considered, the vertical direction deformation obtained was actually the deformation of the surface normal direction.

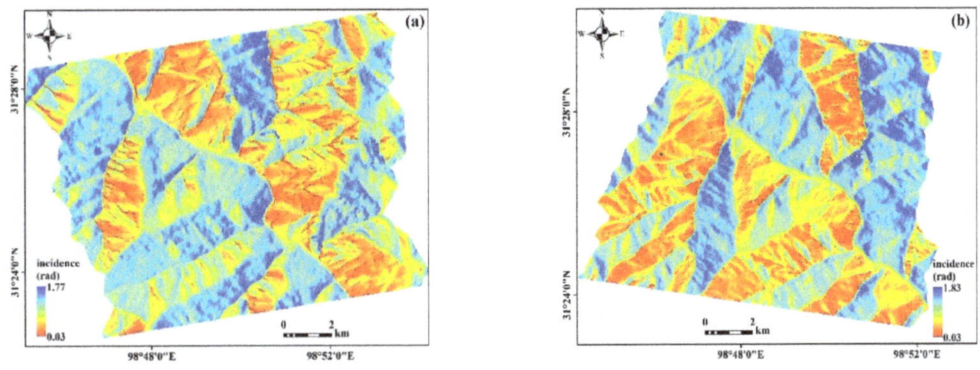

Figure 5. Radar incidence angle distribution map: (**a**) ascending orbit SAR image; (**b**) descending orbit SAR image.

3.3.2. Stability Calculation

We analyzed the stability of the Woda landslide area by the rescaled range (R/S) analysis method [46,47]. The R/S analysis method is a time series fractal statistical method based on fractal theory. It estimates the deformation trend over time and predicts the continuity of future deformation by calculating the Hurst index value H of a non-linear time series. A long time series $\{x_N\}$ consisting of N scenes of SAR images can be divided into m sub-intervals with the length L. For each sub-interval, set

$$X_{t,m} = \sum_{u=1}^{L} (x_u - \overline{x}_u) \tag{15}$$

where $\overline{x}_u$ is the average value of the n-th ($n \leq m$) sub-interval x_u, and $X_{t,n}$ is the cumulative deviation of the n-th sub-interval. Then,

$$R = \max(X_{t,n}) - \min(X_{t,n}) \tag{16}$$

$$S = \sqrt{\frac{1}{L}\sum_{u=1}^{L} (x_u - \overline{x}_u)^2} \tag{17}$$

where R is the range of the interval, and S is the standard deviation of the interval. R/S is

$$R/S = k(n)^H \tag{18}$$

where k is a constant with the empirical value of 0.5; H is the Hurst exponent. Taking the logarithm of both sides of the above equation, we get

$$\log{(R/S)}_n = H\log(n) + \log(k) \tag{19}$$

From the above equation, the Hurst index value H can be estimated by the least square regression. The calculated Hurst index value ranges between 0 and 1. Only when the value is greater than 0.5, the monitoring time series deformation has a long-term correlation, and the current deformation trend will be maintained in the future. The closer the Hurst index value is to 1, the more reliable and stable the correlation and consistency are.

4. Results and Analysis

4.1. Monitoring Results and Analysis

From the 106 ascending and 102 descending SAR images, we derived the surface deformation of the Woda landslide area using the above method. On the basis of the SHPs selected by the HTCI algorithm, we optimized the interferograms by the adaptive nonlocal filtering method. The time coherence threshold was defined as 0.6 to obtain the DS point; the amplitude dispersion index threshold was set as 0.4 to obtain the PS point. Both the PS point and the DS point were adopted for the phase unwrapping and residual phase removal, and finally, the surface deformation during the study period was obtained.

Figure 6 shows the LOS deformation rate of the study area obtained by different methods. The Woda landslide area is mainly mountainous and lacks strong scatterer point targets, such as buildings and bare rock. So, the point targets obtained by the traditional PS-InSAR method are sparse (Figure 6b,d) and cannot be used for analysis. The results of DS-InSAR greatly improve the spatial distribution density of point targets than that of PS-InSAR, especially in the areas with weaker coherence. Specifically, DS-InSAR selected a total of 162,992 point targets from the ascending data and 149,518 point targets from the descending data, with point densities of 25.5% and 23.4%, respectively. The PS-InSAR method selected a total of 2293 point targets from the ascending data and 3088 point targets from the descending data, with densities of 0.4% and 0.5%, respectively. In addition, as Figure 6a,c shows, the results from the ascending and descending data are significantly different in spatial distribution and magnitude. The maximum deformation rate of the ascending data is −156 mm/yr, but for the descending data, it is −129 mm/yr. The reasons for this could be (1) due to the differences in the observation direction, the incidence angle and heading angle of the ascending and descending radar satellites, the projection of the deformation on the LOS of different attitudes at the same location must be different; (2) mountainous areas with great terrain undulations result in geometric distortion in SAR images, which will directly affect the deformation results; (3) due to the changes in slope and aspect, the results of the deformation in the ascending and descending LOS directions of the same location may be opposite.

In this study area, the PS-InSAR method selected no point target at the landslide bodies and could not perform subsequent experimental analysis, indicating that the method has great defects in low-coherence research areas, such as for landslide monitoring. The DS-InSAR method selected dense point targets and provided rich deformation details in the area, which was very helpful for subsequent deformation analysis and discussion.

Due to the absence of leveling and GNSS data, we verified the reliability of the DS-InSAR monitoring results by its correlation with the results of PS-InSAR [48,49]. Figure 7 shows the correlation between the PS-InSAR and DS-InSAR deformation rate results. Since both the PS-InSAR and DS-InSAR results are dependent variables, the correlation coefficient (Corr) proposed by Karl Pearson [50] was used to verify the results of the DS technology. The Corr is a statistical indicator for reflecting the correlation between variables. The closer the Corr value is to 1, the closer the correlation is between the two results. In the ascending orbit, the Corr is 0.83 and the root mean square error (RMSE) is 4 mm/yr; in the descending orbit, the Corr is 0.77 and the RMSE is 5 mm/yr. The results of these two methods are highly correlated, which proves the reliability of DS-InSAR in monitoring the surface deformation of the landslide area.

After analyzing and verifying the deformation rate results, we analyzed the time series deformation in the landslide area to understand the surface deformation characteristics more comprehensively. Figure 8 is a three-dimensional optical image feature map of the landslide area. The yellow curves outline the identification boundaries of the two landslide bodies 1 and 2. Considering the geomorphology and slopes, we selected three points from each landslide body and analyzed their time series deformation trends to understand the law of surface change in the landslide area.

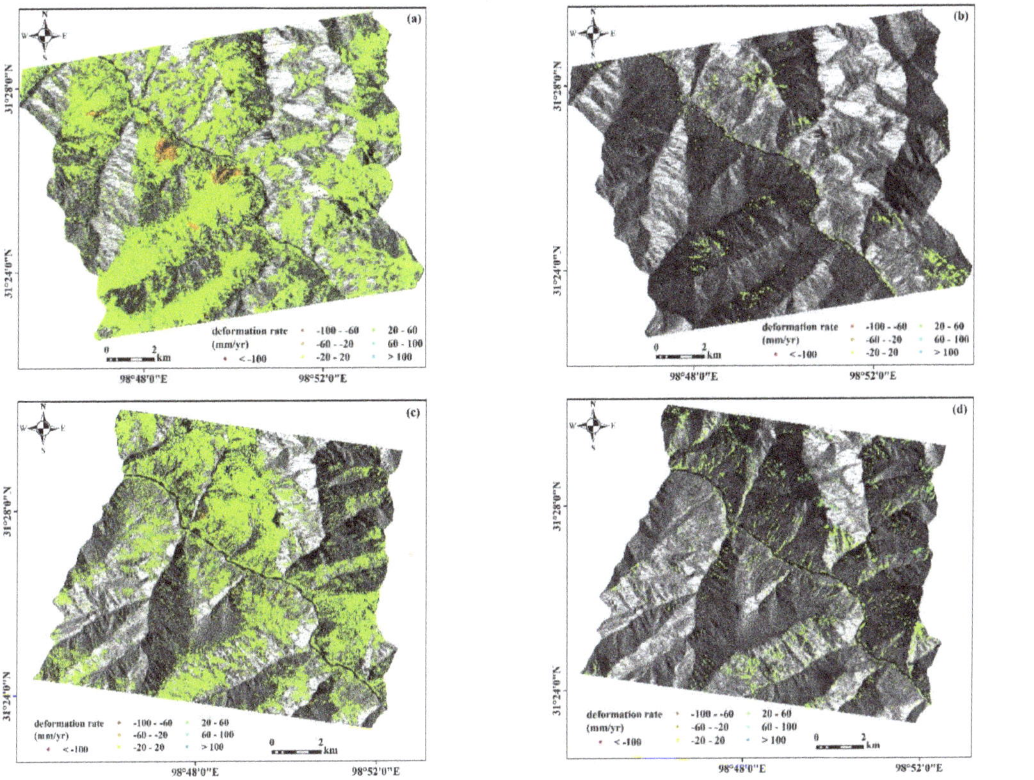

Figure 6. The LOS deformation rate maps of the Woda landslide area (**a**) derived from ascending data using the DS-InSAR method; (**b**) derived from ascending data using the PS-InSAR method; (**c**) derived from descending data using the DS-InSAR method; (**d**) derived from descending data using the PS-InSAR method.

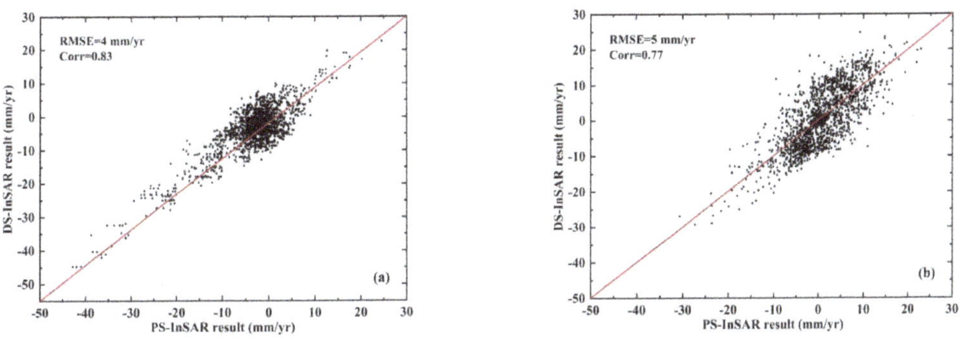

Figure 7. Correlation between PS-InSAR and DS-InSAR deformation rates in the (**a**) ascending orbit and the (**b**) descending orbit.

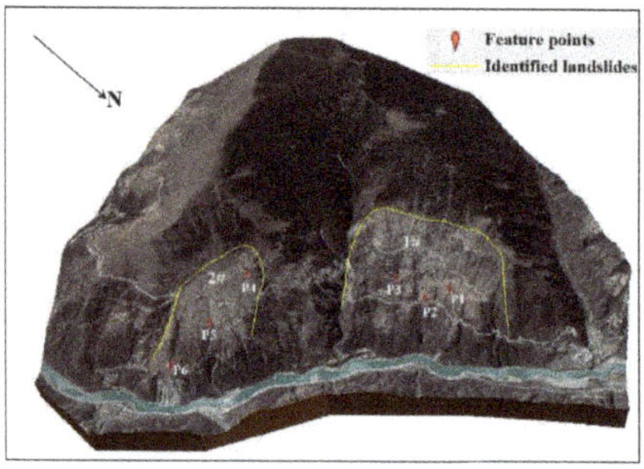

Figure 8. Three-dimensional optical image feature map of the Woda landslide area.

Figure 9 shows the time series deformation trends of points P1–P6 derived from ascending and descending data between 5 November 2014 and 4 September 2019. The six feature points all have varying degrees of deformation and the cumulative deformation statistics of feature points are shown in Table 2. P1 is located in the west-central part of landslide body 1. The ascending result shows great deformation, with a cumulative deformation of −342 mm. The descending result shows a slight uplift (33 mm), which may be a result of the electromagnetic wave emission direction and the radar incidence angle. P2 is located in the south-central part of landslide body 1. The ascending result shows large deformation, with a cumulative deformation of −375 mm. The deformation in the descending result is smaller, with a cumulative deformation of −85 mm. P3 is located in the east-central part of landslide body 1. It has smaller deformation in the ascending result (−163 mm) than P1 and P2, but the change trend is obvious. The descending orbit result is smaller and fluctuates greatly, indicating that the monitoring results were greatly affected by noise. P4 is located in the upper part of landslide body 2. The deformation of the ascending result is large (−209 mm), but the deformation trend is not smooth. The deformation in the descending result is −73 mm. P5 is located in the middle of landslide body 2. The deformations in the ascending and descending results are −223 mm and −220 mm, respectively. P6 is located in the lower part of landslide body 2. The deformation in the ascending orbit result is −284 mm. The deformation in the descending result has a similar trend, but the value is smaller, at −214 mm.

Table 2. Cumulative deformation statistics of feature points.

Feature Points	Cumulative Deformation (mm)	
	Ascending	Descending
P1	−342	33
P2	−375	−85
P3	−163	−42
P4	−209	−73
P5	−223	−220
P6	−284	−214

As Figure 9 shows, the six feature points all experienced deformation acceleration in August 2017, but the deformations in the ascending and descending results of each feature point are quite different. Thus, it was difficult to get a unified summary for the surface deformation law according to the time series deformation trend. Therefore, we established a two-dimensional deformation model using the results of the ascending and descending orbits to further analyze the landslide area.

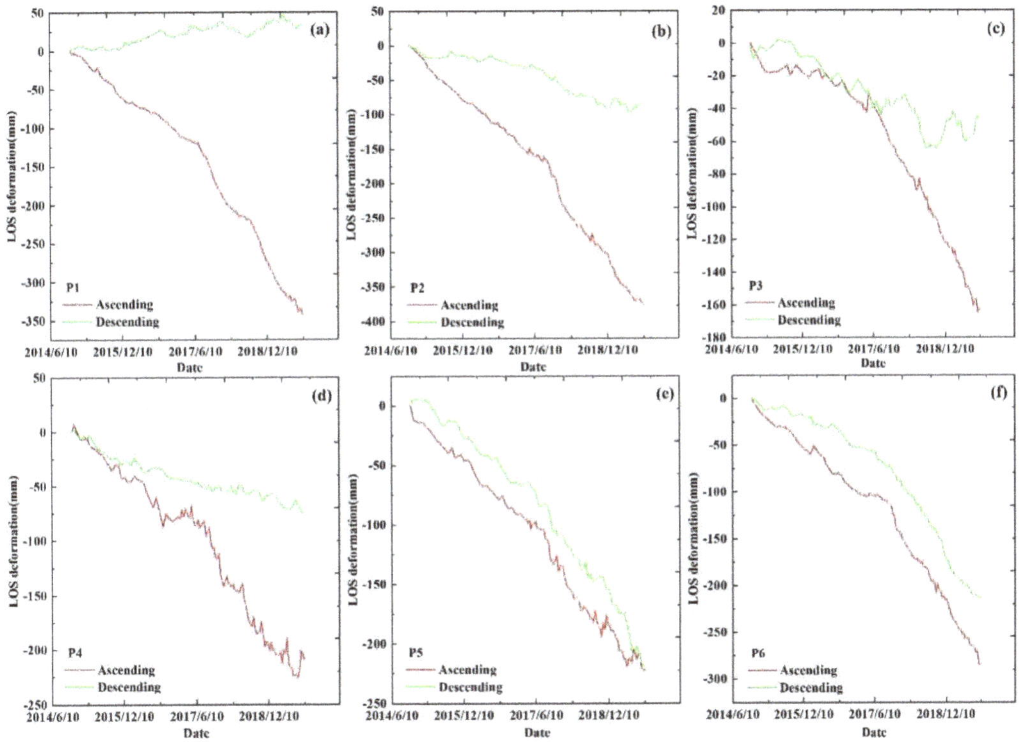

Figure 9. Time series deformation of points P1–P6 (**a**–**f**) based on ascending and descending results.

4.2. Two-Dimensional Deformation Results and Analysis

As the geographic locations of the monitoring points in the ascending orbit SAR image and those in the descending orbit SAR image may not be exactly the same, the homonymy points had to be selected before the two-dimensional deformation modeling. In this study, if the monitoring points on the two orbits had a distance of less than 10 m, they were selected as the homonymy points, and a total of 76,943 monitoring points were obtained (Figure 10).

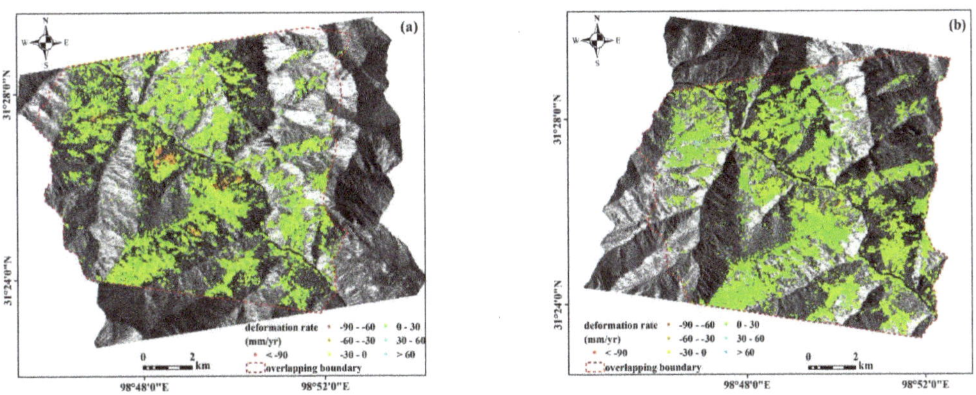

Figure 10. The distribution of the homonymy points: (**a**) ascending orbit results; (**b**) descending orbit results.

According to the distribution results of the homonymy points, we obtained the normal deformation and the east–west deformation by Equations (12) and (14). As shown in Figure 11a,b, the two landslide bodies have obvious deformations, but the other regions are stable. The maximum surface deformation rate in the landslide bodies was −80 mm/yr in the normal direction and 118 mm/yr in the east–west direction. To analyze the surface deformation characteristics more clearly and intuitively, the region in the red dashed box was displayed in a three-dimensional map, as shown in Figure 11c,d. In the normal direction, the deformation rate was between −70mm/yr and −40mm/yr, and higher in the east–west direction. The regions with deformation rates greater than −70mm/yr are concentrated in the middle of landslide body 2. In the east–west direction, the surface deformation rates in the center area of the two landslide bodies were high (60 mm/yr to 90 mm/yr), especially in the middle part of landslide body 1 (>90 mm/yr). The deformation rate of the rest area is about 30 mm/yr and not significant.

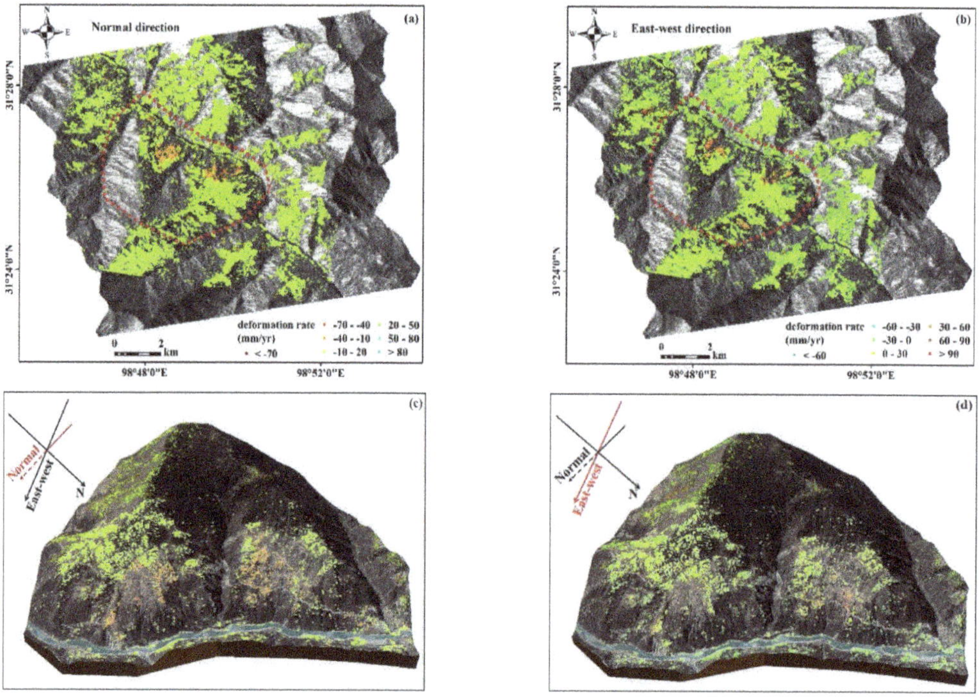

Figure 11. Two-dimensional deformation results (**a**) in the normal direction and (**b**) in the east–west direction. Three-dimensional model of the deformation (**c**) in the normal direction and (**d**) in the east–west deformation.

Through the two-dimensional calculation of the landslide area from the normal and east–west directions, and combined with the actual surface model, the surface deformation analysis was more comprehensive, specific, and vivid. Therefore, compared to the single LOS direction deformation analysis, the two-dimensional deformation is beneficial for comprehensively analyzing the surface deformation characteristics of the landslide area, summarizing the surface deformation laws, as well as providing help for the subsequent discussion of the influence of precipitation and the deformation mechanism.

4.3. Stability Analysis

In order to further study and analyze the surface stability of landslide bodies 1 and 2, we analyzed the stability of the P1–P6 feature points. For the R/S analysis of feature points, we acquired 106 images from 5 November 2014 to 4 September 2019, which imposed a huge calculation burden. We divided the study period into six segments and calculated the Hurst index of each segment separately. The results in the normal direction and east–west direction are shown in Tables 3 and 4, respectively. The distribution of the calculation results is shown in Figure 12.

Table 3. Hurst index calculation results in the normal direction.

No.	Period of Time	Hurst Index					
		P1	P2	P3	P4	P5	P6
1	5 November 2014–18 December 2015	0.934	0.919	0.842	0.938	0.953	0.847
2	18 December 2015–10 February 2017	0.956	0.949	0.907	0.942	0.932	0.939
3	10 February 2017–26 September 2017	0.948	0.818	0.943	0.774	0.726	0.900
4	26 September 2017–12 May 2018	0.932	0.939	0.817	0.932	0.905	0.935
5	12 May 2018–7 January 2019	0.817	0.949	0.956	0.865	0.748	0.924
6	7 January 2019–4 September 2019	0.934	0.945	0.935	0.904	0.937	0.925

Table 4. Hurst index calculation results in the east–west direction.

No.	Period of Time	Hurst Index					
		P1	P2	P3	P4	P5	P6
1	5 November 2014–18 December 2015	0.932	0.938	0.867	0.963	0.840	0.931
2	18 December 2015–10 February 2017	0.934	0.938	0.828	0.945	0.876	0.936
3	10 February 2017–26 September 2017	0.909	0.911	0.956	0.942	0.736	0.748
4	26 September 2017–12 May 2018	0.925	0.929	0.927	0.908	0.959	0.922
5	12 May 2018–7 January 2019	0.947	0.904	0.927	0.881	0.849	0.915
6	7 January 2019–4 September 2019	0.946	0.908	0.935	0.858	0.928	0.905

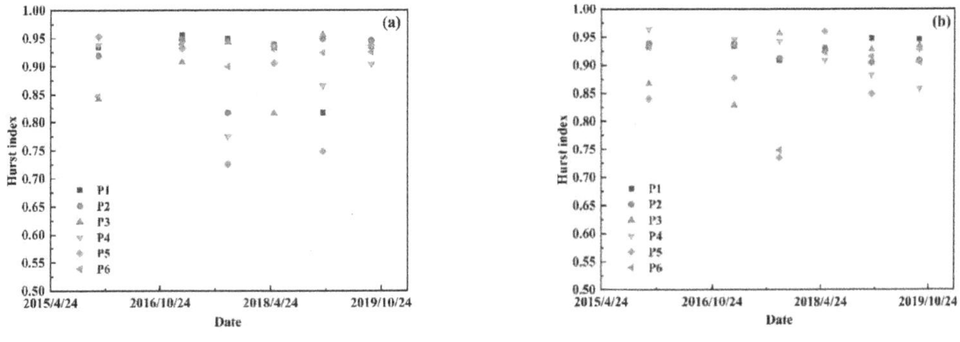

Figure 12. Hurst index values of P1–P6 feature points (**a**) in the normal direction and (**b**) in the east–west direction.

From Table 3 and Figure 12a we found that, in the normal direction, the Hurst index of each feature point is greater than 0.5 at each segment, indicating that the time series deformation results have a long-term correlation. From 10 February 2017 to 26 September 2017, the Hurst index values of each feature point are low, and the Hurst index values of P2, P4, and P5 are all lower than 0.9, which is a little lower than the results of the other periods and also confirms the acceleration trend of deformation after the precipitation peak in 2017 obtained in the previous study. From 7 January 2019 to 4 September 2019, the Hurst index value of each feature point is greater than 0.9, indicating that the normal direction deformation trend will continue for some time. From Table 4 and Figure 12b, we found that in the east–west direction, the Hurst index calculated at each feature point is also greater than 0.5 at each segment, indicating that the time series deformation results have a long-term correlation. From 10 February 2017 to 26 September 2017, the Hurst index values of P5 and P6 are obviously relatively low, at 0.736 and 0.748, respectively, while the other points are greater than 0.9. From 7 January 2019 to 4 September 2019, the Hurst index value of P4 is 0.858, and those of the other points are all above 0.9, indicating that the east–west deformation trend will continue for some time.

As Figure 12 shows, the Hurst index values of P4–P6 have larger fluctuations than those of P1–P3, especially between 7 January 2019 and 4 September 2019. This means that after the peak of precipitation in 2017, the influence of precipitation on landslide body 2 was greater than on landslide body 1. In addition, the Hurst index values of the P1–P6 feature points are higher than 0.5 in both the normal and east–west directions, indicating that the current deformation trend will continue. It is safe to infer that the surface deformation of the two landslide bodies may develop along the current trend. In this study, we adopted the R/S analysis method, which not only analyzes the stability and reliability of the surface deformation trend during the study period but also predicts the deformation trend of the subsequent landslide surface and provides effective theoretical support for local disaster warning and reduction.

5. Discussion

5.1. *Influence of Precipitation*

In order to explore the inducing factors of the surface deformation of the Woda landslide area, we analyzed the relationships between the precipitation and normal deformation and the east–west deformation time series. Since the time span of the ascending orbit SAR images is not completely consistent with that of the descending orbit SAR images, based on the time span of the ascending SAR images (5 November 2014–4 September 2019), we performed time linear interpolation on the time series deformation results of the descending SAR images. We calculated the two-dimensional time series deformation of the ascending and descending data. The precipitation data were processed according to the temporal baseline length between the SAR images of the ascending orbit, which makes the analysis more reasonable.

Figure 13 shows the normal direction time series deformation and precipitation distribution map of P1–P6. Figure 14 shows the east–west direction time series deformation and precipitation distribution map of P1–P6. As the two figures show, the precipitation was mostly concentrated in May to September each year. Therefore, in this study, we focused on the deformation trend after the precipitation peak period from 2016 to 2018, as shown by the green dashed boxes in Figures 13 and 14. As Figures 13a and 14a show, after the precipitation peak period in 2016, the normal direction deformations of P1 and P3 had accelerated trends. After the precipitation peak period in 2017, the normal direction and east–west direction deformations of P1 and P2 had obvious acceleration. The east–west direction deformation of P3 accelerated in May, in the rainy season. After the precipitation peak period in 2018, the east–west deformation of P1 accelerated, and the deformation of the other points did not change significantly. In Figures 13b and 14b, after the precipitation peak period in 2016, only the normal deformation of P6 had an acceleration trend. After the precipitation peak period in 2017, the normal direction deformation and east–west direction deformations of P4–P6 show an obvious acceleration. P6 shows the most obvious acceleration and maintained this trend for some time. After the precipitation peak period in 2018, the deformations of the P4–P6 feature points are not obvious. In summary, the P1–P6 feature points were most affected by the precipitation peak period in 2017.

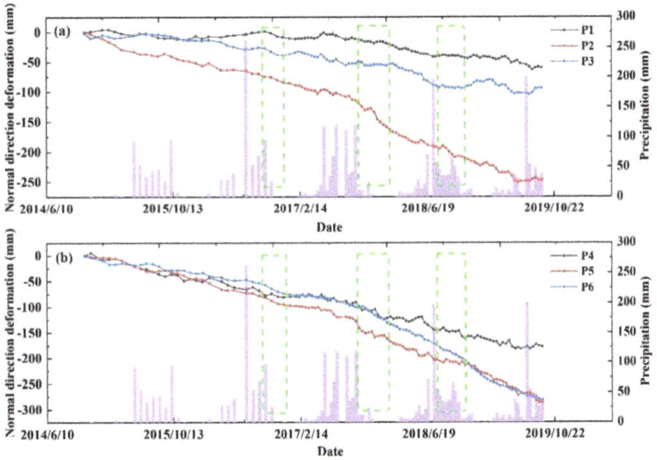

Figure 13. Time series deformation in the normal direction and precipitation distribution map of the P1–P6 feature points. (**a**) includes P1–P3 feature points; (**b**) includes P4–P6 feature points. The green dashed boxes are the deformation trend of the feature points after the period of concentrated precipitation.

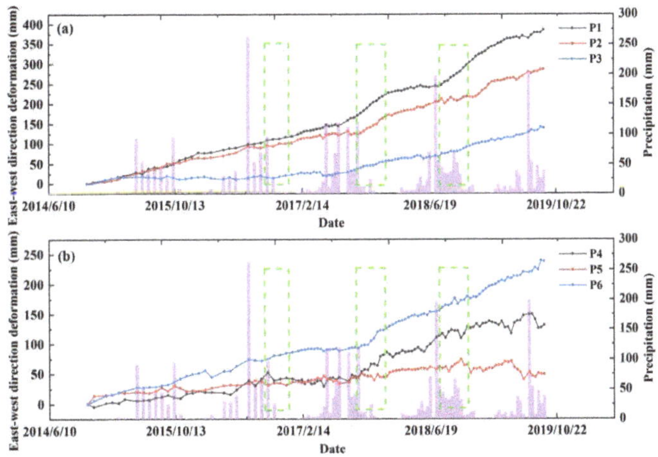

Figure 14. Time series deformation in the east–west direction and precipitation distribution map of the P1–P6 feature points. (**a**) includes P1–P3 feature points; (**b**) includes P4–P6 feature points. The green dashed boxes are the deformation trend of the feature points after the period of concentrated precipitation.

To further analyze the relationship between precipitation and the time series deformation, more comprehensive statistics of precipitation data are listed in Table 5. In May–September from 2015 to 2019, the precipitation in May–September 2017 was the highest, reaching 676.17 mm, which is 28.5% higher than in 2016. Table 5 also lists the maximum daily precipitation rates throughout the whole study period. Three of the five days with the highest precipitation are in 2017. The daily precipitation of 13 May 2017 reached 113.69 mm, which is a heavy rainstorm; the daily precipitation for the two days of 7 September 2017 and 4 July 2017 also reached the heavy rain level. Combining the deformation trends of P1–P6 in the LOS direction (Figure 9), normal direction (Figure 13), and east–west direction (Figure 14), from Table 5, we found that the deformation of each point accelerated during August to November 2017, indicating that precipitation was an important factor leading to the deformation acceleration; the precipitation from May to September in 2015 and 2016 was less than that in 2017 and 2018, so the effect of precipitation on the deformation acceleration is not significant; extremely rainy weather and high precipitation are some of the main reasons for accelerated surface deformation in this landslide area. This provides a useful reference for analyzing and judging the trend of surface deformation in the landslide area.

Table 5. Detailed statistics on precipitation data.

Annual Precipitation from May to September		Precipitation per Day Ranking	
Date	**Precipitation (mm)**	**Date**	**Precipitation (mm)**
2015	597.59	13 May 2017	113.69
2016	526.27	7 September 2017	91.71
2017	676.17	2 July 2019	86.79
2018	642.35	4 July 2017	65.02
2019	559.48	6 August 2015	45.61

5.2. Discussion of Deformation Mechanism

The Woda landslide area is located in a typical alpine and canyon landform, through which the Jinsha River runs. In this area, the Jinsha River is about 115 m wide, and the terrain of the river is gentle. The landslide bodies have circle chair shapes, with multi-level terraces [5]. A large amount of landslide material is accumulated on the slope. The average slope of the landslide bodies is 20°–25°, and the maximum slope at the front edge of the landslide bodies can reach 70°. In addition, the area has a plateau cold temperate semi-humid climate, with uneven precipitation throughout the year. Heavy rains occur occasionally in the area, which has become a significant cause of geological disasters.

As Figure 15 shows, the lithology of this area is mainly composed of schist, slate, and shale of the Upper Triassic Lanashan Formation and of crystalline limestone and feldspar quartz sandstone of the Upper Triassic Tumgou Formation. Under the influence of long-term tectonic movement, weathering, and rainfall, rock mass joints and fissures are abundant and the structure is broken. Therefore, the deformation of the Woda landslide area is controlled by geographical factors, such as stratum lithology and geological structure. In addition, external factors, such as steep terrain, climate environment, and rainfall, promote deformation.

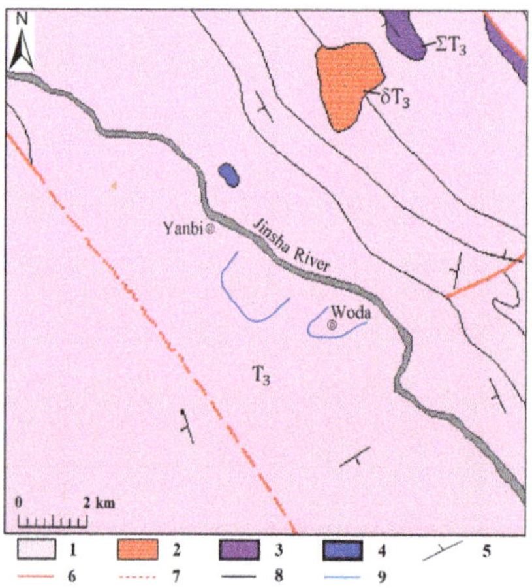

Figure 15. Geological map of the study area. (1) Upper Triassic; (2) Upper Triassic diorite; (3) Upper Triassic ultrabasic rock blocks; (4) diorite vein; (5) stratigraphic occurrence; (6) actual blatt flaws; (7) supposed blatt flaws; (8) stratigraphic boundary; (9) identified landslides.

Once a large-scale slide occurs, the materials may block the Jinsha River, forming a dammed lake and inducing indirect disasters, such as river overtopping, dam breaking, and flooding. Therefore, long-term monitoring of the surface deformation with the InSAR method is necessary, which can be used for analyzing and judging the landslide scale, deformation trends, and influencing factors. It plays an important role in improving emergency management and the prevention and mitigation of landslide disasters.

5.3. Limitations and Prospects

In this study, surface deformation monitoring and stability analysis of the Woda landslide area were successfully carried out. It is undeniable that this study has certain limitations. (1) Due to the influence of geographical location and topography, there is a lack of leveling and GNSS data for support and verification; (2) only SAR data was used, and there was a lack of optical satellite and remote sensing images for collaborative monitoring and analysis; (3) restricted by relevant management regulations, the collection of geological data was not rich enough, and there was no perfect in-depth analysis of the geological structure and lithology.

In future research, with the development of new observation platforms, such as the BeiDou Navigation Satellite System (BDS) and UAV, it is possible to combine navigation and positioning data, UAV photography data, SAR data of different polarization modes, and multi-platform spaceborne SAR data to carry out surface deformation monitoring fieldwork. In the field of landslide surface deformation monitoring, a scientific and reliable monitoring and evaluation system will be formed.

6. Conclusions

This study adopted the state-of-the-art of DS-InSAR method to extract the surface deformation of the Woda landslide area. The HTCI algorithm was used to select SHPs, and the adaptive spatial nonlocal filtering method was combined with the SHP selection result to optimize the phases. Both DS and PS points were used to increase the density of monitoring points in the study area.

From the 106 Sentinel-1A ascending images and 102 Sentinel-1A descending images, we derived the time series deformation of the Woda landslide area for nearly 5 years by the DS-InSAR method. The two landslide bodies have obvious deformations, but the other areas are stable. Compared to the traditional PS-InSAR method, the DS-InSAR method increased the monitoring points density significantly. The results of the monitoring points were verified by the Karl Pearson correlation coefficients, which are 0.83 for the ascending result and 0.77 for the descending result. The obtained two-dimensional deformation model shows that the maximum deformation rate of the normal direction is −80 mm/yr, and for the east–west direction, it is 118 mm/yr. The stability analysis results show that the current deformation trend is sustainable. The time linear interpolation analysis of the six feature points suggests that the deformation is controlled by geographical factors, such as stratum lithology and geological structure, and it is affected by external factors, such as precipitation. The research results and analysis method in this paper provide reference and guidance for the early investigation, monitoring, and timely warning of dense forest landslides.

Author Contributions: Y.L., H.Y., and S.W. conceived of the original design of this paper. Y.L. and H.Y. improve the structure of the paper. L.X. and J.P. provided comments on this paper. All authors have read and agreed to the published version of the manuscript.

Funding: This research was funded by the National Natural Science Foundation of China (42174026), the National Key Research and Development Program of China (2021YFE011004), and the Open Fund of State Key Laboratory of Coal Resources and Safe Mining (Grant No. SKLCRSM20KFA12).

Data Availability Statement: Data sharing is not applicable.

Acknowledgments: The authors would like to thank the European Space Agency for providing Sentinel-1A data and the National Aeronautics and Space Administration for providing SRTM1 DEM data. In addition, thanks to Google for the optical imagery.

Conflicts of Interest: The authors declare that they have no known competing financial interests or personal relationships that could have appeared to influence the work reported in this paper.

References

1. Liu, C.; Li, W.; Wu, H.; Lu, P.; Sang, K.; Sun, W.; Chen, W.; Hong, Y.; Li, R. Susceptibility evaluation and mapping of China's landslides based on multi-source data. *Nat. Hazards* **2013**, *69*, 1477–1495. [CrossRef]
2. Delaney, K.B.; Evans, S.G. The 2000 Yigong landslide (Tibetan Plateau), rockslide-dammed lake and outburst flood: Review, remote sensing analysis, and process modelling. *Geomorphology* **2015**, *246*, 377–393. [CrossRef]
3. Fan, X.; Xu, Q.; Alonso-Rodriguez, A.; Subramanian, S.S.; Li, W.; Zheng, G.; Dong, X.; Huang, R. Successive landsliding and damming of the Jinsha River in eastern Tibet, China: Prime investigation, early warning, and emergency response. *Landslides* **2019**, *16*, 1003–1020. [CrossRef]
4. Sijing, W.; Guohe, L.; Qiang, Z.; Chaoli, L. Engineering Geological Study of the Active Tectonic Region for Hydropower Development on the Jinsha River, Upstream of the Yangtze River. *Acta Geol. Sin.* **2000**, *74*, 353–361. [CrossRef]
5. Feng, W.; Dun, J.; Yi, X.; Zhang, G. Deformation analysis of Woda village old landslide in Jinsha river basin using SBAS-InSAR technology. *J. Eng. Geol.* **2020**, *28*, 384–393. (In Chinese)
6. Ouyang, C.; An, H.; Zhou, S.; Wang, Z.; Su, P.; Wang, D.; Cheng, D.; She, J. Insights from the failure and dynamic characteristics of two sequential landslides at Baige village along the Jinsha River, China. *Landslides* **2019**, *16*, 1397–1414. [CrossRef]
7. Stiros, S.C.; Vichas, C.; Skourtis, C. Landslide Monitoring Based On Geodetically Derived Distance Changes. *J. Surv. Eng.* **2004**, *130*, 156–162. [CrossRef]
8. Du, J.-C.; Teng, H.-C. 3D laser scanning and GPS technology for landslide earthwork volume estimation. *Autom. Constr.* **2007**, *16*, 657–663. [CrossRef]
9. Muntean, A.; Mocanu, V.; Ambrosius, B. A GPS study of land subsidence in the Petrosani (Romania) coal mining area. *Nat. Hazards* **2016**, *80*, 797–810. [CrossRef]
10. Casagli, N.; Frodella, W.; Morelli, S.; Tofani, V.; Ciampalini, A.; Intrieri, E.; Raspini, F.; Rossi, G.; Tanteri, L.; Lu, P. Spaceborne, UAV and ground-based remote sensing techniques for landslide mapping, monitoring and early warning. *Geoenvironmental Disasters* **2017**, *4*, 9. [CrossRef]
11. Bayer, B.; Simoni, A.; Mulas, M.; Corsini, A.; Schmidt, D. Deformation responses of slow moving landslides to seasonal rainfall in the Northern Apennines, measured by InSAR. *Geomorphology* **2018**, *308*, 293–306. [CrossRef]
12. Liu, Y.; Fan, H.; Wang, L.; Zhuang, H. Monitoring of surface deformation in a low coherence area using distributed scatterers InSAR: Case study in the Xiaolangdi Basin of the Yellow River, China. *Bull. Eng. Geol. Environ.* **2021**, *80*, 25–39. [CrossRef]
13. Yin, Y.; Zheng, W.; Liu, Y.; Zhang, J.; Li, X. Integration of GPS with InSAR to monitoring of the Jiaju landslide in Sichuan, China. *Landslides* **2010**, *7*, 359–365. [CrossRef]
14. Chen, Y.; Yu, S.; Tao, Q.; Liu, G.; Wang, L.; Wang, F. Accuracy verification and correction of D-InSAR and SBAS-InSAR in monitoring mining surface subsidence. *Remote Sens.* **2021**, *13*, 4365. [CrossRef]
15. Ferretti, A.; Prati, C.; Rocca, F. Permanent scatterers in SAR interferometry. *IEEE Trans. Geosci. Remote Sens.* **2001**, *39*, 8–20. [CrossRef]
16. Berardino, P.; Fornaro, G.; Lanari, R.; Sansosti, E. A New Algorithm for Surface Deformation Monitoring Based on Small Baseline Differential SAR Interferograms. *IEEE Trans. Geosci. Remote Sens.* **2002**, *40*, 2375–2383. [CrossRef]
17. Werner, C.; Wegmuller, U.; Strozzi, T.; Wiesmann, A. Interferometric point target analysis for deformation mapping. In Proceedings of the IEEE International Geoscience and Remote Sensing Symposium, Toulouse, France, 21–25 July 2003; Institute of Electrical and Electronics Engineers: Muri, Switzerland; pp. 4362–4364.
18. Zhang, L.; Lu, Z.; Ding, X.; Jung, H.-S.; Feng, G.; Lee, C.-W. Mapping ground surface deformation using temporarily coherent point SAR interferometry: Application to Los Angeles Basin. *Remote Sens. Environ.* **2012**, *117*, 429–439. [CrossRef]
19. Osmanoğlu, B.; Sunar, F.; Wdowinski, S.; Cabral-Cano, E. Time series analysis of InSAR data: Methods and trends. *ISPRS J. Photogramm. Remote Sens.* **2016**, *115*, 90–102. [CrossRef]
20. Del Soldato, M.; Riquelme, A.; Bianchini, S.; Tomàs, R.; Di Martire, D.; De Vita, P.; Moretti, S.; Calcaterra, D. Multisource data integration to investigate one century of evolution for the Agnone landslide (Molise, southern Italy). *Landslides* **2018**, *15*, 2113–2128. [CrossRef]
21. Rosi, A.; Tofani, V.; Tanteri, L.; Tacconi Stefanelli, C.; Agostini, A.; Catani, F.; Casagli, N. The new landslide inventory of Tuscany (Italy) updated with PS-InSAR: Geomorphological features and landslide distribution. *Landslides* **2018**, *15*, 5–19. [CrossRef]
22. Intrieri, E.; Raspini, F.; Fumagalli, A.; Lu, P.; Del Conte, S.; Farina, P.; Allievi, J.; Ferretti, A.; Casagli, N. The Maoxian landslide as seen from space: Detecting precursors of failure with Sentinel-1 data. *Landslides* **2018**, *15*, 123–133. [CrossRef]
23. Komac, M.; Holley, R.; Mahapatra, P.; van der Marel, H.; Bavec, M. Coupling of GPS/GNSS and radar interferometric data for a 3D surface displacement monitoring of landslides. *Landslides* **2015**, *12*, 241–257. [CrossRef]
24. Fobert, M.-A.; Singhroy, V.; Spray, J.G. InSAR monitoring of landslide activity in Dominica. *Remote Sens.* **2021**, *13*, 815. [CrossRef]

25. Ferretti, A.; Fumagalli, A.; Novali, F.; Prati, C.; Rocca, F.; Rucci, A. A New Algorithm for Processing Interferometric Data-Stacks: SqueeSAR. *IEEE Trans. Geosci. Remote Sens.* **2011**, *49*, 3460–3470. [CrossRef]
26. Goel, K.; Adam, N. A Distributed Scatterer Interferometry Approach for Precision Monitoring of Known Surface Deformation Phenomena. *IEEE Trans. Geosci. Remote Sens.* **2014**, *52*, 5454–5468. [CrossRef]
27. Fornaro, G.; Verde, S.; Reale, D.; Pauciullo, A. CAESAR: An Approach Based on Covariance Matrix Decomposition to Improve Multibaseline–Multitemporal Interferometric SAR Processing. *IEEE Trans. Geosci. Remote Sens.* **2014**, *53*, 2050–2065. [CrossRef]
28. Jiang, M.; Ding, X.; Hanssen, R.F.; Malhotra, R.; Chang, L. Fast Statistically Homogeneous Pixel Selection for Covariance Matrix Estimation for Multitemporal InSAR. *IEEE Trans. Geosci. Remote Sens.* **2014**, *53*, 1213–1224. [CrossRef]
29. Jiang, M.; Guarnieri, A.M. Distributed Scatterer Interferometry With the Refinement of Spatiotemporal Coherence. *IEEE Trans. Geosci. Remote Sens.* **2020**, *58*, 3977–3987. [CrossRef]
30. Dong, J.; Zhang, L.; Tang, M.; Liao, M.; Xu, Q.; Gong, J.; Ao, M. Mapping landslide surface displacements with time series SAR interferometry by combining persistent and distributed scatterers: A case study of Jiaju landslide in Danba, China. *Remote Sens. Environ.* **2018**, *205*, 180–198. [CrossRef]
31. Wang, J.; Wang, C.; Xie, C.; Zhang, H.; Tang, Y.; Zhang, Z.; Shen, C. Monitoring of large-scale landslides in Zongling, Guizhou, China, with improved distributed scatterer interferometric SAR time series methods. *Landslides* **2020**, *17*, 1777–1795. [CrossRef]
32. Hanssen, R.F. *Radar Interferometry*; Springer: Dordrecht, The Netherlands, 2001. [CrossRef]
33. Fan, H.; Wang, L.; Wen, B.; Du, S. A New Model for three-dimensional Deformation Extraction with Single-track InSAR Based on Mining Subsidence Characteristics. *Int. J. Appl. Earth Obs. Geoinf.* **2021**, *94*, 102223. [CrossRef]
34. Béjar-Pizarro, M.; Notti, D.; Mateos, R.M.; Ezquerro, P.; Centolanza, G.; Herrera, G.; Bru, G.; Sanabria, M.; Solari, L.; Duro, J.; et al. Mapping Vulnerable Urban Areas Affected by Slow-Moving Landslides Using Sentinel-1 InSAR Data. *Remote Sens.* **2017**, *9*, 876. [CrossRef]
35. Confuorto, P.; Di Martire, D.; Centolanza, G.; Iglesias, R.; Mallorqui, J.J.; Novellino, A.; Plank, S.; Ramondini, M.; Thuro, K.; Calcaterra, D. Post-failure evolution analysis of a rainfall-triggered landslide by multi-temporal interferometry SAR approaches integrated with geotechnical analysis. *Remote Sens. Environ.* **2017**, *188*, 51–72. [CrossRef]
36. Sun, Q.; Hu, J.; Zhang, L.; Ding, X. Towards Slow-Moving Landslide Monitoring by Integrating Multi-Sensor InSAR Time Series Datasets: The Zhouqu Case Study, China. *Remote Sens.* **2016**, *8*, 908. [CrossRef]
37. Herrera, G.; Gutiérrez, F.; García-Davalillo, J.C.; Guerrero, J.; Notti, D.; Galve, J.P.; Fernández-Merodo, J.A.; Cooksley, G. Multi-sensor advanced DInSAR monitoring of very slow landslides: The Tena Valley case study (Central Spanish Pyrenees). *Remote Sens. Environ.* **2013**, *128*, 31–43. [CrossRef]
38. Li, T.; Zhang, H.; Fan, H.; Zheng, C.; Liu, J. Position inversion of goafs in deep coal seams based on DS-InSAR data and the probability integral methods. *Remote Sens.* **2021**, *13*, 2898. [CrossRef]
39. Abrahams, J. Probability, Random Variables, and Stochastic Processes by Athanasios Papoulis. *J. Am. Stat. Assoc.* **1984**, *79*, 957. [CrossRef]
40. Neuhäuser, M. Exact tests based on the Baumgartner-Wei-Schindler statistic—A survey. *Stat. Pap.* **2005**, *46*, 1–29. [CrossRef]
41. Jiang, M.; Miao, Z.; Gamba, P.; Yong, B. Application of Multitemporal InSAR Covariance and Information Fusion to Robust Road Extraction. *IEEE Trans. Geosci. Remote Sens.* **2017**, *55*, 3611–3622. [CrossRef]
42. Deledalle, C.A.; Denis, L.; Tupin, F. NL-InSAR: Nonlocal interferogram estimation. *IEEE Trans. Geosci. Remote Sens.* **2011**, *49*, 1441–1452. [CrossRef]
43. Hu, J.; Li, Z.-W.; Ding, X.; Zhu, J.; Zhang, L.; Sun, Q. Resolving three-dimensional surface displacements from InSAR measurements: A review. *Earth Sci. Rev.* **2014**, *133*, 1–17. [CrossRef]
44. Fuhrmann, T.; Garthwaite, M.C. Resolving three-dimensional surface motion with InSAR: Constraints from multi-geometry data fusion. *Remote Sens.* **2019**, *11*, 241. [CrossRef]
45. Colesanti, C.; Wasowski, J. Investigating landslides with space-borne Synthetic Aperture Radar (SAR) interferometry. *Eng. Geol.* **2006**, *88*, 173–199. [CrossRef]
46. Hurst, H.E. Long-Term Storage Capacity of Reservoirs. *Trans. Am. Soc. Civ. Eng.* **1951**, *116*, 770–799. [CrossRef]
47. Pei, L.; Chen, J.; Zhou, J.; Huang, H.; Zhou, Z.; Chen, C.; Yao, F. A Fractal Prediction Method for Safety Monitoring Deformation of Core Rockfill Dams. *Math. Probl. Eng.* **2021**, *2021*, 6655657. [CrossRef]
48. Lv, X.; Yazici, B.; Zeghal, M.; Bennett, V.; Abdoun, T. Joint-Scatterer Processing for Time-Series InSAR. *IEEE Trans. Geosci. Remote Sens.* **2014**, *52*, 7205–7221.
49. Samiei-Esfahany, S.; Martins, J.E.; Van Leijen, F.; Hanssen, R.F. Phase estimation for distributed scatterers in InSAR stacks using integer least squares estimation. *IEEE Trans. Geosci. Remote Sens.* **2016**, *54*, 5671–5687. [CrossRef]
50. Osmani, S.A.; Banik, B.K.; Ali, H. Integrating fuzzy logic with Pearson correlation to optimize contaminant detection in water distribution system with uncertainty analyses. *Environ. Monit. Assess.* **2019**, *191*, 441. [CrossRef]

remote sensing MDPI

Technical Note

Monitoring and Predicting the Subsidence of Dalian Jinzhou Bay International Airport, China by Integrating InSAR Observation and Terzaghi Consolidation Theory

Xianlin Shi [1], Chen Chen [1], Keren Dai [1,2,*], Jin Deng [1], Ningling Wen [1], Yong Yin [1] and Xiujun Dong [2]

[1] College of Earth Science, Chengdu University of Technology, Chengdu 610059, China; shixianlin06@cdut.edu.cn (X.S.); chenchen1@stu.cdut.edu.cn (C.C.); dengjin@stu.cdut.edu.cn (J.D.); wenningling@stu.cdut.edu.cn (N.W.); yinyong@stu.cdut.edu.cn (Y.Y.)
[2] State Key Laboratory of Geological Disaster Prevention and Geological Environmental Protection, Chengdu University of Technology, Chengdu 610059, China; dongxiujun@cdut.cn
* Correspondence: daikeren17@cdut.edu.cn

Abstract: Dalian Jinzhou Bay International Airport (DJBIA) is an offshore artificial island airport, where the reclaimed land is prone to uneven land subsidence due to filling consolidation and construction. Monitoring and predicting the subsidence are essential to assist the subsequent subsidence control and ensure the operational safety of DJBIA. However, the accurate monitoring and prediction of reclaimed subsidence for such a wide area under construction are hard and challenging. This paper utilized the Small Baseline Subset Synthetic Aperture Radar (SBAS-InSAR) technology based on Sentinel-1 images from 2017 to 2021 to obtain the subsidence over the land reclamation area of the DJBIA, in which the results from ascending and descending orbit data were compared to verify the reliability of the results. The SBAS-InSAR results reveal that uneven subsidence is continuously occurring, especially on the runway, terminal, and building area of the airport, with the maximum subsidence rate exceeding 100 mm/year. It was found that there is a strong correlation between the subsidence rate and backfilling time. This study provides important information on the reclaimed subsidence for DJBIA and demonstrates a novel method for reclaimed subsidence monitoring and prediction by integrating the advanced InSAR technology and Terzaghi Consolidation Theory modeling. Moreover, based on the Terzaghi consolidation theory and the corresponding geological parameters of the airport, predicted subsidence curves in this area are derived. The comparison between predicted curves and the actual subsidence revealed by InSAR in 2017–2021 is highly consistent, with a similar trend and falling in a range of ±25 mm/year, which verifies that the subsidence in this area conforms to Terzaghi Consolidation Theory. Therefore, it can be predicted that in the future, the subsidence rate of the new reclamation area in this region will reach about 80 mm/year ± 25 mm/year, and the subsidence rate will gradually slow down with the accumulation of reclamation time. The subsidence rate will slow down to about 30 mm/year ± 25 mm/year after 10 years.

Keywords: Dalian Jinzhou Bay International Airport; SBAS-InSAR; Terzaghi consolidation theory; subsidence monitoring; subsidence prediction

Citation: Shi, X.; Chen, C.; Dai, K.; Deng, J.; Wen, N.; Yin, Y.; Dong, X. Monitoring and Predicting the Subsidence of Dalian Jinzhou Bay International Airport, China by Integrating InSAR Observation and Terzaghi Consolidation Theory. *Remote Sens.* **2022**, *14*, 2332. https://doi.org/10.3390/rs14102332

Academic Editors: Chisheng Wang, Daqing Ge, Guohong Zhang, Wu Zhu and Siting Xiong

Received: 2 April 2022
Accepted: 3 May 2022
Published: 11 May 2022

Publisher's Note: MDPI stays neutral with regard to jurisdictional claims in published maps and institutional affiliations.

1. Introduction

Dalian Jinzhou Bay International Airport (DJBIA), constructed in 2010, is built on a reclaimed artificial island with an overall planned reclamation area of 20.87 km^2, and all the reclaimed land will be used for the airport, which will be the largest offshore airport worldwide after construction. Land reclamation is prone to unstable geological foundations due to sediment consolidation [1–4]. Airports built on reclaimed land commonly suffer from land subsidence, which will threaten the operational safety of the airport, especially the land subsidence occurring in runway areas [5]. Therefore, it is essential to monitor and predict the ground subsidence in this area and assess its impact on the safe operation of the airport.

Reliable surface deformation information can be obtained through traditional point-measurement methods, such as level surveys or global satellite navigation and positioning technology. However, traditional measurement methods was difficult to apply them to ground subsidence monitoring at airports under reclamation. Interferometric synthetic aperture radar (InSAR) is an emerging geodetic technology. It is capable of acquiring high-accuracy and continuously covered surface deformation information, which is widely used in the fields of ground subsidence monitoring [1,3,6,7], and landslide monitoring [8–12]. In recent years, InSAR technology has been widely used in land subsidence monitoring at airports. Such as Hong Kong Chek Lap Kok Airport [13,14], Shanghai Pudong International Airport [15,16], Beijing Capital International Airport [17,18], Shenzhen Baoan International Airport [19,20], Kuala Lumpur International Airport, Malaysia [5], and Japanese airports [21–23]. The land formed by the land reclamation airport is characterized by high compressibility, low strength, and a large porosity ratio, which easily leads to ground subsidence. Studies related to reclaimed land airport monitoring by InSAR have been performed. Lu et al. [19] and Xu et al. [24] monitored ground subsidence at Shenzhen Baoan International Airport based on the SAR data from 2007 to 2010 and 2015 to 2019, revealing that the subsidence in the reclamation area was greater than in the ground, and dynamic loads is an important factor that causing airport subsidence. Zhuo et al. [25] conducted subsidence monitoring of Xiamen Xiang'an New Airport based on SBAS-InSAR from 2018 to 2020 and identified the areas that may be affected in the future based on the type of airport land use. Hong Kong Chek Lap Kok Airport [14] and Xiamen New Airport [26] used SAR data and geological data combined with Terzaghi consolidation theory for long-term subsidence monitoring of the reclamation airports. The above studies used InSAR observation on reclaimed airport subsidence monitoring and made a detailed analysis of subsidence. Integration of InSAR observation and geological condition of the artificial land area to predict the subsidence trend deserves further study.

In this paper, SBAS-InSAR was used to monitor the land subsidence at DJBIA based on 50 Sentinel-1A ascending SAR images between March 2017 and April 2021. The SABS-InSAR monitoring results reveal the regional distribution and time–series characteristics of airport subsidence. Based on the Terzaghi consolidation theory, the geological data of the airport were used to calculate the deformation of the soil layer at any time during the land reclamation process and to predict the subsidence development trend. The InSAR results verified that the subsidence in 2017–2021 is consistent with the predicted curve of the Terzaghi consolidation theory in the area. The integration of InSAR observation and Terzaghi Consolidation Theory works well in this area and could make a great contribution to the subsidence prediction.

2. Study Area and Data Source

2.1. Study Area

Dalian, in Liaoning Province (China), is located between the Yellow Sea and the Bohai Sea, surrounded by the sea, and dominated by mountainous and hilly peninsular landforms. The hills are northeast-southwest direction, with relatively flat terrain on the east and west sides [27]. The study area, DJBIA, is in the sea area of Jinzhou Bay, Dalian City. DJBIA is an offshore artificial island airport on the Chinese mainland, which consists of berms, the runway, and the building area (Figure 1). The western bay of Jinzhou District, where the airport is located, belongs to the high-risk area for geological disasters [28].

DJBIA can serve an average annual passenger throughput of 43 million and 550,000 tons of cargo and mail [29]. Its construction aims to effectively alleviate the tight operating resources and over-saturation of throughput at Dalian Zhoushuizi International Airport, and provide strong support for economic and social development and comprehensive transportation construction in Dalian. DJBIA has a total planned area of 20.87 km^2, including a terminal area of 500,000 m^2, with two 3600 m long teleparallel runways, 45 m, and 60 m wide, respectively [29]. The airport started to fill the land from the beginning of 2010. To better analyze the characteristics of subsidence, the reclaimed area from 2010 to 2021 was defined as six reclamation stages with a period of every two years. The spatial distribution of the reclamation area and reclamation area in different stages is shown in Figure 2a. The road to the airport has been completed in 2011, the building area and part of the terminal in 2013, all of the terminal and part of the heliport filled in 2015, and the focus on the runway and building area filled from 2017 to 2021. As shown in Figure 2b, the largest increase in filled area was 4.96 km^2 in 2015 and the accumulation of filled area will reach 11.29 km^2 by April 2021.

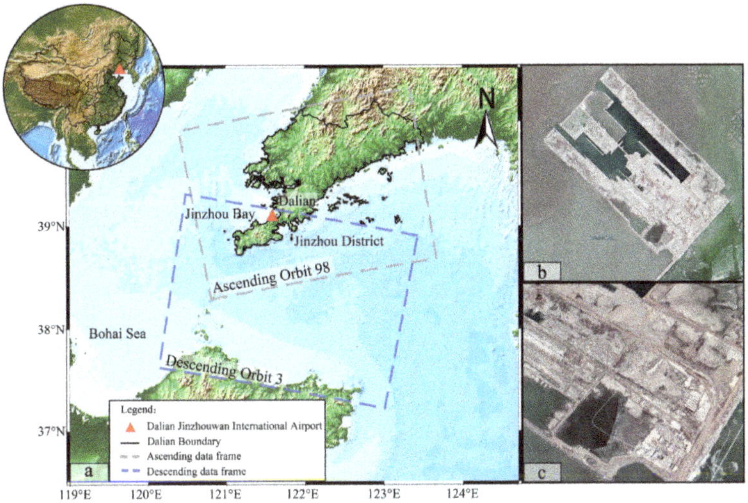

Figure 1. Overview of Dalian Jinzhou Bay International Airport: (a) Location of the study area; (b,c) Satellite map of the study area.

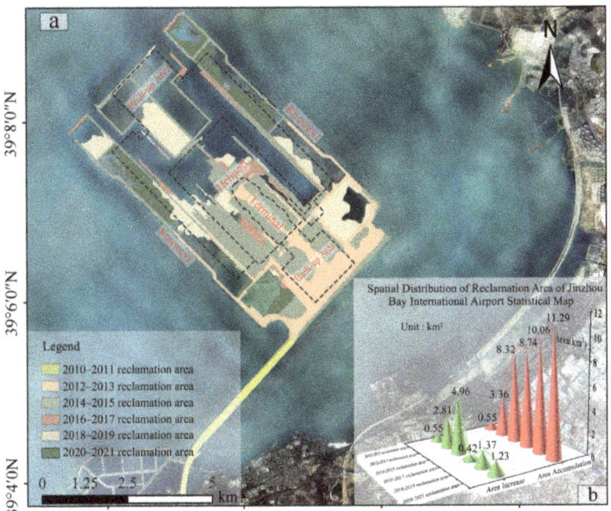

Figure 2. Spatial distribution of different reclamation stages: (a) Map of area change of the airport from 2011 to 2021; (b) Statistical of the increasement and accumulation of reclamation area. Reclamation data source: Google Earth.

2.2. Data Source

In this paper, 50 images of Sentinel-1A ascending orbit and 15 images of Sentinel-1B descending orbit from the European Space Agency (ESA) were acquired to monitor ground subsidence at DJBIA. The main obtained include orbit, resolution, polarization mode, and other parameter information, other detailed parameter information, as shown in Table 1. The 30 m resolution SRTM DEM data provided by the National Aeronautics and Space Administration (NASA) was used in the data processing, to remove the topographic phase. It is worth noting that the study area in this paper is a new reclamation area, and the airport area does not have the latest DEM in the SRTM data or other data. Considering that the entire area is relatively flat (the fluctuation is within 5 m) and the spatial baseline of the sentinel data is controlled within 100 m, the effect of elevation residuals on the deformation of relatively flat terrain in this case of short vertical baseline interference pairs can be neglected. Based on the 50 ascending images, a temporal baseline threshold of 100 days and a spatial baseline threshold of 221 m were set, and 132 pairs of interference pairs were generated by the combination (Figure 3). Spatial and temporal baselines of Sentinel-1 ascending and descending data are shown in Figure 3, Figure 4 respectively.

Table 1. Basic information of the Sentinel-1 images.

Parameter	Value	
Satellite	Sentinel-1A	Sentinel-1B
Orbit	Ascending	Descending
Azimuth/Range pixel spacing	13.99 m/2.33 m	13.99 m/2.33 m
Radar wavelength	5.6 cm	5.6 cm
Polarization mode	VV	VV
Revisit period	12 d	12 d
The angle of incidence	39.2°	39.1°
Temporal coverage	2017.03–2021.04	2019.12.20–2021.04.13
Number of images	50	15

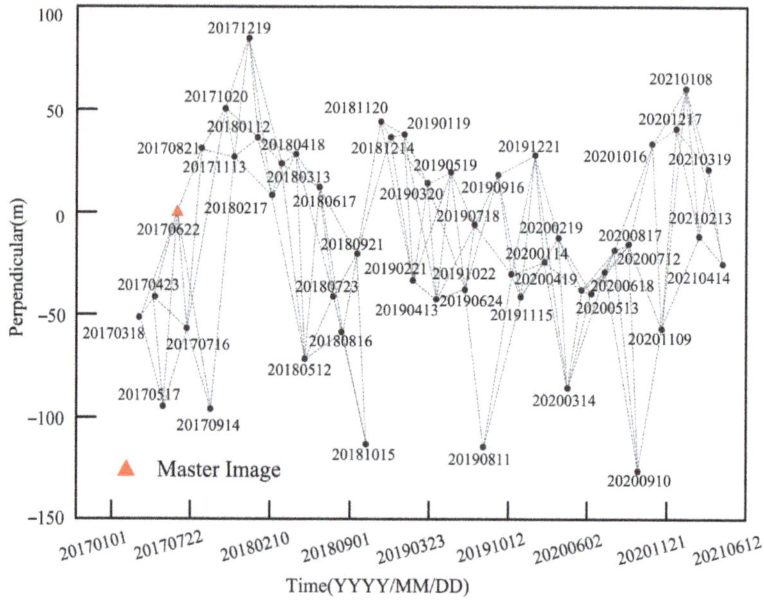

Figure 3. Spatial and temporal baselines of Sentinel-1 ascending data.

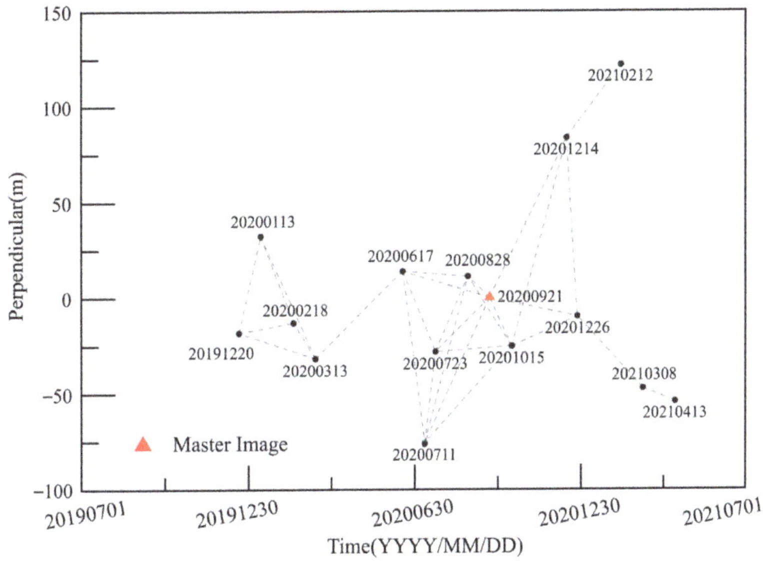

Figure 4. Spatial and temporal baselines of Sentinel-1 descending data.

DJBIA was reclaimed through rock excavation, and the major foundation filled in the site is soft soil, which can be approximately divided into highly compressible silt and clay [30,31]. This kind of foundation condition will result in large foundation subsidence. In order to calculate the subsidence of the reclamation area, geological parameters, such as initial pore ratio, compression coefficient, and the water gravity [30,32] were selected for further analysis in this paper. Detailed information is shown in Table 2.

Table 2. Experiment data in the airport.

Parameter	Corresponding Values
Permeability Coefficient	$\alpha = 0.0342$ m/year
Initial Porosity Ratio	$k_0 = 0.806$
Compression Coefficient	$\beta = 3.52 \times 10^{-4}$ kPa^{-1}
Water Gravity	$G_w = 10.101$ kN/m^3
Maximum Drainage Distance	H = −19.4 m
Solidification Stress	Γ = 250 kPa
Coefficient of Consolidation	C = 17.3959

3. Methodology

In this paper, the subsidence rate of the airport was firstly monitored based on SBAS-InSAR technology, and the subsidence prediction curve of the area was derived by combining the Terzaghi consolidation theory model with the geological sampling data of the airport. Then, the ground subsidence rate results were compared with the subsidence prediction curve to verify the applicability and accuracy of the prediction curve of the Terzaghi consolidation theory. Finally, the development trend of airport subsidence is determined based on this prediction curve. The detailed technology flow chart is shown in Figure 5.

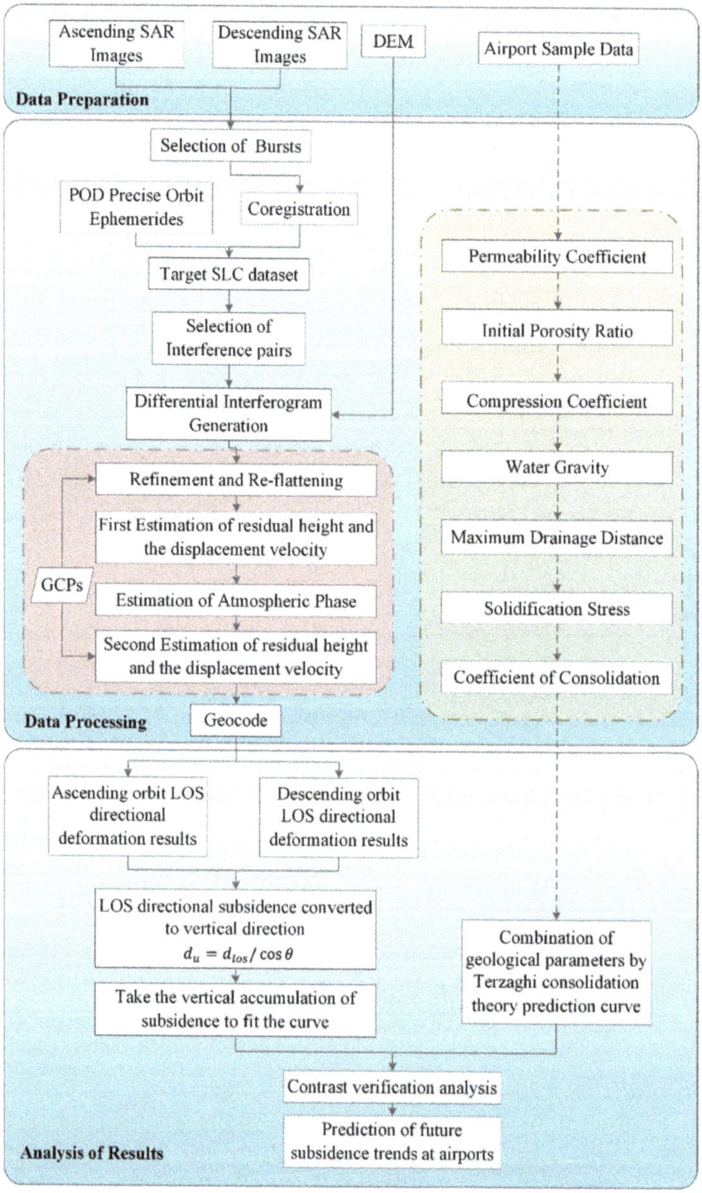

Figure 5. Flowchart of Integration of SBAS-InSAR and Terzaghi consolidation theory.

3.1. SBAS-InSAR Technology

In 2002, Berardino et al. proposed SBAS-InSAR as a novel technology with great prospects for application in the field of monitoring ground subsidence over long-term time series [33,34]. The technology has better applicability and reliability in areas of changing ground cover because it can reduce the negative effects of incoherence and DEM errors. To detect ground subsidence at the airport, the paper used the SBAS plug-in in ENVI software for step-by-step calculation. It includes three main steps: data pre-processing, interferogram generation, and deformation result generation.

The detailed steps of SBAS-InSAR technology are as follows:

Assuming that there are N +1 SAR images covering the same area acquired according to the time series, the image acquisition time series is:

$$t = [t_0, \ldots, t_N]^T \tag{1}$$

After selecting one of them as the master image for registration, in all the free combinations of differential interferometric pairs selected to meet the time baseline and spatial baseline thresholds, $\boldsymbol{M}$ differential interferograms can be obtained, then there are:

$$\frac{N+1}{2} \leq M \leq N\left(\frac{N+1}{2}\right) \tag{2}$$

Combined with the orbit data and external digital elevation model data to remove the flat and terrain effects. The noise is effectively removed by multi-viewing and filtering. In order to obtain the accumulation of deformation on the radar LOS (Line of Sight, LOS), the minimum cost flow (MCF) method is used to perform phase unwinding of M interferometric pairs [35,36]. Suppose that the jth differential interferogram is generated by acquiring the SAR image from the moment of image t_A and the main image t_B, then the corresponding phase unwrapping at any pixel point in the jth interferogram can be expressed as:

$$\delta\varphi^j = \varphi(t_B) - \varphi(t_A) \approx \delta\varphi^j_{def} + \delta\varphi^j_{topo} + \delta\varphi^j_{atm} + \delta\varphi^j_{noise} \tag{3}$$

where $\delta\varphi^j$ is the interference phase at the point; $\varphi(t_A)$ and $\varphi(t_B)$ are the phase values at t_A and t_B moments, respectively, with respect to the initial moment t_0; $\delta\varphi^j_{def}$, $\delta\varphi^j_{topo}$, $\delta\varphi^j_{atm}$, $\delta\varphi^j_{noise}$ indicate the phase difference caused by deformation phase (along with the radar LOS), terrain error, atmospheric delay, and noise, respectively. In order to enhance the accuracy of the deformation monitoring results in the study area, these error terms need to be effectively estimated and eliminated.

Firstly, by selecting more uniform Ground Control Points (Ground Control Points, GCPs) for orbit refinement and re-flattening, not only the orbit parameters can be corrected, but also the residual phase and phase jumps can be estimated and removed [37]. Then, the residual elevation and deformation rate are estimated based on the linear model, and the wrapping phase will be unwrapped twice [38]. Finally, the atmospheric phase is removed by spatial high-pass filtering and temporal low-pass filtering [39]. After removing the above error phase components, the time series deformation information in the LOS direction is obtained by the SVD method.

Due to SAR radar satellite side-view imaging, LOS displacement as a component of real displacement can be acquired [24]. When the horizontal displacement is much smaller than the vertical component, the LOS direction displacement is dominated by vertical subsidence [24,25,40]. As the airport region is relative flat and not like the mountainous area in the InSAR processing, based on the imaging geometry relationship, d_{los} can be converted to ground subsidence d_u based on the incidence angle θ [41], as in Equation (4):

$$d_u = d_{los} / \cos\theta \tag{4}$$

where d_u is the vertical displacement component; d_{los} is the LOS displacement; θ is the radar incidence angle.

3.2. Terzaghi Consolidation Theory

Terzaghi consolidation theory is commonly used to calculate the deformation of a saturated soil layer at any time during infiltration consolidation. The basic assumptions of consolidation theory are as follows: at the point when more than 80% of the pore volume in the soil is filled with water, the soil can be considered saturated although a small amount of gas is present, it is mostly a closed [42]. To find the deformation of the saturated soil layer at any time during infiltration consolidation, the Terzaghi consolidation theory can be established to solve the problem. The Terzaghi theory of consolidation makes many simplifications and assumptions about the changing parameter conditions of the soil during consolidation. Firstly, the external load is applied to the soil instantaneously and at once, and remains constant throughout the consolidation process. Secondly, the soil and water do not produce compression deformation, and the soil compression is caused by drainage. Finally, parameters such as permeability coefficient α and compression coefficient β of the soil during consolidation remain constant during consolidation [30].

In this paper, based on the basic assumptions of the conventional Terzaghi consolidation theory, a differential unit soil body with a height of d_h in the soil body is taken for analysis. The volume compression of the soil is equal to the difference between the inflow and outflow of water from the unit body, which is used to establish the consolidation differential equilibrium equation. Additionally, used Darcy's law and the effective stress principle to solve the equation and obtained the differential equation of Terzaghi consolidation theory [43,44]. As shown in Equation (5):

$$C\frac{\partial^2 P_w}{\partial h^2} = \frac{\partial P_w}{\partial t} \tag{5}$$

where C is the coefficient of consolidation ($m^2/year$); P_w is pore water pressure (kPa); The coefficient of consolidation C and the degree of time consolidation D_c of the soil can be expressed by Equations (6) and (7):

$$C = \alpha\frac{1+k_0}{\beta G_w} \tag{6}$$

$$D_c = 1 - \frac{8e^{-\frac{\pi^2 C}{4H^2}}}{\pi^2} \tag{7}$$

where α is the permeability coefficient (m/year); k_0 is the initial porosity ratio; β is the compression coefficient (kPa^{-1}); G_w is the water gravity (kN/m^3) and H is the maximum drainage distance (m).

Determining and deriving the relevant parameters can calculate any time soil consolidation subsidence S_t(m) [45]. As shown in Equation (8):

$$S_t = D_c\frac{\beta}{1+k_0}\gamma H \tag{8}$$

where γ is the Solidification Stress (kPa).

After obtaining the subsidence prediction curves according to the above calculation method and geological parameters, they were compared with the ground subsidence results monitored by SBAS-InSAR technology. Finally, the curves were used to predict the rate and trend of subsidence in the study area.

4. Results and Discussion

4.1. Analysis of Spatial and Temporal Subsidence Characteristics of DJBIA

The subsidence rate maps of DJBIA were retrieved from the SAR data from March 2017 to April 2021 as shown in Figure 6a, where the negative value denotes the subsidence (i.e., the displacement was far away from the satellite) and the reference point (no subsidence) was chosen away from the reclamation area shown as a red point. Figure 7b–e shows the coherence coefficient, differential interferogram, distribution of elevation residuals, and the error map in the time-series processing, respectively. It can be seen that the area has a high coherence coefficient, a clear interferogram, and small residuals in the time series analysis (maximum less than 4 mm/year). Since the study area is under long-term construction and exploitation, special consideration needs to be given to the possible errors due to inaccurate external elevation information. According to the demand of the landfill here, the elevation error is about 10 m. Based on the results of related studies [46] and the better baseline control of the sentinel data (all vertical baselines are less than 150 m), it is projected that this part of the elevation error will cause the deformation measurement error of interference pair to be less than 2 mm and can be further weakened during the time-series analysis. The inversion results of the elevation residuals in Figure 7d also shows that the elevation residuals of the airport building area, terminal, and heliport are small (between 0.3 and 3 m). Therefore, the measurement errors caused by elevation errors can be controlled. In addition, we introduced the descending orbit data for the same period (December 2019 to April 2021), where the descending orbit data are missing from June-November 2019. It can be found that the deformation results are highly consistent with the ascending orbit results regarding the spatial distribution (Figure 7a). In the runway end area, the number of effective monitoring points becomes more as the backfill progresses, and there is a partial normal decrease in the deformation magnitude. The comparison of the time-series residual results with the ascending and descending orbit results in Figure 6e confirms that the InSAR results of this study are reliable. In summary, the InSAR monitoring results have high accuracy. Because of the lack of data from the descending orbit, the results of the ascending orbit were used for subsequent analysis to ensure data integrity. Since the study area is dominated by vertical ground deformation with minimal horizontal deformation, the absence of horizontal deformation is assumed here to approximate the ascending orbit deformation results to ground subsidence, which is convenient for subsequent 5-year long time series analysis and model building.

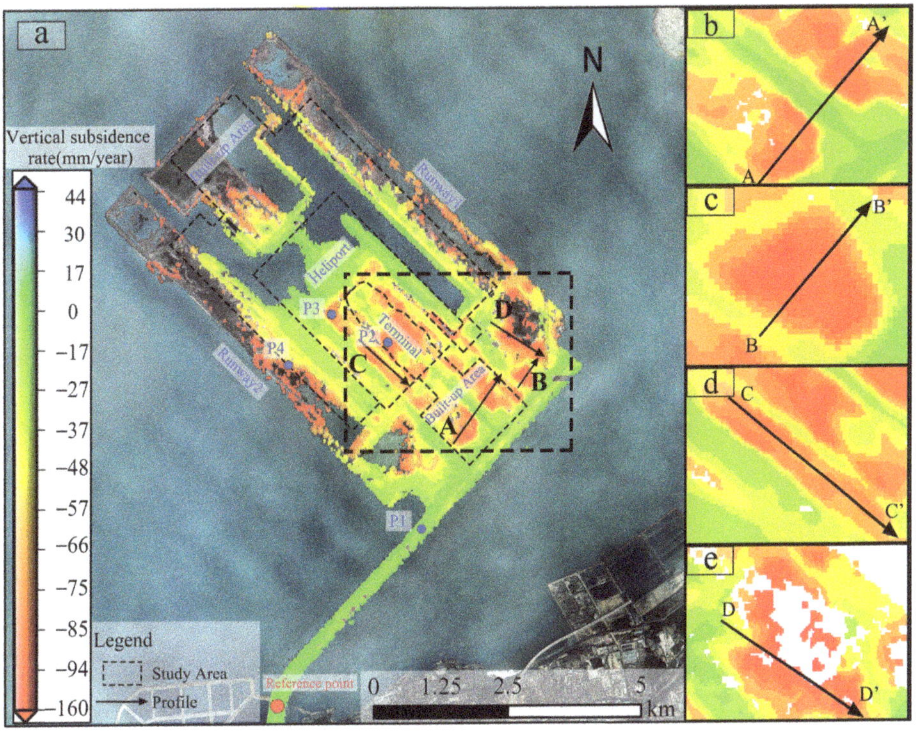

Figure 6. (**a**) Average annual vertical subsidence rate of Jinzhou Bay International Airport; (**b–e**) Subsidence rate along profiles AA′, BB′, CC′, and DD′, respectively; The blue points P1, P2, P3, and P4 in (**a**) were shown in Figure 8 with their accumulated subsidence; The red point in (**a**) denotes the reference point.

As shown in the spatial distribution of subsidence, most areas of the reclamation area are undergoing significant subsidence. From the vertical subsidence results in Figure 6a after ascending orbit decomposition, it can be seen that during the monitoring period from March 2017 to April 2021, some areas of the airport, such as the runway, terminal, and heliport areas of the airport are in a severe subsidence. It is revealed from the combination of Figures 6a and 8 that the accumulation of subsidence in the terminal was over 300 mm and the maximum vertical subsidence rate reached −160 mm/year. Moreover, the time series of subsidence areas in Figure 8 was near linear in the short term, but the subsidence rate was highly associated with the backfilling time, i.e., the longer the reclaimed time, the lower the subsidence rate.

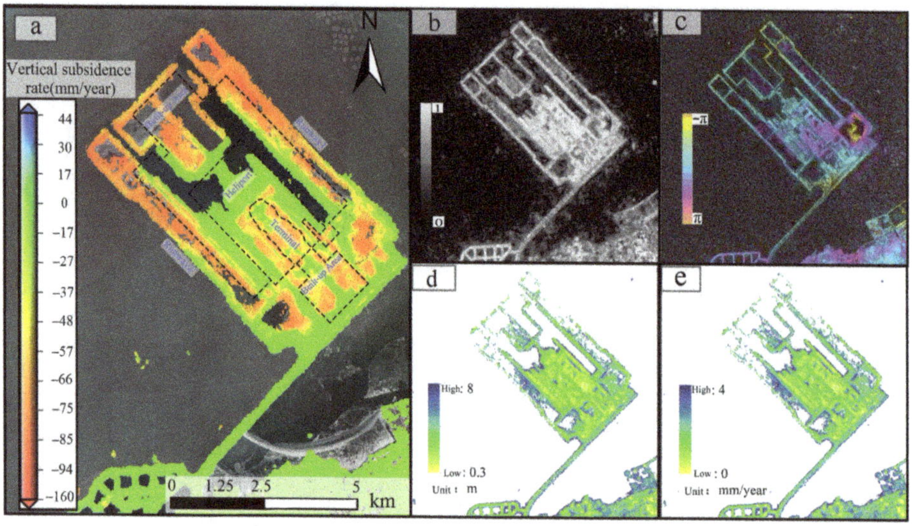

Figure 7. (**a**) Average annual subsidence rate distribution derived from descending track; (**b**) Average coherence coefficient diagram; (**c**) Differential interferogram (20200817–20200910); (**d**) Distribution of elevation residuals; (**e**) The error map of mean velocity results.

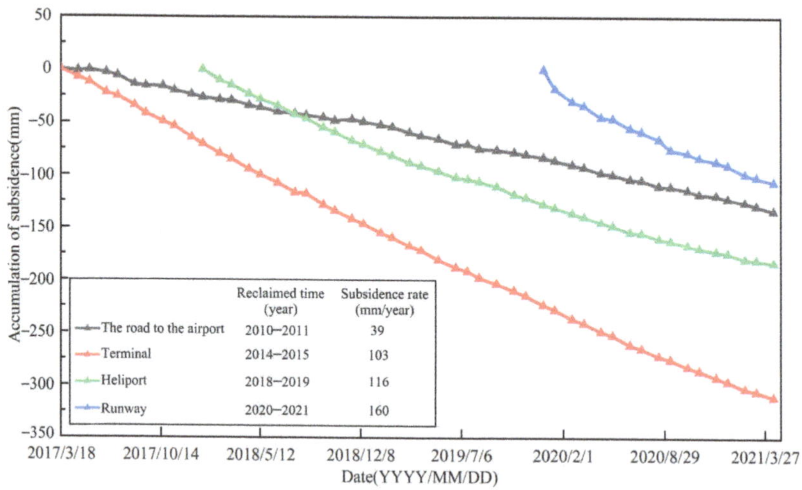

Figure 8. The accumulation subsidence in different areas of the airport. The location of P1–P4 were shown in Figure 6.

In order to further reveal the vertical ground subsidence spatial difference in DJBIA, four typical vertical subsidence areas with complete data, obvious vertical subsidence, and belonging to different reclamation stages were taken for vertical subsidence rate analysis. Figure 6b–e show these four more severe subsidence areas and selected profile lines AA′–DD′, respectively, and the corresponding four average annual vertical subsidence rate profile lines are shown in Figure 9a–d. Ground subsidence can be observed in all four profile line graphs, and there are inconsistent patterns in the curves of the subsidence areas for different reclamation periods and differences in the magnitude of ground subsidence.

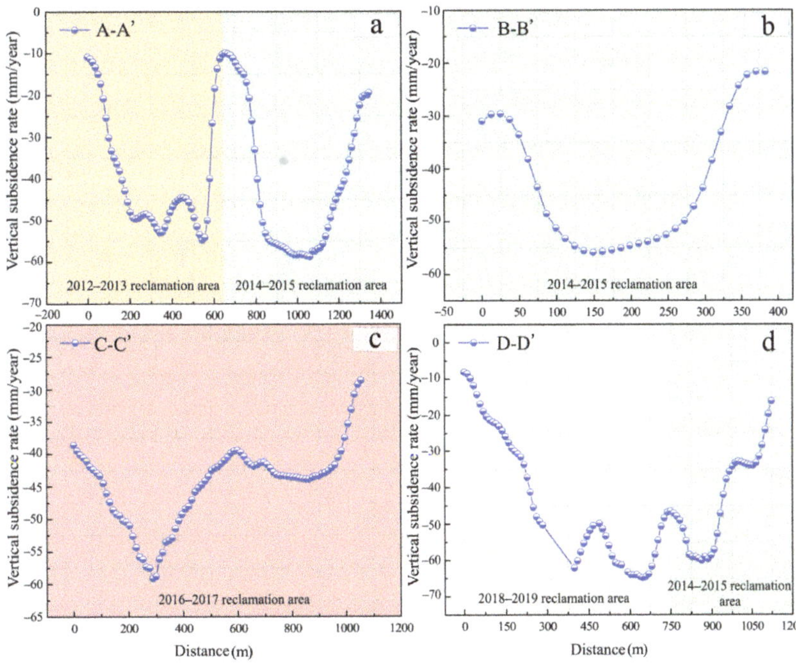

Figure 9. Vertical subsidence rate profiles (the location of the profile are marked in Figure 6). (**a**–**d**) profiles along AA′, BB′,CC′, DD′, respectively.

As shown in Table 3, the earlier the time of reclamation, the lower the vertical subsidence rate. In Figure 9, the different periods of completed reclamation are indicated by the colors corresponding to the periods of reclamation in Figure 2, where two different periods of reclamation are included in Figure 9a,d, respectively. By quantitatively analyzing the correlation between different reclamation periods and ground subsidence, it is found that the dividing line of the drastic change in the vertical subsidence rate coincides with the boundary of reclamation in different periods. The ground subsidence magnitude in the reclaimed area of DJBIA is closely related to its reclamation completion time, which means that significant subsidence occurred in the area where the reclamation was completed in the recent period, and the rate of subsidence decreases with time. This subsidence pattern is more consistent with the subsidence patterns of other offshore reclamation areas, for instance, the Hong Kong Chek Lap Kok Airport [14] and Shanghai Lingang New City, China [15].

The areas with vertical subsidence rates greater than −100 mm/year at DJBIA are located in the corner locations at the airport with the edge locations, which is in the reclamation areas completed in 2018–2019 and 2020–2021. The reclamation of the middle area of the airport was completed in 2012–2013 and 2014–2015, and its vertical subsidence rate was smaller but reached −70 mm/year to −80 mm/year. Overall, the vertical subsidence rate exhibits an emission-like progression from the middle to the edge regions. There is a certain pattern of vertical subsidence rate in the areas where reclamation was completed at different periods. From Figure 10, it can be seen that the maximum, minimum, and median values of vertical subsidence rate are in different reclamation periods. The red points indicate the average vertical subsidence rate. The InSAR subsidence monitoring points in Figure 10 mean that the subsidence rate on the InSAR high-coherence points in Figure 6 shown by counting the magnitude of the rate, and the arcs next to the subsidence points represent the normal distribution curves of these points. The data of this figure are the same as shown in Figure 6 but they are shown in another way to reveal the relationship between reclamation time and vertical subsidence rate. The figure shows that the average vertical subsidence rate of the main body of the airport is: 2020–2021 reclamation area (−80 mm/year) > 2018–2019 reclamation area (−76 mm/year) > 2016–2017 reclamation area (−74 mm/year) > 2014–2015 reclamation area (−66 mm/year) > 2012–2013 reclamation area (−31 mm/year), which shows that the more recent the reclamation completion time is, the greater the average vertical subsidence rate is.

Table 3. Subsidence rate of profile in different periods.

Profile Name	Reclamation Time (Year)		Maximum Subsidence Rate (mm/Year)	
AA′	2012–2013	2014–2015	−91	−97
BB′	2014–2015		−93	
CC′	2016–2017		−99	
DD′	2018–2019	2014–2015	−108	−98

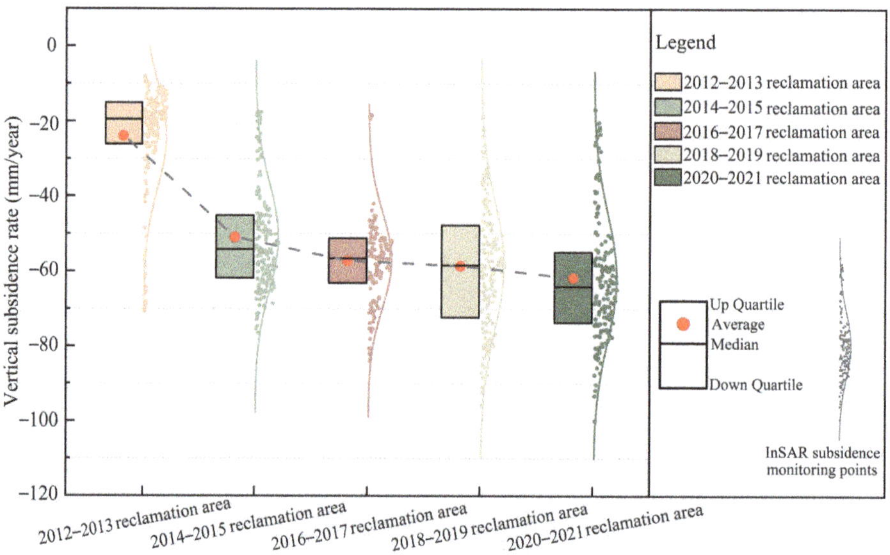

Figure 10. Temporal distribution of average vertical subsidence rates in different reclamation areas.

4.2. *Terzaghi Consolidation Theory Verification and Prediction*

The process stages of land reclamation are coast reclamation, blowing, subsidence, and soft foundation treatment, in which subsidence is an inevitable problem of land reclamation. This subsidence is a characteristic of subsidence exhibited by the compaction of the reclaimed soil layer, which generally tends to stabilize in value after more than 20 years [47]. During this time, the ground will continue to subsidence to different degrees. The ground subsidence related to land reclamation is primarily induced by three main mechanisms, which include the initial consolidation of alluvial clay deposits after land reclamation, the secondary compression of alluvial clay deposits after land reclamation in the long term, and the creep of the fill-in land reclamation [24,26,48,49]. In particular, the initial consolidation of the filling soil in land reclamation accounts for the largest proportion of the total subsidence, which is generally more than 70% in the case of airports in land reclamation [14,50]. Moreover, the subsidence process of secondary compression and filling creep is much slower than the subsidence process of initial consolidation [24]. Terzaghi consolidation theory can provide an excellent explanation for this phenomenon.

From Figure 11, the subsidence rates of different reclamation area subsidence points in the reclamation area of DJBIA are taken during the period from March 2017 to April 2021. The subsidence points of the reclamation area in 2012–2013, 2014–2015, 2016–2017, 2018–2019, and 2020–2021 are indicated by orange, light green, light red, light yellow, and dark green points, respectively. The overall subsidence is undergoing the most significant rapid subsidence phase after the reclamation is completed.

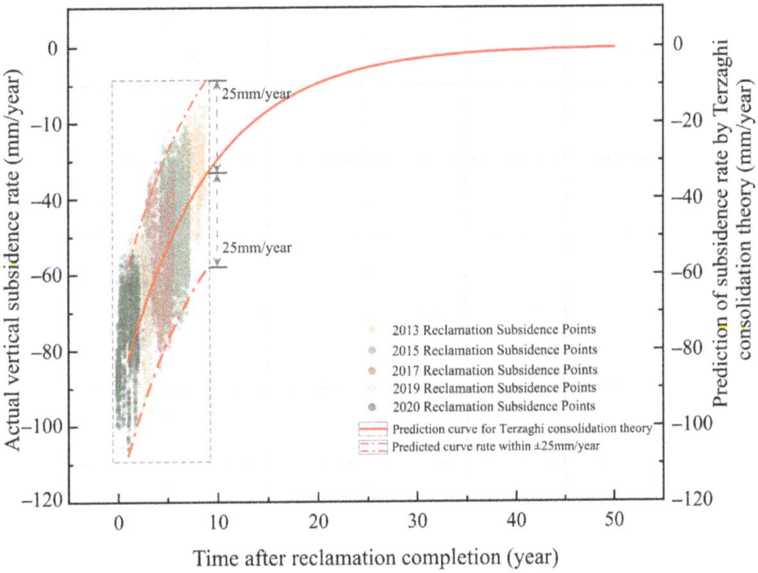

Figure 11. Prediction based on Terzaghi consolidation theory.

Based on the Terzaghi consolidation theory and combined with certain geological sampling data, such as permeability coefficient, initial pore ratio, and compression coefficient of DJBIA [30–32], the subsidence prediction curve of the area is obtained as shown in the red curve. As shown in Figure 11, the red prediction curve obtained by combining the geological parameters of the area with the Terzaghi consolidation theory shows that the reclamation area of the area in 2020 has an annual average vertical subsidence rate of about 82 mm/year in the first year, and the results of InSAR monitoring reveal that the first two years of vertical subsidence are within the range of 80 ± 25 mm/year, which is highly consistent with each other. Additionally, the trend of decreasing deformation in the following years is also highly consistent. The predicted subsidence in the reclaimed area in 2013 was around 32 mm/year, and the real subsidence was about 30 ± 25 mm/year, which confirmed the applicability and feasibility of the Terzaghi consolidation theory and the corresponding parameters used in this paper at Jinzhou Bay Airport. Meanwhile, the results not only show the entire airport is currently in a more significant and rapid subsidence phase, but also the vertical subsidence rate of the airport reclamation area is gradually slowing down over time. In addition, the future vertical subsidence rate of the newly reclaimed area in the region is about 80 ± 25mm/year, and the subsidence rate will gradually slow down with the accumulation of reclamation time, and the vertical subsidence rate will slow down to about 30 ± 25mm/year after 10 years. It is demonstrated that this theory can provide an important reference to the prediction of land reclamation subsidence.

5. Conclusions

In this study, the time-series subsidence results of DJBIA from 2017–2021 were obtained by using the SBAS-InSAR technology based on 50 Sentinel-1A ascending images, and the Terzaghi consolidation theory was integrated to predict the future subsidence of DJBIA in time series. The prediction based on the current stage can generally identify the subsidence rate and trend in the study area, which is important for the control of the current and future construction. The main findings of this paper are listed in the following.

(1) Extensive and significant uneven ground subsidence is occurring throughout the land reclamation area of the airport, especially in the runway, terminal, and building areas, with the maximum subsidence rate exceeding −100 mm/year. By analyzing the maximum subsidence rate in four typical areas and the average regional subsidence rate in six different reclamation periods, it shows that there is a large connection between the ground subsidence rate and the reclamation time in this area, which means that the earlier the reclamation completion, the larger the subsidence rate.

(2) Based on Terzaghi consolidation theory and on-site geological parameters, a predicted subsidence rate curve was derived for the area, which was in high agreement with the InSAR monitored subsidence results in 2017–2021, verifying the applicability of the proposed prediction model based on Terzaghi consolidation theory for this land reclamation airport.

(3) The prediction model reveals that subsidence in the entire reclamation area is in a relatively significant and rapid stage at this stage, and the overall subsidence rate of the reclamation area will be decreasing dramatically as time goes by. The subsidence in the newly reclaimed area in the first two years is within the range of 80 ± 25 mm/year, and it will slow down to about 30 ± 25mm/year after 10 years.

Author Contributions: X.S. and C.C. implemented a combination of both methods. C.C. wrote the draft manuscript. K.D. supervised the experimental analysis and revised the manuscript. J.D., N.W., Y.Y. and X.D. made contributions on the experimental analysis. All authors have read and agreed to the published version of the manuscript.

Funding: This research was funded by the National Natural Science Foundation of China (Grant No. 41941019; No. 42072306; No. 41801391), State Key Laboratory of Geohazard Prevention and Geoenvironment Protection Independent Research Project (SKLGP2020Z012) and the fellowship of China Postdoctoral Science Foundation (2020M673322).

Conflicts of Interest: The authors declare no conflict of interest.

References

1. Cigna, F.; Tapete, D. Urban growth and land subsidence: Multi-decadal investigation using human settlement data and satellite InSAR in Morelia, Mexico. *Sci. Total Environ.* **2022**, *811*, 152211. [CrossRef] [PubMed]
2. Ma, P.; Wang, W.; Zhang, B.; Wang, J.; Shi, G.; Huang, G.; Chen, F.; Jiang, L.; Lin, H. Remotely sensing large- and small-scale ground subsidence: A case study of the Guangdong–Hong Kong–Macao Greater Bay Area of China. *Remote Sens. Environ.* **2019**, *232*, 111282. [CrossRef]
3. Hu, L.; Dai, K.; Xing, C.; Li, Z.; Tomás, R.; Clark, B.; Shi, X.; Chen, M.; Zhang, R.; Qiu, Q.; et al. Land subsidence in Beijing and its relationship with geological faults revealed by Sentinel-1 InSAR observations. *Int. J. Appl. Earth Obs. Geoinf.* **2019**, *82*, 101886. [CrossRef]
4. Aimaiti, Y.; Yamazaki, F.; Liu, W. Multi-Sensor InSAR Analysis of Progressive Land Subsidence over the Coastal City of Urayasu, Japan. *Remote Sens.* **2018**, *10*, 1304. [CrossRef]
5. Marshall, C.; Large, D.J.; Athab, A.; Evers, S.L.; Sowter, A.; Marsh, S.; Sjögersten, S. Monitoring tropical peat related settlement using ISBAS InSAR, Kuala Lumpur International Airport (KLIA). *Eng. Geol.* **2018**, *244*, 57–65. [CrossRef]
6. Jiang, H.; Balz, T.; Cigna, F.; Tapete, D. Land Subsidence in Wuhan Revealed Using a Non-Linear PSInSAR Approach with Long Time Series of COSMO-SkyMed SAR Data. *Remote Sens.* **2021**, *13*, 1256. [CrossRef]

7. Ramirez, R.A.; Lee, G.-J.; Choi, S.-K.; Kwon, T.-H.; Kim, Y.-C.; Ryu, H.-H.; Kim, S.; Bae, B.; Hyun, C. Monitoring of construction-induced urban ground deformations using Sentinel-1 PS-InSAR: The case study of tunneling in Dangjin, Korea. *Int. J. Appl. Earth Obs. Geoinf.* **2022**, *108*, 102721. [CrossRef]
8. Dai, K.; Li, Z.; Xu, Q.; Bürgmann, R.; Milledge, D.G.; Tomas, R.; Fan, X.; Zhao, C.; Liu, X.; Peng, J.; et al. Entering the era of earth observation-based landslide warning systems: A novel and exciting framework. *IEEE Geosci. Remote Sens. Mag.* **2020**, *8*, 136–153. [CrossRef]
9. Shi, X.; Yang, C.; Zhang, L.; Jiang, H.; Liao, M.; Zhang, L.; Liu, X. Mapping and characterizing displacements of active loess slopes along the upstream Yellow River with multi-temporal InSAR datasets. *Sci. Total Environ.* **2019**, *674*, 200–210. [CrossRef]
10. Fan, X.; Xu, Q.; Scaringi, G.; Dai, L.; Li, W.; Dong, X.; Zhu, X.; Pei, X.; Dai, K.; Havenith, H.-B. Failure mechanism and kinematics of the deadly June 24th 2017 Xinmo landslide, Maoxian, Sichuan, China. *Landslides* **2017**, *14*, 2129–2146. [CrossRef]
11. Tomás, R.; Li, Z.; Lopez-Sanchez, J.M.; Liu, P.; Singleton, A. Using wavelet tools to analyse seasonal variations from InSAR time-series data: A case study of the Huangtupo landslide. *Landslides* **2016**, *13*, 437–450. [CrossRef]
12. Wasowski, J.; Bovenga, F. Investigating landslides and unstable slopes with satellite Multi Temporal Interferometry: Current issues and future perspectives. *Eng. Geol.* **2014**, *174*, 103–138. [CrossRef]
13. Jiang, Y.; Liao, M.; Wang, H.; Zhang, L.; Balz, T. Deformation Monitoring and Analysis of the Geological Environment of Pudong International Airport with Persistent Scatterer SAR Interferometry. *Remote Sens.* **2016**, *8*, 1021. [CrossRef]
14. Jiang, L.; Lin, H. Integrated analysis of SAR interferometric and geological data for investigating long-term reclamation settlement of Chek Lap Kok Airport, Hong Kong. *Eng. Geol.* **2010**, *110*, 77–92. [CrossRef]
15. Pepe, A.; Bonano, M.; Zhao, Q.; Yang, T.; Wang, H. The Use of C-/X-Band Time-Gapped SAR Data and Geotechnical Models for the Study of Shanghai's Ocean-Reclaimed Lands through the SBAS-DInSAR Technique. *Remote Sens.* **2016**, *8*, 911. [CrossRef]
16. Zhao, Q.; Lin, H.; Gao, W.; Zebker, H.A.; Chen, A.; Yeung, K. InSAR detection of residual settlement of an ocean reclamation engineering project: A case study of Hong Kong International Airport. *J. Oceanogr.* **2011**, *67*, 415–426. [CrossRef]
17. Dai, K.; Shi, X.; Gou, J.; Hu, L.; Chen, M.; Zhao, L.; Dong, X.; Li, Z. Diagnosing Subsidence Geohazard at Beijing Capital International Airport, from High-Resolution SAR Interferometry. *Sustainability* **2020**, *12*, 2269. [CrossRef]
18. Gao, M.; Gong, H.; Chen, B.; Zhou, C.; Chen, W.; Liang, Y.; Shi, M.; Si, Y. InSAR time-series investigation of long-term ground displacement at Beijing Capital International Airport, China. *Tectonophysics* **2016**, *691*, 271–281. [CrossRef]
19. Lu, M.; Deng, K.; Feng, G.; Li, K.; Xiong, Z.; Wang, Y.; He, S. Reclaimed-Airport Surface-Deformation Monitoring by Improved Permanent-Scatterer Interferometric Synthetic-Aperture Radar: A Case Study of Shenzhen Bao'an International Airport, China. *Photogramm. Eng. Remote Sens.* **2021**, *87*, 105–116.
20. Liu, P.; Chen, X.; Li, Z.; Zhang, Z.; Xu, J.; Feng, W.; Wang, C.; Hu, Z.; Tu, W.; Li, H. Resolving surface displacements in Shenzhen of China from time series InSAR. *Remote Sens.* **2018**, *10*, 1162. [CrossRef]
21. Susaki, J.; Tsujino, M.; Anahara, T. Fusion of different frequency SAR images for DInSAR-based land subsidence monitoring. In Proceedings of the 2017 IEEE International Geoscience and Remote Sensing Symposium (IGARSS), Worth, TX, USA, 23–28 July 2017; pp. 945–948.
22. Ito, H.; Susaki, J.; Anahara, T. Integrating multi-temporal sar images and gps data to monitor three-dimensional land subsidence. *ISPRS Ann. Photogramm. Remote Sens. Spat. Inf. Sci.* **2019**, *4*, 9–16. [CrossRef]
23. Ito, H.; Susaki, J. Land subsidence monitoring by integrating PSI and Geodetic Deformation Measurements. In Proceedings of the IGARSS 2018-2018 IEEE International Geoscience and Remote Sensing Symposium, Valencia, Spain, 23–27 July 2018; pp. 498–501.
24. Xu, B.; Feng, G.; Li, Z.; Wang, Q.; Wang, C.; Xie, R. Coastal Subsidence Monitoring Associated with Land Reclamation Using the Point Target Based SBAS-InSAR Method: A Case Study of Shenzhen, China. *Remote Sens.* **2016**, *8*, 652. [CrossRef]
25. Zhuo, G.; Dai, K.; Huang, H.; Li, S.; Shi, X.; Feng, Y.; Li, T.; Dong, X.; Deng, J. Evaluating Potential Ground Subsidence Geo-Hazard of Xiamen Xiang'an New Airport on Reclaimed Land by SAR Interferometry. *Sustainability* **2020**, *12*, 6991. [CrossRef]
26. Liu, X.; Zhao, C.; Zhang, Q.; Yang, C.; Zhang, J. Characterizing and monitoring ground settlement of marine reclamation land of Xiamen New Airport, China with Sentinel-1 SAR Datasets. *Remote Sens.* **2019**, *11*, 585. [CrossRef]
27. Guo, F.; Zhang, H.; Fan, Y.; Zhu, P.; Wang, S.; Lu, X.; Jin, Y. Environment. Detection and evaluation of a ventilation path in a mountainous city for a sea breeze: The case of Dalian. *Build. Environ.* **2018**, *145*, 177–195. [CrossRef]
28. Zhu, X. Research on Risk Assessment and Disaster Mitigation of Geological Disasters in Dalian. Master's Thesis, Liaoning Normal University, Liaoning, China, 2021.
29. China Civil Aviation Website. Dong Zhiyi went to Dalian New Airport for Research. Available online: http://www.caacnews.com.cn/1/1/201911/t20191107_1285087_wap.html (accessed on 1 April 2022).
30. Bai, W. Improved Algorithm and Application of One-Dimensional Consolidation Theory for Soft Soil Foundation. Master's Thesis, Xi'an University of Technology, Xi'an, China, 2020.
31. Deng, H. Study on Reclamation and Soft Soil Characteristics in Dalian Coastal Zone. Master's Thesis, Jilin University, Jilin, China, 2020.
32. Cong, B. Calculation of the Foundation Settlement of Soft Clay at the DaLian Offshore Airport. Master's Thesis, Dalian University of Technology, Dalian, China, 2013.
33. Lanari, R.; Mora, O.; Manunta, M.; Mallorquí, J.J.; Berardino, P.; Sansosti, E. A small-baseline approach for investigating defor-mations on full-resolution differential SAR interferograms. *IEEE Trans. Geosci. Remote Sens.* **2004**, *42*, 1377–1386. [CrossRef]

34. Osmanoğlu, B.; Sunar, F.; Wdowinski, S.; Cabral-Cano, E. Time series analysis of InSAR data: Methods and trends. *ISPRS J. Photogramm. Remote Sens.* **2016**, *115*, 90–102. [CrossRef]
35. Costantini, M. A novel phase unwrapping method based on network programming. *IEEE Trans. Geosci. Remote Sens.* **1998**, *36*, 813–821. [CrossRef]
36. Hooper, A.; Zebker, H.A. Phase unwrapping in three dimensions with application to InSAR time series. *JOSA A* **2007**, *24*, 2737–2747. [CrossRef]
37. Chen, G.; Zhang, Y.; Zeng, R.; Yang, Z.; Chen, X.; Zhao, F.; Meng, X. Detection of land subsidence associated with land crea-tion and rapid urbanization in the chinese loess plateau using time series insar: A case study of Lanzhou new district. *Remote Sens.* **2018**, *10*, 270. [CrossRef]
38. Liao, M.; Wang, T. *Time Series InSAR Technology and Application*; Science Press: Beijing, China, 2014; pp. 83–86.
39. Goldstein, R.M.; Werner, C. Radar interferogram filtering for geophysical applications. *Geophys. Res. Lett.* **1998**, *25*, 4035–4038. [CrossRef]
40. Dai, K.; Liu, G.; Li, Z.; Li, T.; Yu, B.; Wang, X.; Singleton, A. Extracting Vertical Displacement Rates in Shanghai (China) with Multi-Platform SAR Images. *Remote Sens.* **2015**, *7*, 9542–9562. [CrossRef]
41. Wang, H.; Wright, T.J.; Yu, Y.; Lin, H.; Jiang, L.; Li, C.; Qiu, G. InSAR reveals coastal subsidence in the Pearl River Delta, China. *Geophys. J. Int.* **2012**, *191*, 1119–1128. [CrossRef]
42. Huang, J.; Griffiths, D.J.G. One-dimensional consolidation theories for layered soil and coupled and uncoupled solutions by the finite-element method. *Geotechnique* **2010**, *60*, 709–713. [CrossRef]
43. Chen, Z. *Soil Mechanics*; Tsinghua University Press: Beijing, China, 2006; pp. 100–160.
44. Barnes, G. *Soil Mechanics: Principles and Practice*; Macmillan International Higher Education: London, UK, 2016.
45. Terzaghi, K.; Peck, R.B.; Mesri, G. *Soil Mechanics in Engineering Practice*; John Wiley and Sons: Hoboken, NJ, USA, 1996; pp. 1–592.
46. Dai, K.; Liu, G.; Yu, B.; Jia, H.; Ma, D.; Wang, X. Detecting Subsidence along a High Speed Railway by Ultrashort Baseline TCP-InSAR with High Resolution Images. *Int. Arch. Photogramm. Remote Sens. Spat. Inf. Sci.* **2013**, *XL-7/W2*, 61–65. [CrossRef]
47. Wang, Q.; Zhang, Y.; Dong, Y.; Fu, Y. Subsidence monitoring of land reclamation area dike project based on time-series InSAR technology. *Highway* **2021**, *66*, 276–281.
48. Tosen, R.; Pickles, A.; Jaros, M. Assessment of differential settlement at Chek Lap Kok Airport reclamation site. In *Proceedings of the a Seminar on the Geotechnical Aspects of the Airport Core Projects*; Hong Kong Institution of Civil Engineers: Hong Kong, China, 1998.
49. Plant, G.W.; Covil, C.S.; Hughes, R.A. *Site Preparation for the New Hong Kong International Airport*; Thomas Telford: London, UK, 1998.
50. Pickles, A.; Tosen, R. Settlement of Reclaimed Land for the New Hong Kong International Airport. *Proc. Inst. Civ. Eng.-Geotech. Eng.* **1998**, *131*, 191–209. [CrossRef]

remote sensing

MDPI

Technical Note

Revealing the Morphological Evolution of Krakatau Volcano by Integrating SAR and Optical Remote Sensing Images

Jianming Xiang [1], Shaohua Guo [1], Xianlin Shi [1,*], Daijun Yu [1], Guan Wei [1], Ningling Wen [1], Chen Chen [1] and Keren Dai [1,2]

[1] College of Earth Science, Chengdu University of Technology, Chengdu 610059, China; xiangjianming17@cdut.edu.cn (J.X.); guoshaohua@stu.cdut.edu.cn (S.G.); ydj@cdut.edu.cn (D.Y.); guanwei@stu.cdut.edu.cn (G.W.); wenningling@stu.cdut.edu.cn (N.W.); chenchen1@stu.cdut.edu.cn (C.C.); daikeren17@cdut.edu.cn (K.D.)

[2] State Key Laboratory of Geohazard Prevention and Geoenvironment Protection, Chengdu University of Technology, Chengdu 610059, China

* Correspondence: shixianlin06@cdut.edu.cn

Abstract: On 22 December 2018, volcano Anak Krakatau, located in Indonesia, erupted and experienced a major lateral collapse. The triggered tsunami killed at least 437 people by the 13-m-high tide. Traditional optical imagery plays a great role in monitoring volcanic activities, but it is susceptible to cloud and fog interference and has low temporal resolution. Synthetic aperture radar (SAR) imagery can monitor volcanic activities at a high temporal resolution, and it is immune to the influence of clouds. In this paper, we propose an automatic method to accurately extract the volcano boundary from SAR images by combining multi-polarized water enhancement and the Nobuyuki Otsu (OTSU) method. We extract the area change of the volcano in 2018–2019 from Sentinel-1 images and ALOS-2 imagesThe area change and evolution are verified and analyzed by combing the results from SAR and optical data. The results show that the southeastern part of the volcano expanded significantly after the eruption, and the western part experienced collapse and recovery. The volcano morphology change experienced a slow-fast-slow process in the two years.

Keywords: Anak Krakatau volcano; the Nobuyuki Otsu (OTSU) method; multi-polarization; synthetic aperture radar (SAR)

Citation: Xiang, J.; Guo, S.; Shi, X.; Yu, D.; Wei, G.; Wen, N.; Chen, C.; Dai, K. Revealing the Morphological Evolution of Krakatau Volcano by Integrating SAR and Optical Remote Sensing Images. *Remote Sens.* **2022**, *14*, 1399. https://doi.org/10.3390/rs14061399

Academic Editors: Chisheng Wang, Daqing Ge, Guohong Zhang, Wu Zhu and Siting Xiong

Received: 9 January 2022
Accepted: 7 March 2022
Published: 14 March 2022

1. Introduction

On 22 December 2018, at 21:30 local time, the coastline of the Sunda Strait in Indonesia was stricken by a tsunami, which was brought on by an undersea landslide of Anak Krakatau volcano [1–4]. The tsunami led to 437 casualties, 31,943 injuries and 10 missing people. Over 16,000 people were displaced [5]. Long-term monitoring of volcanic pattern changes may reduce the damage caused by such disasters.

Optical images have high spatial resolution, but they have low temporal resolution and are susceptible to clouds and fog contamination. Synthetic aperture radar (SAR) sensors work with microwave bands that can penetrate clouds and fog, providing better vision than visible light and infrared remote sensing [6,7]. Therefore, SAR images are rapidly developed and widely applied in volcano and landslides monitoring, as well as shorelines and water bodies extraction [8]. The algorithms for extracting water bodies include the sea areas segmentation algorithm based on the Nobuyuki Otsu (OTSU) method and statistical characteristics of sea areas, and the water body extraction algorithm based on thresholds [9–11] or the object-oriented method [12–14]. SAR polarization information is also used to extract the water body [15–19].

The surface deformation caused by volcanic eruptions can be obtained by analyzing the InSAR coherence and backscatter images of Anak Krakatoa [20]. Some ground data and satellite images show that, before the devastating tsunami, flanking motion was evident along with more frequent volcanic activities [4], which may present as morphology changes. The initial landslide volume led by submarine collapse of the Krakatoa volcano ranged from 0.22 to 0.30 km^3, which was estimated by comparing the satellite and aerial photography images before and after the eruption [1]. Thus, using the multi-polarization information of both optical and SAR images to monitor the volcano morphology change may provide useful information for disaster warning.

In this paper, we use Google images and Planet images before and after the volcanic eruption for visual interpretation and comparative analysis of volcanic morphological changes, and use multi-polarization fused SAR images (Sentinel-1 and ALOS-2) to extract the volcano area change during 2018–2019. We analyze the monthly morphological changes of Krakatoa Island, and the changes before and after the eruption. The results show that after the eruption, the southeastern part of the volcano expanded significantly, and the western part experienced collapse and recovery. The volcano morphology experienced slow-fast-slow changes during the two years.

2. Study Area and Data

2.1. Overview of the Study Area

The Sunda Strait of Indonesia, located between Sumatra and Java, bridges the Java Sea with the Indian Ocean and is a shipping lane from North Pacific countries to East and West Africa or around the Cape of Good Hope to Europe. The Sunda Strait region has accommodated many crustal movements including volcanic activities. Krakatoa volcano at the southern entrance is the most famous one (Figure 1a). Anak Krakatoa is the resurgent volcanic island rising from the caldera formed in the historic 1883 eruption, with flat areas covered in dense vegetation prior to the most recent large eruptions (Figure 1b). On 17 December 2018, the Planet satellite observed the ash eruption from the crater. On 22 December 2018, the volcano started the large eruption sequence that was observed by Sentinel-1A (Figure 1d),with a large amount of ash filling the air (Figure 1c). The southwestern part of the volcano cone collapsed, which later formed a crater lake by the materials floating on the sea surface in the southwestern part of the island, as observed by the Google image acquired on 11 January 2019 (Figure 1e).

2.2. Data

In this study, we use Google images, Planet images, Sentinel-1 images and ALOS-2 images (Figure 2). Two Google images and four Planet images before and after the outbreak are adopted to extract the volcanic boundaries, which are then used to obtain the morphological changes before and after the eruption. Google Earth's satellite imagery is not a single source of data, but an integration of satellite imagery and aerial photography. The effective resolution of the global landscape imagery on Google Earth is usually 30 m, but for large cities, famous scenic areas and built-up areas, high precision imagery is provided with a resolution of around 1 m and 0.5 m.

Planet, the world's largest microsatellite constellation, consists of hundreds of Dove satellites (10 cm $\times$ 10 cm $\times$ 30 cm). Each Dove satellite is equipped with an optical system and camera with the resolution of 3–5 m.

The Sentinel-1 satellite, consisting of two polar orbiting satellites (A and B), was launched by the European Space Agency Copernicus program (GMES). The two sun-synchronous satellites each have revisit cycles of 12 days. The C-band Sentinel-1 satellite can provide continuous images in the Vertical Transmit-Vertical Receive and Vertical Transmit-Horizontal Receive(VV + VH) polarization mode and Interferometric Wide swath (IW) mode during the period of 2018–2019.

We use one ALOS-2 image acquired on 24 December 2018 to capture the morphological change of the volcano after the eruption. The ALOS-2 satellite is the only L-band SAR satellite currently operating in orbit. It has a revisit period of 14 days.

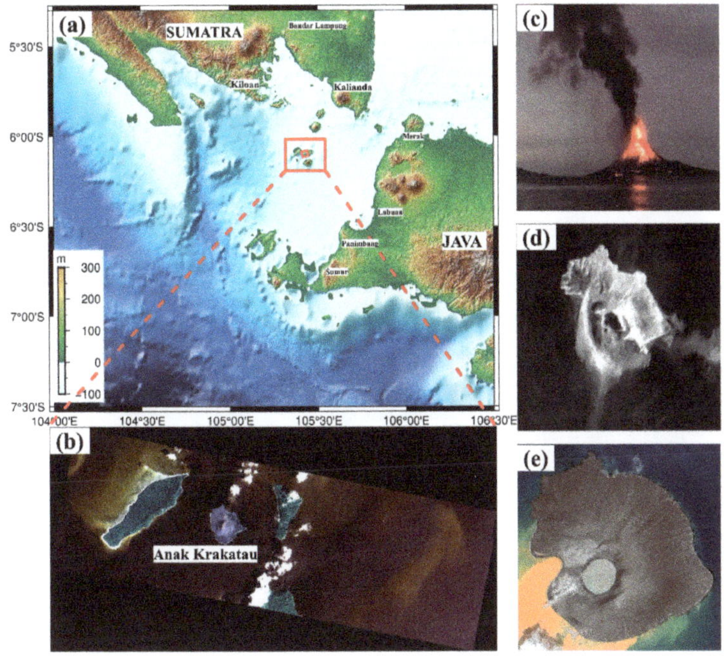

Figure 1. Location map and images of Anak Krakatau volcano. (**a**) Location map; (**b**) volcano before the eruption (Planet image); (**c**) a photo of the volcano eruption (from online source [21]); (**d**) volcano after eruption (synthetic aperture radar (SAR) image); (**e**) volcano after eruption (Google image).

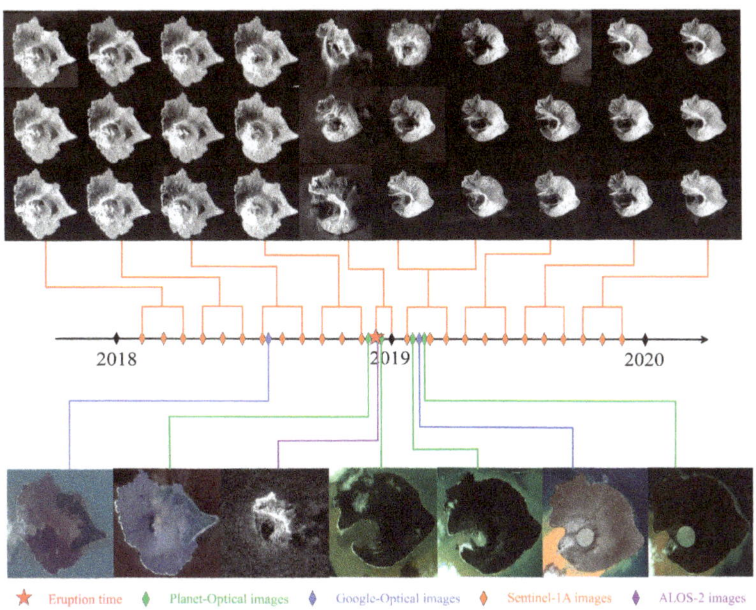

Figure 2. Time chart of the data used in this study, including Planet images, Google images, Sentinel-1A images and ALOS-2 images.

3. Method

We propose a method for volcano morphology extraction on the basis of Sentinel-1 Dual-Polarized Water Index (SDWI) and the Nobuyuki Otsu (OTSU) polarization information enhancement. In this method, the VV and VH polarization of the output SAR image in the IW mode are calculated by band, so as to achieve water body enhancement on the image. The Otsu method is used to search for the best threshold for segmenting the enhanced image. The SAR images can identify water bodies and morphological deformation accurately. Firstly, Sentinel-1A data are preprocessed to obtain the encoded image, which is processed by the SDWI method [15], as shown in Equation (1).

$$k_{sdwi} = \ln(10{*}VV * VH) - 8 \tag{1}$$

Since the backscattering coefficient of water in the SAR image is lower than that of soil and vegetation, the double polarization data is multiplied to enhance the features of water. Then, the Otsu method is used for threshold segmentation using the maximum between class variance of background and target as the criterion. The algorithm is simple, stable and effective [22]. The basic Otsu method is introduced below. Assuming the size of an image is $M * N$, its grayscale range is $0\text{–}L-1$, and the number of the pixels with grayscale i is n_i, then the probability of occurrence of grayscale i is $P_i = n_i/MN$. By setting threshold T to divide the grayscale into target class $C_0 = \{0, 1 \ldots, T-1\}$ and background class $C_1 = \{T-1, \ldots, L-1\}$, we obtain:

The probability of the target part

$$P_0 = \sum_{i=0}^{T-1} P_i \tag{2}$$

The probability of the background part

$$P_1 = \sum_{i=T}^{L-1} P_i = 1 - P_0 \tag{3}$$

The mean value of the target component

$$\mu_0 = \sum_{i=0}^{T-1} iP_i/P_0 \tag{4}$$

The mean value of the background component

$$\mu_1 = \sum_{i=T}^{L-1} iP_i/P_1 \tag{5}$$

The total mean value of image pixels

$$\mu = \sum_{i=0}^{L-1} iP_i + P_0\mu_0 + P_1\mu_1 \tag{6}$$

According to the defining formula of variance as follows:

$$s^2 = \frac{\Sigma(x_i - \overline{x})^2}{n} \tag{7}$$

The interclass variance is

$$\partial^2 = P_0(\mu_0 - \mu)^2 + P_1(\mu_1 - \mu)^2 \tag{8}$$

The formula for the optimal threshold K is

$$K = \text{Arg} \max_{0 \le T \le L-1} (\partial^2) \tag{9}$$

Using the maximum interclass variance method in Python, we get the sea–land binary images, with black being the sea surface and white being the volcanic island. On this basis, the classification and post-processing are done in ArcGIS platform to eliminate redundant invalid map spots. The monthly variation of the volcanic area in 2018–2019 is finally calculated. The flowchart of this method is shown in Figure 3.

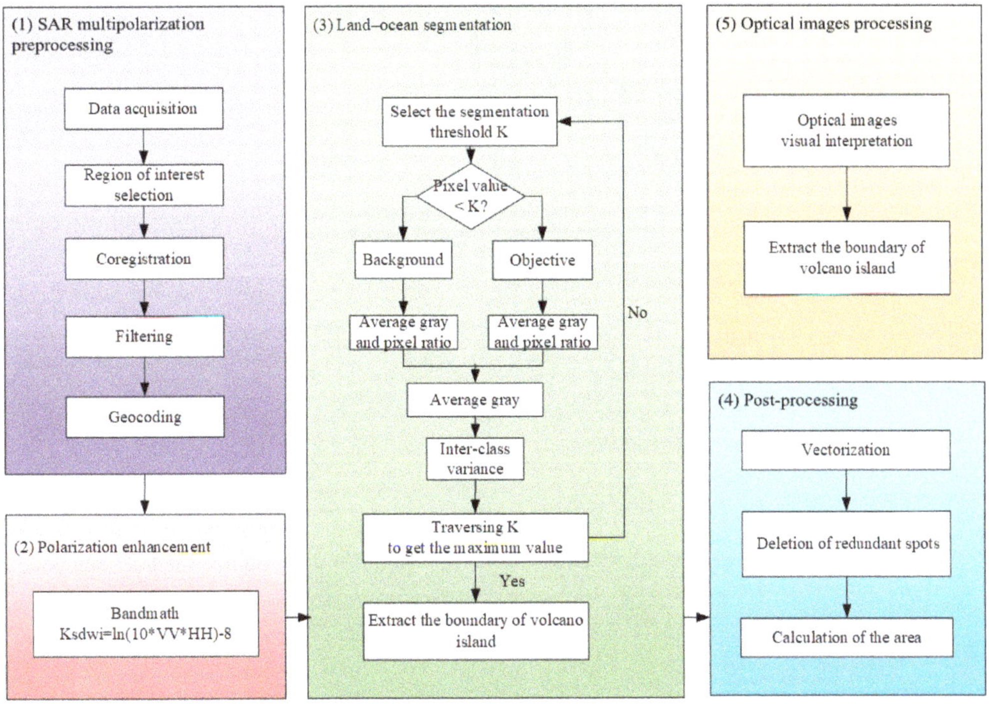

Figure 3. Flowchart of the proposed method.

4. Optical Image and SAR Image Results

4.1. Optical Image Results

After visually interpreting the Planet images (Figure 4a–c) and Google images (Figure 4d–f), we extracted the volcanic morphological changes by vectorization. The eruption smoke is obvious on the Planet image from 17 December 2018 (Figure 4a), indicating that the volcano was in an active period. Similarly, Figure 4c,f shows that the volcano has significant morphological changes after the eruption. The southwestern part of the cone collapsed, the crater formed a concave crater lake and the boundaries clearly expanded eastward and northward.

4.2. SAR Time Series Analysis Results

We calculate the area change of Krakatoa volcano from 2018 to 2019 on the basis of the Sentinel-1A data. The area of the volcano in January 2018 is taken as the initial area (2.86 km^2). The volcanic changes can be divided into the following stages (Figure 5).

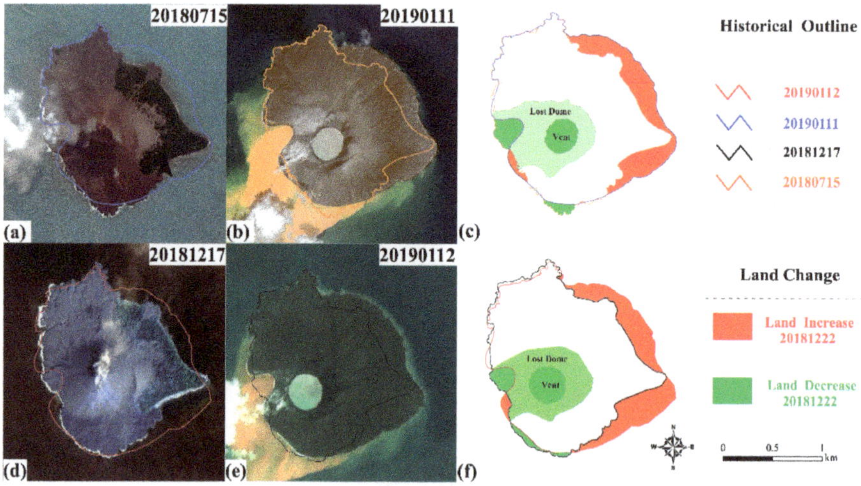

Figure 4. Comparative analysis of optical images: (**a**) the Google image on 15 July 2018; (**b**) the Google image on 11 January 2019; (**c**) morphological changes after eruptions (Google images); (**d**) the Planet image on 17 December 2018; (**e**) the Planet image on 12 January 2019; (**f**) morphological changes after eruptions (Planet images).

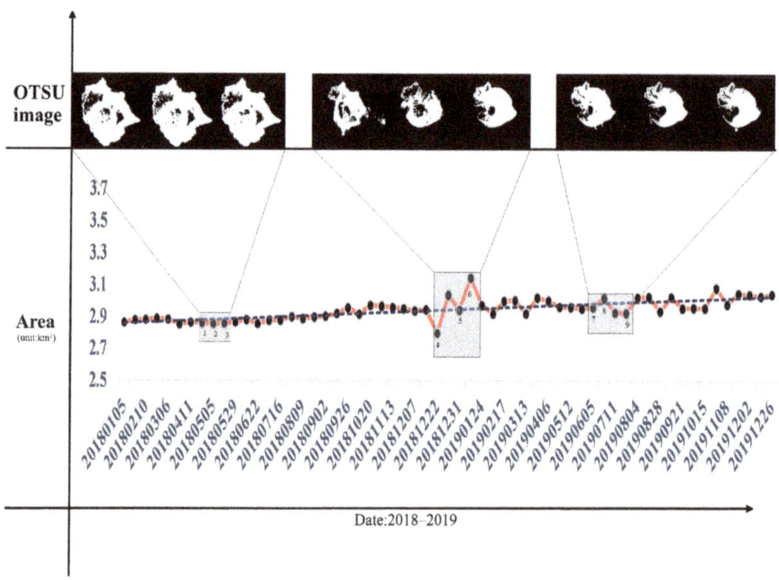

Figure 5. The change of volcanic area from 2018 to 2019 and the comparison of OTSU images and SDWI images at different stages. 1, 2, 3 represent the images of 20180423, 20180505, 20180529, respectively; 4, 5, 6 represent the images of 20181222, 20181227, 20910112; 7, 8, 9 represent the images of 20190629, 20190711, 20190804.

a. Slow volcanic active phase (5 January to 22 December 2018): The area (about 2.85 to 2.97 km^2) grew slowly (before the eruption).

b. Volcano eruption phase (22 December 2018 to 28 December 2018 [23]: On the Sentinel-1A image acquired at 22:34 on 22 December 2018, there are ripples on the sea surface around the island, and the volcanic plume covering the sky. Accordingly, the volcano was still in a sustained phase of activity until 22:34 p.m. The area of the volcano at that time was minimum due to the collapse of the southwest side and the loss of material. The area was 2.79 km^2, and decreased by 0.15 km^2. According to the analysis in [23], the volcano was active from 16:55 on 22 December to 5:00 on 28 December, which is confirmed by our combined analysis of area, optical images and SAR images.

c. Stable phase after eruption (since January 2019): After the eruption, the eastern and northern parts of the volcano stabilized gradually. The crater formed a closed lake by the material floated back. This crater lake has small changes due to subsequent volcanic activities [24] and ocean currents. During this process, the area of the volcano fluctuated, as shown in the images acquired on NO. 7, 8 and 9 in Figure 5.

The volcano morphological changes occurred in the pre-eruption stage (stage a), the eruption stage (stage b) and the post-eruption stage (stage c). We select the nine images (Figure 5) processed by the SDWI and OTSU methods to show the volcano morphological changes.

4.3. Combined Optical and SAR Image Analysis

For the eruption phase, we combined optical images and SAR images to restore the morphological change process of the volcano (Figure 6).

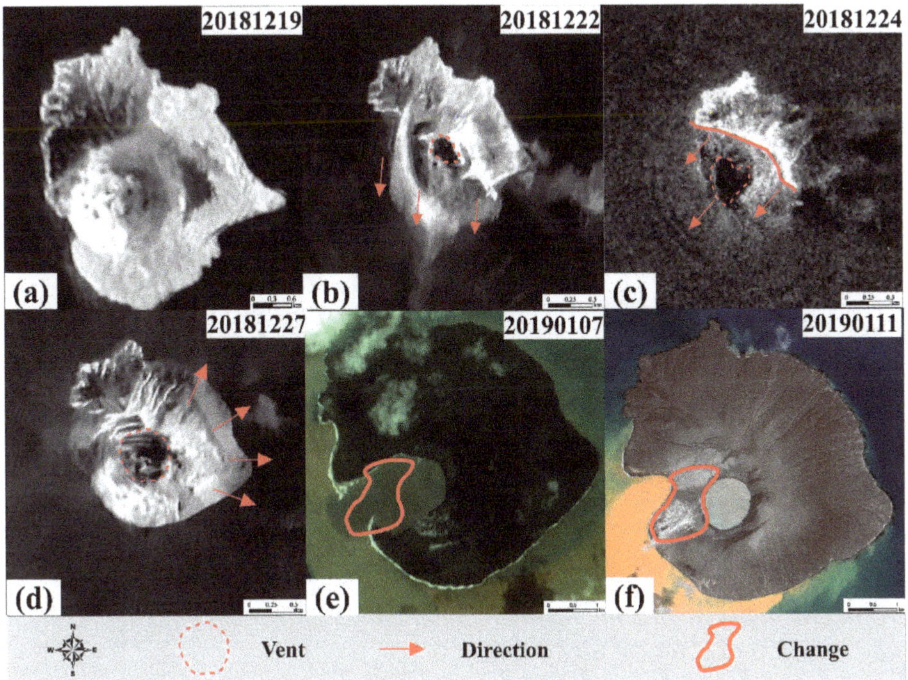

Figure 6. The volcano morphologic changes during the eruption stage analyzed by optical and SAR images. (**a**) the Sentinel-1A image on 19 December 2018; (**b**) the Sentinel-1A image on 22 December 2018; (**c**) the Alos-2 image on 24 December 2018; (**d**) the Sentinel-1A image on 27 December 2018; (**e**) the Planet image on 7 January 2019; (**f**) the Google image on 11 January 2019.

The Sentinel-1A image on 19 December 2018 shows the state of the volcano before the eruption (Figure 6a). The Sentinel-1A image acquired at 22:34 on 22 December 2018 shows that the volcanic cone has collapsed. The lost material moved to the southwest side (Figure 6b). On 24 December 2018, the volcano could be divided into two parts (highlighted in red line in Figure 6c), with the southwest side showing signs of overall collapse (Figure 6c). There are some ripples on the sea surface, so the volcano was still active. The volcano shape in the descending orbit image on 27 December 2018 is not much different from that in the subsequent time image (Figure 5), which proves that the eastern part of the island has stabilized on 27 December 2018 (Figure 6d). The Planet image on 7 January 2019 shows an open crater lake with a collapsed gap in the southwestern part of the island (Figure 6e). By 11 January 2019 the gap had closed, as observed by both Google images (Figure 6f).

In general, the eruption started at 13:55 on 22 December 2018, and the volcanic material was being lost until 22:34 on 22 December 2018 [4]. The southwestern part of the volcano collapsed and became a crescent-shaped gap. A large amount of collapsed material piled up on the eastern and northern sides and expanded the volcano area (Figure 6b,c). Some material was driven back by the ocean currents and finally filled up the crescent gap by 11 January 2019. Since then, the volcano area became stable.

5. Accuracy Analysis and Discussion

5.1. Accuracy Analysis

In this paper, the volcanic boundaries are automatically identified by band polarization enhancement combined with the Otsu method. To test the boundary accuracy, optical images and SAR images acquired at the same time are selected for comparison. The optical images are analyzed by visual interpretation (Figure 7a,d), and the SAR images are analyzed by visual interpretation (Figure 7b,e) and automatic identification interpretation separately (Figure 7c,f). As Table 1 shows, extraction accuracy before volcanic eruption can reach 99.14%, and after volcanic eruption can reach 96.25%, which confirms the feasibility of automatically extracting volcanic boundaries from SAR images.

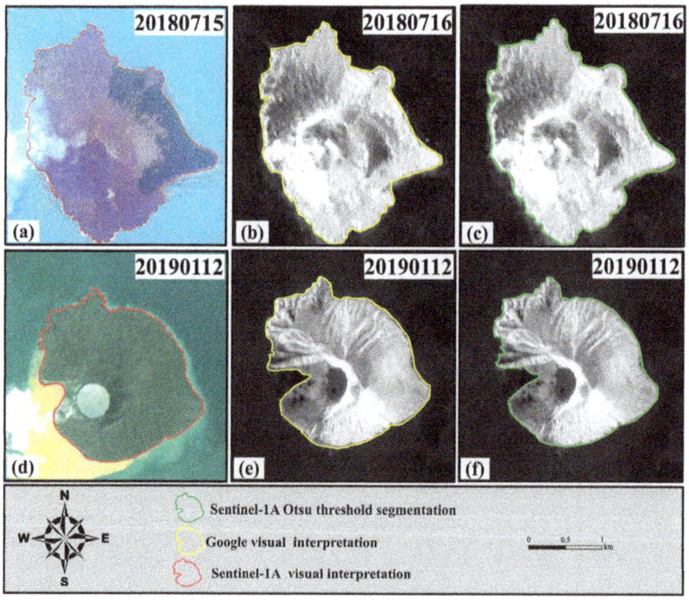

Figure 7. Volcano boundaries extracted from (**a**) the Google image on 15 July 2018 by visual interpretation; (**b**) the Sentinel-1A image on 16 July 2018 by visual interpretation; (**c**) the Sentinel-1A image on 16 July 2018 by the Otus threshold segmentation method; (**d**) the Planet image on 12 January 2019 by visual interpretation; (**e**) the Sentinel-1A image on 12 January 2019 by visual interpretation; (**f**) the Sentinel-1A image on 12 January 2019 by the Otus threshold segmentation method.

Table 1. The volcano area obtained by different methods.

Time	Otus Threshold	Visual Interpretation (km^2)		Difference (km^2)	Relative Accuracy	Average Accuracy
20180716	2.88	SAR image	2.90	0.02	99.31%	99.14%
		Optical image	2.85	0.03	98.96%	
20190112	3.16	SAR image	3.21	0.15	98.44%	96.25%
		Optical image	3.36	0.20	94.05%	

5.2. Discussion

In this paper, four pairs of optical images and SAR images with close acquisition time (maximum interval of one day) are selected to explore the advantages and disadvantages between them, taking Krakatoa Island as an example.

Figure 8a1,a2 shows the images acquired before the eruption. In Figure 8a1, there was vegetation in the eastern part of the volcano, which was covered by a large amount of lava after the eruption. In Figure 8a2, ground objects have sensitive polarization information and high classification accuracy, so the image is suitable for the segmentation of the sea–land boundary. Figure 8b1,b2 shows the post-eruption images, which have the same information. The shadow in the eastern part of the volcanic island in the SAR image is due to the geometric distortion. Both the optical image (Figure 8c1,c2) and SAR image (Figure 8d1,d2) show the formation of the closed crater lake in the southwest. Since SAR images have high temporal resolution and are free of charge, they can be better applied to analyze the morphological changes of long time series. In summary, optical images and SAR images have mutual respective advantages and disadvantages (Table 2), and how to better integrate the advantages of optical images and SAR images for volcano monitoring is worthy of further study.

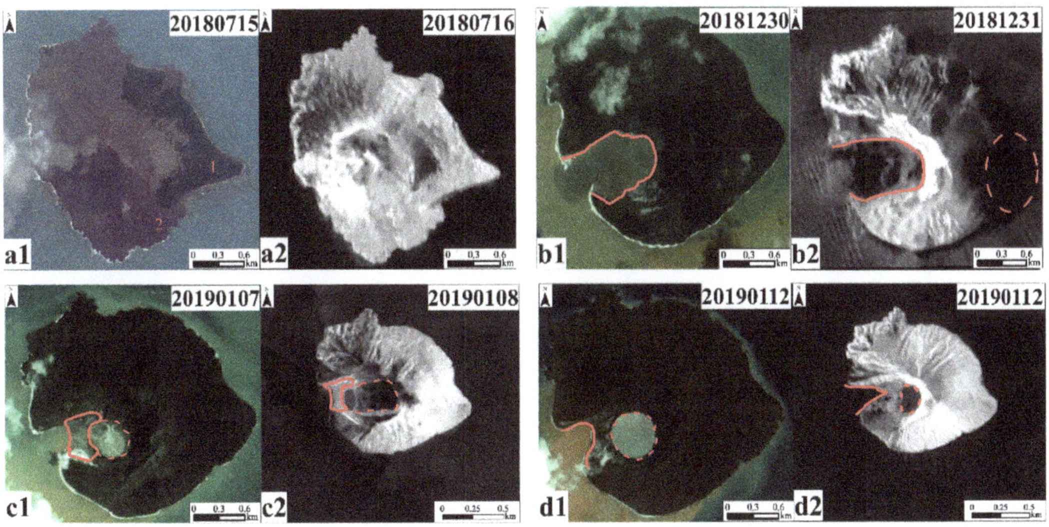

Figure 8. Comparison of optical and SAR images at the same time; optical images for (**a1**,**b1**,**c1**,**d1**); SAR images for (**a2**,**b2**,**c2**,**d2**).

Table 2. Comparison of optical and SAR images.

Data Source	Optical Images	SAR Images
Advantages	1. Rich in texture and spectral information, visually reflecting geomorphological features 2. High spatial resolution, up to 0.3 m	1. All-day, all-weather detection 2. Unaffected by clouds and fog 3. Short revisit period
Disadvantages	1. Effective data constrained by clouds and fog 2. Long revisit period	1. Include geometric distortion areas 2. Geomorphological features are not intuitive
Interpretation method	1. Combine texture, color, shape and other features to interpret 2. Different combinations of bands can be interpreted according to different features	1. Interpretation by amplitude information 2. Fusion interpretation based on polarization information
Applicability	More suitable for areas with lush vegetation and more complex terrain	Suitable for a wide range of flat terrain and perennial cloudy areas

6. Conclusions

In this paper, we use the optical images and SAR images to analyze the morphological changes of Krakatoa volcano before and after the eruption in 2018, by visual interpretation and polarization enhancement combined with the maximum interclass variance method. The polarization enhancement of water body information improves the identification accuracy, and distinguishes water bodies and non-water bodies clearly. The area of Krakatoa was stable before the eruption in December. It grew between 22 December 2018 and 11 January 2019, during which the volcanic morphology also changed greatly. After 11 January 2019, the volcanic morphology became stable. Krakatoa volcano experienced slow to fast and then slow changes in 2018–2019. The volcano eruption had significant impacts on the local climate and people's lives. Timely monitoring of volcanoes is important for volcanic disaster early warning and post-disaster assessment.

Author Contributions: Conceptualization, J.X. and X.S.; methodology, S.G. and D.Y.; software, S.G.; investigation, G.W.; resources, G.W. and K.D.; writing—original draft preparation, S.G. and N.W.; writing—review and editing, K.D., J.X., X.S. and C.C.; funding acquisition, J.X., K.D. and X.S. All authors have read and agreed to the published version of the manuscript.

Funding: This research received no external funding.

Institutional Review Board Statement: Not applicable.

Informed Consent Statement: Not applicable.

Data Availability Statement: Not applicable.

Conflicts of Interest: The authors declare no conflict of interest.

References

1. Grilli, S.T.; Tappin, D.R.; Carey, S.; Watt, S.F.L.; Ward, S.N.; Grilli, A.R.; Engwell, S.L.; Zhang, C.; Kirby, J.T.; Schambach, L.; et al. Modelling of the tsunami from the December 22, 2018 lateral collapse of Anak Krakatau volcano in the Sunda Straits, Indonesia. *Sci. Rep.* **2019**, *9*, 11946. [CrossRef] [PubMed]
2. Robertson, I.; Head, M.; Roueche, D.; Wibowo, H.; Kijewski-Correa, T.; Mosalam, K.; Prevatt, D. *Steer-Sunda Strait Tsunami (Indonesia): Preliminary Virtual Assessment Team (p-Vat) Report*; DesignSafe-CI: Seattle, WA, USA, 2018.
3. Takabatake, T.; Shibayama, T.; Esteban, M.; Achiari, H.; Nurisman, N.; Gelfi, M.; Tarigan, T.A.; Kencana, E.R.; Fauzi, M.A.R.; Panalaran, S.; et al. Field survey and evacuation behaviour during the 2018 Sunda Strait tsunami. *Coast. Eng. J.* **2019**, *61*, 423–443. [CrossRef]

4. Walter, T.R.; Haghshenas Haghighi, M.; Schneider, F.M.; Coppola, D.; Motagh, M.; Saul, J.; Babeyko, A.; Dahm, T.; Troll, V.R.; Tilmann, F.; et al. Complex hazard cascade culminating in the Anak Krakatau sector collapse. *Nat. Commun.* **2019**, *10*, 4339. [CrossRef] [PubMed]
5. Tsunami Sunda Strait (Update 14 January 2019). Available online: https://bnpb.go.id/ (accessed on 6 May 2019).
6. Dai, K.; Li, Z.; Xu, Q.; Bürgmann, R.; Milledge, D.; Tomás, R.; Fan, X.; Zhao, C.; Liu, X.; Peng, J.; et al. Entering the era of Earth observation-based landslide warning systems: A novel and exciting framework. *IEEE Geosci. Remote Sens. Mag.* **2020**, *8*, 136–153. [CrossRef]
7. Deng, H. Application of High Precision Satellite Remote Sensing Technology in Geological Disaster Investigation and Evaluation. Ph.D. Thesis, Chengdu University of Technology, Chengdu, China, 2007.
8. Wu, Y.; Liu, Z. Research progress on methods of automatic coastline extraction based on remote sensing images. *J. Remote Sens.* **2019**, *23*, 582–602.
9. Zhai, Q.; Zhang, D.; Tian, H.; Wu, K.; Wang, X.; Liang, P. Optimal Threshold Segmentation Method for Flood Inundation Range of Multi-exponential Shadow Sentinel-1 SAR Data: Take Shaoyang County as an example. *Mapp. Spat. Geogr. Inf.* **2020**, *43*, 79–83, 86.
10. Chen, L.; Liu, Z.; Zhang, H. SAR image Water Extraction based on Scattering Characteristics. *Remote Sens. Technol. Appl.* **2014**, *29*, 963–969.
11. Jing, B.; Ning, W. Water Body Information Extraction of SAR Images Based on Threshold Segmentation and Decision tree. *Geospat. Inf.* **2021**, *19*, 46–49.
12. Zhu, J.; Guo, H.; Fan, X.; Ding, C.; Ligang, L.; Li, L. Water detection with high-resolution SAR image based on texture and imaging knowledge. *Adv. Water Sci.* **2006**, *04*, 525–530.
13. Wang, J.; Wang, S.; Wang, F.; Zhou, Y.; Ji, J.; Xiong, Y.; Wang, Z.; Zhao, Q. Flood Inundation Region Extraction Method based on Sentinel-1 SAR data. *Disaster Sci.* **2021**, *36*, 7.
14. Wang, H. Water Body Change Detection Based on Ascending and Descending SAR Wavelet Fusion Imagery. Master's Thesis, Nanjing University, Nanjing, China, 2012.
15. Jia, S.; Xue, D.; Li, C.; Zheng, J.; Li, W. Study on new method for water area information extraction based on Sentinel-1 date. *Peoples Yangtze River* **2019**, *50*, 213–217.
16. Zeng, L.; Li, L.; Wan, L. Fast extraction of flood inundation extent based on Sentinel-1 satellite SAR data. *Geogr. World* **2015**, *22*, 100–103, 107.
17. Jiang, J.; Zhou, J.; Yang, B. Automatic Extraction of water Boundary based on Radar Image. *Mar. Inf.* **2020**, *35*, 20–28.
18. Song, H. Research on Extraction of Typical Land Cover Types with PolSAR—A case Research of Water Body Extraction. Master's Thesis, PLA Information Engineering University, Zhengzhou, China, 2012.
19. Deng, Y.; Zhang, H.; Wang, C.; Liu, M. An Object-oriented Water Extraction Method based on Texture andPolarization Decomposition Feature. *Remote Sens. Technol. Appl.* **2016**, *31*, 714–723.
20. Babu, A.; Kumar, S. InSAR coherence and backscatter images based analysis for the Anak Krakatau volcano eruption. *Multidiscip. Digit. Publ. Inst. Proc.* **2019**, *24*, 21.
21. Garcia, G. Volcano Discovery. Available online: https://www.volcanodiscovery.com/krakatoa/2018/dec/eruption-tsunami/updates.html (accessed on 22 December 2018).
22. Chen, X.; Sun, J.; Yin, K.; Yu, J. Sea-Land Segmentation Algorithm for SAR image Based on Otsu Method and Statistical Characteristic of sea Area. *Data Acquis. Process.* **2014**, *29*, 603–608.
23. Perttu, A.; Caudron, C.; Assink, J.D.; Metz, D.; Tailpied, D.; Perttu, B.; Hibert, C.; Nurfiani, D.; Pilger, C.; Muzli, M.; et al. Reconstruction of the 2018 tsunamigenic flank collapse and eruptive activity at Anak Krakatau based on eyewitness reports, seismo-acoustic and satellite observations. *Earth Planet. Sci. Lett.* **2020**, *541*, 116268. [CrossRef]
24. Darmawan, H.; Worosuprojo, S.; Wahyudi, W.; Harijoko, A. Available online: https://www.researchgate.net/publication/349578183_Morphological_Changes_of_Anak_Krakatau_after_the_22_December_2018_Flank_Collapsed (accessed on 24 February 2021).